Nonlinear System Identification – Input-Output Modeling Approach

MATHEMATICAL MODELLING:
Theory and Applications

VOLUME 7/1

This series is aimed at publishing work dealing with the definition, development and application of fundamental theory and methodology, computational and algorithmic implementations and comprehensive empirical studies in mathematical modelling. Work on new mathematics inspired by the construction of mathematical models, combining theory and experiment and furthering the understanding of the systems being modelled are particularly welcomed.

Manuscripts to be considered for publication lie within the following, non-exhaustive list of areas: mathematical modelling in engineering, industrial mathematics, control theory, operations research, decision theory, economic modelling, mathematical programming, mathematical system theory, geophysical sciences, climate modelling, environmental processes, mathematical modelling in psychology, political science, sociology and behavioural sciences, mathematical biology, mathematical ecology, image processing, computer vision, artificial intelligence, fuzzy systems, and approximate reasoning, genetic algorithms, neural networks, expert systems, pattern recognition, clustering, chaos and fractals.

Original monographs, comprehensive surveys as well as edited collections will be considered for publication.

The titles published in this series are listed at the end of this volume.

Nonlinear System Identification – Input-Output Modeling Approach

Volume 1: Nonlinear System Parameter Identification

by

Robert Haber
Laboratory for Process Control,
Department of Plant and Process Engineering,
University of Applied Sciences Cologne (Fachhochschule Köln),
Köln, Germany

and

László Keviczky
Computer and Automation Institute,
Hungarian Academy of Sciences,
Budapest, Hungary

KLUWER ACADEMIC PUBLISHERS
DORDRECHT / BOSTON / LONDON

A C.I.P. Catalogue record for this book is available from the Library of Congress.

ISBN 0-7923-5856-2
ISBN 0-7923-5859-9 (Set)

Published by Kluwer Academic Publishers,
P.O. Box 17, 3300 AA Dordrecht, The Netherlands.

Sold and distributed in North, Central and South America
by Kluwer Academic Publishers,
101 Philip Drive, Norwell, MA 02061, U.S.A.

In all other countries, sold and distributed
by Kluwer Academic Publishers,
P.O. Box 322, 3300 AH Dordrecht, The Netherlands.

Printed on acid-free paper

To Agnes and Csilla
and to our readers

CONTENTS

VOLUME 1
NONLINEAR SYSTEM PARAMETER ESTIMATION

VOLUME 2
NONLINEAR SYSTEM STRUCTURE IDENTIFICATION

Preface

The subject of the book is to present the modeling, parameter estimation and other aspects of the identification of nonlinear dynamic systems. The treatment is restricted to the input–output modeling approach. Because of the widespread usage of digital computers discrete time methods are preferred. Time domain parameter estimation methods are dealt with in detail, frequency domain and power spectrum procedures are described shortly. The theory is presented from the engineering point of view, and a large number of examples of case studies on the modeling and identifications of real processes illustrate the methods.

Almost all processes are nonlinear if they are considered not merely in a small vicinity of the working point. To exploit industrial equipment as much as possible, mathematical models are needed which describe the global nonlinear behavior of the process. If the process is unknown, or if the describing equations are too complex, the structure and the parameters can be determined experimentally, which is the task of identification.

The book is divided into seven chapters dealing with the following topics:

1. Nonlinear dynamic process models
2. Test signals for identification
3. Parameter estimation methods
4. Nonlinearity test methods
5. Structure identification
6. Model validity tests
7. Case studies on identification of real processes

Chapter 1 summarizes the different model descriptions of nonlinear dynamical systems. They are the nonparametric models which are the nonlinear extensions of the linear weighting functions; the block oriented models which consist of linear dynamic and static nonlinear parts; the explicit and implicit models linear-in-parameters; the quasi-linear models having signal dependent parameters; and the multi-models that have different equations under different working modes.

Chapter 2 describes the usual test signals used in active identification experiments. First the classical (but non-practical) Gaussian white noise is treated as test signal. Then the generating algorithms and features of pseudo-random binary, ternary and five-level signals are summarized. Time sequences, auto-correlation and power density spectra of the usual sequences are derived and shown. Recommendations for the choice of the parameters of a pseudo-random multilevel signal generator are given, as well.

Chapter 3 covers many parameter estimation methods for nonlinear systems. The manifold of methods arises from the variety of process models and from the diversity of parameter estimation methods: grapho-analytical, correlation, frequency, least squares and modified least squares ones.

Chapter 4 deals with the problem of how a decision of whether a process is linear or not dependent on input–output measurements can be made without long computations. The different methods need various test signals, e.g., steps, Gaussian, stationary stochastic or arbitrary but persistently exciting ones. The methods cover time domain, steady state, frequency, correlation, dispersion methods and some procedures that are based on the parameter estimation of simple structures.

Chapter 5 is devoted to structure identification methods. This topic is much more important where the inner structure of the model is usually unknown than for linear systems where mostly the order of the system is only searched for. Some of the methods are nonparametric which means that no *a priori* assumption of the structure is needed.

Methods are given for the structure detection of block oriented models, cascade systems consisting of linear dynamic and nonlinear static parts, and quasi-linear models with signal dependent parameters. Different techniques of regression analysis are used for the selection of the components of models linear-in-parameters. Finally, the group method of data handling is dealt with.

Chapter 6 discusses the problem of whether the estimated model is adequate or not. Because a comparison with the true parameters is not possible, the residual is analyzed instead. Different techniques applicable also for linear systems are reviewed first then special correlation techniques are introduced for the nonlinear case.

Chapter 7 presents nine different case studies of the modeling and identification of real processes. The processes are from different application areas: a biological membrane, fermentation, a heat exchanger, a distillation column, a flood process and a cement mill.

References are given at the end of each chapter.

Each chapter contains several examples. Some of them are pencil and paper problems and others are simulations that illustrate the theoretical statements.

At the time of writing the book there was no comprehensive textbook dealing with the identification of nonlinear dynamic processes available. There are hundreds of articles in journals, preprints, and proceedings of symposia and some books deal with a small band of the methods only. Our aim was to collect the diverse methods scattered in the literature into a unified form when some of the original publications are written in a mathematical way, too. The methods are presented in the book rather as recipes, and the assumptions of their usage are given exactly.

The backgrounds are probability theory and statistics, the theory of parameter estimation of linear dynamic systems, and the representation of linear systems by differential and difference equations. In other words, it is assumed that the reader knows the foundations of the identification of linear dynamic systems.

The book is intended to be a textbook for graduate and postgraduate students and as a handbook for engineers. Identification is not only related to control, everybody who would like to obtain a deeper knowledge of an unknown complex process can use it.

Robert Haber
Cologne
(hr@fh-av.av.fh-koeln.de)

László Keviczky
Budapest
(keviczky@sztaki.hu)

Glossary

Notations

$A(s)$	denominator polynomial of a linear transfer function
$A(p)$	polynomial of the differential operator
$A(p, x)$	polynomial of the differential operator with $x(t)$ signal dependent parameters
$A(z^{-1})$	denominator polynomial of a linear pulse transfer function
$A(q)$	polynomial of the forward shift operator
$A(q^{-1})$	polynomial of the backward shift operator
$\tilde{A}(q^{-1})$	$A(q^{-1})-1$
$A(q^{-d}, x)$	autoregressive polynomial of the backward shift operator with $x(k)$ signal dependent parameters
a_i	coefficient of the polynomial $A(\ldots)$
$a_i(x)$	coefficient of the polynomial $A(\ldots, x)$
$B(s)$	numerator polynomial of a linear transfer function
$B(p)$	polynomial of the differential operator
$B(p, x)$	moving average polynomial of the differential operator with $x(t)$ signal dependent parameters
$B(z^{-1})$	numerator polynomial of a linear pulse transfer function
$B(q)$	polynomial of the forward shift operator
$B(q^{-1})$	polynomial of the backward shift operator
$B(q^{-1}, x)$	polynomial of the backward shift operator with $x(k)$ signal dependent parameters
b_i	coefficient of the polynomial $B(\ldots)$
$b_i(x)$	coefficient of the polynomial $B(\ldots, x)$
c_0	constant term in a static polynomial
c_1	coefficient of the linear term in a static polynomial
c_2	coefficient of the quadratic term in a static polynomial
[°C]	temperature degrees Celsius
C_p	Mallow's statistic
$C^{(r)}(q^{-1})$	modulo-*r* polynomial of the backward shift operator
d	integer time delay relative in the unit of the sampling period
[dB]	decibel
$D_{uy}(\kappa)$	discrete time cross-dispersion function of the sequences $\{u(k)\}$ and $\{y(k)\}$

$D^{(r)}\left(q^{-1}\right)$	modulo-r polynomial of the backward shift operator
$\deg\{\ldots\}$	degree
$\exp\{\ldots\}$	exponential
$e(t)$	continuous time source noise
$\{e(k)\}$	discrete time source noise sequence
$\mathrm{E}\{\ldots\}$	expected value
f	frequency
$f(\ldots)$	function
F	Fisher distribution value
$\mathcal{F}\{\ldots\}$	Fourier transform
g_0	constant term of the continuous time Volterra kernel
$g(\tau)$	linear weighting function
$g_1(\tau_1)$	degree-1 (first degree) continuous time Volterra kernel
$g_1^{\mathrm{o}}(\tau_1)$	degree-1 (first degree) continuous time Wiener kernel
$g_n(\tau_1,\ldots,\tau_n)$	degree-n continuous time Volterra kernel
$g_n^{\mathrm{sym}}(\tau_1,\ldots,\tau_n)$	degree-n continuous time symmetrical Volterra kernel
$g_n^{\mathrm{o}}(\tau_1,\ldots,\tau_n)$	degree-n continuous time Wiener kernel
$G(s)$	linear transfer function
$G(p)$	operator form of linear differential equation
$G(s, x)$	operator form of a quasi-linear differential equation whose parameters depend on the signal $x(t)$
$G(j\omega)$	frequency function of the transfer function
$\arg G(j\omega)$	phase function
$G(j\omega_1,\ldots,j\omega_n)$	frequency function of the multi-dimensional transfer function
[h]	hour
h_0	constant term of the discrete time Volterra kernel
$h(\kappa)$	linear weighting function series
$h_1(\kappa)$	degree-1 (first-degree) discrete time Volterra kernel
$h_n(\kappa_1,\ldots,\kappa_n)$	degree-n discrete time Volterra kernel
$h_n^{\mathrm{sym}}(\kappa_1,\ldots,\kappa_n)$	degree-n discrete time symmetrical Volterra kernel
$H(z)$	linear pulse transfer function
$H\left(q^{-1}\right)$	operator form of a linear difference equation
$H\left(q^{-1}, x\right)$	operator form of a quasi-linear difference equation whose parameters depend on the signal $x(k)$
$J(\ldots)$	loss function
k	discrete time unit
K	static gain

[kcal]	kilocalorie
[kg]	kilogram
[kW]	kilowatt
[ℓ]	litre
$\mathscr{L}\{\ldots\}$	Laplace transform
lim	limit
m	memory length (order) of a nonparametric model
[m]	metre
M	number of input signals
[mA]	milliampere
$\max\{\ldots\}$	maximum value
$\min\{\ldots\}$	minimum value
[min]	minute
[mm]	millimetre
[mol]	mole
[ms]	millisecond
[mS]	millisiemens
$\overline{(\ldots)}$	mean value
$\otimes_i$	modulo-i polynomial multiplication
$\oplus_i$	modulo-i polynomial addition
$\ominus_i$	modulo-i polynomial subtraction
$\odot_i$	modulo-i polynomial division
n	degree
na	degree of the autoregressive polynomial $A(q^{-1})$
nb	degree of the moving average polynomial $B(q^{-1})$
n_e	memory of the error terms in the memory vector
n_u	memory of the input terms in the memory vector
n_y	memory of the output terms in the memory vector
n_r	number of registers in a shift register
$n_{\boldsymbol{\theta}}$	number of parameters
$n(t)$	continuous time output error
$\{n(k)\}$	discrete time output error sequence
N	number of samplings
N_q	number of equal intervals in an amplitude range
N_p	period
p	differential operator
[Pa]	pascal
p_i	ith pole
$p(x)$	probability density function
$P(x)$	cumulative probability density function
$P[\ldots]$	static polynomial
plim	probability limit

q	forward shift operator
q^{-1}	backward shift operator
$Q_i[\ldots]$	ith gate function
r	number of levels of a PRMS
$r_{uu}(\tau)$	continuous time auto-correlation function of signal $u(t)$
$r_{uy}(\tau)$	continuous time cross-correlation function of signals $u(t)$ and $y(t)$
$r_x(\tau_1)$	continuous time degree-1 correlation function of the signal $x(t)$
$r_x(\tau_1,\ldots,\tau_n)$	continuous time degree-n correlation function of the signal $x(t)$
$r_{uu}(\kappa)$	discrete time auto-correlation function of sequence $\{u(k)\}$
$r_{uy}(\kappa)$	discrete time cross-correlation function of sequences $\{u(k)\}$ and $\{y(k)\}$
$r_x(\kappa_1)$	discrete time degree-1 correlation function of the sequence $\{x(k)\}$
$r_x(\kappa_1,\ldots,\kappa_n)$	discrete time degree-n correlation function of the sequence $\{x(k)\}$
R^2	multiple correlation coefficient
R_a^2	adjusted multiple correlation coefficient
s	variable of the continuous time Laplace transformation
[s]	second
sgn(...)	sign function
$S_{uu}(\omega)$	auto-spectral density function of signal $u(t)$
$S_{uy}(\omega)$	cross-spectral density function of signals $u(t)$ and $y(t)$
[t]	tonne
t	continuous time
T	time constant
T_d	time delay
T_p	time period
T_{95}	settle time of transients reaching the 5% bias region compared to the stationary value
ΔT	sampling period
ΔT_e	minimum switching time of an exciting PRMS
$u(t)$	continuous time input signal
$\{u(k)\}$	discrete time input sequence
U	steady state value or amplitude of the input signal
U_{extr}	input value belonging to the extremum value of the steady state output
$U(j\omega)$	Fourier transform of the input signal
$\overline{U}(j\omega)$	complex conjugate of the Fourier transform of the input signal
Y_{extr}	extremum value of the steady state output
$Y(j\omega)$	Fourier transform of the output signal
$\overline{Y}(j\omega)$	complex conjugate of the Fourier transform of the output signal
$v(t)$	state variable, auxiliary signal
$V_n[\ldots]$	degree-n homogeneous Volterra operator

$V_n^{\circ}[\ldots]$	degree-*n* homogeneous Wiener operator
$V_{nc}[\ldots]$	degree-*n* (complete) Volterra operator
$\mathrm{var}\{\ldots\}$	variance
$x(t)$	state variable, auxiliary signal, multiplier
$\dot{x}(t)$	differential value of $x(t)$
$\ddot{x}(t)$	second-order differential value of $x(t)$
$\hat{x}$	estimated quantity of x
$\bar{x}$	average value of $x(t)$ or $\{x(k)\}$
$\Delta x(k)$	difference value of $x(k)$
$y(t)$	continuous time output signal
$\{y(k)\}$	discrete time output sequence
Y	steady state value of the output signal
[V]	volt
z	variable of the discrete time Laplace transformation
z^{-1}	inverse of the discrete time Laplace operator
z_i	*i*th zero
$\mathcal{Z}\{\ldots\}$	z-transform
$w(t)$	continuous time noise-free output signal
$\{w(k)\}$	discrete time noise-free output sequence
$\delta(t)$	Dirac impulse
Δ	first-order difference
∇	second-order difference
$\{\varepsilon(k)\}$	discrete time residual sequence
γ	scalar factor
ν	nonlinearity index
ξ	damping factor
$\boldsymbol{\theta}$	parameter vector
κ	discrete shifting time
η	forgetting factor
Π	product
τ	continuous shifting time
$\sigma\{\ldots\}$	standard deviation
σ_ε	estimated standard deviation of the residuals
$\sigma\{\theta_i\}$	standard deviation of the parameter θ_i
Σ	sum
$\boldsymbol{\phi}$	memory vector
$\boldsymbol{\Phi}$	memory matrix
φ	phase
ω	angular frequency
ψ	noise/signal ratio
Ξ^2	chi-square value
π	3.141592

$\forall$	all
$1(t)$	unit step function
$\|...\|$	absolute value
$(...)^{o}$	orthogonal
∞	infinity

Abbreviations

A/D	analog/digital (converter)
AIC	Akaike's information criterion
B	bilinear
BIC	Bayesian information criterion
COR-LS	correlation analysis with least squares parameter estimation
D/A	digital/analog (converter)
DC	direct current component
DE	direct estimation
EGM	elementary gate function model
ELS	extended least squares
F	partial *F* value
FPE	final prediction error
GH	generalized Hammerstein
GMDH	group method of data handling
H	Hammerstein
H	Hermite
L	linear
LD	linear dynamic
LHS	left hand side
LILC	Khinchin's law of iterated logarithm criterion
LS	least squares
MIMO	multi-input multi-output
MISO	multi-input single-output
MSE	mean square error
NARMAX	nonlinear auto regressive moving average with exogenous input
NRSS	normalized residual sum of squares
NS	nonlinear static
OVF	overall *F*-test
P	general, polynomial model
PE	prediction error
PRBS	pseudo-random binary signal
PRESS	prediction sum of squares
PRMS	pseudo-random multi-level signal
PRQS	pseudo-random quinary signal
PRTS	pseudo-random ternary signal
PV	parametric Volterra
RHS	right hand side
RMSE	relative mean square error
RLS	recursive least squares
RSS	residual sum of squares

SISO	single-input single-output
SR	static regression
SS	sum of squares due to regression
SRE	step response equivalent
TSS	total sum of squares
V	Volterra, parametric Volterra

Notational conventions

entier(...)	integer part
::=	assignment operator
*	convolution

1. Nonlinear Dynamic Process Models

1.1 INTRODUCTION

The topic of the book is restricted to single-input single-output (SISO) systems. Any process can be described either

- by a state space or
- by an input–output equivalent model.

The latter structure is usually chosen if

- identification and/or control methods elaborated for linear input–output models should be extended to the nonlinear case,
- the inner structure of the process is unknown or not of interest.

As is very often the case, only input–output equivalent process models are treated.

Assume that there is a single noise source which disturbs the process. Then the noisy process is a two-input single-output model, where the inputs are the real input signal and the source noise. The process and the noise model have similar structures and the measured output is the sum of the output signals of the two models for linear systems. The situation is more complex for nonlinear systems because the superposition law is not valid any more and two-variable nonlinear functions (e.g., product terms) of the input signal and the noise may occur. A general nonlinear multi-input single-output (MISO) model is rarely set up because the noisy model would be very complex and would contain too many parameters. How a noise model can be parametrized in respect of an easy parameter estimation will not be dealt with in this chapter but in Chapter 3 together with the parameter estimation methods.

Before starting with the classification of nonlinear process models let us investigate a simple example which shows the complexity of a nonlinear model.

Example 1.1.1 *Cascade model containing a linear dynamic part and a square element*

Figure 1.1.1 shows two simple models. The input signal is denoted by $u(t)$, the output signal by $y(t)$, where t is the continuous time, $g_1(\tau)$ is the weighting function of the linear parts and $\prod$ means a product. The equation of the linear dynamic part is

$$y(k) = \int_{\tau=0}^{t} g_1(\tau)u(t-\tau)d\tau$$

Because of the different positions of the linear and nonlinear elements in the two processes, the input–output descriptions are very different.

1. The square element is followed by a linear dynamic part (Figure 1.1.1a)

$$y(k) = \int_{\tau=0}^{t} g_1(\tau)u^2(t-\tau)d\tau = \int_{\tau_1=0}^{t} \int_{\tau_2=0}^{t} g_2(\tau_1,\tau_2)u(t-\tau_1)u(t-\tau_2)d\tau_1 d\tau_2 \qquad (1.1.1)$$

with

$$g_2(\tau_1,\tau_2) = \begin{cases} g_1^2(\tau_1) & \text{if } \tau_1 = \tau_2 = \tau \\ 0 & \text{if } \tau_1 \neq \tau_2 \end{cases} \qquad (1.1.2)$$

2. The square element is preceded by a linear dynamic part (Figure 1.1.1b)

$$y(k)=\left[\int_{\tau=0}^{t} g_1(\tau)u(t-\tau)d\tau\right]^2=\int_{\tau_1=0}^{t}\int_{\tau_2=0}^{t} g_1(\tau_1)g_1(\tau_2)u(t-\tau_1)u(t-\tau_2)d\tau_1 d\tau_2$$
$$=\int_{\tau_1=0}^{t}\int_{\tau_2=0}^{t} g_2(\tau_1,\tau_2)u(t-\tau_1)u(t-\tau_2)d\tau_1 d\tau_2 \qquad (1.1.3)$$

with

$$g_2(\tau_1,\tau_2)=g_1(\tau_1)\,g_1(\tau_2) \qquad (1.1.4)$$

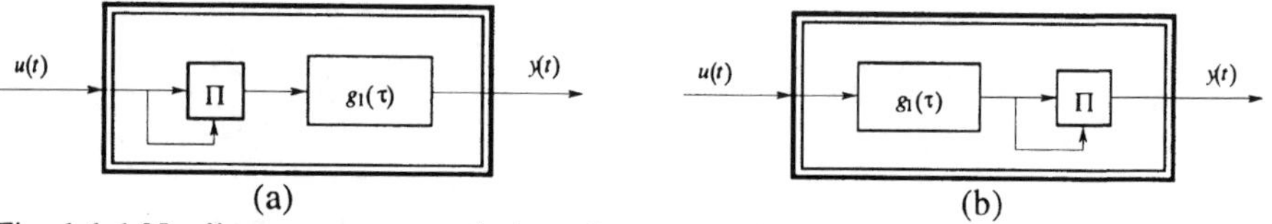

Fig. 1.1.1 Nonlinear system consisting of a square element and a linear dynamic term: (a) square element preceeds the linear dynamic part; (b) square element follows the linear dynamic part

The equations (1.1.1) and (1.1.3) represent a so-called quadratic continuous time Volterra weighting function model with a Volterra kernel $g_2(\tau_1,\tau_2)$ that is different in both cases. Figure 1.1.2 shows the weighting function $g_1(\tau)$ (a), and the two quadratic kernels ((b) and (c)). As is seen, they depend on the structure of the process.

Assume that the input and output signals are sampled and they are denoted by $u(k)$ and $y(k)$, respectively. In the linear case the relation between them can be described by the weighting function series $h_1(\kappa)$

$$y(k)=\sum_{\kappa=0}^{k} h_1(\kappa)u(k-\kappa)$$

Both nonlinear models have the same weighting function series form

$$y(k)=\sum_{\kappa_1=0}^{k}\sum_{\kappa_2=0}^{k} h_2(\kappa_1,\kappa_2)\,u(k-\kappa_1)u(k-\kappa_2) \qquad (1.1.5)$$

but have different kernels:

- in the case of Figure 1.1.1a:

$$h_2(\kappa_1,\kappa_2)=\begin{cases} h_1^2(\kappa_1) & \text{if } \kappa_1=\kappa_2 \\ 0 & \text{if } \kappa_1\neq\kappa_2 \end{cases} \qquad (1.1.6)$$

- in the case of Figure 1.1.1b:

$$h_2(\kappa_1,\kappa_2)=h_1(\kappa_1)h_1(\kappa_2) \qquad (1.1.7)$$

■

The example showed the different parametrizations of the nonlinear systems:

- in respect of the measured signals:
  - continuous time,
  - discrete time.
- in respect of the parametrization:
  - parametric models (which can be described by finite number of parameters)
  - nonparametric models (which need theoretically infinite number of parameters)

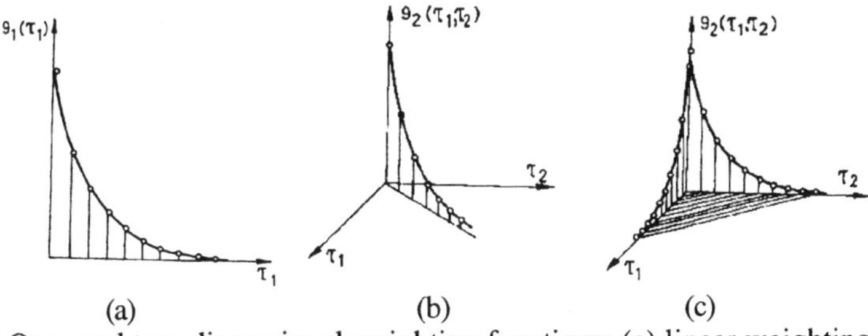

(a) (b) (c)

Fig. 1.1.2 One- and two-dimensional weighting functions: (a) linear weighting function; (b) quadratic kernel of Figure 1.1.1a; (c) quadratic kernel of Figure 1.1.1b

Some definitions are summarized now.

Definition 1.1.1 A model is called parametric if it can be described by a finite number of parameters. ■

Definition 1.1.2 A model is called nonparametric if an infinitely large number of parameters is needed to describe it. ■

Definition 1.1.3 A model is called a continuous time model if the equation describing it is valid in all time points. ■

Definition 1.1.4 A model is called a discrete time model if the equation describing it is valid only in dedicated (mostly equidistant) sampling points. ■

Definition 1.1.5 A noise-free process model is called linear-in-parameters if the output signal can be described by a scalar product of two vectors

$$y(k) = \boldsymbol{\phi}^{\mathrm{T}}\boldsymbol{\theta} \tag{1.1.8}$$

where $\boldsymbol{\theta}$ is the parameter vector, and the memory vector $\boldsymbol{\phi}$ contains measured values or computed values based on the measurements. (Here T stands for the transposition.) ■

Definition 1.1.6 A noise-free process model is called nonlinear-in-parameters if the output signal cannot be described by the scalar product (1.1.8). ■

Definition 1.1.7 A noisy model is called linear-in-parameters if the source noise can be computed from the measurements by a scalar product of two vectors

$$\varepsilon(k) = \boldsymbol{\phi}_n^{\mathrm{T}}\boldsymbol{\theta}_n \tag{1.1.9}$$

where $\varepsilon(k)$ is the computed source noise, the so-called residual. $\boldsymbol{\theta}_n$ is the parameter vector of the noisy model, and the memory vector $\boldsymbol{\varphi}_n$ contains measured values or computed ones based on the measurements. ■

Definition 1.1.8 A noisy model is called nonlinear-in-parameters if the source noise cannot be computed from the measurements by the scalar product (1.1.9). ■

Definition 1.1.9 A model is called strictly causal if the output signal at a time instant depends only on the actual and past value of the input signal. ■

In view of computer control applications, discrete time parametric models are at the center of our interest. In the case of continuos time models their discretization procedure will be presented.

The simple nonlinear parametric models can be arranged into the following groups:

- block oriented models
- models linear-in-parameters
- quasi-linear models with signal dependent parameters
- quasi-linear models with piecewise constant parameters (multi-models).

In this chapter no care will be given to the dead time. It will be treated with the parameter estimation methods in Chapter 3 in details.

1.2 NONPARAMETRIC POLYNOMIAL DYNAMIC MODELS

The nonparametric models can be described by infinite number of parameters. Therefore they are restricted suitably for identification and control purposes. Historically the nonparametric models were used earlier for modeling and identification than the parametric models. Several methods that are commonly used today with parametric models were developed first for nonparametric models.

Most of the researchers in these topics were mathematicians, and the original references are in mathematical journals. The style of the papers is too mathematical. The engineering aspects were clarified later in the publications of application engineers. Some good survey books and papers on the subject are those of Marmarelis and Marmarelis (1978), Billings (1980), Schetzen (1980), Rugh (1981).

As the number of parameters may be very large, it is of great importance to develop orthogonal models. Then the parameters can be estimated independently from each other and the model can be improved by including further components without re-estimating the already known parameters.

1.2.1 The Volterra series model

A static mild nonlinear system with input signal $u(t)$ and output signal $y(t)$ can be described by its Taylor series

$$y(t) = c_0 + c_1 u(t) + c_2 u^2(t) + \ldots = \sum_{i=0}^{n} c_i u^i(t) \tag{1.2.1}$$

A mild nonlinear system is analytic; i.e., it has a single-valued characteristic without any break point, and can be differentiated any times. The approximation is better and better with increasing the degree n.

A linear dynamic system can be described by its weighting function $g(t)$. The output is given by the convolution integral

$$y(t)=\int_{\tau=0}^{t} g(\tau)u(t-\tau)d\tau \tag{1.2.2}$$

If the signals are sampled then the linear dynamic relation between them is described by the convolution sum

$$y(k)=\sum_{\kappa=0}^{k} h(\kappa)u(k-\kappa) \tag{1.2.3}$$

where $u(k)$ and $y(k)$ are the equidistant samplings of the input and output signals, respectively, and $h(k)$ is the linear weighting function series.

A mild nonlinear dynamic system can be approximated in the vicinity of the working point at least by the so-called Volterra integral (Frechet, 1910; Volterra, 1959) or the Volterra weighting function model

$$\begin{aligned} y(t)&=g_0+\int_{\tau_1=0}^{t} g_1(\tau_1)u(t-\tau_1)d\tau_1+\int_{\tau_1=0}^{t}\int_{\tau_2=0}^{t} g_2(\tau_1,\tau_2)u(t-\tau_1)u(t-\tau_2)d\tau_1 d\tau_2+\ldots \\ &=\sum_{i=0}^{n}\int_{\tau_1=0}^{t}\ldots\int_{\tau_i=0}^{t} g_i(\tau_1,\ldots,\tau_i)\prod_{j=1}^{i} u(t-\tau_j)d\tau_1\ldots d\tau_i \end{aligned} \tag{1.2.4}$$

Similarly to the discrete time description of the linear systems, a multi-dimensional convolution sum describes the relation between the sampled input and output signals (Alper, 1965):

$$\begin{aligned} y(k)&=h_0+\sum_{\kappa_1=0}^{k} h_1(\kappa_1)u(k-\kappa_1)+\sum_{\kappa_1=0}^{k}\sum_{\kappa_2=0}^{k} h_2(\kappa_1,\kappa_2)u(k-\kappa_1)u(k-\kappa_2)+\ldots \\ &=\sum_{i=0}^{n}\ldots\sum_{\kappa_i=0}^{k} h_i(\kappa_1,\ldots,\kappa_i)\prod_{j=1}^{i} u(k-\kappa_j) \end{aligned} \tag{1.2.5}$$

We call (1.2.5) the Volterra weighting function series. Another name for it is the Gabor–Kolmogorov series (Eykhoff, 1974). Figure 1.2.1 shows the scheme of the model.

The process is characterized by its Volterra kernels:

- $g_n(\tau_1,\ldots,\tau_n)$ $n=0,1,2,\ldots$ in the continuous time case and
- $h_n(\kappa_1,\ldots,\kappa_n)$ $n=0,1,2,\ldots$ in the discrete time case.

The relation between the continuous and the discrete time kernels of a system is given by the formulas, which are based on Barker (1969):

1. The input signal is a pulse series:

$$
\begin{aligned}
h_0 &= g_0 \\
h_1(\kappa_1) &= g_1(\kappa_1 \Delta T) \\
h_2(\kappa_1, \kappa_2) &= g_2(\kappa_1 \Delta T, \kappa_2 \Delta T) \\
&\vdots \\
h_n(\kappa_1, \dots, \kappa_n) &= g_n(\kappa_1 \Delta T, \dots, \kappa_n \Delta T)
\end{aligned}
\tag{1.2.6}
$$

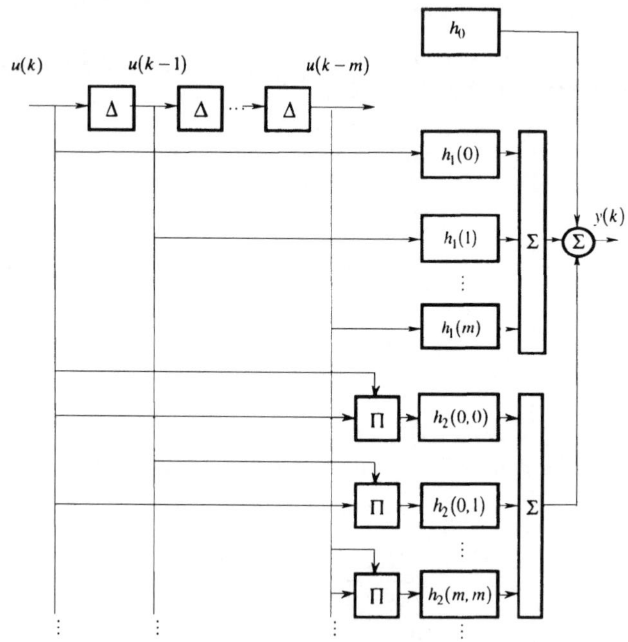

Fig. 1.2.1 Discrete time quadratic Volterra model

2. The input signal is constant in the sampling instances:

$$
\begin{aligned}
h_0 &= g_0 \\
h_1(\kappa_1) &= \int_{\tau_1 = \kappa_1 \Delta T}^{(\kappa_1 + 1)\Delta T} g_1(\tau_1)\, d\tau_1 \\
h_2(\kappa_1, \kappa_2) &= \int_{\tau_1 = \kappa_1 \Delta T}^{(\kappa_1 + 1)\Delta T} \int_{\tau_2 = \kappa_2 \Delta T}^{(\kappa_2 + 1)\Delta T} g_2(\tau_1, \tau_2)\, d\tau_1 d\tau_2 \\
&\vdots \\
h_n(\kappa_1, \dots, \kappa_2) &= \int_{\tau_1 = \kappa_1 \Delta T}^{(\kappa_1 + 1)\Delta T} \dots \int_{\tau_n = \kappa_n \Delta T}^{(\kappa_n + 1)\Delta T} g_n(\tau_1, \dots, \tau_n)\, d\tau_1 \dots d\tau_n
\end{aligned}
\tag{1.2.7}
$$

An n-dimensional kernel is called degree- n Volterra kernel. A Volterra model including only one degree- n part is called a homogeneous Volterra weighting function model:

$$y(t) = V_n[u(t)] = \int_{\tau_1=0}^{t} \cdots \int_{\tau_n=0}^{t} g_n(\tau_1, \ldots, \tau_n) \prod_{j=1}^{n} u(t-\tau_j) d\tau_1 \ldots d\tau_n$$

or homogeneous Volterra weighting function series model

$$y(k) = V_n[u(k)] = \sum_{\kappa_1=0}^{k} \cdots \sum_{\kappa_n=0}^{k} h_n(\kappa_1, \ldots, \kappa_n) \prod_{j=1}^{n} u(k-\kappa_j) \tag{1.2.8}$$

The Volterra model is a degree- n polynomial system

$$y(t) = V_{nc}[u(t)] = \sum_{i=0}^{n} V_i[u(t)]$$

or

$$y(k) = V_{nc}[u(k)] = \sum_{i=0}^{n} V_i[u(k)]$$

where $V_{nc}[\ldots]$ denotes the complete (not homogeneous) system of degree- n.

The Volterra kernel is called a homogeneous polynomial kernel because if the input signal becomes γ times then the outputs of the degree- i subsystems become γ^i times

$$y(t) = \sum_{i=0}^{n} V_i[\gamma u(t)] = \sum_{i=0}^{n} \gamma^i \, V_i[u(t)]$$

The Volterra kernels have the following properties:

- *causality:*

$$g_n(\tau_1, \ldots, \tau_n) = 0 \qquad \text{if} \quad \tau_i < 0 \quad i = 1, \ldots n$$
$$h_n(\kappa_1, \ldots, \kappa_n) = 0 \qquad \text{if} \quad \kappa_i < 0 \;\; i = 1, \ldots n \tag{1.2.9}$$

- *stability:*

$$\int_{\tau_1=0}^{t} \cdots \int_{\tau_n=0}^{t} \left| g_n(\tau_1, \ldots, \tau_n) \right| < \infty$$
$$\sum_{\kappa_1=0}^{k} \cdots \sum_{\kappa_n=0}^{k} \left| h_n(\kappa_1, \ldots, \kappa_n) \right| < \infty \tag{1.2.10}$$

Proof (discrete time case). Assume the input signal is bounded. Then the output signal is also bounded if

$$|y(t)| = \left| \sum_{n=0}^{\infty} \sum_{\kappa_1=0}^{k} \cdots \sum_{\kappa_n=0}^{k} h_n(\kappa_1, \ldots, \kappa_n) \prod_{j=1}^{n} u(k-\kappa_j) \right|$$

$$\leq \sum_{n=0}^{\infty} \left| \sum_{\kappa_1=0}^{k} \cdots \sum_{\kappa_n=0}^{k} h_n(\kappa_1, \ldots, \kappa_n) \right| \left| \prod_{j=1}^{n} u(k-\kappa_j) \right| \leq \infty$$

which is fulfilled through (1.2.10).

- *decay in the infinite time:*

$$\lim_{\tau_i \to \infty} g_n(\tau_1, \ldots, \tau_n) = 0$$

$$\lim_{\kappa_i \to \infty} h_n(\kappa_1, \ldots, \kappa_n) = 0$$

This property is generally valid for physical systems. However, systems with integrating character do not decay at infinity.

The Volterra kernels can be represented in different forms: symmetrical, triangular and regular (Rugh, 1981). A triangular form can be an upper or lower form depending on the fact the elements are zero below or above the main diagonal $\kappa_1 = \ldots = \kappa_n$. In the sequel the different forms are presented in details.

- *symmetrical kernels:*
Each kernel can be expressed in a symmetric form, which is a consequence of the definition of the Volterra model

$$h_n^{\text{sym}}(\kappa_1, \ldots, \kappa_n) = \frac{1}{n!} \sum_{i=1}^{n!} h_n(\kappa_1', \ldots, \kappa_n') \tag{1.2.11}$$

where the summations have to be performed over all permutations of the variables $\kappa_1' = \ldots = \kappa_n'$.

Proof. The proof is given here for $n = 3$, an extension to the higher (and lower) degree case is trivial. The product $u(k-\kappa_1)u(k-\kappa_2)u(k-\kappa_3)$ occurs $3! = 6$ times in the kernels. Therefore

$$\left[h_3(\kappa_1', \kappa_2', \kappa_3') + h_3(\kappa_1', \kappa_3', \kappa_2') + h_3(\kappa_2', \kappa_1', \kappa_3') + h_3(\kappa_2', \kappa_3', \kappa_1')\right.$$
$$\left. + h_3(\kappa_3', \kappa_1', \kappa_2') + h_3(\kappa_3', \kappa_2', \kappa_1')\right] u(k-\kappa_1')u(k-\kappa_2')u(k-\kappa_3')$$
$$= h_3^{\text{sym}}(\kappa_1, \kappa_2, \kappa_3) u(k-\kappa_1) u(k-\kappa_2) u(k-\kappa_3).$$

- *upper triangular kernels:*
In order to avoid that the same product term of the input signal appears many times in the Volterra series, the sum can be defined as

$$V_n[u(k)] = \sum_{\kappa_1=0}^{k} \cdots \sum_{\kappa_n=\kappa_{n-1}}^{k} h^{\mathrm{tri}}(\kappa_1,\ldots,\kappa_n) \prod_{j=1}^{n} u(k-\kappa_j) \tag{1.2.12}$$

As is seen, only those values of the kernels differ from zero where $\kappa_1 \le \ldots \le \kappa_n$. The upper triangular kernels can be calculated from the symmetrical kernels by the formula (Rugh, 1981):

$$h^{\mathrm{tri}}(\kappa_1,\ldots,\kappa_n) = \frac{1}{n!}\sum^{n!} h^{\mathrm{sym}}(i_1,\ldots,i_n) \tag{1.2.13}$$

where $\overset{n!}{\Sigma}$ means that the summation has to be performed for all possible permutations of the arguments of the symmetrical kernels. If some arguments of the triangular kernels are equal to each other then the number of summations can be reduced (Rugh, 1981).

- *lower triangular kernels:*
  Here only those kernels are used where $\kappa_1 \ge \ldots \ge \kappa_n$. Now the Volterra series is

$$V_n[u(k)] = \sum_{\kappa_1=\kappa_2}^{k} \sum_{\kappa_2=\kappa_3}^{k} \cdots \sum_{\kappa_n=0}^{k} h^{\mathrm{tri}}(\kappa_1,\ldots,\kappa_n) \prod_{j=1}^{n} u(k-\kappa_j) \tag{1.2.14}$$

The lower triangular kernels can be calculated from the symmetrical kernels in a similar way as the upper triangular kernels. Further on, the upper triangular kernels will be meant under triangular kernels unless it is stated otherwise.

- *upper regular kernels:*
  The upper regular kernels are defined by the following modified version of the Volterra series

$$V_n[u(k)] = \sum_{\kappa_1=0}^{k} \cdots \sum_{\kappa_n=0}^{k} h^{\mathrm{reg}}(\kappa_1,\ldots,\kappa_n) \prod_{j=1}^{n} u\left(k-\sum_{\ell=1}^{j}\kappa_\ell\right) \tag{1.2.15}$$

The relation between the upper regular and the upper triangular forms is given by

$$h^{\mathrm{reg}}(\kappa_1,\kappa_2,\ldots,\kappa_n) = h^{\mathrm{tri}}(\kappa_1,\kappa_2+\kappa_1,\ldots,\kappa_n+\kappa_{n-1}+\ldots+\kappa_1)$$

Proof. Introduce the distances $\kappa_i' = \kappa_i - \kappa_{i-1};\ i=2,\ldots,n$ between the κ_i-s as new variables. Substitute $\kappa_i = \kappa_i' + \kappa_{i-1};\ i=2,\ldots,n$ into the upper triangular form of the Volterra series (1.2.12)

$$V_n[u(k)] = \sum_{\kappa_1=0}^{k} \sum_{\kappa_2'=0}^{k-\kappa_1} \cdots \sum_{\kappa_n'=0}^{k-\kappa_1-\ldots-\kappa_n} h^{\mathrm{tri}}(\kappa_1,\kappa_2'+\kappa_1,\ldots,\kappa_n'+\ldots+\kappa_2'+\kappa_1) \\ \times u(k-\kappa_1)u(k-\kappa_2'-\kappa_1)\ldots u(k-\kappa_n'-\ldots-\kappa_2'-\kappa_1) \tag{1.2.16}$$

Substitute κ_i' by κ_i in (1.2.16)

$$V_n[u(k)] = \sum_{\kappa_1=0}^{k} \sum_{\kappa_2=0}^{k-\kappa_1} \cdots \sum_{\kappa_n=0}^{k-\kappa_1-\ldots-\kappa_n} h^{\mathrm{reg}}(\kappa_1, \kappa_2+\kappa_1, \ldots, \kappa_n+\ldots+\kappa_2+\kappa_1) \times u(k-\kappa_1)u(k-\kappa_2-\kappa_1)\ldots u(k-\kappa_n-\ldots-\kappa_2-\kappa_1) \tag{1.2.17}$$

■

- *lower regular kernels:*
  The lower regular kernels are defined by the following modified version of the Volterra series

$$V_n[u(k)] = \sum_{\kappa_1=0}^{k} \cdots \sum_{\kappa_n=0}^{k} h^{\mathrm{reg}}(\kappa_1, \ldots, \kappa_n) \prod_{j=1}^{n} u\left(k - \sum_{\ell=1}^{n-j+1} \kappa_\ell\right)$$

The relation between the lower regular and the lower triangular forms is given by

$$h^{\mathrm{reg}}(\kappa_1, \kappa_2, \ldots, \kappa_n) = h^{\mathrm{tri}}(\kappa_1+\kappa_2+\ldots+\kappa_n, \kappa_2+\kappa_3+\ldots+\kappa_n, \ldots, \kappa_n)$$

Proof. The proof is similar to that of the upper regular kernels. ■
Further on, by upper regular kernels will be meant regular kernels unless it is stated otherwise.

Example 1.2.1 *Degree-3 kernel with memory 3.*
The different representation forms will be illustrated by means of a strictly causal degree-3 kernel with memory 3. The strict causality means that the actual input does not influence the actual output signal, which means $h_3(\kappa_1, \kappa_2, \kappa_3) = 0$, if one of the κ_i-s is zero. The 27 non-zero kernels are

$$\begin{array}{lll} h_3(1,1,1), & h_3(1,1,2), & h_3(1,1,3), \\ h_3(1,2,1), & h_3(1,2,2), & h_3(1,2,3), \ldots \\ h_3(3,3,1), & h_3(3,3,2), & h_3(3,3,3). \end{array}$$

We shall calculate the special forms of the kernels for the three possible cases

- all shifting times are different (e.g., $\kappa_1 = 1,\ \kappa_2 = 2,\ \kappa_3 = 3$),
- two shifting times are equal and the third differs from them (e.g., $\kappa_1 = 2$, $\kappa_2 = 2,\ \kappa_3 = 3$),
- all three shifting times are equal (e.g., $\kappa_1 = 2,\ \kappa_2 = 2,\ \kappa_3 = 2$),

only once.

1. Symmetrical kernels

$$h_3^{\mathrm{sym}}(1,2,3) = \tfrac{1}{6}\left[h_3(1,2,3) + h_3(1,3,2) + h_3(2,1,3) + h_3(2,3,1) + h_3(3,1,2) + h_3(3,2,1)\right]$$

$$\begin{aligned} h_3^{\mathrm{sym}}(2,2,3) &= \tfrac{1}{6}\left[h_3(2,2,3) + h_3(2,3,2) + h_3(2,2,3) + h_3(2,3,2) + h_3(3,2,2) + h_3(3,2,2)\right] \\ &= \tfrac{1}{3}\left[h_3(2,2,3) + h_3(2,3,2) + h_3(3,2,2)\right] \end{aligned}$$

$$h_3^{\text{sym}}(2,2,2) = \tfrac{1}{6}\left[h_3(2,2,2)+h_3(2,2,2)+h_3(2,2,2)+h_3(2,2,2)+h_3(2,2,2)+h_3(2,2,2)\right]$$
$$= h_3(2,2,2)$$

2. Upper triangular kernels

$$h_3^{\text{tri}}(1,2,3) = 6h_3^{\text{sym}}(1,2,3)$$
$$= \left[h_3(1,2,3)+h_3(1,3,2)+h_3(2,1,3)+h_3(2,3,1)+h_3(3,1,2)+h_3(3,2,1)\right]$$
$$h_3^{\text{tri}}(2,2,3) = \tfrac{6}{2}h_3^{\text{sym}}(2,2,3)$$
$$= 3\tfrac{1}{6}\left[h_3(2,2,3)+h_3(2,3,2)+h_3(2,2,3)+h_3(2,3,2)+h_3(3,2,2)+h_3(3,2,2)\right]$$
$$= h_3(2,2,3)+h_3(2,3,2)+h_3(3,2,2)$$
$$h_3^{\text{tri}}(2,2,2) = \tfrac{6}{6}h_3^{\text{sym}}(2,2,2)$$
$$= \tfrac{1}{6}\left[h_3(2,2,2)+h_3(2,2,2)+h_3(2,2,2)+h_3(2,2,2)+h_3(2,2,2)+h_3(2,2,2)\right]$$
$$= h_3(2,2,2)$$

3. Upper regular kernels
The triangular kernels calculated above correspond to the following regular kernels

$$h_3^{\text{tri}}(1,2,3) = h_3^{\text{reg}}(1,1,1)$$
$$h_3^{\text{tri}}(2,2,3) = h_3^{\text{reg}}(2,0,1)$$
$$h_3^{\text{tri}}(2,2,2) = h_3^{\text{reg}}(2,0,0)$$

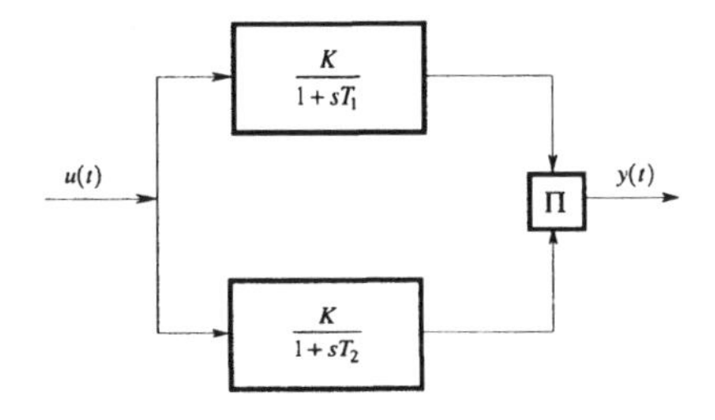

Fig. 1.2.2 Product of two first-order systems

Example 1.2.2 *Product of two linear first-order systems*
Figure 1.2.2 shows the product of the output signals of two first-order systems with the same input signal. The transfer functions are

$$G_1(s) = \frac{K_1}{1+sT_1}$$

and

$$G_2(s) = \frac{K_2}{1+sT_2}$$

respectively.

The degree-2 (second-degree or second-order, as also used in the literature) continuous time Volterra kernel is the product of the weighting functions

$$g_2(\tau_1,\tau_2)=g_1(\tau_1)g_2(\tau_2)=\frac{K_1}{T_1}\exp\left(-\frac{\tau_1}{T_1}\right)\frac{K_2}{T_2}\exp\left(-\frac{\tau_2}{T_2}\right)$$

The discrete time kernel can be calculated according to (1.2.7) if a zero-order holding device is before the system

$$h_2(\kappa_1,\kappa_2)=\int\limits_{\tau_1=\kappa_1\Delta T}^{(\kappa_1+1)\Delta T}\int\limits_{\tau_2=\kappa_2\Delta T}^{(\kappa_2+1)\Delta T}g_2(\tau_1,\tau_2)=\int\limits_{\tau_1=\kappa_1\Delta T}^{(\kappa_1+1)\Delta T}g_1(\tau_1)d\tau_1\int\limits_{\tau_2=\kappa_2\Delta T}^{(\kappa_2+1)\Delta T}g_2(\tau_2)d\tau_2$$

$$=K_1\left[1-\exp\left(-\frac{\Delta T}{T_1}\right)\right]\exp\left(-\frac{\Delta T}{T_1}\kappa_1\right)K_2\left[1-\exp\left(-\frac{\Delta T}{T_2}\right)\right]\exp\left(-\frac{\Delta T}{T_2}\kappa_2\right)$$

$$\kappa_1\geq 1,\ \kappa_2\geq 1$$

The discrete time kernels are zero for zero shifting time. As is seen, both the continuous and the discrete time kernels have the same exponential decay.

The different realization forms of the kernel exist:

1. Symmetrical kernel:

$$g_2^{\text{sym}}(\tau_1,\tau_2)=\tfrac{1}{2}\left[g_2(\tau_1,\tau_2)+g_2(\tau_2,\tau_1)\right]$$

which reduces on the main diagonal $(\tau_1=\tau_2)$ to

$$g_2^{\text{sym}}(\tau_1,\tau_1)=\tfrac{1}{2}\left[g_2(\tau_1,\tau_1)+g_2(\tau_1,\tau_1)\right]=g_2(\tau_1,\tau_1)$$

and

$$h_2^{\text{sym}}(\kappa_1,\kappa_2)=\tfrac{1}{2}\left[h_2(\kappa_1,\kappa_2)+h_2(\kappa_2,\kappa_1)\right]$$

which reduces on the main diagonal $(\kappa_1=\kappa_2)$ to

$$h_2^{\text{sym}}(\kappa_1,\kappa_1)=\tfrac{1}{2}\left[h_2(\kappa_1,\kappa_1)+h_2(\kappa_1,\kappa_1)\right]=h_2(\kappa_1,\kappa_1)$$

2. Upper triangular kernel:

$$g_2^{\text{tri}}(\tau_1,\tau_2)=2g_2^{\text{sym}}(\tau_1,\tau_2)=g_2(\tau_1,\tau_2)+g_2(\tau_2,\tau_1)\qquad\text{if}\quad\tau_1\neq\tau_2$$

$$g_2^{\text{tri}}(\tau_1,\tau_1)=\tfrac{2}{2}g_2^{\text{sym}}(\tau_1,\tau_1)=g_2(\tau_1,\tau_1)$$

$$h_2^{\text{tri}}(\kappa_1,\kappa_2)=2h_2^{\text{sym}}(\kappa_1,\kappa_2)=h_2(\kappa_1,\kappa_2)+h_2(\kappa_2,\kappa_1)\qquad\text{if}\quad\kappa_1\neq\kappa_2$$

$$h_2^{\text{tri}}(\kappa_1,\kappa_1)=\tfrac{2}{2}h_2^{\text{sym}}(\kappa_1,\kappa_1)=h_2(\kappa_1,\kappa_1)$$

3. Upper regular kernel:

$$g_2^{\text{reg}}(\tau_1,\tau_2)=g_2^{\text{tri}}(\tau_1,\tau_2+\tau_1)=g_2(\tau_1,\tau_2+\tau_1)+g_2(\tau_2+\tau_1,\tau_1) \quad \text{if} \quad \tau_2\neq 0$$
$$g_2^{\text{reg}}(\tau_1,0)=g_2^{\text{tri}}(\tau_1,\tau_1)=g_2(\tau_1,\tau_1)$$
$$h_2^{\text{reg}}(\kappa_1,\kappa_2)=h_2^{\text{tri}}(\kappa_1,\kappa_2+\kappa_1)=h_2(\kappa_1,\kappa_2+\kappa_1)+h_2(\kappa_2+\kappa_1,\kappa_1) \quad \text{if} \quad \kappa_2\neq 0$$
$$h_2^{\text{reg}}(\kappa_1,0)=h_2^{\text{tri}}(\kappa_1,\kappa_1)=h_2(\kappa_1,\kappa_1)$$

■

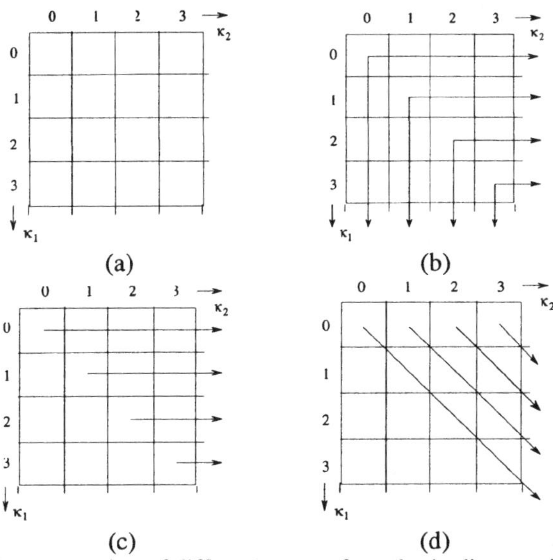

Fig. 1.2.3 Demonstration of different types of quadratic discrete time kernels: (a) general; (b) symmetrical; (c) upper triangular; (d) upper regular

The physical meaning of the different representations of the Volterra kernels will be demonstrated now for the homogeneous quadratic case. For the reason of better illustration the discrete time case is treated. The corresponding kernels are drawn in Figure 1.2.3. The *original* definition is

$$y_2(k)=V_2[u(k)]=\sum_{\kappa_1=0}^{k}\sum_{\kappa_2=0}^{k}h_2(\kappa_1,\kappa_2)u(k-\kappa_1)u(k-\kappa_2)$$

The summation is the same with the symmetrical kernels

$$y_2(k)=\sum_{\kappa_1=0}^{k}\sum_{\kappa_2=0}^{k}h_2^{\text{sym}}(\kappa_1,\kappa_2)u(k-\kappa_1)u(k-\kappa_2)$$

but in this case it would be enough to sum up first the terms on the main diagonal and then all other terms on one side of the main diagonal and multiply them by two

$$y_2(k)=\sum_{\kappa_1=0}^{k}\left[h_2^{\text{sym}}(\kappa_1,\kappa_2)u(k-\kappa_1)+\sum_{\kappa_2=\kappa_1+1}^{k}2h_2^{\text{sym}}(\kappa_1,\kappa_2)u(k-\kappa_2)\right]u(k-\kappa_1) \quad (1.2.18)$$

The triangular kernels are defined in such a way that the sum (1.2.18) can be written in a single double sum

$$y_2(k) = \sum_{\kappa_1=0}^{k} \sum_{\kappa_2=\kappa_1}^{k} h_2^{\text{tri}}(\kappa_1, \kappa_2) u(k-\kappa_1) u(k-\kappa_2) \tag{1.2.19}$$

The triangular kernels are summarized parallel to the axis κ_2 as shown in Figure 1.2.3c. The same kernels (area) can be covered by lines parallel to the main diagonal. This is done with the regular kernels. Introduce κ' as the difference between κ_2 and κ_1 as $\kappa' = \kappa_2 - \kappa_1$, then (1.2.19) becomes

$$y_2(k) = \sum_{\kappa_1=0}^{k} \sum_{\kappa'=0}^{k-\kappa_1} h_2^{\text{tri}}(\kappa_1, \kappa_1 + \kappa') u(k-\kappa_1) u(k-\kappa_1-\kappa')$$

A change about the summation shows now

$$y_2(k) = \sum_{\kappa'=0}^{k} \sum_{\kappa_1=0}^{k} h_2^{\text{tri}}(\kappa_1, \kappa_1 + \kappa') u(k-\kappa_1) u(k-\kappa_1-\kappa') \tag{1.2.20}$$

that the triangular kernels are considered in lines parallel to the main diagonal and κ' is the distance of the off-diagonals from the main diagonal.

A degree- n homogeneous Volterra kernel is called factorable or separable if it can be produced as the product of n degree-1 kernels

$$g_n(\tau_1, \ldots, \tau_n) = \prod_{i=1}^{n} g_i^{\text{f}}(\tau_i) \quad ; \quad h_n(\kappa_1, \ldots, \kappa_n) = \prod_{i=1}^{n} h_i^{\text{f}}(\kappa_i)$$

A factorable quadratic kernel was treated in Example 1.2.2.

A Volterra series model has too many parameters. It is rarely used in real applications. Some applications are now listed herebelow:

1. series of four *stirred tank reactors* (Bard and Lapidus, 1970);
2. *rainfall run off process* (Amorocho and Hart, 1964; Diskin and Boneh, 1973; Hoffmeyer–Zlotnik *et al.*, 1979);
3. *nerve cell* (Kitajima and Hara, 1987),
4. *human controller* (Taylor and Balakrishnan, 1967);
5. *blood pressure of a dog* (Hubbell, 1968).

1.2.2 Wiener model using Laguerre functions and Taylor series

The disadvantage of a Volterra series model is that too many parameters are needed to describe it. Even in the special case, if the system is linear, at least 10–20 points of the weighting function series have to be used. An alternative to it can be a function series, which at least approximates the weighting functions of the low order linear systems with few terms. Wiener (1958) recommended using the Laguerre functions.

The Laguerre functions $\ell_i(t)$ are orthonormal

$$\int_{\tau=0}^{\infty} \ell_i(\tau)\ell_j(\tau)d\tau = \begin{cases} 0 & \text{if} \quad i \neq j \\ 1 & \text{if} \quad i = j \end{cases}$$

and build a complete set

$$\sum_{i=0}^{\infty} \ell_i^2(\tau) < \infty \qquad \forall\tau$$

Then any weighting function can be approximated by the weighted sum of the Laguerre functions

$$g_1(\tau_1) = \int_{i=0}^{\infty} c_i^{\ell} \ell_i(\tau_1)$$

As mentioned before, usually a rapid convergence can be achieved with few Laguerre functions.

Wiener (1958) has shown that the multi-dimensional kernels can be approximated by the weighted sum of the multi-dimensional Laguerre functions

$$g_n(\tau_1,\ldots,\tau_n) = \sum_{i_1=0}^{\infty} \cdots \sum_{i_n=0}^{\infty} c_{i_1\ldots i_n}^{\ell} \ell_{i_1\ldots i_n}(\tau_1,\ldots,\tau_n)$$

where the multi-dimensional Laguerre function is the product of the one-dimensional ones

$$\ell_{i_1\ldots i_n}(\tau_1,\ldots,\tau_n) = \prod_{j=1}^{n} \ell_{i_j}(\tau_j) \tag{1.2.21}$$

The approximation lies on the fact that the multi-dimensional Laguerre function builds a complete set and is orthonormal

$$\int_{\tau_1=0}^{\infty} \cdots \int_{\tau_n=0}^{\infty} \ell_{i_1\ldots i_n}(\tau_1,\ldots,\tau_n)\ell_{i_1\ldots i_n}(\tau_1,\ldots,\tau_n) = \begin{cases} 1 & \text{if} \quad i_1 = \ldots = i_n \\ 0 & \text{otherwise} \end{cases}$$

which is a consequence of the choice of the multi-dimensional Laguerre functions (1.2.21).

The continuous time linear Laguerre function is defined by

$$\ell_i(\tau) = \sqrt{2\gamma}\, \tfrac{1}{i!} e^{\gamma\tau} \frac{\partial^i}{\partial\tau^i}\left[(-\tau)^i e^{-2\gamma\tau}\right]$$

where γ is a free selectable parameter (e.g., Schetzen, 1980). The first three functions are

$$\ell_0(\tau) = \sqrt{2\gamma}\; e^{-\gamma\tau}$$
$$\ell_1(\tau) = \sqrt{2\gamma}\; [2\gamma\tau - 1]\; e^{-\gamma\tau}$$
$$\ell_2(\tau) = \sqrt{2\gamma}\; \left[2(\gamma\tau)^2 - 4\gamma\tau + 1\right] e^{-\gamma\tau}$$

They are plotted in Figure 1.2.4a for $\gamma = 0.5$. All functions have infinite memory.

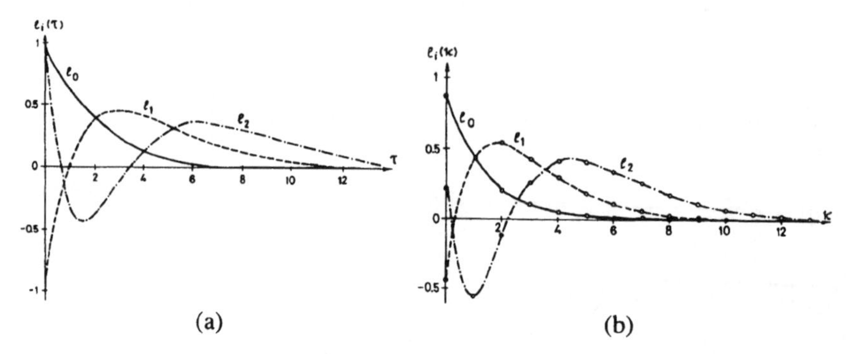

Fig. 1.2.4 The first three Laguerre functions for $\gamma = 0.5$:
(a) continuous time; (b) discrete time

The Laplace transform of the Laguerre functions is given by

$$L_i(s) = \sqrt{2\gamma}\,\frac{(\gamma - s)^i}{(\gamma + s)^{i+1}}$$

which allows a recursive computation of the Laguerre function.

$$L_i(s) = \frac{(\gamma - s)}{(\gamma + s)} L_{i-1}(s) \qquad i = 1, 2, \ldots \qquad \text{and} \qquad L_0(s) = \sqrt{2\gamma}\,\frac{1}{(\gamma + s)}$$

The discrete time Volterra kernel can be approximated by the discrete time Laguerre functions in a similar way as the continuous time one with the continuous time Laguerre functions. The only difference is that now the integrals have to be replaced by sums

$$h_n(\kappa_1, \ldots, \kappa_n) = \sum_{i_1=0}^{\infty} \cdots \sum_{i_n=0}^{\infty} c^{\ell}_{i_1 \ldots i_n} \ell_{i_1 \ldots i_n}(\kappa_1, \ldots, \kappa_n)$$

where the multi-dimensional Laguerre function is the product of the one-dimensional discrete time Laguerre functions

$$\ell_{i_1 \ldots i_n}(\kappa_1, \ldots, \kappa_n) = \prod_{j=1}^{n} \ell_{i_j}(\kappa_j)$$

The z-transform of the discrete time linear Laguerre function is defined by (Simpson, 1965; King and Paraskevopoulos, 1979)

$$L_i(z) = \sqrt{1 - \gamma^2}\,\frac{\left(z^{-1} - \gamma\right)^i}{\left(1 - \gamma z^{-1}\right)^{i+1}} \qquad i = 0, 1, 2, \ldots$$

The discrete time Laguerre functions are also orthonormal as can be seen (Thathachar and Ramaswamy, 1973)

$$\sum_{\kappa=0}^{\infty} \ell_i(\kappa)\ell_j(\kappa) = \begin{cases} 0 & \text{if } i \neq j \\ 1 & \text{if } i = j \end{cases}$$

Figure 1.2.5 shows how the discrete time Volterra kernels can be synthesized from the Laguerre functions.

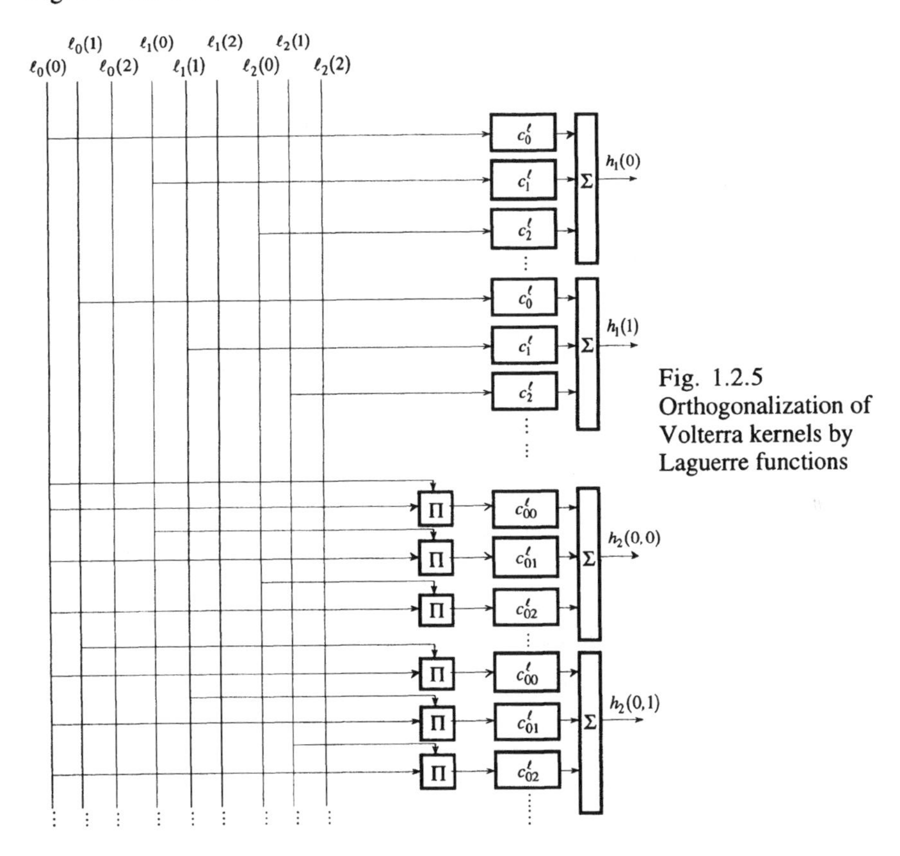

Fig. 1.2.5 Orthogonalization of Volterra kernels by Laguerre functions

Write the degree- n Volterra series model with the one-dimensional Laguerre functions in details

$$y(t) = \sum_{i=0}^{n} V_i[u(t)] = c_0 + \sum_{i=1}^{n} \int_{\tau_1=0}^{t} \cdots \int_{\tau_n=0}^{t} g_n(\tau_1,\ldots,\tau_n) \prod_{j=1}^{i} u(t-\tau_j)d\tau_1 \ldots d\tau_n$$

$$= c_0 + \sum_{i=1}^{n} \int_{\tau_1=0}^{t} \cdots \int_{\tau_n=0}^{t} \sum_{i_1=0}^{\infty} \cdots \sum_{i_n=0}^{\infty} c_{i_1\ldots i_n}^{\ell} \prod_{j=1}^{n} \ell_{i_j}(\tau_j) u(t-\tau_j)d\tau_1 \ldots d\tau_n$$

$$= c_0 + \sum_{i=1}^{n} \sum_{i_1=0}^{\infty} \cdots \sum_{i_n=0}^{\infty} c_{i_1 \ldots i_n}^{\ell} \int_{\tau_1=0}^{t} \cdots \int_{\tau_n=0}^{t} \prod_{j=1}^{n} \ell_{i_j}(\tau_j) u(t-\tau_j) d\tau_1 \ldots d\tau_n$$

$$= c_0 + \sum_{i=1}^{n} \sum_{i_1=0}^{\infty} \cdots \sum_{i_n=0}^{\infty} c_{i_1 \ldots i_n}^{\ell} \prod_{j=1}^{n} \int_{\tau_j=0}^{t} \ell_{i_j}(\tau_j) u(t-\tau_j) d\tau_j \qquad (1.2.22)$$

$$= c_0 + \sum_{i=1}^{n} \sum_{i_1=0}^{\infty} \cdots \sum_{i_n=0}^{\infty} c_{i_1 \ldots i_n}^{\ell} \prod_{j=1}^{n} v_{i_j}(t)$$

where the signal $v_i(t)$ is the output of the ith order Laguerre filter whose input is the input signal of the process

$$v_{i_j}(t) = \int_{\tau_j=0}^{t} \ell_{i_j}(\tau_j) u(t-\tau_j) d\tau_j$$

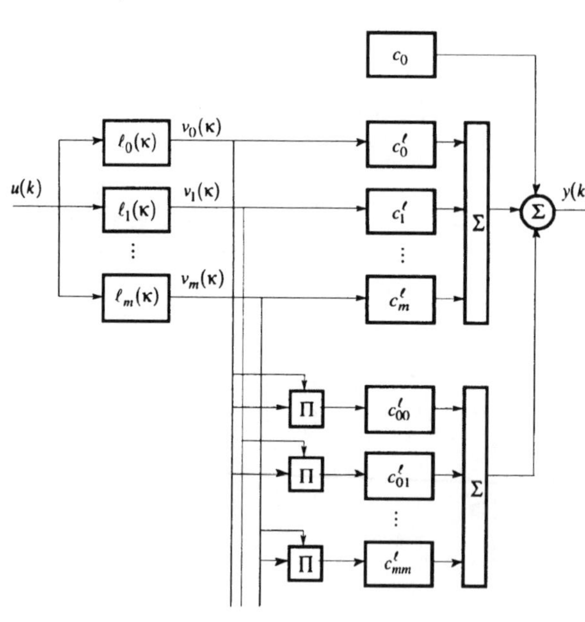

Fig. 1.2.6
Discrete time quadratic Wiener model using Laguerre functions

The output signal of the model is the sum of all possible weighted products of the $v_i(t)$-s. All possible products mean all products of the $v_i(t)$-s, which have no more terms than n.

The discrete time model has the same structure

$$y(k) = \sum_{i=0}^{n} V_i[u(k)] = c_0 + \sum_{i=1}^{n} \sum_{i_1=0}^{\infty} \cdots \sum_{i_n=0}^{\infty} c_{i_1 \ldots i_n}^{\ell} \prod_{j=1}^{n} v_{i_j}(k)$$

and $v_i(k)$ is the output of the ith order discrete time Laguerre filter. Figure 1.2.6 shows the scheme of the model.

1.2.3 Wiener model using Laguerre and Hermite functions

(1.2.22) has all possible combination of the $v_i(t)$-s upto degree n. This can also be expressed by another form. Denote a weighted polynomial of $v_i(t)$ by

$$P^{\mathrm{H}}[v_i(t)] = \sum_{j=0}^{n} c_j^{\mathrm{H}} v_i^j(t)$$

Then (1.2.22) can be rewritten as

$$y(t) = \sum_{i_1=0}^{\infty} \cdots \sum_{i_n=0}^{\infty} c_{i_1 \ldots i_n}^{\mathrm{H}} \prod_{j=0}^{n} P_1^{\mathrm{H}}\left[v_{i_j}(t)\right] \qquad \left(P_1^{\mathrm{H}}[v_i(t)]\right)^j \to P_j^{\mathrm{H}}[v_i(t)] \tag{1.2.23}$$

which means that the products of the degree-1 functionals of the same argument have to be replaced by a higher degree functional with the same argument and with degree equal to the multiple occurrence. The weighting coefficients in (1.2.22) and (1.2.23) differ from each other, therefore now the upper index H is used. The discrete time model is built up analogously:

$$y(k) = \sum_{i_1=0}^{\infty} \cdots \sum_{i_n=0}^{\infty} c_{i_1 \ldots i_n}^{\mathrm{H}} \prod_{j=0}^{n} P_1^{\mathrm{H}}\left[v_{i_j}(k)\right] \qquad \left(P_1^{\mathrm{H}}[v_i(k)]\right)^j \to P_j^{\mathrm{H}}[v_i(k)]$$

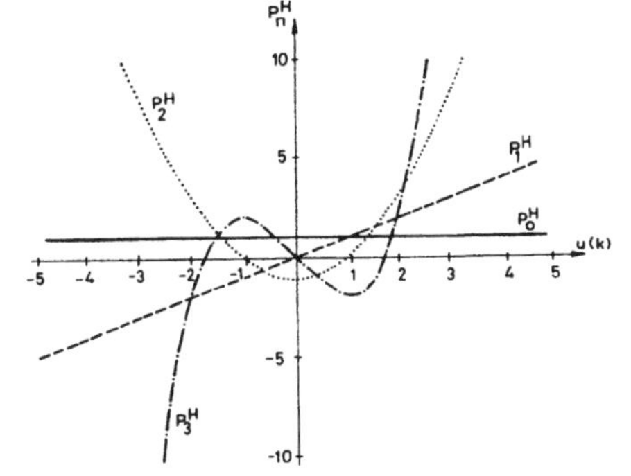

Fig. 1.2.7
The first four Hermite functions $\sigma_u = 1$

Wiener proposed using the Hermite polynomials because of some orthogonal properties of the model that make the estimation of the unknown coefficients easy. The degree-n Hermite polynomials are (e.g., Schetzen, 1980) as follows

$$P_n^{\mathrm{H}}[u(k)] = \sum_{i=0}^{\mathrm{entier}[n/2]} \frac{(-1)^i n!}{i!(n-2i)!} \left(\frac{\sigma_u^2}{2}\right)^i u^{n-2i}(k) \tag{1.2.24}$$

and the first four terms are

$$P_0^{\mathrm{H}}[u(k)] = 1 \qquad P_1^{\mathrm{H}}[u(k)] = u(k)$$

$$P_2^{\mathrm{H}}[u(k)] = u^2(k) - \sigma_u^2 \qquad P_3^{\mathrm{H}}[u(k)] = u^3(k) - 3\sigma_u^2 u(k)$$

They are plotted in Figure 1.2.7 for $\sigma_u = 1$.

The Hermite polynomials are orthogonal when the argument is a Gaussian random variable with zero mean and standard deviation σ_u

$$\mathrm{E}\left\{P_i^{\mathrm{H}}[u(k)]P_j^{\mathrm{H}}[u(k)]\right\} = \begin{cases} 0 & \text{if} \quad i \neq j \\ i!\,\sigma_u^{2i} & \text{if} \quad i = j \end{cases}$$

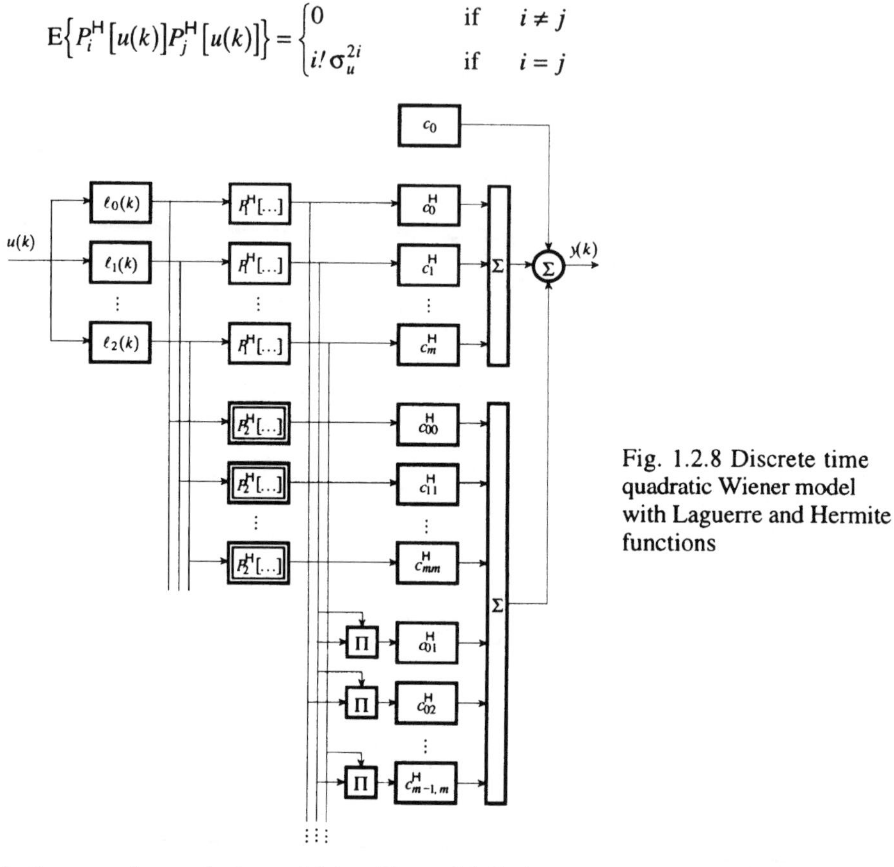

Fig. 1.2.8 Discrete time quadratic Wiener model with Laguerre and Hermite functions

In the case of a quadratic model (see Figure 1.2.8) the components are as follows

$$\begin{aligned} y(k) = c_0 &+ c_0^{\mathrm{H}} P_1^{\mathrm{H}}(v_0(t)) + c_1^{\mathrm{H}} P_1^{\mathrm{H}}(v_1(t)) + \ldots + c_m^{\mathrm{H}} P_1^{\mathrm{H}}(v_m(t)) \\ &+ c_{00}^{\mathrm{H}} P_2^{\mathrm{H}}(v_0(t)) + c_{11}^{\mathrm{H}} P_2^{\mathrm{H}}(v_1(t)) + \ldots + c_{mm}^{\mathrm{H}} P_2^{\mathrm{H}}(v_m(t)) \\ &+ c_{01}^{\mathrm{H}} P_1^{\mathrm{H}}(v_0(t)) P_1^{\mathrm{H}}(v_1(t)) + c_{02}^{\mathrm{H}} P_1^{\mathrm{H}}(v_0(t)) P_1^{\mathrm{H}}(v_2(t)) + \ldots \\ &+ c_{m-1,m}^{\mathrm{H}} P_1^{\mathrm{H}}(v_{m-1}(t)) P_1^{\mathrm{H}}(v_m(t)) \end{aligned}$$

The advantage of selecting the Hermite polynomials is that the components of the model are orthogonal to each other (Schetzen, 1980). This is true because the outputs of the Laguerre filters are also Gaussian random signals with the same standard deviation as the input signal.

1.2.4 The orthogonal Wiener series model

The degree-n Volterra series model contains n parallel homogeneous terms. Wiener

(1958) has shown how these submodels can be orthogonalized to a Gaussian random input signal.

Before dealing with the dynamical case consider the orthogonalization of a static polynomial

$$y(t)=\sum_{i=0}^{\infty} y_i(t)=c_0 1+c_1 u(t)+c_2 u^2(t)+c_3 u^3(t)+\ldots$$

Denote the orthogonalized model by

$$y(t)=\sum_{i=0}^{\infty} y_i^{\circ}(t)=c_0^{\circ}[1]^{\circ}+c_1^{\circ}[u(t)]^{\circ}+c_2^{\circ}\left[u^2(t)\right]^{\circ}+c_3^{\circ}\left[u^3(t)\right]^{\circ}+\ldots$$

where the index $^{\circ}$ means the orthogonalized values. The new components can be derived by using the Gram–Schmidt orthogonalization procedure. With the assumption made to the input signal the elements of the orthogonal polynomial are the Hermite polynomials. We expect a similar result for the dynamical case because the Volterra series are also polynomial functions.

The constant term is equal to the constant term of the Volterra series

$$y_0^{\circ}=V_0^{\circ}[u(t)]=g_0^{\circ}=g_0=V_0[u(t)]=y_0$$

The most general form of the linear term is

$$y_1^{\circ}(t)=V_1^{\circ}(t)=\int_{\tau=0}^{t} g_1^{\circ}(\tau)u(t-\tau)d\tau+g_{10}^{\circ} \tag{1.2.25}$$

(1.2.25) has to be orthogonal to any constant term $y_0=g_0$

$$\mathrm{E}\left\{y_1^{\circ}(t)y_0\right\}=\mathrm{E}\left\{\int_{\tau=0}^{t} g_0 g_1^{\circ}(\tau)u(t-\tau)d\tau+g_0 g_{10}^{\circ}\right\}=0$$

if

$$g_{10}^{\circ}=0$$

The degree-1 (first-order) term of the Wiener series is therefore

$$y_1^{\circ}(t)=V_1^{\circ}[u(t)]=\int_{\tau=0}^{t} g_1^{\circ}(\tau)u(t-\tau)d\tau$$

The most general form of the quadratic term is

$$\begin{aligned} y_2^{\circ}(t)&=V_2^{\circ}(t)\\ &=\int_{\tau_1=0}^{t}\int_{\tau_2=0}^{t} g_2^{\circ}(\tau_1,\tau_2)u(t-\tau_1)u(t-\tau_2)d\tau_1 d\tau_2+\int_{\tau_1=0}^{t} g_{21}^{\circ}(\tau_1)u(t-\tau_1)d\tau_1+g_{20}^{\circ} \end{aligned} \tag{1.2.26}$$

(1.2.26) has to be orthogonal to any linear and constant terms. The first condition is

$$\begin{aligned} \mathrm{E}\left\{y_2^{\mathrm{o}}(t)y_1(t)\right\} &= \mathrm{E}\left\{y_2^{\mathrm{o}}(t)\int_{\tau=0}^{t} g_1(\tau)u(t-\tau)d\tau\right\} \\ &= \mathrm{E}\left\{\int_{\tau_1=0}^{t} g_{21}^{\mathrm{o}}(\tau_1)u(t-\tau_1)d\tau_1 \int_{\tau=0}^{t} g_1(\tau)u(t-\tau)d\tau\right\} = 0 \end{aligned} \quad (1.2.27)$$

The quadratic and the constant terms do not contribute to (1.2.27) because the expected values of the odd order moments of a Gaussian random process are zero. (1.2.27) can be put zero by selecting

$$g_{21}^{\mathrm{o}}(\tau_1) = 0$$

The second condition gives

$$\begin{aligned} \mathrm{E}\left\{y_2^{\mathrm{o}}(t)y_0(t)\right\} &= \mathrm{E}\left\{y_2^{\mathrm{o}}(t)g_0\right\} \\ &= \mathrm{E}\left\{g_0 \int_{\tau_1=0}^{t}\int_{\tau_2=0}^{t} g_2^{\mathrm{o}}(\tau_1,\tau_2)u(t-\tau_1)u(t-\tau_2)d\tau_1\, d\tau_2 + g_0 g_{20}^{\mathrm{o}}\right\} \\ &= g_0\left[\sigma_u^2 \int_{\tau=0}^{\infty}\int_{\tau=0}^{\infty} g_2^{\mathrm{o}}(\tau,\tau)\, d\tau + g_{20}^{\mathrm{o}}\right] = 0 \end{aligned}$$

which leads to

$$g_{20}^{\mathrm{o}} = -\sigma_u^2 \int_{\tau=0}^{\infty}\int_{\tau=0}^{\infty} g_2^{\mathrm{o}}(\tau,\tau)\, d\tau$$

The total form of the quadratic Wiener kernel is

$$y_2^{\mathrm{o}}(t) = V_2^{\mathrm{o}}[u(t)] = \int_{\tau_1=0}^{t}\int_{\tau_2=0}^{t} g_2^{\mathrm{o}}(\tau_1,\tau_2)u(t-\tau_1)u(t-\tau_2)d\tau_1\, d\tau_2 - \sigma_u^2 \int_{\tau=0}^{\infty} g_2^{\mathrm{o}}(\tau,\tau)d\tau$$

The cubic term of the Wiener series can be derived in a similar manner (Schetzen, 1980; Rugh, 1981):

$$\begin{aligned} y_3^{\mathrm{o}}(t) &= V_3^{\mathrm{o}}[u(t)] \\ &= \int_{\tau_1=0}^{t}\int_{\tau_2=0}^{t}\int_{\tau_3=0}^{t} g_3^{\mathrm{o}}(\tau_1,\tau_2,\tau_3)u(t-\tau_1)u(t-\tau_2)\, u(t-\tau_3)d\tau_1\, d\tau_2 d\tau_3 \\ &\quad -3\sigma_u^2 \int_{\tau_1=0}^{t}\int_{\tau_1^{\mathrm{o}}=0}^{\infty} g_3^{\mathrm{o}}(\tau_1,\tau_1^{\mathrm{o}},\tau_1^{\mathrm{o}})u(t-\tau_1)d\tau_1^{\mathrm{o}}d\tau_1 \end{aligned}$$

The general form of the Wiener series is (Rugh, 1981):

$$y_i^o(t) = V_i^o[u(t)]$$

$$= \sum_{i=0}^{\text{entier}[n/2]} \frac{(-1)^i n!}{i!(n-2i)!}\left(\frac{\sigma_u^2}{2}\right)^i \int_{\tau_1=0}^{t} \cdots \int_{\tau_{n-2i}=0}^{t} \int_{\tau_1^o=0}^{\infty} \cdots \int_{\tau_i^o=0}^{\infty} g_n^o\left(\tau_1,\ldots,\tau_{n-2i},\tau_1^o,\tau_1^o,\ldots,\tau_i^o,\tau_i^o\right)$$

$$\times d\tau_1^o \ldots d\tau_i^o \prod_{j=1}^{n-2i} u\left(t-\tau_j\right)d\tau_1 \ldots d\tau_{n-2i}$$

The degree- n Wiener kernel can be calculated from the Volterra kernels as (e.g., Rugh, 1981)

$$g_n^o\left(\tau_1,\ldots,\tau_n\right) = \sum_{i=0}^{\infty} \frac{(n+2i)!}{i!\,n!}\left(\frac{\sigma_u^2}{2}\right)^i$$

$$\times \int_{\tau_1^o=0}^{\infty} \cdots \int_{\tau_i^o=0}^{\infty} g_{n+2i}^{\text{sym}}\left(\tau_1,\ldots,\tau_n,\tau_1^o,\tau_1^o,\ldots,\tau_i^o,\tau_i^o\right)d\tau_1^o \ldots d\tau_i^o$$

The backward transformation is when the degree- n Volterra kernel can be calculated from the Wiener kernels as (e.g., Rugh, 1981)

$$g_n^{\text{sym}}\left(\tau_1,\ldots,\tau_n\right) = \sum_{i=0}^{\infty} \frac{(-1)^i (n+2i)!}{i!\,n!}\left(\frac{\sigma_u^2}{2}\right)^i$$

$$\times \int_{\tau_1^o=0}^{\infty} \cdots \int_{\tau_i^o=0}^{\infty} g_{n+2i}^o\left(\tau_1,\ldots,\tau_n,\tau_1^o,\tau_1^o,\ldots,\tau_i^o,\tau_i^o\right)d\tau_1^o \ldots d\tau_i^o$$

The proof of the above transformation is given in (Rugh, 1981).

Remarks: (Marmarelis and Marmarelis, 1978; Schetzen, 1980; Rugh, 1981)

1. The Volterra kernels depend only on the system (and not on the input signal);
2. The Wiener kernels depend on both the system and the input signal;
3. Higher degree Wiener functionals contain also lower degree terms. The contributing terms have the degrees $n-2, n-4, n-6, \ldots$;
4. As because of *Remark 3* the degree-n homogeneous Volterra kernel is a function of the Wiener kernels of degree-n, $n+2, n+4, \ldots$;
5. As a further consequence of *Remark 3* the Wiener functionals are not homogeneous;
6. The Wiener functionals are orthogonal to any functionals of lower degree if the input signal is a Gaussian white noise;
7. The Wiener functionals of different degrees are orthogonal to the Gaussian random input signal and build a complete set (Ahmed, 1970). This has the following consequences:
   - The Wiener kernels can be estimated independently form each other;
   - The Wiener series have a better approximate. Even a truncated model with less degree than the degree of the real system would

approximate the process better than the Volterra series of the same degree;

- An inclusion of a higher degree term into the already estimated model would not effect the already estimated parameters in the modeling with the Wiener series. Using, however, the Volterra series approach, any increase of the degree of the fitting model would follow that the already estimated parameters are invalid.

8. The Volterra series approximation of a system exists only if the steady state input–output relation of the system can be described by Taylor series. This is not the case for non-analytic functions, like saturation with cut points. As the Wiener series are orthogonal, those processes can also be described by them, if the assumption for the input signal is fulfilled.

The discrete time kernels have the similar structure as the continuous time ones.

Orthogonal series can be developed for any type of input signals. The orthogonalization for the following input signals is known:

- constant switching pace symmetric random signals (Marmarelis and Marmarelis, 1978; Marmarelis, 1979);
- multi-level, uniformly distributed random signal (Melton and Kroeker, 1982);
- non-Gaussian random white noise (Klein and Yasui, 1979);
- Gaussian white noise (Barrett, 1963);
- Gaussian colored noise (Barrett, 1963).

An application of the Wiener model is

- neuron chain (Marmarelis and Naka, 1972).

1.2.5 Other nonparametric models

1. The Hammerstein series model

The Hammerstein operator (Hammerstein, 1930) is the simplest extension of the linear weighting function to the nonlinear case; the nonlinear relation exists only between the input signal of the process and the input signal of the linear dynamic part of the system

$$y(t) = \int_{\tau=0}^{t} g(\tau)P[u(t-\tau)]d\tau \tag{1.2.28}$$

Here $P[u(t)]$ is usually a polynomial

$$P[u(t)] = \sum_{i=0}^{n} c_i u^i(t) \tag{1.2.29}$$

Putting (1.2.28) into (1.2.29) leads to

$$y(t) = \int_{\tau=0}^{t} g(\tau)\sum_{i=0}^{n} c_i u^i(t-\tau) = \sum_{i=0}^{n} c_i \int_{\tau=0}^{t} g(\tau)u^i(t-\tau)d\tau \tag{1.2.30}$$

(1.2.30) is equivalent to a Volterra series model with the kernels

$$g_n(\tau_1,\ldots,\tau_n) = \begin{cases} 0 & \text{if} \quad \tau_1 \neq \ldots \neq \tau_n \\ g(\tau) & \text{if} \quad \tau_1 = \ldots = \tau_n = \tau \end{cases} \tag{1.2.31}$$

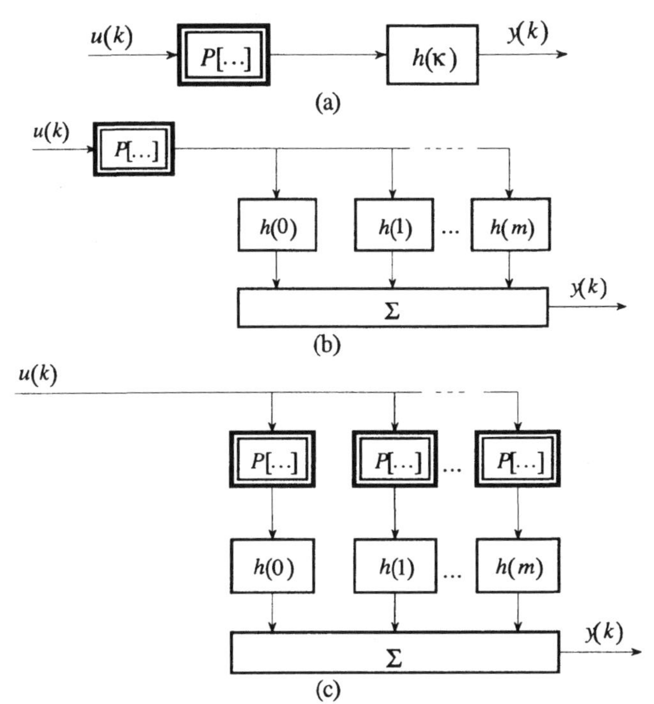

Fig. 1.2.9 Discrete time Hammerstein series model:
(a) cascade structure; (b) equivalent parallel form 1; (c) equivalent parallel form 2

(1.2.31) means that all kernels of the Volterra series are diagonal, i.e., the crossproduct terms of the differently shifted input signal do not influence the output signal.

The discrete time Hammerstein series model (see Figure 1.2.9) has the equation

$$y(k) = \sum_{\kappa=0}^{k} h(\kappa) P[u(k-\kappa)]$$

2. The Uryson series model

The Uryson series model is the extension of the Hammerstein model (Gallman, 1971; Gallman and Narendra, 1976). Its continuous time form is

$$\begin{aligned} y(k) &= g_0 + \sum_{i_1=0}^{n} \int_{\tau_1=0}^{t} g_{1i_1}(\tau_1) P_{1i_1}[u(t-\tau_1)] d\tau_1 \\ &+ \sum_{i_1=0}^{n} \sum_{i_2=0}^{n} \left[\int_{\tau_1=0}^{t} g_{21i_1}(\tau_1) P_{21i_1}[u(t-\tau_1)] d\tau_1 \right] \left[\int_{\tau_2=0}^{t} g_{22i_2}(\tau_2) P_{22i_2}[u(t-\tau_2)] d\tau_2 \right] + \dots \\ &= \sum_{i=0}^{n} \sum_{i_1=0}^{n} \dots \sum_{i_n=0}^{n} \prod_{j=1}^{i} \int_{\tau_j=0}^{t} g_{iji_j}(\tau_j) P_{iji_j}\left[u(t-\tau_j)\right] d\tau_j \end{aligned} \tag{1.2.32}$$

and the discrete time equation is

$$y(k)=\sum_{i=0}^{n}\sum_{i_1=0}^{n}\cdots\sum_{i_n=0}^{n}\prod_{j=1}^{i}\sum_{\kappa_n=0}^{n}h_{iji_j}(\kappa_j)P_{iji_j}\left[u(k-\kappa_j)\right] \qquad (1.2.33)$$

The scheme of the Uryson series model is seen in Figure 1.2.10. The degree-0 Uryson model is the constant term. The degree-1 model is the sum of different Hammerstein models. The degree-2 model is the sum of the products of the output signals of different Hammerstein models. Clearly the degree of the model is not equal to the degree of the nonlinear power of the polynomial steady state relation of the system, i.e., the degree of the power of the steady state nonlinearity is higher than the degree of the model.

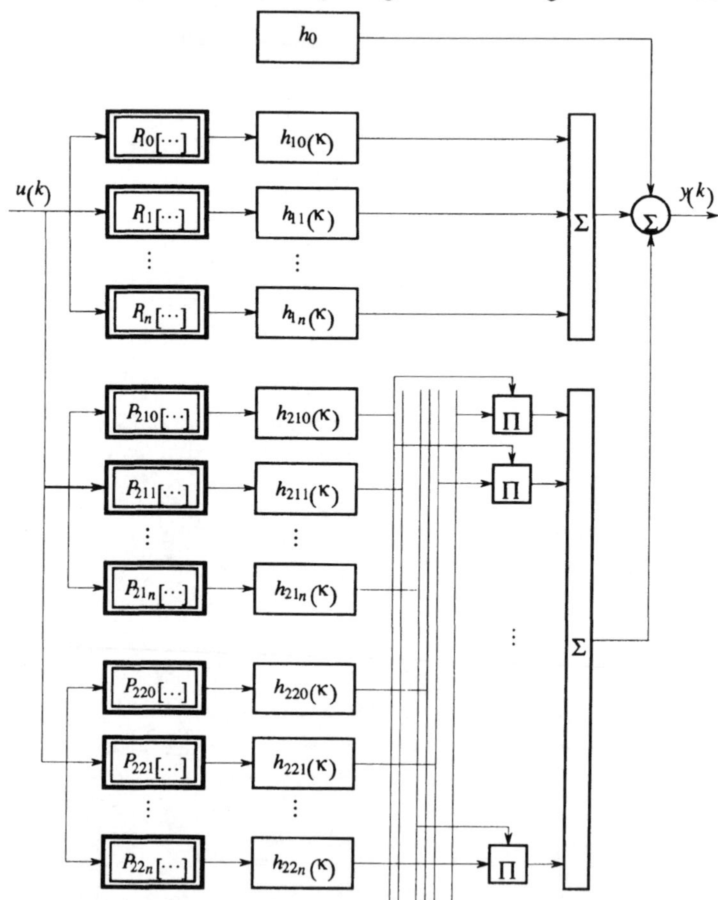

Fig. 1.2.10 Discrete time degree-2 Uryson series model

3. The Zadeh series model

The Uryson series model can be rewritten to

$$y(t)=g_0+\sum_{i_1=0}^{n}\int_{\tau_1=0}^{t}g_{1i_1}(\tau_1)P_{1i_1}\left[u(t-\tau_1)\right]d\tau_1$$

$$+\sum_{i_1=0}^{n}\sum_{i_2=0}^{n}\int_{\tau_1=0}^{t}\int_{\tau_2=0}^{t} g_{21i_1}(\tau_1)g_{21i_2}(\tau_2)P_{22i_1}\left[u(t-\tau_1)\right]P_{22i_2}\left[u(t-\tau_2)\right]d\tau_1 d\tau_2+\ldots$$

$$=g_0+\sum_{i_1=0}^{n}\int_{\tau_1=0}^{t} g_{1i_1}(\tau_1)P_{i_1}\left[u(t-\tau_1)\right]d\tau_1$$

$$+\sum_{i_1=0}^{n}\sum_{i_2=0}^{n}\int_{\tau_1=0}^{t}\int_{\tau_2=0}^{t} g_{2i_1i_2}(\tau_1,\tau_2)P_{21i_1}\left[u(t-\tau_1)\right]P_{22i_2}\left[u(t-\tau_2)\right]d\tau_1 d\tau_2+\ldots$$

$$=g_0+\int_{\tau_1=0}^{t}\sum_{i_1=0}^{n} g_{1i_1}(\tau_1)P_{1i_1}\left[u(t-\tau_1)\right]d\tau_1$$

$$+\int_{\tau_1=0}^{t}\int_{\tau_2=0}^{t}\sum_{i_1=0}^{n}\sum_{i_2=0}^{n} g_{2i_1i_2}(\tau_1,\tau_2)P_{21i_1}\left[u(t-\tau_1)\right]P_{22i_2}\left[u(t-\tau_2)\right]d\tau_1 d\tau_2+\ldots$$

$$=\sum_{i=0}^{n}\int_{\tau_{i_1}=0}^{t}\cdots\int_{\tau_{i_n}=0}^{t}\sum_{i_1=0}^{n}\cdots\sum_{i_n=0}^{n} g_{1i_1\ldots i_n}\left(\tau_{i_1},\ldots,\tau_{i_n}\right)\prod_{j=1}^{i} P_{iji_j}\left[u\left(t-\tau_j\right)\right]d\tau_j$$

$$=\sum_{i=0}^{n}\int_{\tau_{i_1}=0}^{t}\cdots\int_{\tau_{i_n}=0}^{t} f_{ii_1\ldots i_n}\left(u\left(t-\tau_{i_1}\right),\ldots,u\left(t-\tau_{i_n}\right)\right)d\tau_{i_1}\ldots d\tau_{i_n} \tag{1.2.34}$$

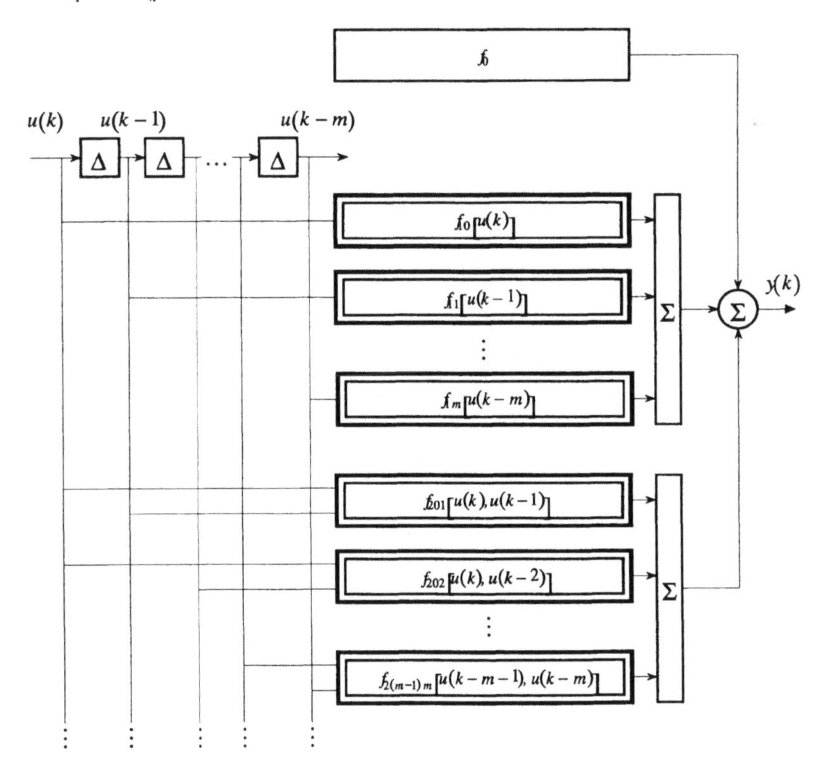

Fig. 1.2.11 Discrete time degree-2 Zadeh series model

which is called the continuous time Zadeh series model (Zadeh, 1953a, 1953b; Gallman and Narendra, 1976). A comparison with the Volterra series model shows that the degree-n term of the Volterra series $g_n(\tau_1,\dots,\tau_n)u(t-\tau_1)\dots u(t-\tau_n)$ is replaced now by $f_n(u(t-\tau_1),\dots,u(t-\tau_n))$. The kernel $f_n(.)$ is an arbitrary function of the shifted input signals $u(t-\tau_i)$. Similarly to the case of the Uryson model, the degree of the Zadeh series model is not equal to the degree (or power) of the nonlinear steady state relation of the system. In other words, the Volterra series are special cases of the Zadeh series.

The discrete time Zadeh series model has the form

$$y(k)=\sum_{i=0}^{n}\sum_{\kappa_{i_1}=0}^{k}\dots\sum_{\kappa_{i_n}=0}^{k} f_{ii_1\dots i_n}\left(u\left(k-\kappa_{i_1}\right),\dots,u\left(k-\kappa_{i_n}\right)\right) \tag{1.2.35}$$

Its scheme is given in Figure 1.2.11.

1.3 BLOCK ORIENTED MODELS

The block oriented models are separable models consisting of linear dynamic and static nonlinear elements. They may or may not have feedbacks. In feedback-free systems the highest degree of the power of the nonlinearity is the maximum of the sums of the powers of the nonlinear polynomial functions in each input–output channel. In feedback case the resulting power is infinite. That means even higher degree nonlinear systems can be built up by low degree terms.

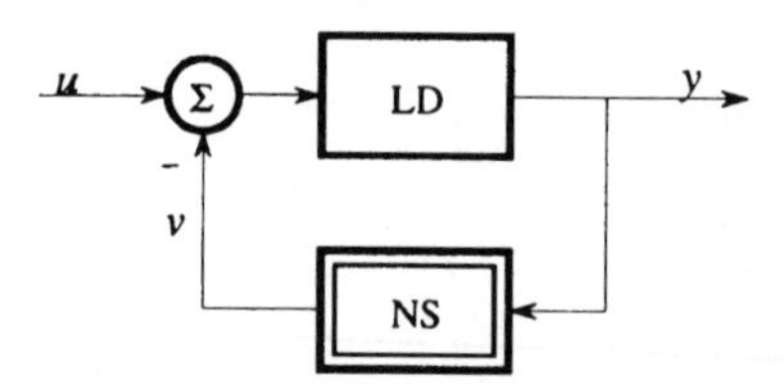

Fig. 1.3.1 Linear dynamic system with a static nonlinear term in the feedback

Example 1.3.1 *Linear dynamic system feedbacked by a static term with parabolic characteristics (Figure 1.3.1)*

Have the linear dynamic (LD) part a unity static gain and the nonlinear static (NS) element a parabolic characteristic:

$$v(k)=c_0+c_1y(k)+c_2y^2(k)$$

Then the steady state equation of the closed loop system is

$$Y=1\times\left[U-\left(c_0+c_1Y+c_2Y^2\right)\right]$$

which is a quadratic polynomial $U=U(Y)$ and can be approximated by a Taylor series $Y=Y(U)$ of infinite degree. ■

First we deal with those block oriented systems that have no feedback.

1.3.1 Classification according to the number of the multipliers

Schetzen (1965a) classified the block oriented systems (without feedback) according to the number of multipliers occurring. He observed that any homogeneous nonlinear system of a given degree may have only degree minus one multipliers in a path (channel) between the input and the output. He called such a channel a basis system. A homogeneous nonlinear system may consist of more basis systems of the same degree.

TABLE 1.3.1 Number of possible basis system of the same degree (Schetzen, 1965)

Degree	0	1	2	3	4	5	6	7	8	9	10
Number of basis system	1	1	1	1	2	3	6	11	24	47	103

The lower degree basis systems are as follows:

1. A constant system is a constant term;
2. A linear system consists of a single linear dynamic term;
3. A quadratic system may have only one multiplier between the input and the output signals. Since a multiplier has two inputs and one output, two linear dynamic terms may precede and a third one may follow the multiplier;
4. A cubic system may have two multipliers. Since there are six connection possibilities to the two multipliers, and they are connected with each other, there can be "3·2-1=5" linear dynamic terms in the system;
5. A fourth-degree basis system has three multipliers. Since there are nine connection possibilities to the three multipliers, and they are connected with each other, there can be "3·3-2=7" linear dynamic terms in the system. There are two ways how the basis system can be built up.

Figure 1.3.2 shows the basis systems till degree four. The number of possible different basis systems of the same degree is given up to degree-10 in Table 1.3.1.

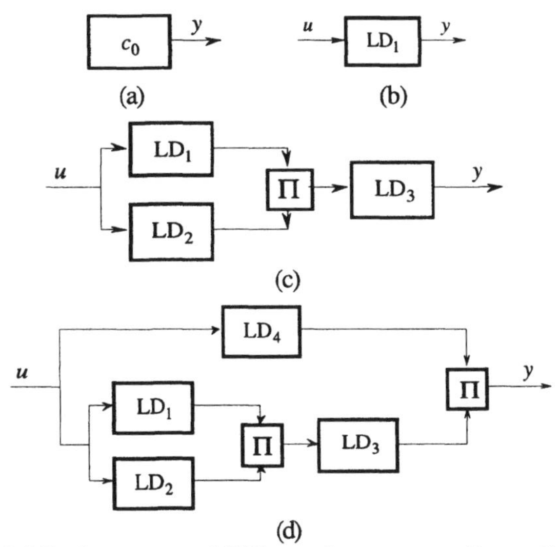

Fig. 1.3.2 Basic systems of different degrees according to Schetzen: (a) degree-0; (b) degree-1; (c) degree-2; (d) degree-3

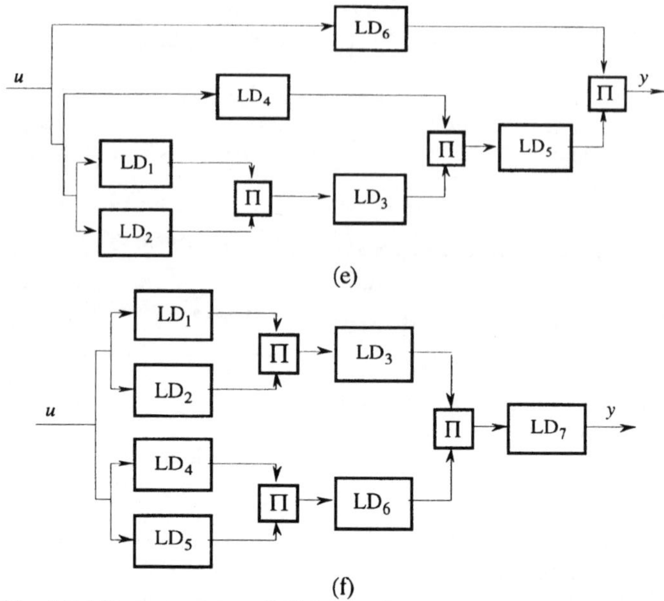

Fig. 1.3.2 Basic systems of different degrees according to Schetzen: (e) degree-4 version 1; (f) degree-4 version 2

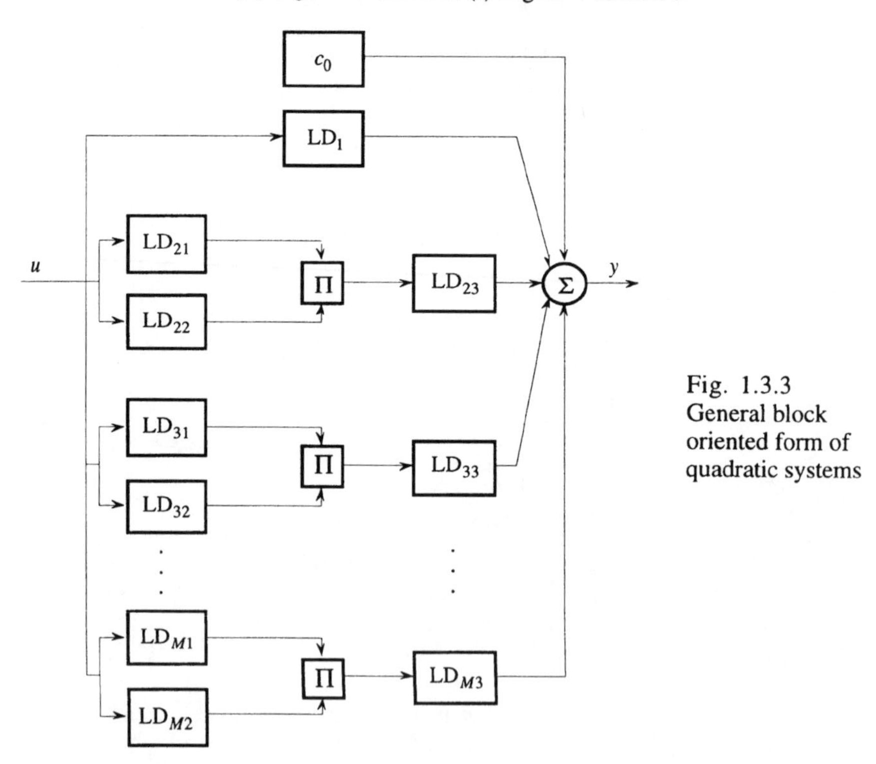

Fig. 1.3.3 General block oriented form of quadratic systems

Figure 1.3.3 shows the general form of a discrete time quadratic system; it consists of a constant term, a linear dynamic channel, and infinite number of parallel degree-2 basis systems. In practical cases only a few, mostly one quadratic channel is used.

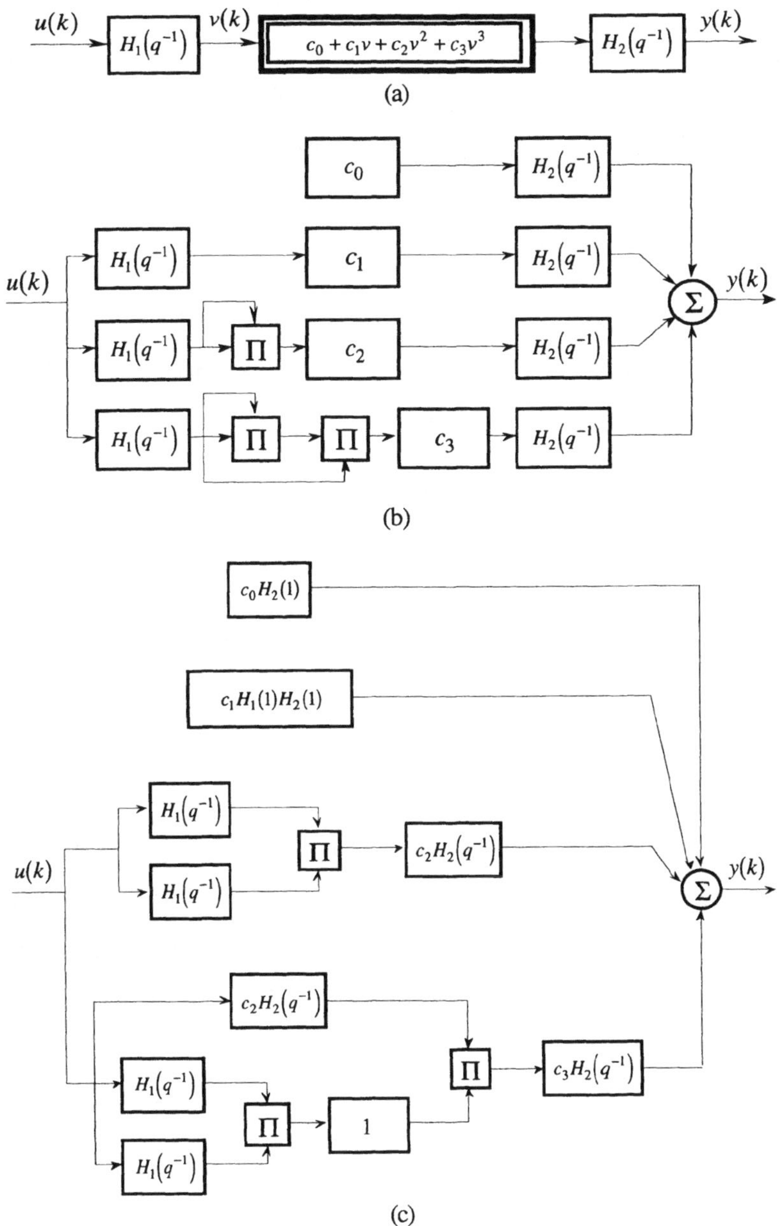

Fig. 1.3.4 Cubic simple Wiener–Hammerstein cascade model: (a) cascade structure; (b) parallel scheme; (c) Schetzen model

Example 1.3.2 *Cubic simple Wiener–Hammerstein cascade system*
A discrete time cascade model consisting of an element with cubic nonlinear characteristic and preceded and followed by linear dynamic terms (Figure 1.3.4a) has to be redrawn into a canonical form according to Schetzen. The system can be redrawn into a parallel form, where each channel has systems of different degree among 0 and 3, as seen in Figure 1.3.4b. The output signals of a degree-*n* system are denoted by $y_i(k)$. Now, each channel can be reconstructed into a form suggested by Schetzen (Figure 1.3.4c). The following special features of the model can be observed:

1. There is only one parallel quadratic and cubic channels;
2. The dynamic parts do not differ from each other;
3. The pulse transfer functions of all linear dynamic parts originate from the parameters (five terms) of the cascade model.

1.3.2 Quadratic block oriented models
In practice processes are usually approximated by simple structures. The simplest nonlinear model is the quadratic model. By means of these models a better input–output approximation can be achieved than by a linear one. Further on an extremum control can be realized based on quadratic models.

In Figure 1.3.5 the different quadratic block oriented models are summarized. The discrete time description has been chosen.

1. Simple Hammerstein model (Figure 1.3.5a)

It consists of a static parabolic characteristic

$$v(k) = c_0^{\mathrm{H}} + c_1 u(k) + c_2 u^2(k)$$

followed by a linear dynamic term

$$y(k) = \frac{B(q^{-1})}{A(q^{-1})} v(k)$$

Here

$$B(q^{-1}) = b_0 + b_1 q^{-1} + \ldots + b_{nb} q^{-nb} \text{ and } A(q^{-1}) = 1 + a_1 q^{-1} + \ldots + a_{na} q^{-na}$$

are the polynomials of the numerator and the denominator of the pulse (discrete time) transfer function $B(q^{-1})/A(q^{-1})$, respectively. The simple model can be rewritten in a form in which the constant, linear and quadratic terms are individually filtered by the linear dynamic term.

$$y(k) = c_0^{\mathrm{H}} \frac{B(1)}{A(1)} + c_1 \frac{B(q^{-1})}{A(q^{-1})} u(k) + c_2 \frac{B(q^{-1})}{A(q^{-1})} u^2(k)$$

The simple Hammerstein model is always linear in the parameters since

$$A(q^{-1}) y(k) = c_0^{\mathrm{H}} B(1) + c_1 B(q^{-1}) u(k) + c_2 B(q^{-1}) u^2(k)$$

but this form is redundant in the parameters.

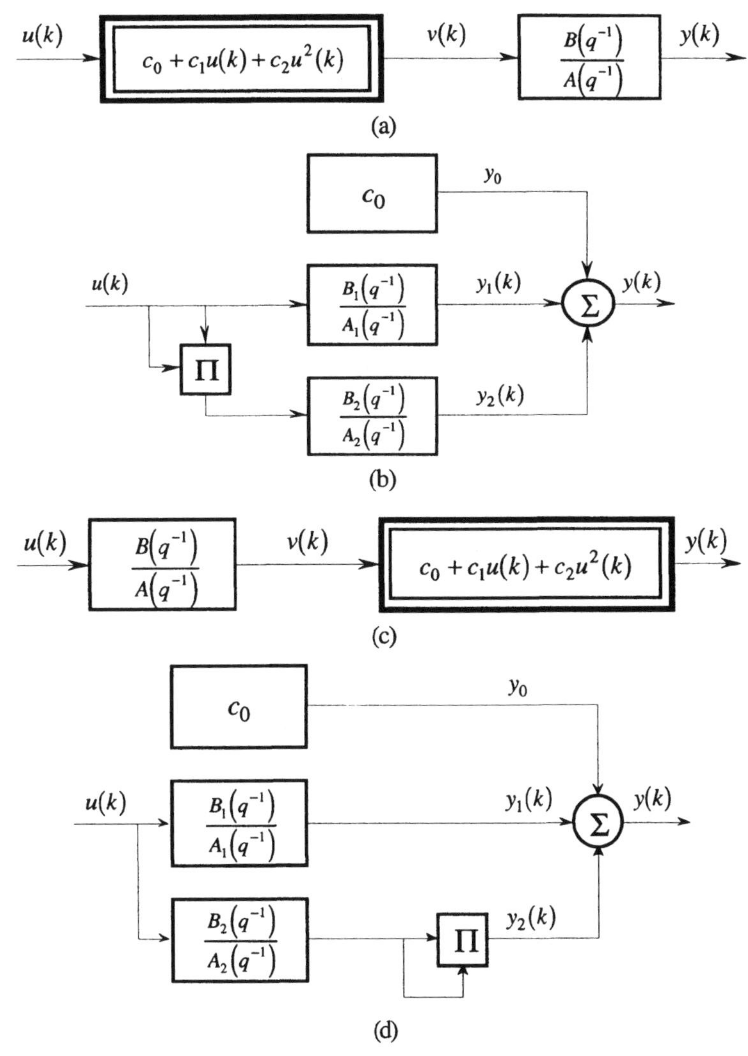

Fig. 1.3.5 Quadratic block oriented models: (a) simple Hammerstein; (b) generalized Hammerstein; (c) simple Wiener; (d) generalized Wiener

2. *Generalized Hammerstein model (Figure 1.3.5b)*
   A generalization of the model is if the filters of the linear and quadratic channels differ from each other:

$$y(k) = c_0 + \frac{B_1(q^{-1})}{A_1(q^{-1})} u(k) + \frac{B_2(q^{-1})}{A_2(q^{-1})} u^2(k)$$

The generalized Hammerstein model is linear in the parameters if $A_1(q^{-1}) = A_2(q^{-1})$

because $A(q^{-1})y(k) = c_0 + B_1(q^{-1})u(k) + B_2(q^{-1})u^2(k)$.

A form that is linear-in-parameters can always be achieved by using the common denominator $A(q^{-1}) = A_1(q^{-1})A_2(q^{-1})$. In this case, however, the difference equation is redundant in the unknown parameters.

3. *Simple Wiener model (Figure 1.3.5c)*
The simple Wiener model has the same parts as the simple Hammerstein model, except when the sequence is reversed. Its equations are

$$v(k) = \frac{B(q^{-1})}{A(q^{-1})}u(k) \tag{1.3.1}$$

$$y(k) = c_0 + c_1 v(k) + c_2 v^2(k) \tag{1.3.2}$$

Substituting (1.3.1) into (1.3.2) leads to the equation

$$y(k) = c_0 + c_1 \frac{B(q^{-1})}{A(q^{-1})}u(k) + c_2 \left[\frac{B(q^{-1})}{A(q^{-1})}u(k) \right]^2$$

which is nonlinear-in-parameters.

4. *Generalized Wiener model (Figure 1.3.5d)*
Assuming different linear filters in the linear and in the quadratic channels leads to the generalized Wiener model

$$y(k) = c_0 + \frac{B_1(q^{-1})}{A_1(q^{-1})}u(k) + \left[\frac{B_2(q^{-1})}{A_2(q^{-1})}u(k) \right]^2$$

5. *Extended Wiener model (Figure 1.3.5e)*
A further generalization is if different linear dynamic filters are allowed in the quadratic channel

$$y(k) = c_0 + \frac{B_1(q^{-1})}{A_1(q^{-1})}u(k) + \left[\frac{B_2(q^{-1})}{A_2(q^{-1})}u(k) \right] \left[\frac{B_3(q^{-1})}{A_3(q^{-1})}u(k) \right]$$

This model is called the extended Wiener model.

6. *Simple Wiener–Hammerstein cascade model (Figure 1.3.5f).*
The most compounded quadratic cascade model is obtained if a static quadratic element is preceded and followed by (different) linear filters

$$v_1(k) = \frac{B_1(q^{-1})}{A_1(q^{-1})}u(k) \qquad v_2(k) = c_0 + c_1 v_1(k) + c_2 v_1^2(k)$$

$$y(k) = \frac{B_2(q^{-1})}{A_2(q^{-1})} v_2(k)$$

A separation of the model output into constant, linear and quadratic channels results in

$$y_0 = c_0 \frac{B_2(1)}{A_2(1)}$$

$$y_1(k) = c_1 \frac{B_1(1)}{A_1(1)} \frac{B_2(1)}{A_2(1)} u(k)$$

$$y_2(k) = c_2 \frac{B_2(q^{-1})}{A_2(q^{-1})} \left[\frac{B_1(q^{-1})}{A_1(q^{-1})} u(k) \right]^2$$

7. *Generalized Wiener–Hammerstein cascade model (Figure 1.3.5g)*
   Denote further on the constant term by c_0 and the pulse transfer function of the linear channel by $B_1(q^{-1})/A_1(q^{-1})$. Then

$$y_0 = c_0$$

$$y_1(k) = \frac{B_1(1)}{A_1(1)} u(k)$$

   The present model may have different linear filters in the quadratic channel

$$y_2(k) = \frac{B_3(q^{-1})}{A_3(q^{-1})} \left[\frac{B_2(q^{-1})}{A_2(q^{-1})} u(k) \right]^2$$

8. *Extended Wiener–Hammerstein cascade model (Figure 1.3.5h)*
   Finally, if the linear filters before the multiplier may differ from each other then the most complex quadratic structure is achieved

$$y_2(k) = \frac{B_4(q^{-1})}{A_4(q^{-1})} \left[\frac{B_2(q^{-1})}{A_2(q^{-1})} u(k) \right] \left[\frac{B_3(q^{-1})}{A_3(q^{-1})} u(k) \right]$$

   Observe that this structure is the quadratic basis system suggested by Schetzen (1965).

9. *Schetzen's canonical form*
   Schetzen showed that any quadratic dynamic system can be built up from parallel connected basis systems. This means an extension of the quadratic channel of the

$$y_2(k) = \sum_{i=1}^{M} \frac{B_{4i}(q^{-1})}{A_{4i}(q^{-1})} \left[\frac{B_{2i}(q^{-1})}{A_{2i}(q^{-1})} u(k) \right] \left[\frac{B_{3i}(q^{-1})}{A_{3i}(q^{-1})} u(k) \right]$$

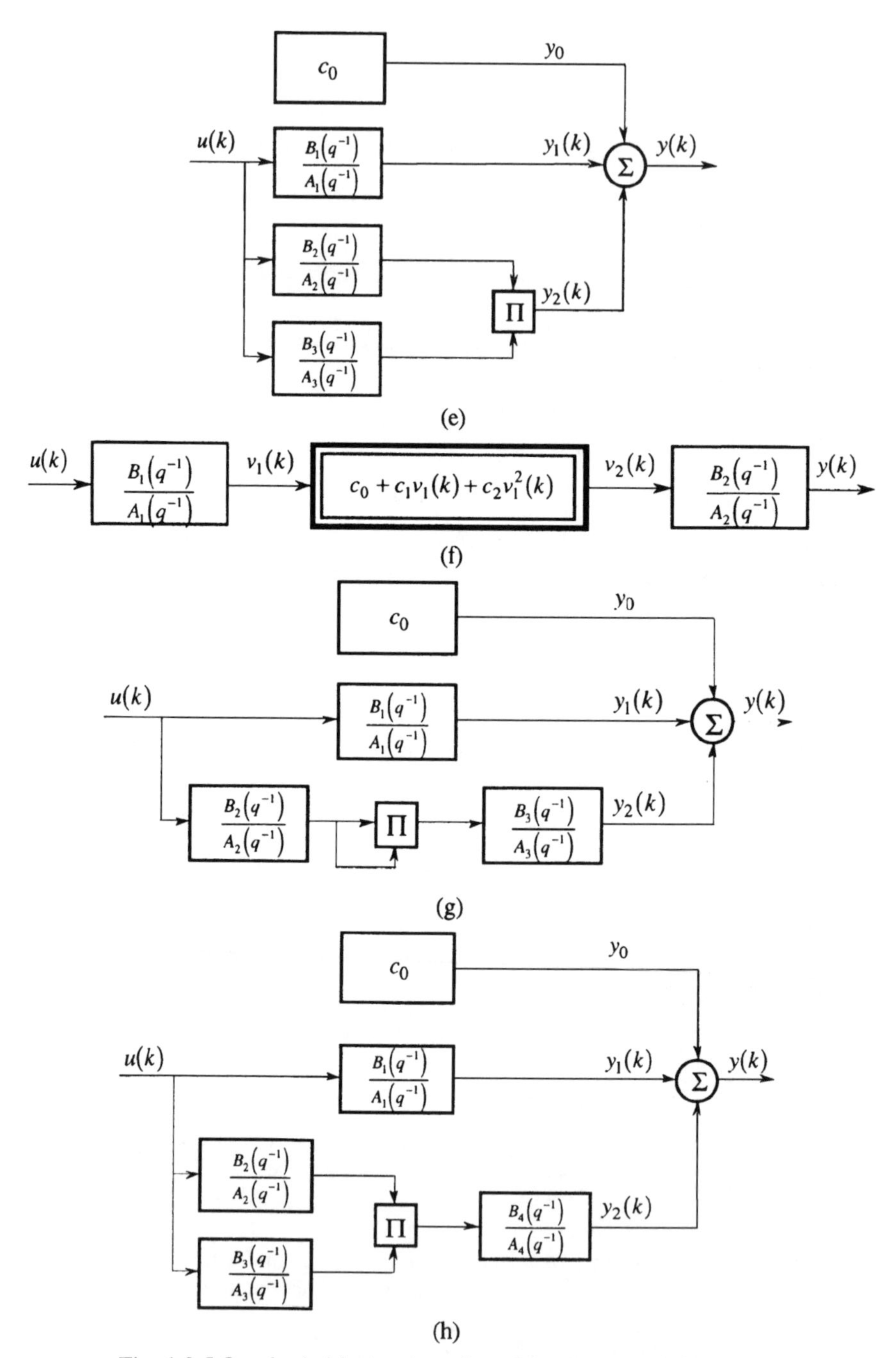

Fig. 1.3.5 Quadratic block oriented models: (e) extended Wiener; (f) simple Wiener–Hammerstein cascade; (g) generalized Wiener–Hammerstein cascade; (h) extended Wiener–Hammerstein cascade

Only the Hammerstein models can be expressed in a form linear-in-parameters. They are widely used in control applications. Some Wiener type of models are also used in adaptive control because the recursive parameter estimation is generally a nonlinear iterative procedure. There are of course also block oriented models of degree higher than 2. Their structure is easily imagined on the basis of the quadratic structures. Some applications that use cascade models for approximating physical processes are listed in the next section.

1.3.3 Cascade systems

Cascade structure systems consist of linear dynamic and nonlinear static terms connected in series. The degree of the power of the nonlinearity is arbitrary. A term with non-polynomial characteristics may also appear.

Figure 1.3.6 shows the simple Wiener–Hammerstein cascade model, which is known already as one of the quadratic block oriented models. Now the limitation regarding the form of the nonlinearity is annulled. The simple Wiener and simple Hammerstein cascade models are obtained if only the first two or the second two terms exist, respectively. These models are also marked in Figure 1.3.6. From an identification point of view the inner variable is not measurable in a cascade model. If it were measurable then only the parameters of a single linear or a nonlinear term had to estimated.

Representatives of this model class are often to be found in the practice, as the following examples show.

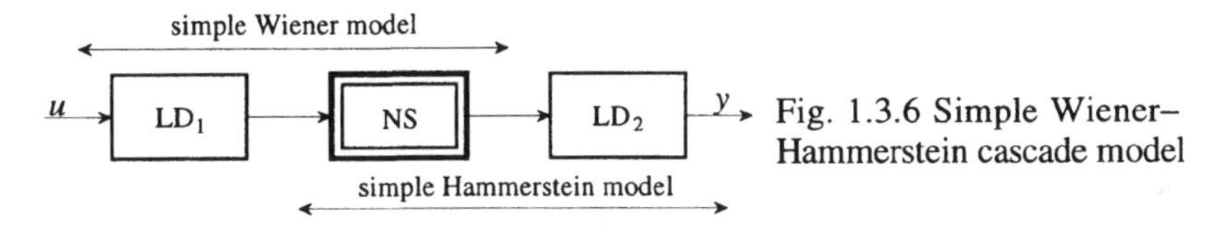

Fig. 1.3.6 Simple Wiener–Hammerstein cascade model

Example 1.3.3 *Valve actuator before a linear process*

Figure 1.3.7 shows a linear dynamic process whose input signal (e.g., the flow) is the output of a valve with a nonlinear shape. If the input signal (flow) of the process is not measurable, then the whole system has the structure of a simple Hammerstein model.

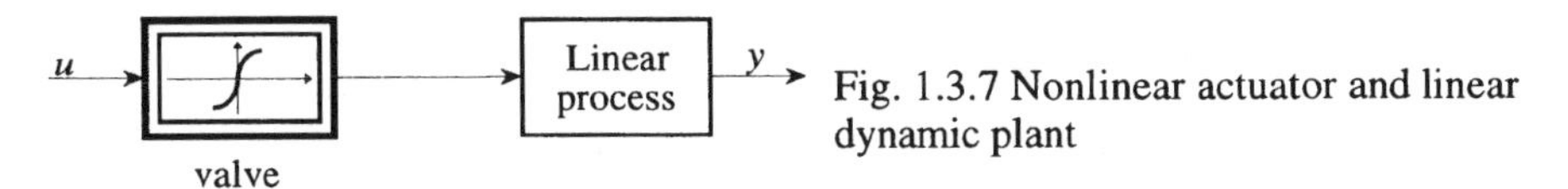

Fig. 1.3.7 Nonlinear actuator and linear dynamic plant

Example 1.3.4 *Linear controller with saturation*

Figure 1.3.8 shows a linear regulator (e.g., PID) followed by a saturation element. This structure is a simple Wiener model.

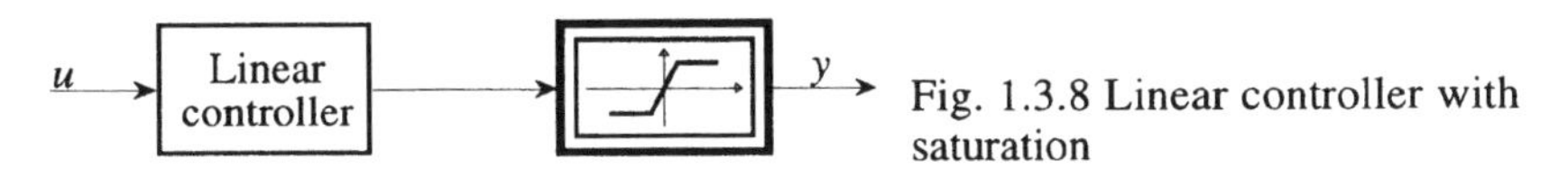

Fig. 1.3.8 Linear controller with saturation

Example 1.3.5 *Water neutralization process*

Figure 1.3.9 shows the scheme of a neutralization plant. The input signal is either the

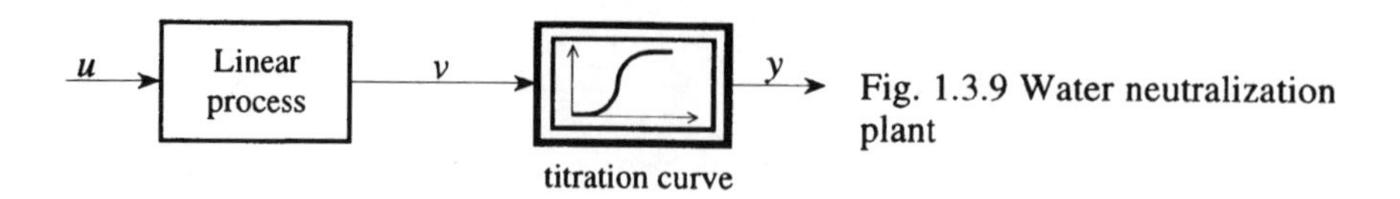

Fig. 1.3.9 Water neutralization plant

concentration or the flow of the waste water, the inner variable is the concentration of the outflow and the output signal is the pH of the outflow.

Cascade models occur frequently in the practice. Some cases are listed here:

1. Simple Hammerstein model:
- cement kiln (Li and Chen, 1988);
- heat exchanger (Corlis and Luus, 1969).

2. Simple Wiener model:
- pH-process (Pajunen, 1985a; 1985b; Proudfoot *et al.*, 1983; 1985; Lachmann and Goedecke, 1982);
- neuron chain in the retina of the catfish (Marmarelis and Naka, 1974);
- active frog tibias anterior muscle fiber (Hunter, 1985a; 1985b);
- gas chromatography with process (Moss).

3. Simple Wiener-Hammerstein model:
- frequency control loop of a water-power plant (Tuis, 1976);
- electrohydraulic servo-motor (Parker and Moore, 1982);
- encoding part of the cochlear system of mammals (Lammers, Verbruggen and Boer, 1979);
- peripheral auditory system (Weiss, 1966);
- neural chain in the primary visual pathway (Spekreijse, 1969; Spekreijse and Oosting, 1970);
- parts of the model of the pupil system (Stark, 1969).

We have shown that cascade models can be transformed into parallel representations. The reverse transformation from the parallel structure to the cascade one is generally not possible. It is only performable in some special cases, e.g., the parallel structure of Figure 1.3.4b is equivalent to the cascade structure of Figure 1.3.4a, as shown in Example 1.3.2.

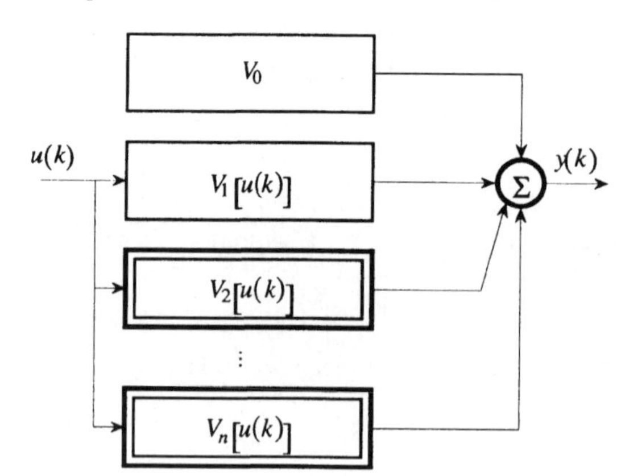

Fig. 1.3.10 Parallel structure of the degree-n Volterra model

1.3.4 Description of systems with polynomial nonlinearities by parallel channels of different nonlinear degree

Any nonlinear dynamical system can be separated to homogeneous parallel channels of different degree. If the degree of the highest power of the system is a finite integer value then the number of parallel channels is finite and equal to this value. Figure 1.3.10 shows the parallel structure.

Example 1.3.6 *Cubic simple Wiener–Hammerstein cascade system (Part of Example 1.3.2)*

A discrete time cascade model consisting of an element with cubic nonlinear characteristic and preceded and followed by linear dynamic terms (Figure 1.3.4a) was redrawn into a parallel form (Figure 1.3.4b) in Example 1.3.2. ■

A systematic method for calculating the output signals of the parallel channels was suggested by Gardiner (1973). His method will be presented under the structure search methods in Chapter 5. In the cited paper a further example is also shown of how a degree-4 cascade system can be transformed into a parallel form.

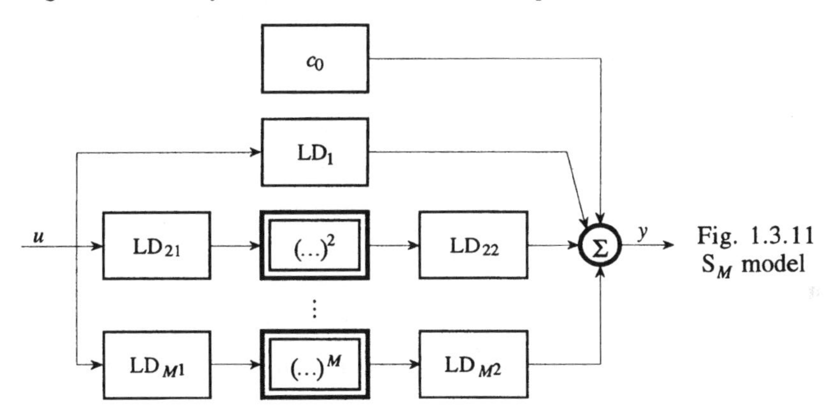

Fig. 1.3.11 S_M model

1.3.5 The S_M and the factorable Volterra models

There are two special parallel structures, which have their own names.

1. S_M model (Figure 1.3.11)

The *i*th channel consists of a static homogeneous nonlinear term of *i*th power both preceded and followed by linear dynamic systems. In other words we can say that each channel consists of a Wiener–Hammerstein cascade model with a homogeneous power form nonlinear term. (M in S_M denotes the highest power of the nonlinearity.) This model type was studied by Baumgarten and Rugh (1975), Wysocki and Rugh (1976), and Billings and Fakhouri (1979a).

The parallel form of any simple Wiener–Hammerstein cascade model with polynomial nonlinearity belongs to the class of S_M systems. This fact was shown in Economakos (1971) and can be verified in Example 1.3.2, as well.

2. Factorable Volterra model (Figure 1.3.12)

This model consists of linear dynamic terms, whose inputs are the inputs of the process. The *i*th parallel channel consists of the product of the outputs of *i* linear dynamic systems. This model type was studied by Harper and Rugh (1976), and Billings and Fakhouri (1979b).

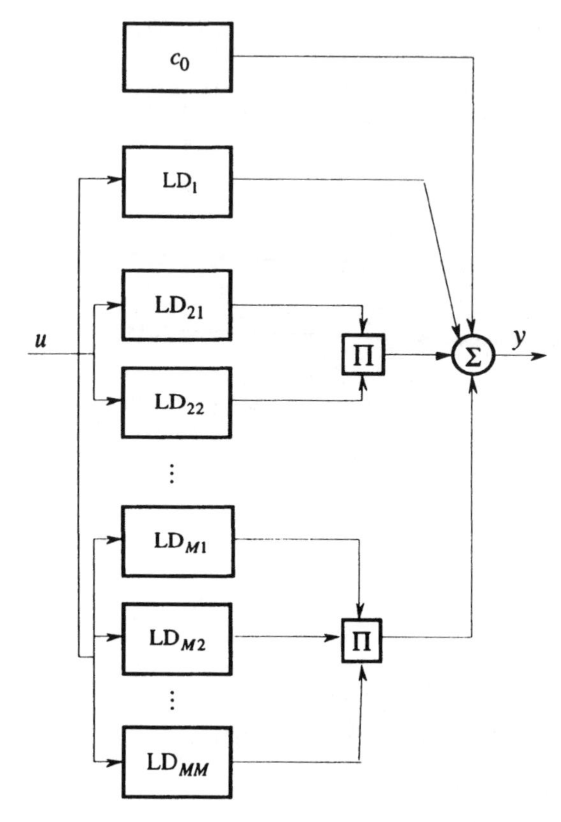

Fig. 1.3.12
Factorable degree- M Volterra model

The two forms cannot be generally transformed into each other. The transformation can be performed

- if the linear dynamic terms $(\mathrm{LD}_{i2}; i = 1, \ldots, M)$ behind the nonlinear powers are missing in the S_M model, or
- if the linear dynamic terms before the multipliers are equal to each other in the same channel in the factorable Volterra model.

The simple Wiener cascade model with polynomial static nonlinearity can be transformed into both forms. This is illustrated by the next example.

Example 1.3.7 *Simple quadratic Wiener cascade model*
Figure 1.3.13a shows the scheme of the simple Wiener model. Its transformation into the S_M form is given in Figure 1.3.13b. This is also the form of the factorable Volterra model (Figure 1.3.13c).

1.3.6 Nonlinear systems having only one nonlinear term and feedback and/or feedforward paths

Physical systems rarely have nonlinear parts in form of pure multipliers. There are often static nonlinear terms with any characteristic in the processes. A possible classification of such systems is according to the number of the individual static nonlinear parts in the systems.

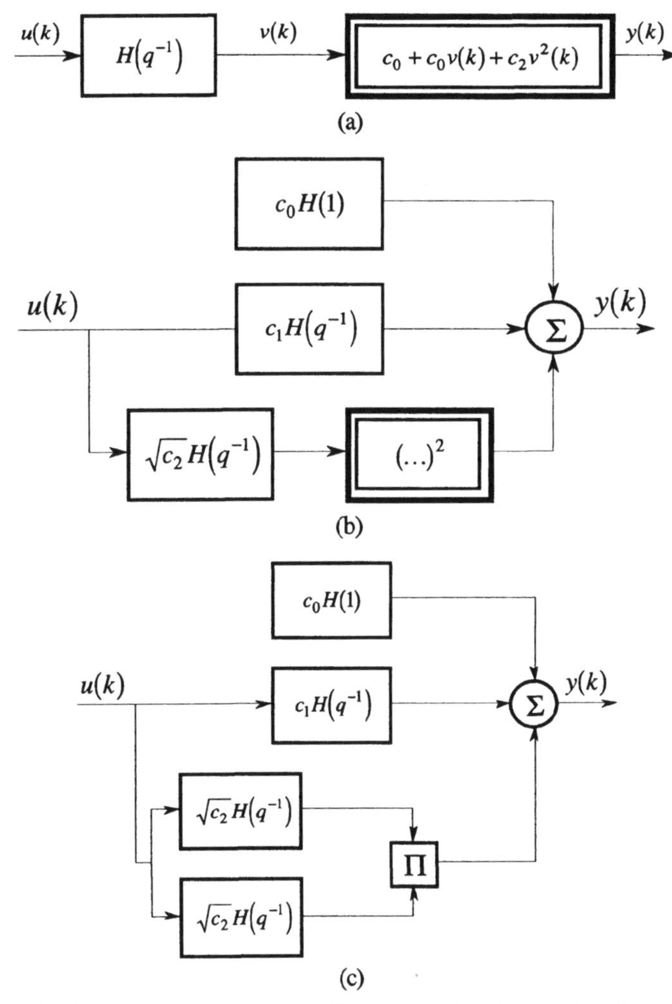

Fig. 1.3.13 Quadratic simple Wiener model: (a) cascade scheme; (b) parallel structure (S_M model); (c) factorable Volterra model

The three simple cascade models are as follows:

1. *the simple Hammerstein model*: a nonlinear static element is followed by a linear dynamic term;
2. *the simple Wiener model*: a linear dynamic term is followed by a nonlinear static element;
3. *the simple Wiener–Hammerstein model*: a nonlinear static element is both preceded and followed by linear dynamic terms.

Each of the above models can be:

- feedbacked by a linear dynamic term (Figure 1.3.14a);
- feedforwarded by a linear dynamic term (Figure 1.3.14b);
- both feedbacked and feedforwarded by linear dynamic terms (Figure 1.3.14c).

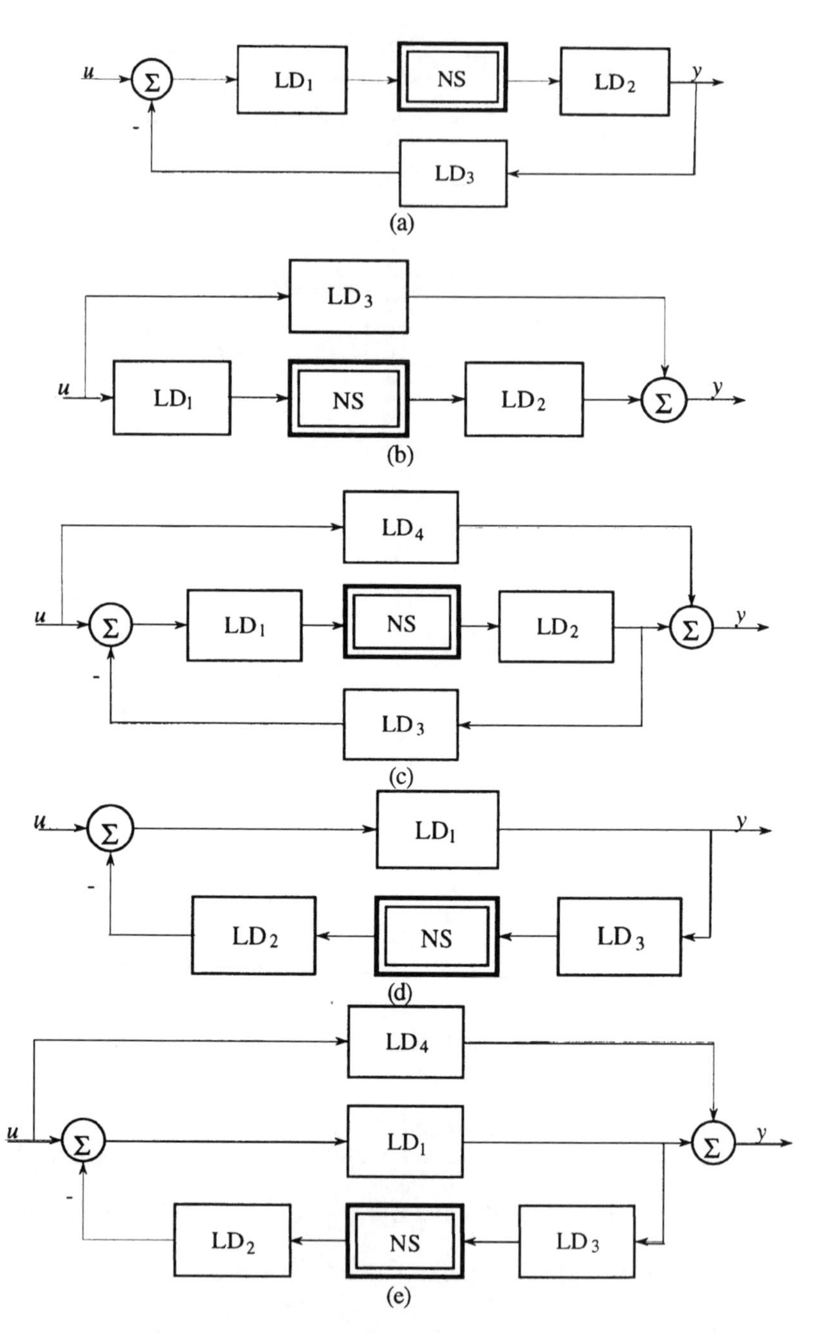

Fig. 1.3.14 Feedback and feedforward structures with one nonlinearity: (a) linear dynamic feedback; (b) linear dynamic feedforward; (c) linear dynamic feedback and feedforward; (d) nonlinear feedback; (e) nonlinear feedback and linear dynamic feedforward

In Figure 1.3.14 the feedback and feedforward structures are given only for the simple Wiener–Hammerstein cascade model. By omitting the corresponding linear dynamic terms, the cases of the simple Hammerstein and Wiener models can be obtained as special cases. The most simple feedback and/or feedforward structures are if both linear dynamic terms are omitted from the cascade model. Further on the nonlinearity can be placed either in the feedback or in the feedforward path, as seen in Figure 1.3.14d and Figure 1.3.14e, respectively.

Two applications are listed below:

linear dynamic systems are feedbacked by a static nonlinear term:

- jet engine (Rault *et al.*);
- distillation column (Fronza *et al.*, 1969).

1.4 THE GENERALIZED HAMMERSTEIN AND THE PARAMETRIC VOLTERRA MODELS

There are two models that pay an important role in the identification and control of nonlinear dynamic processes. They are the generalized Hammerstein and the parametric Volterra models. If they are expressed by a form linear-in-parameters the estimation of the parameters is an easy task. Since both models are parametric they give information with a few parameters about both the dynamics and the steady state characteristics of the process. In the sequel we restrict the nonlinear degree of the models to two. An extension to higher than quadratic model is trivial. The quadratic models approximate the processes in a given neighborhood of the working point with enough accuracy, and they give proper information for an extremum control. Since the models are widely used for discrete time (control) purposes these descriptions will be presented here.

1.4.1 The generalized Hammerstein model

As mentioned with the block oriented models, the generalized Hammerstein model may have different linear dynamic filters in the linear and quadratic channels. Supply the polynomials of the pulse transfer function with the index L (linear) and H (Hammerstein), then

$$y(k+d) = c_0 + \frac{B_1^{\mathrm{L}}\left(q^{-1}\right)}{A_1^{\mathrm{L}}\left(q^{-1}\right)} u(k) + \frac{B_2^{\mathrm{H}}\left(q^{-1}\right)}{A_2^{\mathrm{H}}\left(q^{-1}\right)} u^2(k) \tag{1.4.1}$$

Introduce the weighting function series $H_1\left(q^{-1}\right)$ and $H_{20}\left(q^{-1}\right)$ as

$$H_1\left(q^{-1}\right) = \frac{B_1^{\mathrm{L}}\left(q^{-1}\right)}{A_1^{\mathrm{L}}\left(q^{-1}\right)} \tag{1.4.2}$$

and

$$H_{20}\left(q^{-1}\right) = \frac{B_2^{\mathrm{H}}\left(q^{-1}\right)}{A_2^{\mathrm{H}}\left(q^{-1}\right)} \tag{1.4.3}$$

and substitute the pulse transfer functions in (1.4.1) by (1.4.2) and (1.4.3)

$$y(k+d) = c_0 + H_1\left(q^{-1}\right)u(k) + H_{20}\left(q^{-1}\right)u^2(k)$$

which corresponds to the difference equation

$$y(k+d) = c_0 + \sum_{\kappa_1=0}^{\infty} h_1(\kappa_1)u(k-\kappa_1) + \sum_{\kappa_1=0}^{\infty} h_{20}(\kappa_1)u^2(k-\kappa_1) \tag{1.4.4}$$

A comparison of (1.4.4) with the Volterra series model shows that the quadratic generalized Hammerstein model covers only those terms of the second-degree Volterra kernels that lie on the main diagonal ($\kappa_1 = \kappa_2$) (Figure 1.4.1).

κ_1 \ κ_2	0	1	2	3	4
0	GH	PV	PV		
1		GH	PV	PV	
2			GH	PV	PV
3				GH	PV
4					GH

Fig. 1.4.1
Discrete time kernels approximated by the generalized Hammerstein (GH) and the parametric Volterra model (PV), the latter for $M = 2$

Introduce the common denominator

$$A\left(q^{-1}\right) = A_1^{\mathrm{L}}\left(q^{-1}\right)A_2^{\mathrm{H}}\left(q^{-1}\right)$$

and rearrange (1.4.1)

$$y(k+d) = c_0 + \frac{B_1\left(q^{-1}\right)}{A\left(q^{-1}\right)}u(k) + \frac{B_2\left(q^{-1}\right)}{A\left(q^{-1}\right)}u^2(k) \tag{1.4.5}$$

where

$$B_1\left(q^{-1}\right) = B_1^{\mathrm{L}}\left(q^{-1}\right)A_2^{\mathrm{H}}\left(q^{-1}\right), \qquad B_2\left(q^{-1}\right) = B_2^{\mathrm{H}}\left(q^{-1}\right)A_1^{\mathrm{L}}\left(q^{-1}\right)$$

The advantage of (1.4.5) is that it is linear in the parameters that can be seen either from the difference equation

$$A\left(q^{-1}\right)y(k) = c_0A(1) + B_1\left(q^{-1}\right)u(k) + B_2\left(q^{-1}\right)u^2(k)$$

or from the scalar product

$$y(k+d) = \boldsymbol{\phi}^{\mathrm{T}}(k)\boldsymbol{\theta}$$

with

$$\boldsymbol{\phi}^{\mathrm{T}}(k)=\left[1, u(k), \ldots, u\left(k-nb_1\right), u^2(k), \ldots, u^2\left(k-nb_2\right), -y(k-1), \ldots, -y(k-n)\right]$$

$$\boldsymbol{\theta}^{\mathrm{T}}(k)=\left[c_0, b_{10}, \ldots, b_{1nb_1}, b_{20}, \ldots, b_{2nb_2}, a_1, \ldots, a_n\right]$$

The parameters of the generalized Hammerstein model can be estimated with program packages elaborated for linear MISO systems. Depending on whether different or equal denominators are assumed in the pulse transfer functions of the linear and quadratic channels an appropriate identification package has to be used.

1.4.2 The parametric Volterra model

The Volterra series model has also terms of the type $h_2(\kappa_1, \kappa_2)u(k-\kappa_1)u(k-\kappa_2)$; $\kappa_1 \neq \kappa_2$. To cover these terms the generalized Hammerstein model (1.4.1) has to be extended by further terms

$$y(k+d)=c_0+\frac{B_1^{\mathrm{L}}\left(q^{-1}\right)}{A_1^{\mathrm{L}}\left(q^{-1}\right)}u(k)+\frac{B_2^{\mathrm{H}}\left(q^{-1}\right)}{A_2^{\mathrm{H}}\left(q^{-1}\right)}u^2(k)+\sum_{j=0}^{\infty}\frac{B_{2j}^{\mathrm{V}}\left(q^{-1}\right)}{A_{2j}^{\mathrm{V}}\left(q^{-1}\right)}u(k)u(k-j) \tag{1.4.6}$$

Denote the linear weighting function series by

$$H_{2j}\left(q^{-1}\right)=\frac{B_{2j}^{\mathrm{V}}\left(q^{-1}\right)}{A_{2j}^{\mathrm{V}}\left(q^{-1}\right)} \tag{1.4.7}$$

then (1.4.6) becomes with (1.4.2), (1.4.3) and (1.4.7)

$$\begin{aligned}y(k+d)&=c_0+H_1\left(q^{-1}\right)u(k)+H_{20}\left(q^{-1}\right)u^2(k)+\sum_{j=1}^{\infty}H_{2j}\left(q^{-1}\right)u(k)u(k-j)\\&=c_0+H_1\left(q^{-1}\right)u(k)+\sum_{j=0}^{\infty}H_{2j}\left(q^{-1}\right)u(k)u(k-j)\end{aligned}$$

which corresponds to the difference equation

$$\begin{aligned}y(k+d)&=c_0+\sum_{\kappa_1=0}^{\infty}h_1(\kappa_1)u(k-\kappa_1)+\sum_{\kappa_1=0}^{\infty}h_{20}(\kappa_1)u^2(k-\kappa_1)\\&\quad+\sum_{j=1}^{\infty}\sum_{\kappa_2=0}^{\infty}h_{2j}(\kappa_1)u(k-\kappa_1)u(k-\kappa_1-j)\\&=c_0+\sum_{\kappa_1=0}^{\infty}h_1(\kappa_1)u(k-\kappa_1)+\sum_{j=0}^{\infty}\sum_{\kappa_2=0}^{\infty}h_{2j}(\kappa_1)u(k-\kappa_1)u(k-\kappa_1-j)\end{aligned} \tag{1.4.8}$$

A comparison of (1.4.8) with the Volterra series model shows that (1.4.8) covers all terms of the second-degree Volterra kernels with the only restriction, that any off-diagonal (that means parallel line to the main diagonal) is approximated by a linear weighting function series:

$$h_2(\kappa_1, \kappa_1 + j) \approx h_{2j}^{\mathrm{V}}(\kappa_1) \approx \frac{B_{2j}^{\mathrm{V}}(q^{-1})}{A_{2j}^{\mathrm{V}}(q^{-1})} \qquad j = 0, 1, 2, \ldots$$

Here the quadratic channel in the Hammerstein model is used as the first ($j = 0$) channel of the parametric Volterra model

$$\frac{B_{20}^{\mathrm{V}}(q^{-1})}{A_{20}^{\mathrm{V}}(q^{-1})} \equiv \frac{B_2^{\mathrm{H}}(q^{-1})}{A_2^{\mathrm{H}}(q^{-1})}$$

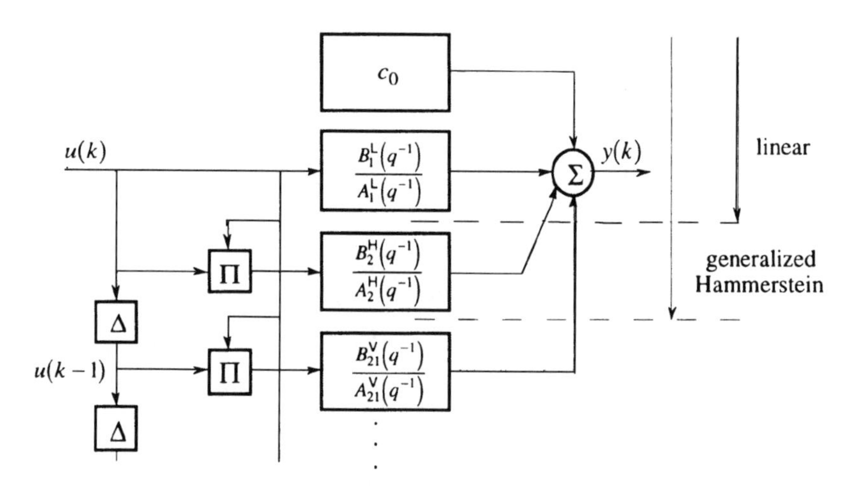

Fig. 1.4.2 Parametric Volterra model nonlinear-in-parameters

The model is called the parametric Volterra model if both the number of the parallel channels (M) and the order of the pulse transfer functions are finite

$$y(k+d) = c_0 + \frac{B_1^{\mathrm{L}}(q^{-1})}{A_1^{\mathrm{L}}(q^{-1})} u(k) + \sum_{j=0}^{M} \frac{B_{2j}^{\mathrm{V}}(q^{-1})}{A_{2j}^{\mathrm{V}}(q^{-1})} u(k) u(k-j) \tag{1.4.9}$$

The scheme of the model is shown in Figure 1.4.2.

The disadvantage of (1.4.3) is that it is not linear in the parameters. Using, however, a common denominator in all pulse transfer functions leads to the parametric Volterra model linear-in-parameters (Figure 1.4.3):

$$y(k+d) = c_0 + \frac{B_1(q^{-1})}{A(q^{-1})} u(k) + \sum_{j=0}^{M} \frac{B_{2j}(q^{-1})}{A(q^{-1})} u(k) u(k-j) \tag{1.4.10}$$

The linearity in the parameters can be seen from the form

$$A\left(q^{-1}\right)y(k) = c_0 A(1) + B_1\left(q^{-1}\right)u(k) + \sum_{j=0}^{M} B_{2j}\left(q^{-1}\right)u(k)u(k-j)$$

Figure 1.4.1 shows in the (κ_1, κ_2)-plane that those kernels are covered by the model that lies either on the main diagonal or on parallel lines whose distance is not more than M from the main diagonal $(\kappa_2 \le \kappa_1 + M)$.

The parameters of the parametric Volterra model can be estimated by program packages elaborated for linear MISO systems. The approximation of an unknown process is usually better when different denominators in the pulse transfer functions of the channels are assumed.

1.4.3 The generalized and extended parametric Volterra models

A parametric linear model can be obtained formally by equaling the weighting function models of the input signal $B\left(q^{-1}\right)u(k)$ and that of the output signal $A\left(q^{-1}\right)y(k)$:

$$A\left(q^{-1}\right)y(k) = B\left(q^{-1}\right)u(k)$$

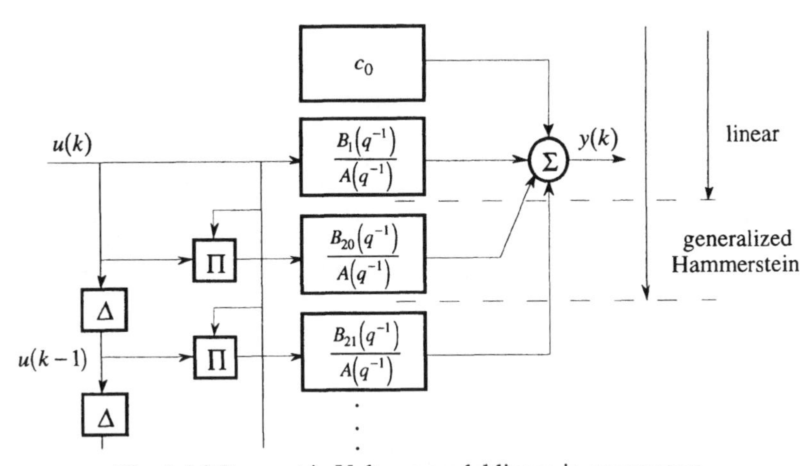

Fig. 1.4.3 Parametric Volterra model linear-in-parameters

In a similar way we can equal the finite-order Volterra series models of the input and output signals

$$\begin{aligned} A_1\left(q^{-1}\right)y(k) + \sum_{j=0}^{MA} A_{2j}\left(q^{-1}\right)y(k-1)y(k-1-j) = c_0 + B_1\left(q^{-1}\right)u(k) \\ + \sum_{j=0}^{MB} B_{2j}\left(q^{-1}\right)u(k)u(k-j) \end{aligned} \tag{1.4.11}$$

As is seen, the two constant terms are combined, and to obtain a recursive model the $y(k)$ term is omitted from the quadratic term. This model can be considered as the generalized parametric Volterra model. (Here MA and MB are properly chosen structural constants.)

(1.4.11) contains quadratic terms of the input and output signals. However, the steady state characteristic is of high nonlinear degree if $A_2\left(q^{-1}\right) \neq 0$. (1.4.11) can be extended with the cross-product terms of the input and the output signals. The extended parametric Volterra model is defined as

$$\begin{aligned} & A_1\left(q^{-1}\right)y(k)+\sum_{j=0}^{MA} A_{2j}\left(q^{-1}\right)y(k-1)y(k-1-j)=c_0+B_1\left(q^{-1}\right)u(k) \\ & +\sum_{j=0}^{MB} B_{2j}\left(q^{-1}\right)u(k)u(k-j)+\sum_{j=1}^{MC} C_{2j}\left(q^{-1}\right)u(k)y(k-j) \\ & +\sum_{j=2}^{MD} D_{2j}\left(q^{-1}\right)y(k-1)u(k-j) \end{aligned} \tag{1.4.12}$$

(Here MA, MB, MC and MD are properly chosen structural constants.) These extensions were proposed by Haber and Keviczky (1976). Another, widely used classification of models linear-in-parameters is given in the next section.

1.5 PARAMETRIC DISCRETE TIME MODELS LINEAR-IN-PARAMETERS

A model is linear in the parameters if the output signal can be expressed as a scalar product of a vector of the parameters and of a vector of the measured data. If the process and/or the measurements are disturbed by noise then the equation used with the parameter estimation should be expressed not for the output signal but for the error. It can occur that this equation is not linear in the parameters, although the noise-free output signal is linear in the parameters. In this section we deal only with deterministic, i.e., noise-free models being linear in the parameters. The stochastic case is treated with the parameter estimation in Chapter 3.

1.5.1 Polynomial difference equations

Definition 1.5.1 A polynomial difference equation is of type

$$f\left(y(k),\ldots,y(k-ny),u(k),\ldots,u(k-nu)\right)=0 \tag{1.5.1}$$

where f is any polynomial function and $n=\max\{nu,ny\}$ is the order of the model (e.g., Diaz and Desrochers, 1988). (Here nu and ny are properly chosen structural constants.) ■

Example 1.5.1 *First-order polynomial difference equation*
A first-order dynamic model has the components $y(k), y(k-1)$ and $u(k-1)$. Consider all terms up to the quadratic ones:

$$\begin{aligned} & \theta_1+\theta_2 y(k)+\theta_3 y(k-1)+\theta_4 u(k-1)+\theta_5 y^2(k)+\theta_6 y(k)y(k-1) \\ & +\theta_7 y(k)u(k-1)+\theta_8 y^2(k-1)+\theta_9 y(k-1)u(k-1)+\theta_{10}u^2(k-1)=0 \end{aligned} \tag{1.5.2}$$

(1.5.2) is an implicit function of the output signal $y(k)$. ■

Definition 1.5.2 A polynomial difference equation is an explicit function of the output signal, if

$$y(k) = f\big(y(k-1), \ldots, y(k-ny), u(k), \ldots, u(k-nu)\big) \tag{1.5.3}$$

where f is any polynomial function and $n = \max\{nu, ny\}$ is the order of the model. This type of equation is called recursive (Leontaritis and Billings, 1985a) or of regression type (Diaz and Desrochers, 1988). ■

Example 1.5.2 *First-order polynomial difference equation*
A first-order dynamic model has the components $y(k), y(k-1)$ and $u(k-1)$. Consider all terms up to the quadratic ones, then we get (1.5.2). Let $\theta_5 = \theta_6 = \theta_7 = \theta_8 = 0$ then

$$\theta_1 + \theta_2 y(k) + \theta_3 y(k-1) + \theta_4 u(k-1) + \theta_9 y(k-1)u(k-1) + \theta_{10} u^2(k-1) = 0$$

or

$$y(k) = -\frac{\theta_1}{\theta_2} - \frac{\theta_3}{\theta_2} y(k-1) - \frac{\theta_4}{\theta_2} u(k-1) - \frac{\theta_9}{\theta_2} y(k-1)u(k-1) - \frac{\theta_{10}}{\theta_2} u^2(k-1)$$

which is a recursive difference equation.

With strictly causal systems $u(k)$ fails in the difference equations (1.5.1) or (1.5.3).

1.5.2 Recursive rational difference equation

Definition 1.5.3 The recursive rational difference equation was defined by Sontag (1979b) as

$$\begin{aligned} &a\big(y(k-1), \ldots, y(k-ny), u(k), \ldots, u(k-nu)\big)y(k) \\ &= b\big(y(k-1), \ldots, y(k-ny), u(k), \ldots, u(k-nu)\big) \end{aligned} \tag{1.5.4}$$

a(.) and *b(.)* are polynomials of the actual and past input signals and of the past output signals. (Here nu and ny are properly chosen structural constants.) ■

Remarks:

1. Sontag (1979b) defined (1.5.4) with $ny = nu$.
2. With strictly causal systems $u(k)$ fails in the difference equation.
3. Both *a(.)* and *b(.)* may be polynomials of the input and the output signals, including cross product terms, as well. Therefore the model is formally equivalent with the extended parametric Volterra model, which was introduced in a heuristic way by Haber and Keviczky (1976).
4. The model is recursive because $y(k)$ can be expressed explicitly from (1.5.4).
5. $y(k)$ can be explicitly expressed from (1.5.4) by polynomial division only if $a(.) \neq 0$,

$$y(k)=\frac{b(y(k-1),\ldots,y(k-ny),u(k),\ldots,u(k-nu))}{a(y(k-1),\ldots,y(k-ny),u(k),\ldots,u(k-nu))} \tag{1.5.5}$$

Sontag (1979b) proved that if the system can be modeled by a discrete time finite-order nonlinear state space model then the input–output equivalent difference equation is a recursive rational function with finite order.

Example 1.5.3 *Quasi-linear first-order continuous time model with output signal dependent reciprocal gain*

The differential equation of the system is

$$T\dot{y}(t)+y(t)=Ku(t) \tag{1.5.6}$$

Assume an inverse linear dependence of the static gain on the output signal

$$K^{-1}(t)=\overline{K}_0^{-1}+\overline{K}_1^{-1}y(t) \tag{1.5.7}$$

The time constant is not time dependent, i.e., $T=T_0$. If the sampling time ΔT is small then the Euler transformation can be used while discretizing (1.5.6). The equivalent difference equation is

$$y(k)+\left[\frac{\Delta T}{T_0}-1\right]y(k-1)=K(t)\frac{\Delta T}{T_0}u(k-1) \tag{1.5.8}$$

Divide (1.5.8) by the gain

$$K^{-1}(t)y(k)+K^{-1}(t)\left[\frac{\Delta T}{T_0}-1\right]y(k-1)=\frac{\Delta T}{T_0}u(k-1) \tag{1.5.9}$$

Now, substitute the discrete time form of (1.5.7) into (1.5.9)

$$\begin{aligned}&\left[\overline{K}_0^{-1}+\overline{K}_1^{-1}y(k-1)\right]y(k)+\left[\overline{K}_0^{-1}+\overline{K}_1^{-1}y(k-1)\right]\left[\frac{\Delta T}{T_0}-1\right]y(k-1)\\&=\frac{\Delta T}{T_0}u(k-1)\end{aligned} \tag{1.5.10}$$

(1.5.10) is a recursive rational difference equation with the polynomials

$$\begin{aligned}a(.)&=a(y(k-1))=\overline{K}_0^{-1}+\overline{K}_1^{-1}y(k-1)\\b(.)&=b(y(k-1),u(k-1))\\&=-\left[\overline{K}_0^{-1}+\overline{K}_1^{-1}y(k-1)\right]\left[\frac{\Delta T}{T_0}-1\right]y(k-1)+\frac{\Delta T}{T_0}u(k-1)\end{aligned}$$

■

1.5.3 Output–affine difference equations

Definition 1.5.4 The output–affine polynomial difference equation was defined by Sontag (1979a, 1979b) as

$$\sum_{i=0}^{na} a_i\left(u(k),\ldots,u(k-nu)\right)y(k-i) = b_0\left(u(k),\ldots,u(k-nu)\right). \quad (1.5.11)$$

$a_i(.)$ and $b_0(.)$ are polynomials of the actual and shifted input signals and $a_0(.) \neq 0$. (Here nu and na are properly chosen structural constants.) ■

Remarks:

1. Sontag (1979a) defined (1.5.11) with $na = nu$.
2. With strictly causal systems $u(k)$ fails in the difference equation.
3. Sontag (1979a) called the model output linear if $b_0 = 0$.
4. The RHS of (1.5.11) is a polynomial of the input signal, similar to the terms including the input signal of the parametric Volterra model.
5. The LHS of (1.5.11) is similar to the LHS of a linear difference equation

   $$\sum_{i=0}^{na} a_i\, y(k-i) = \sum_{i=0}^{nb} b_i u(k-i)$$

   with the difference that the coefficients a_i may depend on the input signal.
6. The LHS of (1.5.11) contains cross product terms between the shifted input and the output signals. Such terms were recommended also in the extended parametric Volterra model by Haber and Keviczky (1976).
7. $a_i(.) \neq 0$ is the condition for expressing the output signal from (1.5.11)

   $$y(k) = \frac{1}{a_0\left(u(k),\ldots,u(k-nu)\right)} \times\left[b_0\left(u(k),\ldots,u(k-nu)\right) - \sum_{i=1}^{na} a_i\left(u(k),\ldots,u(k-nu)\right)y(k-i)\right] \quad (1.5.12)$$

The importance of the output–affine models lies in the fact that a very wide range of nonlinear systems can be approximated by them. This is enlightened by the following two theorems:

1. Assume that:
   - the output signal of a system depends continuously on the input signal;
   - the input signal is bounded;
   - the model output should be fitted to finite samplings of the outputs;

   then the system can be modeled by a discrete time finite-order state–affine state space model (Fliess and Normand–Cyrot, 1982).

2. If a discrete time state–affine state space model is finitely realizable then it is equivalent to an output–affine polynomial difference equation with finite order (Sontag, 1979a).

Example 1.5.4 *Quasi-linear first-order continuous time model with input signal dependent time constant*

The differential equation of the system is given by (1.5.6). Assume a linear dependence of the time constant on the input signal

$$T(t) = T_0 + T_1 u(t) \tag{1.5.13}$$

and the static gain is constant $(K = K_0)$. If the sampling time ΔT is small the Euler transformation can be used while discretizing (1.5.6). The equivalent difference equation is

$$y(k) + \left[\frac{\Delta T}{T(t)} - 1\right] y(k-1) = K_0 \frac{\Delta T}{T(t)} u(k-1) \tag{1.5.14}$$

Now, multiply (1.5.14) by $T(t)$

$$T(t)y(k) + [\Delta T - T(t)]y(k-1) = K_0 \, \Delta T u(k-1) \tag{1.5.15}$$

and substitute the discrete time form of (1.5.13) into (1.5.15)

$$[T_0 + T_1 u(k-1)]y(k) + [\Delta T - T_0 - T_1 u(k-1)]y(k-1) = K_0 \, \Delta T u(k-1) \tag{1.5.16}$$

As is seen, (1.5.16) has an output–affine form. ■

A physical application of this model class is reported in the literature:
binary distillation column (Beghelli and Guidorzi, 1976)

1.5.4 Recursive polynomial difference equations

The definition of this model class was given in *Definition 1.5.2.*

Leontaritis and Billings (1985a) gave sufficient but not necessary conditions for the existence of the recursive nonlinear difference equations. This means that they allowed nonlinear functions other than polynomials, which may lead to a model nonlinear-in-parameters. The conditions are repeated here in a somewhat loosely stated form:

1. the system can be described by a finite-dimensional nonlinear discrete time state space model, and
2. there exist linearized models of the system in the whole region, where the recursive polynomial difference equation is defined, and the order of the models is equal to the order of the state space model.

Leontaritis and Billings (1985a) called the recursive nonlinear difference equation model NARMAX model (Nonlinear AutoRegressive Moving-Average with eXogenous inputs).

Remarks:

1. With strictly causal systems $u(k)$ fails in the difference equation.
2. Having performed the polynomial division in the output–affine difference equation we obtain a recursive difference equation with infinite components. However, stopping the division after certain terms, the output–affine model can be approximated by a parametric recursive polynomial difference equation. Of

course, the condition $a(.) \neq 0$ should be fulfilled, furthermore (Diaz and Desrochers, 1988).

3. If $a_0(.) = a_0$ does not depend on the input signal in (1.5.11) then the output–affine model is equivalent to the recursive polynomial difference equation model (Diaz and Desrochers, 1988).
4. Having performed the polynomial division in the rational difference equation we obtain a recursive difference equation with infinite components. However, stopping the division after certain terms, the rational difference equation model can be approximated by a parametric recursive polynomial difference equation. Of course, the condition $a(.) \neq 0$ should be fulfilled, furthermore.
5. If $a(.)$ does not depend either on the input nor on the output signal in (1.5.4) then the rational difference equation model is equivalent to the recursive polynomial difference equation model.
6. Leontaritis and Billings (1985b) have shown that the recursive nonlinear difference equation model reduces itself to the linear model if the amplitude of the input signal is small. This is not the case with the output–affine model.
7. The recursive polynomial difference equation model is a special case of the recursive nonlinear difference equation model.
8. The nonrecursive polynomial difference equation model can be approximated by a recursive polynomial difference equation if the nonlinear function is approximated by finite number of terms of its Taylor series.
9. The nonrecursive polynomial difference equation model is not necessarily linear in the parameters. The recursive polynomial model is always linear in the parameters.

Example 1.5.5 *Quasi-linear first-order continuous time model with input signal dependent reciprocal time constant*

The differential equation of the system is given in (1.5.6). Assume an inverse linear dependence of the time constant on the input signal

$$T^{-1}(t) = \overline{T}_0^{-1} + \overline{T}_1^{-1}u(t) \tag{1.5.17}$$

The static gain is constant $K = K_0$. If the sampling time ΔT is small then the Euler transformation can be used while discretizing (1.5.6). The equivalent difference equation is (1.5.14). Substitute the discrete time form of (1.5.17) into (1.5.14)

$$y(k) + \left[\Delta T\left[\overline{T}_0^{-1} + \overline{T}_1^{-1}u(k-1)\right] - 1\right]y(k-1) = K_0\Delta T\left[\overline{T}_0^{-1} + \overline{T}_1^{-1}u(k-1)\right]u(k-1) \tag{1.5.18}$$

Equation (1.5.18) is a recursive polynomial difference equation. ■

Example 1.5.6 *Quadratic, first-order simple Hammerstein model*

Let the equation of the static part of the model be

$$v(k) = c_0 + c_1u(k) + c_2u^2(k) \tag{1.5.19}$$

and that of the dynamic part

$$y(k) = b_1v(k-1) - a_1y(k-1) \tag{1.5.20}$$

The input–output equivalent recursive difference equation is obtained by substituting (1.5.19) into (1.5.20)

$$y(k) = b_1c_0 + b_1c_1u(k-1) + b_1c_2u^2(k-1) - a_1y(k-1) \tag{1.5.21}$$

Equation (1.5.21) is an input–output difference equation if $a_1 \neq 0$ and

$$b_1\left[c_0 + c_1u(k-1) + c_2u^2(k-1)\right] \neq 0$$

The latter is fulfilled if $b_1 \neq 0$ and

$$c_0 + c_1u(k-1) + c_2u^2(k-1) \neq 0$$

Now we shall investigate whether the sufficient conditions introduced by Leontaritis and Billings (1985b) are fulfilled. The first condition is that a state space model exists. This can be set up from (1.5.19) and (1.5.20). The second is that a linearized model exists with the same order as (1.5.20), i.e., with order one.

Assume the input signal varies around a working point u_0

$$u(k) = u_0 + \Delta u(k)$$

and the varying part of the input signal $(\Delta u(k))$ is small. Then the output signal can be separated into a constant (y_0) and a varying part $(\Delta y(k))$

$$y(k) = y_0 + \Delta y(k)$$

The globally valid equation is

$$\begin{aligned} y(k) &= b_1c_0 + b_1c_1\left[u_0 + \Delta u(k-1)\right] + b_1c_2\left[u_0 + \Delta u(k-1)\right]^2 - a_1y(k-1) \\ &= b_1c_0 + b_1c_1u_0 + b_1c_1\Delta u(k-1) + b_1c_2u_0^2 + 2b_1c_2u_0\Delta u(k-1) \\ &\quad + b_1c_2\Delta u^2(k-1) - a_1y(k-1) \end{aligned} \tag{1.5.22}$$

The expected value of the output signal is determined by

$$\mathrm{E}\{y(k)\} + a_1\mathrm{E}\{y(k)\} = b_1c_0 + b_1c_1u_0 + b_1c_2u_0^2 + b_1c_2\mathrm{E}\left\{\Delta u^2(k-1)\right\}$$

and the linearized equation is

$$\Delta y(k) = \left[b_1\left(c_1 + 2c_2u_0\right)\right]\Delta u(k-1) - a_1y(k-1) \tag{1.5.23}$$

As is seen, the order of the linearized model is one, as expected. The linearized equation vanishes if and only if

$$c_1 + 2c_2u_0 = 0$$

which is the case if the working point is in the extremum of the polynomial.

(1.5.23) can be obtained by another way, as well. The linearized equation of the nonlinear part is

$$\Delta v(k) = c_1 \Delta u(k) + 2c_2 u_0 \Delta u(k) \tag{1.5.24}$$

Substitution of (1.5.24) into (1.5.20) leads again to (1.5.23).

Example 1.5.7 *Quadratic, first-order simple Wiener model*
Let the equation of the linear dynamic part of the model be

$$v(k) = b_1 u(k-1) - a_1 v(k-1) \tag{1.5.25}$$

and that of the dynamic part be

$$y(k) = c_0 + c_1 v(k) + c_2 v^2(k) \tag{1.5.26}$$

Substitute (1.5.25) into (1.5.26)

$$\begin{aligned} y(k) &= c_0 + c_1\left[b_1 u(k-1) - a_1 v(k-1)\right] + c_2\left[b_1 u(k-1) - a_1 v(k-1)\right]^2 \\ &= c_0 + c_1\, b_1 u(k-1) - c_1 a_1 v(k-1) + c_2 b_1^2 u^2(k-1) \\ &\quad -2c_2 b_1 a_1 u(k-1) v(k-1) + c_2 a_1^2 v^2(k-1) \end{aligned} \tag{1.5.27}$$

Substitute now $c_2 v^2(k)$ from (1.5.26) into (1.5.28)

$$\begin{aligned} y(k) &= c_0 + c_1\, b_1 u(k-1) - c_1 a_1 v(k-1) + c_2 b_1^2 u^2(k-1) \\ &\quad -2c_2 b_1 a_1 u(k-1) v(k-1) + a_1^2 y(k-1) - a_1^2 c_0 - a_1^2 c_1 v(k-1) \\ &= c_0' + c_1\, b_1 u(k-1) + c_2 b_1^2 u^2(k-1) + a_1^2 y(k-1) \\ &\quad -a_1 v(k-1)\left[c_1' + 2c_2 b_1 u(k-1)\right] \end{aligned} \tag{1.5.28}$$

with

$$c_0' = c_0\left(1 - a_1^2\right)$$
$$c_1' = c_1\left(1 + a_1\right)$$

Substitute $v(k-1)$ from the one step delayed (1.5.26) into (1.5.28)

$$\begin{aligned} y(k) &= c_0' + c_1 b_1 u(k-1) + c_2 b_1^2 u^2(k-1) + a_1^2 y(k-1) \\ &\quad -a_1\left[c_1' + 2c_2 b_1 u(k-1)\right]\left[-\frac{c_1}{2c_2} \pm \sqrt{\left(\frac{c_1}{2c_2}\right)^2 - \frac{c_0 - y(k-1)}{c_2}}\right] \end{aligned} \tag{1.5.29}$$

The sign before the root depends on which side of the polynomial the working point is. The condition for having a real root is that the term under the root in (1.5.29) is positive:

$$y(k-1) > c_0 - \frac{c_1^2}{4c_2}$$

which is fulfilled if the output signal is greater than the minimum of the parabola. If the square root is approximated by its truncated Taylor series then (1.5.29) has only polynomial terms.

Assume the input signal varies around a working point u_0

$$u(k) = u_0 + \Delta u(k)$$

and the varying part of the input signal $\Delta u(k)$ is small. The linearized equation of the static nonlinear part is from (1.5.24)

$$\Delta y(k) = [c_1 + 2c_2 v_0]\Delta v(k)$$

where

$$\Delta v(k) = b_1 \Delta u(k) - a_1 \Delta v(k-1)$$

and

$$v_0 = \frac{b_1}{1+a_1} u_0$$

The linearized equation has the order one if

$$a_1 \neq 0 \quad \text{and} \quad b_1 \neq 0 \quad \text{and} \quad c_1 + 2c_2 v_0 \neq 0$$

The latter condition means that the working point is not in the extremum of the static polynomial. This example was derived for the special case $c_0 = 0$, $c_1 = c_2 = b_1 = -a_1 = 1$ in Leontaritis and Billings (1985b). ■

1.5.5 Classification of the polynomial difference equations

A parametric polynomial nonlinear model linear-in-parameters restricted up to the quadratic terms in the memory vector, is given by

$$\begin{aligned} &A_1(q^{-1})y(k) + A_2(q_1^{-1}, q_2^{-1})y^2(k-\ell) \\ &= c_0 + B_1(q^{-1})u(k-d) + B_2(q_1^{-1}, q_2^{-1})u^2(k-d) + F_2(q_1^{-1}, q_2^{-1})u(k-d)y(k-\ell) \end{aligned} \tag{1.5.30}$$

where the polynomials of the shifting (delaying) operator q^{-1} are

$$B_1(q^{-1}) = \sum_{i=0}^{nb_1} b_{1i} q^{-i} \tag{1.5.31}$$

$$A_1(q^{-1}) = 1 + \tilde{A}_1(q^{-1}) = \sum_{j=0}^{na_1} a_{1j} q^{-j} \tag{1.5.32}$$

$$B_2\left(q_1^{-1},q_2^{-1}\right)u^2(k)=\sum_{i=0}^{nb_2}\sum_{j=i}^{nb_2}b_{2ij}u(k-i)u(k-j) \tag{1.5.33}$$

$$A_2\left(q_1^{-1},q_2^{-1}\right)y^2(k)=\sum_{i=0}^{na_2}\sum_{j=i}^{na_2}a_{2ij}y(k-i)y(k-j) \tag{1.5.34}$$

$$F_2\left(q_1^{-1},q_2^{-1}\right)u(k)y(k)=\sum_{i=0}^{nf_2}\sum_{j=0}^{nf_2}f_{2ij}u(k-i)y(k-j) \tag{1.5.35}$$

In (1.5.30) d is the integer dead time in the unit of the sampling time. The equation is implicit in the output signal if $\ell=0$ and it can also be rearranged to an explicit recursive form if $\ell=1$. The orders of the one- and two-dimensional polynomials are as follows:

$$\deg\left\{B_1\left(q^{-1}\right)\right\}=nb_1,\ \deg\left\{A_1\left(q^{-1}\right)\right\}=na_1,\ \deg\left\{B_2\left(q_1^{-1},q_2^{-1}\right)\right\}=\left[nb_2,nb_2\right],$$
$$\deg\left\{A_2\left(q_1^{-1},q_2^{-1}\right)\right\}=\left[na_2,na_2\right],\ \deg\left\{F_2\left(q_1^{-1},q_2^{-1}\right)\right\}=\left[nf_2,nf_2\right]$$

In special cases the two-dimensional polynomials $B_2\left(q_1^{-1},q_2^{-1}\right)$ or $F_2\left(q_1^{-1},q_2^{-1}\right)$ have zero off-diagonal values (which means the coefficients are zero if the delays in the two dimensions are different). Then they can be replaced by one dimensional polynomials as

$$B_2\left(q_1^{-1},q_2^{-1}\right)\to B_2\left(q_1^{-1}\right)\qquad\left(b_{2ij}\to b_{2i}\right)\qquad\text{if}\quad b_{2ij}=0\quad i\neq j \tag{1.5.36}$$

and

$$F_2\left(q_1^{-1},q_2^{-1}\right)\to F_2\left(q_1^{-1}\right)\qquad\left(f_{2ij}\to f_{2i}\right)\qquad\text{if}\quad f_{2ij}=0\quad i\neq j \tag{1.5.37}$$

In the followings we deal with the recursive equations. The general quadratic description (1.5.30) with $\ell=1$ includes the following special cases:

1. Linear pulse transfer function model:

$$A_1\left(q^{-1}\right)y(k)=c_0+B_1\left(q^{-1}\right)u(k-d) \tag{1.5.38}$$

or

$$y(k)=-\sum_{j=1}^{na_1}a_{1j}y(k-j)+c_0+\sum_{i=0}^{nb_1}b_{1i}u(k-d-i) \tag{1.5.39}$$

2. Generalized Hammerstein model:

$$A_1\left(q^{-1}\right)y(k)=c_0+B_1\left(q^{-1}\right)u(k-d)+B_2\left(q^{-1}\right)u^2(k-d) \tag{1.5.40}$$

or

$$y(k)=-\sum_{j=1}^{na_1}a_{1j}y(k-j)+c_0+\sum_{i=0}^{nb_1}b_{1i}u(k-d-i)+\sum_{i=0}^{nb_2}b_{2i}u^2(k-d-i) \tag{1.5.41}$$

3. Parametric Volterra model (Haber and Keviczky, 1974; 1976)

$$A_1(q^{-1})y(k) = c_0 + B_1(q^{-1})u(k-d) + B_2(q_1^{-1}, q_2^{-1})u^2(k-d) \tag{1.5.42}$$

or

$$y(k) = -\sum_{j=1}^{na_1} a_{1j}y(k-j) + c_0 + \sum_{i=0}^{nb_1} b_{1i}u(k-d-i) + \sum_{i=0}^{nb_2}\sum_{j=i}^{nb_2} b_{2ij}u(k-d-i)u(k-d-j) \tag{1.5.43}$$

4. Simple bilinear model (e.g., Beghelli and Guidorzi, 1976)

$$A_1(q^{-1})y(k) = c_0 + B_1(q^{-1})u(k-d) + F_2(q^{-1})u(k-d)y(k-1) \tag{1.5.44}$$

or

$$y(k) = -\sum_{j=1}^{na_1} a_{1j}y(k-j) + c_0 + \sum_{i=0}^{nb_1} b_{1i}u(k-d-i) + \sum_{i=0}^{nf_2} f_{2i}u(k-d-i)\,y(k-1-i) \tag{1.5.45}$$

5. General bilinear (multi-linear) model (e.g., Beghelli and Guidorzi, 1976)

$$A_1(q^{-1})y(k) = c_0 + B_1(q^{-1})u(k-d) + F_2(q_1^{-1}, q_2^{-1})u(k-d)y(k-1) \tag{1.5.46}$$

or

$$y(k) = -\sum_{j=1}^{na_1} a_{1j}y(k-j) + c_0 + \sum_{i=0}^{nb_1} b_{1i}u(k-d-i) + \sum_{i=0}^{nf_2}\sum_{j=0}^{nf_2} f_{2ij}u(k-d-i)\,y(k-1-j) \tag{1.5.47}$$

6. Linear input signal nonlinear output signal model (Lachmann, 1983)

$$A_1(q^{-1})y(k) + A_2(q_1^{-1}, q_2^{-1})y^2(k-1) = c_0 + B_1(q^{-1})u(k-d) \tag{1.5.48}$$

or

$$y(k) = -\sum_{j=1}^{na_1} a_{1j}y(k-j) - \sum_{i=0}^{na_2}\sum_{j=i}^{na_2} a_{2ij}y(k-1-i)y(k-1-j) + c_0 + \sum_{i=0}^{nb_1} b_{1i}u(k-d-i) \tag{1.5.49}$$

Table 1.5.1 shows which model components the different models include.

TABLE 1.5.1 Possible model components of recursive polynomial difference equations limited to the quadratic terms

Model \ Component	1	$u(k-d-i)$ $i \geq 1$	$y(k-i)$	$u^2(k-d-i)$	$u(k-d-i)u(k-d-j)$ $i \neq j$	$u(k-d-i)y(k-i)$ $i \geq 1$	$u(k-d-i)y(k-j)$ $j \geq 1; i \neq j$	$y^2(k-i)$ $i \geq 1$	$y(k-i)y(k-j)$ $i \geq 1; j \geq 1; i \neq j$
generalized Hammerstein	×	×	×	×					
parametric Volterra	×	×	×	×	×				
simple bilinear	×	×	×	×		×			
generalized bilinear	×	×	×	×		×	×		
linear-in-parameters	×	×	×	×				×	×
general polynomial	×	×	×	×	×	×	×	×	×

The natural question arises which model should be applied for the approximation of real processes. One possibility is to allow all components of (1.5.30) compete to become a model component, and to use a structure search. There are some advantages, however, if only model components of a certain model class are used. Some standpoints are given now.

1. *Generalized Hammerstein model*
   - Because the output signal occurs only in the form $A_1(q^{-1})y(k)$, the filtering of a linear additive noise at the output can be done similarly as with the linear systems;
   - The prediction of the output signal can be calculated by a similar algorithm as with the linear systems because the only term where the output signal occurs is $A_1(q^{-1})y(k)$. This is important in predictive control algorithms;
   - The nonlinear degree of the steady state characteristics of the model is equal to the highest power of the input signal used as a model component;
   - The Volterra kernels $[h_2(\kappa_1, \kappa_2), \ \kappa_1 \neq \kappa_2]$belonging to the cross product terms are not approximated by the model.

2. *Parametric Volterra model*
   - Because the output signal occurs only in the form $A_1(q^{-1})y(k)$, the filtering of a linear additive noise at the output can be done similarly to the linear systems;
   - The prediction of the output signal can be calculated by a similar algorithm as with the linear systems because the only term where the output signal occurs is $A_1(q^{-1})y(k)$. This is important in predictive control algorithms;
   - The nonlinear degree of the steady state characteristics of the model is equal to the highest power of the input signal used as a model component;
   - The Volterra kernels $[h_2(\kappa_1, \kappa_2), \ \kappa_1 \neq \kappa_2]$ belonging to the cross-product terms are also approximated by the model;
   - All components of the generalized Hammerstein model are covered.

3. *Bilinear models (simple and general)*
   - Any continuous time analytic nonlinear dynamic system can be approximated

by an equivalent input–output continuous time bilinear model. This is, however, not true for discrete time models (Chen and Billings, 1989);
- The nonlinear degree of the steady state characteristics of the model is infinite;
- All Volterra kernels are somehow approximated by the model;
- There is no function of the output signal, like $y^2(k)$, which would cause a problem while filtering the noise;
- The derivation of a control algorithm by minimizing a quadratic function of the control error is an easy task, because the model is linear in the control signal. The control algorithms are, therefore, similar to those of the linear systems;
- The prediction of the future output signals cannot be calculated analytically only by recursive substitution;
- The inverse model is single-valued and the input signal belonging to a given desired output signal can be computed unambiguously.

4. *Model linear in the input signal and nonlinear in the output signal*
   - The model describes continuous time systems having on the LHS only linear and nonlinear functions of the output signal and on the RHS only linear functions of the input signal. Such a system occurs often in the practice. An example is a quasi-linear second-order system with working point dependent damping factor and/or time constant;
   - The nonlinear degree of the steady state characteristics of the model is infinite;
   - All Volterra kernels are somehow approximated by the model;
   - The derivation of a control algorithm by minimizing a quadratic function of the control error is an easy task, because the model is linear in the control signal;
   - The prediction of the future output signals cannot be calculated analytically only by recursive substitution;
   - The inverse model is single-valued and the input signal belonging to a given desired output signal can be computed unambiguously.

5. *General polynomial difference equation model*
   - Almost any continuous time or discrete time analytic dynamic system can be approximated by an equivalent input–output discrete time general polynomial model;
   - The nonlinear degree of the steady state characteristics of the model is infinite;
   - All Volterra kernels are somehow approximated by the model;
   - Using a general model and a structure search leads to the self-selection of the best fitting model components;
   - Simulation experiences have shown that usually a proper input–output equivalent model can be fitted to the measured data if the linear and quadratic functions of the input and output signals inclusive the cross product terms $\left[u(k), y(k), u^2(k), y^2(k), u(k)y(k)\right]$ are chosen as the model components and the order of the memory is one or two;
   - The prediction of the future output signals cannot be calculated analytically only by recursive substitution.

In the following applications where different models were used for approximating the real processes are listed:

1. *Generalized Hammerstein model*
   - electrical power unit (Kaminskas, Sidlauskas and Tallat–Kelpsa, 1988);
   - thermal power plant (Bamberger, 1978; Bamberger and Isermann, 1976, 1977, 1978);
   - heat exchanger (Haber, 1979a);
   - stream flow process in hydrology (Hu and Yuan, 1988);
   - cement kiln (Li and Chen, 1988);
   - closed circuit cement ball grinding mill (Keviczky, 1976).

2. *Parametric Volterra model*
   - two heat exchangers working parallel (Chantre, 1987, 1989);
   - pH process (Lachmann, 1983; 1985b);
   - air conditioner (Lachmann, 1983, 1985b);
   - water tank (Kortmann and Unbehauen, 1988a).

3. *Bilinear models*
   - fermentation (Dochain and Bastin, 1984);
   - waste water treatment (Goodwin *et al.*, 1982; Goodwin and Sin, 1984);
   - continuously stirred reactor (Yi *et al.*, 1989);
   - blood pressure (McInnis *et al.*, 1985);
   - pH process (Gilles and Laggoune, 1985a, 1985b);
   - nerve cell (Kitajima and Hara, 1987);
   - distillation column (Haber and Zierfuss, 1991).

4. *Linear input signal nonlinear output signal model*
   - water tank (Lachmann and Goedecke, 1982; Lachmann, 1983, 1985b; Kortmann and Unbehauen, 1987, 1988a);
   - electrically heated heat exchanger (Haber and Unbehauen, 1990).

5. *General polynomial model*
   - robot manipulator (Agarwal and Seborg, 1985);
   - two-link robot manipulator (Agarwal and Seborg, 1986);
   - flood process (Haber, 1982);
   - turbo generator (Kortmann and Unbehauen, 1988b);
   - blast furnace (Kortmann and Unbehauen, 1988b; Kortmann *et al.*, 1988);
   - reheater of a thermal plant (Diaz and Desrochers, 1988);
   - heat exchanger (Billings and Fadzil, 1985);
   - water tank (Chen *et al.*, 1989).

1.5.6 Calculation of the Volterra kernels

The property that a Volterra series model is a polynomial model in the input signal makes the calculation of the Volterra kernels from a nonlinear difference equation possible.

Algorithm 1.5.1 (Diaz, 1986; Diaz and Desrochers, 1988): The homogeneous subsystems

$$y_i(k) = y_i\big(u(k)\big) = V_i\big[u(k)\big] \qquad i = 0, 1, 2, \ldots$$

can be determined by the following algorithm:

1. Make the following substitution in the difference equation:

$$u(k) \to \gamma u(k)$$
$$y(k) \to y_0 + \gamma y_1(k) + \gamma^2 y_2(k) + ...$$

2. Since the equation is true for arbitrary γ the terms belonging to the same powers of γ are to be equated. Then a set of equations for $y_i(k)$, $i = 0, 1, 2, \ldots$ can be set up.
3. Solve the equations in turn starting with $i = 0$. The resulting expressions for $y_i(k)$ should contain only terms of the input signal.
4. The coefficients of $\prod_{j=1}^{i} u(k - \kappa_j)$ in $y_i(k)$ are the discrete time Volterra kernels.

Example 1.5.8 *Generalized Hammerstein model*
Consider the difference equation of a Hammerstein model (1.5.41) with $d = 1, na_1 = 1, nb_1 = nb_2 = 0$

$$y(k+1) = -a_1 y(k) + b_0 + b_{10} u(k) + b_{20} u^2(k)$$

or shifted backwards by one

$$y(k) = -a_1 y(k-1) + b_0 + b_{10} u(k-1) + b_{20} u^2(k-1) \qquad (1.5.50)$$

Replace $u(k)$ by $\gamma u(k)$ and $y(k)$ by $\Sigma \gamma^i y_i(k)$ in (1.5.50)

$$y_0 + \gamma y_1(k) + \gamma^2 y_2(k) + \ldots = -a_1 y_0 - a_1 \gamma y_1(k-1) - a_1 \gamma^2 y_2(k-1)$$
$$\ldots + b_0 + b_{10} \gamma u(k-1) + b_{20} \gamma^2 u^2(k-1)$$

Equate the coefficients of equal powers of γ:

$$y_0 = -a_1 y_0 + b_0$$
$$y_1(k) = -a_1 y_1(k-1) + b_{10} u(k-1)$$
$$y_2(k) = -a_1 y_2(k-1) + b_{20} u^2(k-1)$$
$$y_i(k) = -a_1 y_i(k) \qquad i = 3, 4, 5, \ldots$$

The solutions of the difference equations are as follows

$$y_0 = \frac{b_0}{1 + a_1}$$
$$y_1(k) = \sum_{\kappa_1 = 1}^{k} b_{10} (-a_1)^{\kappa_1 - 1} u(k - \kappa_1)$$
$$y_2(k) = \sum_{\kappa_1 = 1}^{k} b_{10} (-a_1)^{\kappa_1 - 1} u^2(k - \kappa_1)$$
$$y_i(k) = 0 \qquad i = 3, 4, 5 \ldots$$

Finally, the kernels are

$$h_0 = b_0/(1+a_1)$$

$$h_1(\kappa_1) = \begin{cases} b_{10}(-a_1)^{\kappa_1-1} & \kappa_1 = 1,2,3,\dots \\ 0 & \kappa_1 = 0 \end{cases}$$

$$h_2(\kappa_1,\kappa_2) = \begin{cases} b_{10}(-a_1)^{\kappa_1-1} & \kappa_1 = \kappa_2 = 1,2,3,\dots \\ 0 & \kappa_1 \neq \kappa_2 \text{ and } \kappa_1 = \kappa_2 = 0 \end{cases}$$

$$h_i(\kappa_1,\kappa_2,\dots,\kappa_i) = 0 \qquad i = 3,4,5,\dots$$

Example 1.5.9 *Parametric Volterra model*
Consider the difference equation of a parametric Volterra model (1.5.43) with $d = 1, na_1 = 1, nb_1 = nb_2 = 1$

$$y(k+1) = -a_1 y(k) + b_0 + b_{10}u(k) + b_{11}u(k-1) + b_{200}u^2(k) \\ +b_{201}u(k)u(k-1) + b_{211}u^2(k-1)$$

or shifted backwards by one

$$y(k) = -a_1 y(k-1) + b_0 + b_{10}u(k-1) + b_{11}u(k-2) + b_{200}u^2(k-1) \\ +b_{201}u(k-1)u(k-2) + b_{211}u^2(k-2) \tag{1.5.51}$$

Replace $u(k)$ by $\gamma u(k)$ and $y(k)$ by $\Sigma\gamma^i y_i(k)$ in (1.5.51)

$$y_0 + \gamma y_1(k) + \gamma^2 y_2(k) + \dots = -a_1 y_0 - a_1\gamma y_1(k-1) - a_1\gamma^2 y_2(k-1) \\ \dots + b_0 + b_{10}\gamma u(k-1) + b_{11}\gamma u(k-2) + b_{200}\gamma^2 u^2(k-1) \\ +b_{201}\gamma^2 u(k-1)u(k-2) + b_{211}\gamma^2 u^2(k-2)$$

Equate the coefficients of equal powers of γ:

$$y_0 = -a_1 y_0 + b_0$$

$$y_1(k) = -a_1 y_1(k-1) + b_{10}u(k-1) + b_{11}u(k-2)$$

$$y_2(k) = -a_1 y_2(k-1) + b_{200}u^2(k-1) + b_{201}u(k-1)u(k-2) + b_{211}u^2(k-2)$$

$$y_i(k) = -a_1 y_i(k) \qquad i = 3,4,5,\dots$$

The kernels are as follows

$$h_0 = \frac{b_0}{1+a_1}$$

$$h_1(\kappa_1) = \begin{cases} 0 & \kappa_1 = 0 \\ b_{10} & \kappa_1 = 1 \\ (b_{11} - a_1 b_{10})(-a_1)^{\kappa_1-2} & \kappa_1 = 2,3,4,\dots \end{cases}$$

$$h_2(\kappa_1, \kappa_1) = \begin{cases} 0 & \kappa_1 = 0 \\ b_{200} & \kappa_1 = 1 \\ (b_{211} - a_1 b_{200})(-a_1)^{\kappa_1 - 2} & \kappa_1 = 2, 3, 4, \ldots \end{cases}$$

$$h_2(\kappa_1, \kappa_1 + 1) = \begin{cases} 0 & \kappa_1 = 0 \\ b_{201}(-a_1)^{\kappa_1 - 1} & \kappa_1 = 1, 2, 3, \ldots \end{cases}$$

$$h_2(\kappa_1, \kappa_1 + \kappa) = 0 \qquad \kappa = 2, 3, 4, \ldots \text{ or } \kappa = -1, -2, -3, \ldots$$

$$h_i(\kappa_1, \ldots, \kappa_i) = 0 \qquad i = 3, 4, 5, \ldots$$

Example 1.5.10 *Simple bilinear model*
Consider the difference equation of a simple bilinear model (1.5.45) with $d = 1, na_1 = 1, nb_1 = nc_1 = 0$

$$y(k+1) = -a_1 y(k) + b_0 + b_{10} u(k) + c_{20} u(k) y(k)$$

or shifted backwards by one

$$y(k) = -a_1 y(k-1) + b_0 + b_{10} u(k-1) + c_{20} u(k-1) y(k-1) \tag{1.5.52}$$

Replace $u(k)$ by $\gamma u(k)$ and $y(k)$ by $\Sigma \gamma^i y_i(k)$ in (1.5.52)

$$\begin{aligned} y_0 + \gamma y_1(k) + \gamma^2 y_2(k) + \ldots = {} & -a_1 y_0 - a_1 \gamma y_1(k-1) - a_1 \gamma^2 y_2(k-1) - \ldots \\ & \ldots + b_0 + b_{10} \gamma u(k-1) + c_{20} \gamma u(k-1) y_0 + c_{20} \gamma^2 u(k-1) y_1(k-1) \\ & + c_{20} \gamma^3 u(k-1) y_2(k-1) + \ldots \end{aligned}$$

Equate the coefficients of equal powers of γ :

$$\begin{aligned} y_0 &= -a_1 y_0 + b_0 \\ y_1(k) &= -a_1 y_1(k-1) + b_{10} u(k-1) + c_{20} u(k-1) y_0 \\ y_i(k) &= -a_1 y_i(k-1) + c_{20} u(k-1) y_{i-1}(k-1) \qquad i = 2, 3, 4, \ldots \end{aligned}$$

The solutions of the difference equations are as follows

$$y_0 = \frac{b_0}{1 + a_1}$$

$$y_1(k) = \sum_{\kappa_1 = 1}^{k} \left(b_{10} + c_{20} \frac{b_0}{1 + a_1} \right) (-a_1)^{\kappa_1} u(k - \kappa_1)$$

$$y_2(k)=\sum_{\kappa_1=1}^{k} c_{20}(-a_1)^{\kappa_1} u(k-\kappa_1) y_1(k-\kappa_1)$$

$$=\sum_{\kappa_1=1}^{k} c_{20}(-a_1)^{\kappa_1} u(k-\kappa_1)\left[\sum_{\kappa_2=1}^{k-\kappa_1}\left(b_{10}+c_{20}\frac{b_0}{1+a_1}\right)(-a_1)^{\kappa_2} u(k-\kappa_1-\kappa_2)\right]$$

$$=\sum_{\kappa_1=1}^{k}\sum_{\kappa_2=1}^{k-\kappa_1} c_{20}\left(b_{10}+c_{20}\frac{b_0}{1+a_1}\right)(-a_1)^{\kappa_1+\kappa_2} u(k-\kappa_1)u(k-\kappa_1-\kappa_2)$$

$$y_3(k)=\sum_{\kappa_1=1}^{k} c_{20}(-a_1)^{\kappa_1} u(k-\kappa_1) y_2(k-\kappa_1)=\sum_{\kappa_1=1}^{k} c_{20}(-a_1)^{\kappa_1} u(k-\kappa_1)$$

$$\times\sum_{\kappa_2=1}^{k-\kappa_1}\sum_{\kappa_3=1}^{k-\kappa_1-\kappa_2} c_{20}\left(b_{10}+c_{20}\frac{b_0}{1+a_1}\right)(-a_1)^{\kappa_2+\kappa_3} u(k-\kappa_1-\kappa_2)u(k-\kappa_1-\kappa_2-\kappa_3)$$

$$=\sum_{\kappa_1=1}^{k}\sum_{\kappa_2=1}^{k-\kappa_1}\sum_{\kappa_3=1}^{k-\kappa_1-\kappa_2} c_{20}\left(b_{10}+c_{20}\frac{b_0}{1+a_1}\right)$$

$$\times(-a_1)^{\kappa_1+\kappa_2+\kappa_3} u(k-\kappa_1)u(k-\kappa_1-\kappa_2)u(k-\kappa_1-\kappa_2-\kappa_3)$$

The successive terms $y_i(k)$, $i=4,5,6,\ldots$ can be calculated similarly. As is seen, a bilinear system has infinite-degree Volterra series. Finally, the first four kernels are

$$h_0=\frac{b_0}{1+a_1}$$

$$h_1(\kappa_1)=\begin{cases}\left(b_{10}+c_{20}\frac{b_0}{1+a_1}\right) b_{10}(-a_1)^{\kappa_1-1} & \kappa_1=1,2,3,\ldots\\ 0 & \kappa_1=0\end{cases}$$

$$h_2(\kappa_1,\kappa_1+\kappa_2)=\begin{cases}c_{20}\left(b_{10}+c_{20}\frac{b_0}{1+a_1}\right)(-a_1)^{\kappa_1+\kappa_2-2} & \kappa_1=1,2,3,\ldots\ \kappa_2=1,2,3,\ldots\\ 0 & \kappa_1\kappa_2=0\end{cases}$$

$$h_3(\kappa_1,\kappa_1+\kappa_2,\kappa_1+\kappa_2+\kappa_3)=$$

$$=\begin{cases}c_{20}\left(b_{10}+c_{20}\frac{b_0}{1+a_1}\right)(-a_1)^{\kappa_1+\kappa_2+\kappa_3-3} & \kappa_1=1,2,3,\ldots\quad \kappa_2=1,2,3,\ldots\\ & \kappa_3=1,2,3,\ldots\\ 0 & \kappa_1\kappa_2\kappa_3=0\end{cases}$$

The special case of this example with $b_0=0$ is in (Diaz and Desrochers, 1988).

Example 1.5.11 *Linear input signal nonlinear model*
Consider the difference equation of a linear input signal nonlinear model (1.5.49) with $d=1, na_1=1, nb_1=na_2=0$

$$y(k)=-a_1y(k-1)+b_0+b_{10}u(k-1)+a_{20}y^2(k-1) \tag{1.5.53}$$

Replace $u(k)$ by $\gamma u(k)$ and $y(k)$ by $\Sigma\gamma^i y_i(k)$ in (1.5.53)

$$\begin{aligned}y_0+\gamma y_1(k)+\gamma^2y_2(k)+\ldots=&-a_1y_0-a_1\gamma y_1(k-1)-a_1\gamma^2y_2(k-1)-\ldots\\&\ldots+b_0+b_{10}\gamma u(k-1)+a_{20}y_0^2+2a_{20}\gamma y_0y_1(k-1)\\&+2a_{20}\gamma^2\left[y_0y_2(k-1)+y_1^2(k-1)\right]+\ldots\end{aligned}$$

Equate the coefficients of equal powers of γ:

$$\begin{aligned}y_0&=-a_1y_0+b_0+a_{20}y_0^2\\y_1(k)&=-a_1y_1(k-1)+b_{10}u(k-1)+2a_{20}y_0y_1(k-1)\\y_2(k)&=-a_1y_2(k-1)+2a_{20}\left[y_0y_2(k-1)+y_1^2(k-1)\right]\end{aligned}$$

etc. The solutions of the difference equations are as follows

$$y_0=\frac{1}{2a_{20}}\left[(a_1+1)\pm\sqrt{(a_1+1)^2-4a_{20}b_0}\right]$$

$$y_1(k)=\sum_{\kappa_1=1}^{k}b_{10}\left(-a_1+2a_{20}y_0\right)^{\kappa_1-1}u(k-\kappa_1)$$

$$\begin{aligned}y_2(k)&=\sum_{\kappa_3=1}^{k}2a_{20}\left(-a_1+2a_{20}y_0\right)^{\kappa_3-1}y_1^2(k-\kappa_3)\\&=\sum_{\kappa_3=1}^{k}\sum_{\kappa_1=1}^{k}\sum_{\kappa_2=1}^{k}2a_{20}\left(-a_1+2a_{20}y_0\right)^{\kappa_3}b_{10}^2\left(-a_1+2a_{20}y_0\right)^{\kappa_1-1}\\&\times\left(-a_1+2a_{20}y_0\right)^{\kappa_2-1}u(k-\kappa_3-\kappa_1)u(k-\kappa_3-\kappa_2)\end{aligned}$$

The successive terms $y_i(k)$, $i=3,4,5,\ldots$ can be calculated similarly. As is seen, the quadratic term in the output signal causes an infinite-degree Volterra series. Finally, the kernels are

$$h_0=-\frac{1}{2a_{20}}\left[(a_1+1)\pm\sqrt{(a_1+1)^2-4a_{20}b_0}\right]$$

$$h_1(\kappa_1)=\begin{cases}b_{10}\left(-a_1+2a_{20}h_0\right)^{\kappa_1-1} & \kappa_1=1,2,3,\ldots\\0 & \kappa_1=0\end{cases}$$

$$h_2(\kappa_1,\kappa_2)=\begin{cases}\sum\limits_{\kappa_3=1}^{k} 2a_{20}\left(-a_1+2a_{20}h_0\right)^{\kappa_3} b_{10}^2\left(-a_1+2a_{20}h_0\right)^{\kappa_1-1}\left(-a_1+2a_{20}h_0\right)^{\kappa_2-1} & \kappa_1=1,2,3,\ldots\ \kappa_2=1,2,3,\ldots \\ 0 & \kappa_1\kappa_2=0\end{cases}$$

etc. ■

1.6 QUASI-LINEAR MODELS HAVING SIGNAL DEPENDENT PARAMETERS

1.6.1 Physical meaning

A nonlinear process can be described by a quasi-linear system with signal dependent parameters in a domain of the input signal

- if the system can be linearized for small excitations around all possible working points, and
- if the parameters are functions of any measurable or computable signal.

We distinguish between

- continuous time and
- discrete time

quasi-linear systems with signal dependent parameters. For both kinds of models their difference equations will be presented, which are the bases of the usage of the effective discrete time parameter estimation methods. All parameters of a quasi-linear model may depend on a signal, e.g.:

- static gain;
- time constant;
- damping factor;
- zero of the (pulse) transfer function;
- pole of the (pulse) transfer function;
- coefficient of the numerator of the (pulse) transfer function;
- coefficient of the denominator of the (pulse) transfer function.

Any parameter may depend on the following signals:

- input signal;
- output signal;
- external signal;
- computed signal based on the measured ones.

The signal(s), the parameter(s) depends on, can be:

- time varying signal;
- change in an arbitrary signal;
- direction of the change in a signal.

The most frequent occurrences of the dependency are the

- working-point-dependent or
- direction dependent ones.

The advantage of this description over other nonlinear model classes (e.g., parametric

Volterra model, block oriented models) is as follows:

1. It is clear enough to be understood for the application engineers;
2. Although it is an input–output model it gives an insight into the structure of the process;
3. It is understandable even for those who are familiar only with the linear terminology;
4. It allows the application of the identification programs already elaborated for linear systems with a pre-transformation of the measured data;
5. It makes the usage of linear controller design methods possible.

In practice numerous processes can be modeled by quasi-linear models with signal dependent parameters. Here are some examples.

Example 1.6.1 *Horizontal drum boiler (Figure 1.6.1a)*
The relation between the water flow $u(t)$ and the level $y(t)$ is described by

$$y(t) = \int_{\tau=0}^{t} \frac{1}{A_c} u(\tau) d\tau = \int_{\tau=0}^{t} \frac{1}{T_I} u(\tau) d\tau$$

where T_I is the integrating time constant that is equal to the cross-section (A_c) of the cylinder. This depends on the level according to the formula

$$T_I(y(t)) = 2L\sqrt{y(t)[2R - y(t)]}$$

where R is the radius of the cylinder and L is the length of the boiler. The maximum is achieved at $y(t) = R$. Figure 1.6.1b shows the relation

$$\frac{T_I(Y)}{T_{Imax}} = \frac{T_I}{T_I(R)} = \sqrt{\frac{Y}{R}\left(2 - \frac{Y}{R}\right)}$$

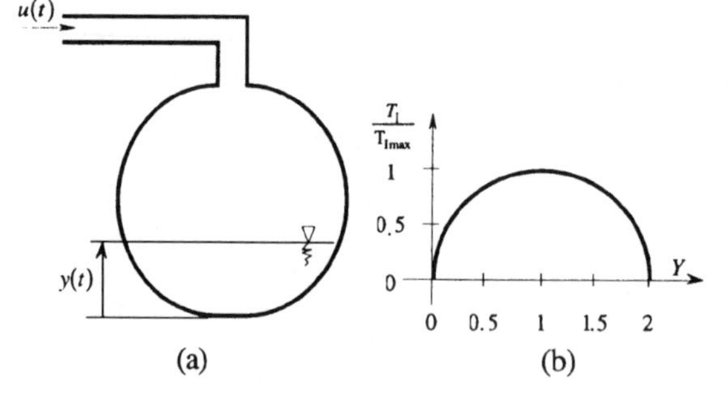

Fig. 1.6.1 Horizontal drum boiler: (a) scheme; (b) integrating time constant as a function of the level

Example 1.6.2 *Gravity tank (Figure 1.6.2)*
The following notations are used:

- the flow rate of the inlet liquid is the input signal $[u(t)]$;
- the flow rate of the outlet is the output signal $[y(t)]$;
- the level in the tank is a measured state variable $[x(t)]$.

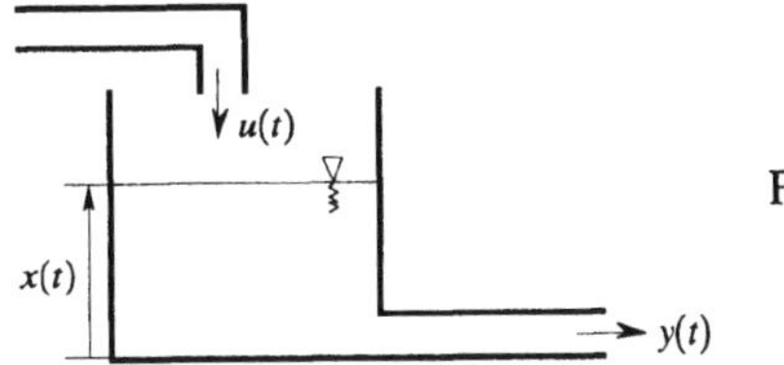

Fig. 1.6.2 Gravity tank

The following assumptions are made:

- the outlet flow is proportional to the level;
- the frictional force in the pipe is proportional to the square of the velocity (turbulent flow).

They lead to the following state space equation (Luyben, 1973):

$$\begin{aligned} \dot{y}(t) &= k_1 x(t) - k_2 y^2(t) \\ \dot{x}(t) &= -k_4 y(t) + k_3 u(t) \end{aligned} \tag{1.6.1}$$

(1.6.1) can be rewritten to an input–output equivalent differential equation

$$\ddot{y}(t) + 2k_2 y(t)\dot{y}(t) + k_1 k_4 y(t) = k_1 k_3 u(t) \tag{1.6.2}$$

which is similar to the differential equation of a second-order system

$$T^2 \ddot{y}(t) + 2T\xi\dot{y}(t) + y(t) = Ku(t) \tag{1.6.3}$$

A comparison of the coefficients of (1.6.2) with (1.6.3) gives the parameters

- static gain: $K = k_3/k_4$;
- time constant: $T = 1/\sqrt{k_1 k_4}$;
- damping factor: $\xi = \left(k_2/\sqrt{k_1 k_4}\right) y(t)$

Consequently the gravity tank behaves as a second-order linear system and the damping factor is proportional to the output signal.

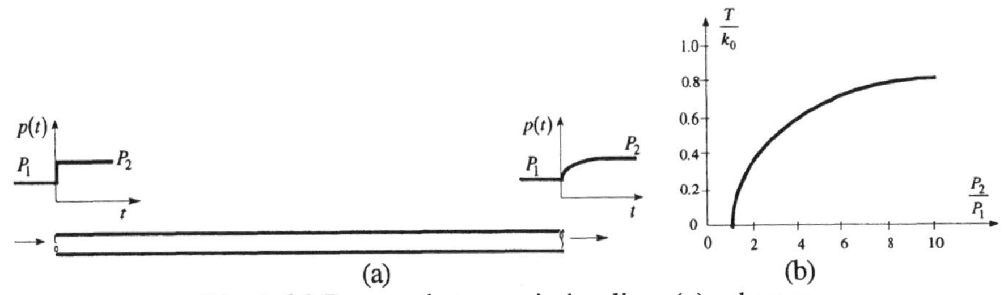

Fig. 1.6.3 Pneumatic transmission line: (a) scheme; (b) time constant as a function of the ratio of the pressures

Example 1.6.3 *Pneumatic transmission line (Figure 1.6.3a)*
If the pressure is changed from P_1 to P_2 at the front of a pneumatic transmission line, then the pressure at the end of the pipe achieves its new value with a delay. The time constant of the equivalent first-order quasi-linear system has the formula (Grabbe *et al.*,

1958)

$$T = k_0 \ln\left(\frac{0.63 + 0.37(P_2/P_1) + \sqrt{[0.63 + 0.37(P_2/P_1)]^2 - 1}}{(P_2/P_1) + \sqrt{(P_2/P_1)^2 - 1}}\right)$$

Figure 1.6.3b shows the dependence of T_1/k_0 on the ratio (P_2/P_1).

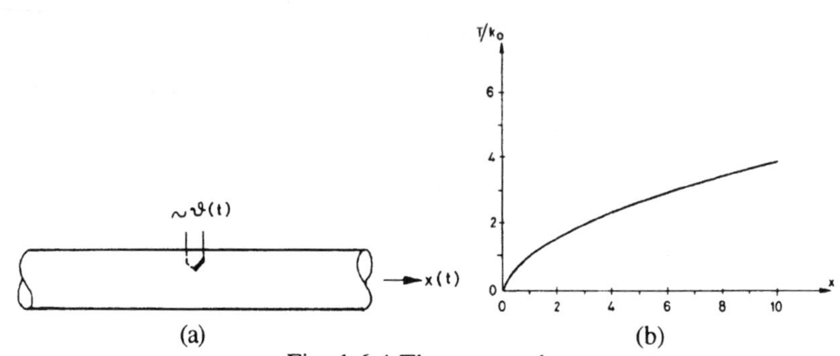

Fig. 1.6.4 Thermocouple:
(a) scheme; (b) time constant as a function of the flow rate

Example 1.6.4 *Thermocouple (Figure 1.6.4a)*
A thermocouple does not transmit the temperature of a flow without any delay. The transfer function of the sensor is linear, fist-order with constant static gain and an equivalent time constant *(T)* depending on the flow rate $(x(t))$ (Grabbe *et al.*, 1958)

$$T = k_0 \frac{1}{x(t)^{0.6}}$$

Here k_0 is a constant factor. The relation $T/k_0 = f(X)$ is shown in Figure 1.6.4b.

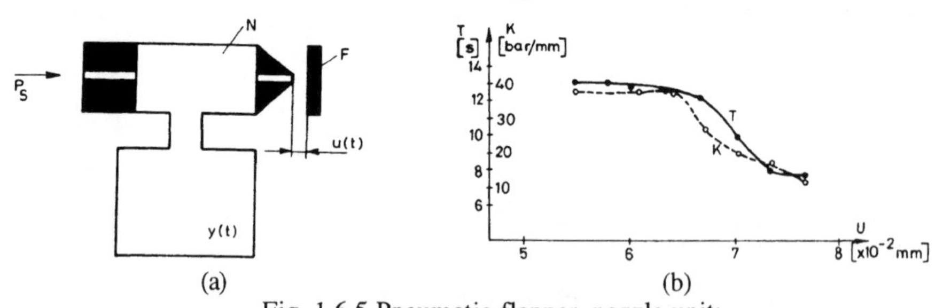

Fig. 1.6.5 Pneumatic flapper–nozzle unit:
(a) scheme; (b) gain and time constant as a function of the distance between the flapper and the nozzle

Example 1.6.5 *Pneumatic flapper–nozzle unit*
A pneumatic flapper–nozzle unit (Figure 1.6.5a) is a distance–pressure transmitter. The

input signal (u) is the distance between the nozzle and the flapper and the output signal (y) is the pressure. The linearized input–output relation can be modeled by a first-order model. Figure 1.6.5b shows the gains and the time constants as a function of the distance. The measurements were taken on a real transmitter with small changes near the working points marked in the plots (Haber, 1979b).

Example 1.6.6 *Distillation column*
Several authors report that the relation between the change of the reflux flow and the top temperature or the concentration is highly nonlinear, i.e., both signal and direction dependent (DeLorenzo *et al.*, 1972; Mizuno *et al.*, 1972; Weigand *et al.*, 1972; Tuschák *et al.*, 1982; Haber and Zierfuss, 1991, etc.).

Assume that first-order linear models are fitted to the temperature responses of the top plate to reflux steps. The static gain is inversely proportional to the reflux rate because the separation becomes increasingly difficult if the top product is almost pure alcohol. This is the reason that the time constant increases with the increase of the reflux rate. At the same operating point the time constant is less with increasing flow rates than with decreasing rates, because the reflux is cooler than the reflux plate and cools the temperature faster than the vapor warms it when decreasing the reflux rate. (A case study deals with this problem in Chapter 7.) ∎

1.6.2 Continuous time description of continuous time models with signal dependent parameters

A quasi-linear continuous time system with signal dependent parameters can be described by two types of equations:

- the differential equation of the linear dynamic system with constant parameters, and
- the static equations that describe how those *constant* parameters depend on (a) measurable (or computable) signal(s).

Although the principle is valid for any linear systems, we restrict ourselves to simple processes that are commonly used in the engineering practice. They are

- first-order process:

$$T\dot{y}(t) + y(t) = Ku(t) \tag{1.6.4}$$

- second-order aperiodic process:

$$T_1 T_2 \ddot{y}(t) + (T_1 + T_2)\dot{y}(t) = Ku(t) \tag{1.6.5}$$

- second-order under damped process:

$$T^2 \ddot{y}(t) + 2T\xi\dot{y}(t) = Ku(t) \tag{1.6.6}$$

Here K is the static gain, T the time constant and ξ the damping factor.

Usually a polynomial static function is assumed for the parameters

$$\theta_i(t) = \sum_{j=0}^{m} \theta_{ij} x_{\theta_i}^j(t)$$

or for their reciprocal values

$$\theta_i^{-1}(t) = \sum_{j=0}^{\overline{m}} \overline{\theta}_{ij}^{-1} x_{\theta_i}^j(t)$$

Here θ_i denotes a parameter of the differential equation and $x_{\theta_i}(t)$ is the signal the parameters depend on, furthermore θ_{ij}, m and $\overline{\theta}_{ij}$, $\overline{m}$ are the parameters of these relationships. The reason for using the parameter dependence of the reciprocal values of the parameters is that highly nonlinear functions of the parameter dependence can be approximated with only a few unknown parameters. Functions other than polynomial functions for the parameter dependence can also be handled easily.

The most simple signal dependence is the linear one, which is assumed either for the parameters

$$\begin{aligned} K(t) &= K_0 + K_1 x_K(t) \\ T(t) &= T_0 + T_1 x_T(t) \\ \xi(t) &= \xi_0 + \xi_1 x_\xi(t) \end{aligned} \tag{1.6.7}$$

or for their reciprocal values:

$$\begin{aligned} K^{-1}(t) &= \overline{K}_0^{-1} + \overline{K}_1^{-1} x_K(t) \\ T^{-1}(t) &= \overline{T}_0^{-1} + \overline{T}_1^{-1} x_T(t) \\ \xi^{-1}(t) &= \overline{\xi}_0^{-1} + \overline{\xi}_1^{-1} x_\xi(t) \end{aligned} \tag{1.6.8}$$

The signals x_K, x_T and x_ξ denote those (not necessarily different) signals the parameters depend on.

Example 1.6.7 *First-order system with a time constant being a linear function of the output signal*

The process is described by two equations

$$\begin{aligned} T(t)\dot{y}(t) + y(t) &= K_0 u(t) \\ T(t) &= T_0 + T_1 y(t) \end{aligned} \tag{1.6.9}$$

and three unknown parameters K_0, T_0 and T_1 (1.6.9) can be rewritten into one equation

$$T_0\dot{y}(t) + T_1 y(t)\dot{y}(t) + y(t) = K_0 u(t) \tag{1.6.10}$$

which is linear in the parameters. Therefore the form (1.6.10) is more advantageous for parameter estimation than (1.6.9).

Example 1.6.8 *Second-order under damped system with a time constant being a linear function of the output signal*

The process is described by two equations

$$T^2(t)\ddot{y}(t) + 2T\xi_0\dot{y}(t) + y(t) = K_0u(t)$$
$$T(t) = T_0 + T_1y(t) \qquad (1.6.11)$$

and four unknown parameters K_0, ξ, T_0 and T_1. (1.6.11) can be rewritten into one equation

$$T_0^2\ddot{y}(t) + 2T_0T_1y(t)\ddot{y}(t) + T_1^2y^2(t)\ddot{y}(t) + 2T_0\xi_0\dot{y}(t)$$
$$+2T_1\xi_0y(t)\dot{y}(t) + y(t) = K_0u(t) \qquad (1.6.12)$$

which is linear in the parameters. If estimating the parameters in a linear-in-parameters form (1.6.12) six unknowns have to be estimated, which are redundant in the four unknown parameters. The form (1.6.11) has only four unknowns, but it is not linear in the parameters.

Applications that report about the continuous time identification of processes with signal dependent parameters are listed below:

1. *Reflux flow-rate and top concentration model of a distillation column* (Fronza *et al.*, 1969);
2. *Gas-turbine engine* (Godfrey and Moore, 1974);
3. *Heat-exchanger* (Franck and Rake, 1985);
4. *Reactor using for the production of acetone from isopropanol* (Godfrey and Briggs, 1973);
5. *Steam raising plant* (Godfrey and Briggs, 1973);
6. *Electrically excited biological membrane* (Haber and Wernstedt, 1979).

Since digital computers work in a discrete time mode, we deal with the discrete time modeling further on.

1.6.3 Discrete time description of continuous time models with signal dependent parameters

Several methods are known for the discretization of the differential equations. They differ from each other in integrating of the continuous derivative between t and $(t+\Delta T)$:

- *Euler transformation*:

$$y(t+\Delta T) \approx y(t) + \Delta T\dot{y}(t)$$

- *bilinear transformation*:

$$y(t+\Delta T) \approx y(t) + \Delta T\frac{\dot{y}(k)+\dot{y}(k-1)}{2}$$

- *step response equivalent transformation*:
  By this method a zero order holding device is assumed for the input signal in the interval between two samplings

$$y(t+\Delta T) \approx y(t) + \int_{\tau=t}^{t+\Delta T} \dot{y}(\tau)d\tau\Big|_{u(\tau)=u(t);\ t\le\tau<t+\Delta T}$$

In the difference equation form $y(t)$ is denoted by $y(k)$ and $y(t+\Delta T)$ by $y(k+1)$, i.e., the sampling is given in sampling time units.

A differential equation with constant parameters can be described in the Laplace domain, as well. The above time domain criteria can be used for transforming a transfer function $G(s)$ to a pulse transfer function $H(z^{-1})$ by the following procedures:

- *Euler transformation*:

$$s=\frac{z-1}{\Delta T} \tag{1.6.13}$$

- *bilinear transformation*:

$$s=\frac{2}{\Delta T}\frac{1-z^{-1}}{1+z^{-1}} \tag{1.6.14}$$

- *step-response equivalent transformation*:

$$G(z^{-1})=(1-z^{-1})\mathscr{L}^{*}\left\{\frac{1}{s}G(s)\right\}$$

($\mathscr{L}^{*}\{\ldots\}$ means the discrete time Laplace transformation)

Lemma 1.6.1.
The difference equation of a first-order continuous time system

$$T\dot{y}(t)+y(t)=Ku(t) \tag{1.6.15}$$

is

$$T\Delta y(k)+\tilde{y}(k)=K\tilde{u}(k) \tag{1.6.16}$$

if the Euler transformation is used. The variables are

$$\begin{aligned}\tilde{u}(k)&=u(k-1)\\ \tilde{y}(k)&=y(k-1)\\ \Delta y(k)&=\frac{1}{\Delta T}[y(k)-y(k-1)]\end{aligned} \tag{1.6.17}$$

Proof. The Laplace transform of the differential equation (1.6.15) is

$$Tsy(s)+y(s)=Ku(s) \tag{1.6.18}$$

Substitution of (1.6.13) into the Laplace transform (1.6.18) leads to

$$T\frac{1}{\Delta T}\frac{1-z^{-1}}{z^{-1}}y(s)+y(s)=Ku(s) \tag{1.6.19}$$

Multiply (1.6.19) by z^{-1}:

$$T\frac{1}{\Delta T}\left[1-z^{-1}\right]y(s)+z^{-1}y(s)=Kz^{-1}u(s) \tag{1.6.20}$$

The transformation of (1.6.20) back to the time domain results in (1.6.16) with (1.6.17). ■

Lemma 1.6.2
The difference equation of a first-order continuous time system (1.6.15) has the form (1.6.16) if the bilinear transformation was used. The variables are

$$\begin{aligned}\tilde{u}(k)&=\tfrac{1}{4}\left[u(k)+u(k-1)\right]\\ \tilde{y}(k)&=\tfrac{1}{4}\left[y(k)+y(k-1)\right]\\ \Delta y(k)&=\tfrac{1}{\Delta T}\left[y(k)+y(k-1)\right]\end{aligned} \tag{1.6.21}$$

Proof. The Laplace transform of the differential equation (1.6.15) is (1.6.18). The substitution of (1.6.21) into (1.6.18) leads to

$$T\frac{2}{\Delta T}\frac{1-z^{-1}}{1+z^{-1}}y(s)+y(s)=Ku(s) \tag{1.6.22}$$

Multiply (1.6.22) by $\left(1+z^{-1}\right)/2$:

$$T\frac{1}{\Delta T}\left[1-z^{-1}\right]y(s)+\frac{1}{2}\left[1+z^{-1}\right]y(s)=K\frac{1}{2}\left[1+z^{-1}\right]u(s) \tag{1.6.23}$$

The transformation of (1.6.23) back to the time domain results in (1.6.16) with (1.6.21). ■

Lemma 1.6.3
The difference equation of a second-order continuous time system

$$a_2\ddot{y}(t)+a_1\dot{y}(t)+y(t)=Ku(t) \tag{1.6.24}$$

is

$$a_2\nabla y(k)+a_1\Delta y(k)+\tilde{y}(k)=K\tilde{u}(k) \tag{1.6.25}$$

if the Euler transformation is used. The variables are

$$\begin{aligned}&\tilde{u}(k)=u(k-2),\quad \tilde{y}(k)=y(k-2)\\ &\Delta y(k)=\frac{1}{\Delta T}\left[y(k-1)-y(k-2)\right]\\ &\nabla y(k)=\frac{1}{\Delta T^2}\left[y(k)-2y(k-1)+y(k-2)\right]\end{aligned} \tag{1.6.26}$$

(here ∇ stands for the second-order difference).

Proof. The Laplace transform of the differential equation (1.6.24) is

$$a_2 s^2 y(s) + a_1 s y(s) + y(s) = Ku(s) \tag{1.6.27}$$

The substitution of (1.6.13) into (1.6.27) leads to

$$a_2 \frac{1}{\Delta T^2}\left[\frac{1-z^{-1}}{z^{-1}}\right]^2 y(s) + a_1 \frac{1}{\Delta T}\frac{1-z^{-1}}{z^{-1}} y(s) + y(s) = Ku(s) \tag{1.6.28}$$

Multiply (1.6.28) by z^{-2}:

$$a_2 \frac{1}{\Delta T^2}\left[1 - 2z^{-1} + z^{-2}\right]y(s) + a_1 \frac{1}{\Delta T} z^{-1}\left[1 - z^{-1}\right]y(s) + z^{-2}y(s) = Kz^{-2}u(s) \tag{1.6.29}$$

The transformation of (1.6.29) back to the time domain results in (1.6.25) with (1.6.26). ■

Lemma 1.6.4
The difference equation of a second-order continuous time system (1.6.24) has the form (1.6.25) if the bilinear transformation was used. The variables are

$$\begin{aligned}
\tilde{u}(k) &= \tfrac{1}{4}\left[u(k) + 2u(k-1) + u(k-2)\right] \\
\tilde{y}(k) &= \tfrac{1}{4}\left[y(k) + 2y(k-1) + y(k-2)\right] \\
\Delta y(k) &= \frac{1}{\Delta T}\left[y(k) - 2y(k-2)\right] \\
\nabla y(k) &= \frac{1}{\Delta T^2}\left[y(k) - 2y(k-1) + y(k-2)\right]
\end{aligned} \tag{1.6.30}$$

Proof. Substitution of (1.6.14) into (1.6.25) leads to

$$a_2 \frac{4}{\Delta T^2}\left[\frac{1-z^{-1}}{1+z^{-1}}\right]^2 y(s) + \frac{2}{\Delta T}\frac{1-z^{-1}}{1+z^{-1}} y(s) + y(s) = Ku(s) \tag{1.6.31}$$

Multiply (1.6.31) by $\left(1+z^{-1}\right)^2/4$:

$$\begin{aligned}
&a_2 \frac{1}{\Delta T^2}\left[1 - 2z^{-1} + z^{-2}\right]y(s) + \frac{1}{2\Delta T}\left[1 - z^{-2}\right]y(s) + \tfrac{1}{4}\left[1 - 2z^{-1} + z^{-2}\right]y(s) \\
&= \tfrac{1}{4}\left[1 - 2z^{-1} + z^{-2}\right]Ku(s)
\end{aligned} \tag{1.6.32}$$

The transformation of (1.6.32) back to the time domain results in (1.6.25) with (1.6.30). ■

The results of Lemmas 1.6.1 to 1.6.4 are summarized in Table 1.6.1.

Example 1.6.9 *First-order system with constant parameters*
We apply the three different discretization methods in turn.

1. *Euler transformation:*
   The difference equation is

$$T\frac{y(k)-y(k-1)}{\Delta T}+y(k-1)=Ku(k-1)$$

and after having rearranged it to $y(k)$ we obtain

$$y(k)=-\left(\frac{\Delta T}{T}-1\right)y(k-1)+K\frac{\Delta T}{T}u(k-1)$$

TABLE 1.6.1 Discretization of first-order systems

Model			First-order	Second-order
Continuous time	transfer function		$\frac{K}{1+sT}$	$\frac{K}{1+a_1s+a_2s^2}$
	differential equation		$T\dot{y}(t)+y(t)=Ku(t)$	$a_2\ddot{y}(t)+a_1\dot{y}(t)+y(t)=Ku(t)$
Discrete time	difference equation		$T\Delta y(k)+y(k)=Ku(k)$	$a_2\nabla y(k)+a_1\Delta y(k)+y(k)=Ku(k)$
	Euler transformation	$u(k)$	$u(k-1)$	$u(k-2)$
		$y(k)$	$y(k-1)$	$y(k-2)$
		$\Delta y(k)$	$\frac{1}{\Delta T}[y(k)-y(k-1)]$	$\frac{1}{\Delta T}[y(k)-y(k-1)]$
		$\nabla y(k)$		$\frac{1}{\Delta T^2}[y(k)-2y(k-1)+y(k-1)]$
	bilinear transformation	$u(k)$	$\frac{1}{2}[u(k)+u(k-1)]$	$\frac{1}{4}[u(k)+2u(k-1)+u(k-2)]$
		$y(k)$	$\frac{1}{2}[y(k)+y(k-1)]$	$\frac{1}{4}[y(k)+2y(k-1)+y(k-2)]$
		$\Delta y(k)$	$\frac{1}{\Delta T}[y(k)-y(k-1)]$	$\frac{1}{2\Delta T}[y(k)-y(k-2)]$
		$\nabla y(k)$		$\frac{1}{\Delta T^2}[y(k)-2y(k-1)+y(k-1)]$

2. *Bilinear transformation:*
   The difference equation is

$$T\frac{y(k)-y(k-1)}{\Delta T}+\frac{1}{2}[y(k)+y(k-1)]=K\frac{1}{2}[u(k)+u(k-1)]$$

and after having rearranged it to $y(k)$ we obtain

$$y(k)=-\frac{\Delta T-2T}{\Delta T+2T}y(k-1)+K\frac{\Delta T}{\Delta T+2T}[u(k)+u(k-1)]$$

3. *Step-response equivalent transformation:*
   The difference equation is

$$y(k) = K\left[1-\exp\left(-\tfrac{\Delta T}{T}\right)\right]u(k-1) + \exp\left(-\tfrac{\Delta T}{T}\right)y(k-1)$$ ■

Example 1.6.10 *First-order system with constant gain and linearly signal dependent time constant*

The difference equations of Example 1.6.9 have to be extended by

$$K = K_0$$
$$T(t) = T_0 + T_1 x(t)$$

1. *Euler transformation:*

$$\left[T_0 + T_1 x(k)\right]\frac{y(k)-y(k-1)}{\Delta T} + y(k-1) = K_0 u(k-1)$$

and after having rearranged it to $y(k)$ we obtain

$$y(k) = -\left(\frac{\Delta T}{T_0} - 1\right)y(k-1) + K_0\frac{\Delta T}{T_0}u(k-1) - \frac{T_1}{T_0}x(k)\left[y(k)-y(k-1)\right]$$

2. *Bilinear transformation:*

$$\left[T_0 + T_1 x(k)\right]\frac{y(k)-y(k-1)}{\Delta T} + \frac{1}{2}\left[y(k)+y(k-1)\right] = K_0\frac{1}{2}\left[u(k)+u(k-1)\right]$$

and after having rearranged it to $y(k)$ we obtain

$$y(k) = -\frac{\Delta T - 2T_0}{\Delta T + 2T_0}y(k-1) + K_0\frac{\Delta T}{\Delta T + 2T_0}\left[u(k)+u(k-1)\right]$$
$$-2\frac{T_1}{\Delta T + 2T_0}x(k)\left[y(k)-y(k-1)\right]$$

3. *Step-response equivalent transformation:*

$$y(k) = K_0\left[1-\exp\left(-\frac{\Delta T}{T_0 + T_1 x(k)}\right)\right]u(k-1) + \exp\left(-\frac{\Delta T}{T_0 + T_1 x(k)}\right)y(k-1)$$ ■

From Examples 1.6.9 and 1.6.10 we can draw the following conclusions:

- Each difference equation is linear in the parameters of the difference equation;
- Only the Euler and bilinear transformations lead to a form which can be rewritten as a form linear-in-parameters of the differential equation.

Therefore we deal hereafter with the Euler and the bilinear transformation methods. In the practice it is usual that the input signal of the process is constant between two samplings. In this case only the step response equivalent transformation gives a difference equation that describes accurately the continuous system in the sampling instants. Using, however, a small sampling time (e.g., less than a fifth of the least time constant of the process) both the Euler and the bilinear transformations present approximations good enough.

Example 1.6.11 *Comparison of the different transformation methods at a first-order system*

The coefficients of the pulse transfer function of a first-order system (1.6.15) have the general form

$$H\left(z^{-1}\right)=\frac{b_0+b_1z^{-1}}{1+a_1z^{-1}}$$

for all three transformation methods. The coefficients were calculated in Example 1.6.9:

- Euler-transformation:

$$b_0=0,\quad b_1=K\frac{\Delta T}{T},\quad a_1=\frac{\Delta T}{T}-1$$

- bilinear-transformation:

$$b_0=b_1=K\frac{\Delta T}{\Delta T+2T},\quad a_1=\frac{\Delta T-2T}{\Delta T+2T}$$

- step-response equivalent transformation:

$$b_0=0,\quad b_1=K\left[1-\exp\left(\frac{-\Delta T}{T}\right)\right],\quad a_1=-\exp\left(\frac{-\Delta T}{T}\right)$$

The continuous time first-order system with $K=K_0=2$ and $T=T_0=10$ was chosen. Figure 1.6.6 shows the coefficients of the denominator and the sum of the coefficients in the numerator of the pulse transfer function, respectively, in the domain $0<\Delta T/T\leq 2$. As is seen, the coefficients, when compared, hardly differ from each other if $\Delta T/T<0.1$ and the bilinear transformation almost coincides with the step response equivalent transformation up to $\Delta T/T<0.5$. ■

In considering Lemmas 1.6.1 to 1.6.4 and Table 1.6.1 the difference equations using the Euler and the bilinear transformations have the same structure, and only the variables depend on the method used.

In the sequel we analyze the relation between the number of the parameters of the continuous and the discrete time forms.

Example 1.6.12 *First-order system with constant gain and linearly signal dependent time constant*

The difference equation using either the Euler or the bilinear transformation has the common form (1.6.16). Assume the following parameter dependence on the signal $x_{\mathrm{T}}(t)$ in the form

$$K=K_0,\quad T(t)=T_0+T_1x_{\mathrm{T}}(t)$$

then the input–output difference equation becomes

$$\left[T_0 + T_1 x_T(t)\right]\Delta y(k) + \tilde{y}(k) = K\tilde{u}(k)$$

or after rearranging to $\Delta y(k)$

$$\Delta y(k) = -\frac{T_1}{T_0} x_T \Delta y(k) - \frac{1}{T_0}\tilde{y}(k) + \frac{K_0}{T_0}\tilde{u}(k)$$

which is a form linear-in-parameters T_1/T_0, $1/T_0$ and K_0/T_0. The number of parameters in the form linear-in-parameters is equal to the unknown parameters of the continuous time model.

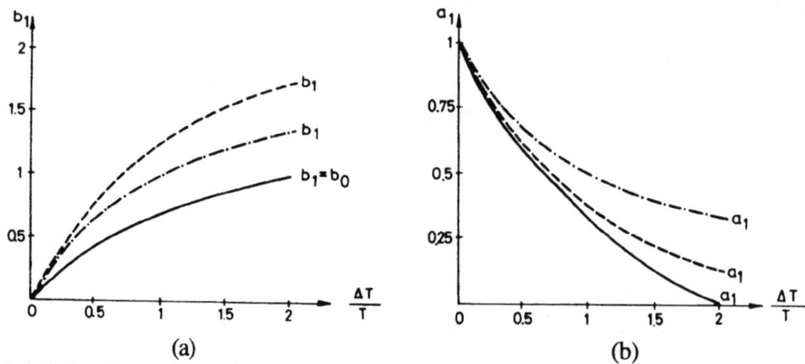

(a) (b)

Fig. 1.6.6 Coefficients of the pulse transfer function as a function of the sampling time: (a) numerator; (b) denominator
(Step response equivalent: - - - , bilinear: — , Euler: – · –)

Example 1.6.13 *First-order system with constant gain and inverse linearly signal dependent time constant*

The parameter dependence is

$$K = K_0, \quad T^{-1}(t) = \overline{T}_0^{-1} + \overline{T}_1^{-1} x_T(t) \tag{1.6.33}$$

Rearrange (1.6.16) to a form where the parameters occur similarly to (1.6.33)

$$\Delta y(k) = -T^{-1}\tilde{y}(k) + KT^{-1}\tilde{u}(k) \tag{1.6.34}$$

and put (1.6.33) into (1.6.34)

$$\Delta y(k) = -\left[\overline{T}_0^{-1} + \overline{T}_1^{-1} x_T(t)\right]\tilde{y}(k) + K_0\left[\overline{T}_0^{-1} + \overline{T}_1^{-1} x_T(t)\right]\tilde{u}(k)$$

or

$$\Delta y(k) = -\overline{T}_0^{-1}\tilde{y}(k) - \overline{T}_1^{-1} x_T(t)\tilde{y}(k) + K_0\overline{T}_0^{-1}\tilde{u}(k) + K_0\overline{T}_1^{-1} x_T(t)\tilde{u}(k) \tag{1.6.35}$$

(1.6.35) is a form linear-in-parameters $1/\overline{T}_0$, $1/\overline{T}_1$, $K_0/\overline{T}_0$, $K_0/\overline{T}_1$. The number of parameters is four, one more than those of the original continuous time form. ■

From Examples 1.6.12 and 1.6.13 we can draw the following conclusions:

1. Using the Euler or bilinear transformation the difference equations of the systems with signal dependent parameters are linear in the parameters;
2. The number of parameters in the models linear-in-parameters is equal to or greater than the number of unknowns in the (physical) continuous time model; If more parameters than necessary are estimated, the model is called redundant;
3. It is the decision of the user to decide which model he uses:
   - one that is linear-in-parameters but may be redundant, or
   - a model that has only a few parameters but is not linear in the parameters.

TABLE 1.6.2 Difference equations of first-order systems with signal dependent parameters

Components	Δy	$-x_K\Delta y$	$-x_T\Delta y$	$-x_Kx_T\Delta y$	$-\tilde{y}$	$-x_K\tilde{y}$	$-x_T\tilde{y}$	$-x_Kx_T\tilde{y}$	$\tilde{u}$	$x_K\tilde{u}$	$x_T\tilde{u}$	$x_Kx_T\tilde{u}$
Parameter	1	$\frac{K_0}{K_1}$	$\frac{T_1}{T_0}$	$\frac{K_0}{K_1}\frac{T_1}{T_0}$	$\frac{1}{T_0}$	$\frac{K_0}{K_1}\frac{1}{T_0}$	$\frac{1}{T_1}$	$\frac{K_0}{K_1}\frac{1}{T_1}$	$\frac{K_0}{T_0}$	$\frac{K_1}{T_0}$	$\frac{K_0}{T_1}$	$\frac{K_1}{T_1}$
K_0T_0	×				×				×			
K_1T_0	×				×				×	×		
K_0T_1	×		×		×				×			
K_1T_1	×		×		×				×	×		
$\bar{K}_1^{-1}T_0$	×	×			×	×			×			
$\bar{K}_1^{-1}T_1$	×	×	×	×	×	×			×			
$K_0\bar{T}_1^{-1}$	×				×		×		×		×	
$K_1\bar{T}_1^{-1}$	×				×		×		×	×	×	×
$\bar{K}_1^{-1}\bar{T}_1^{-1}$	×	×			×	×	×	×	×		×	

Tables 1.6.2 to 1.6.4 summarize the form linear-in-parameters of the difference equations of the first- and second-order over- and under-damped systems. All difference equations are arranged to $\Delta y(k)$ or $\nabla y(k)$. The derivations were made in a similar way as in Examples 1.6.12 and 1.6.13 (Haber and Keviczky, 1985; Haber *et al.*, 1986).

TABLE 1.6.3a Difference equations of second-order aperiodic systems with signal dependent parameters

Component	*Parameter*		A	B	C	D	E	F	G	H	I
∇y	1		×	×	×	×	×	×	×	×	×
$-x_K\nabla y$	K_0/K_1								×	×	×
$-x_{T_1}\nabla y$	T_{11}/T_{10}				×	×	×	×		×	×
$-x_{T_2}\nabla y$	T_{21}/T_{20}					×		×			×
$-x_Kx_{T_1}\nabla y$	K_0T_{11}/K_1T_{10}									×	×
$-x_Kx_{T_2}\nabla y$	K_0T_{21}/K_1T_{20}										×
$-x_{T_1}x_{T_2}\nabla y$	$T_{11}T_{21}/T_{10}T_{20}$					×		×			×
$-x_Kx_{T_1}x_{T_2}\nabla y$	$K_0T_{11}T_{21}/K_1T_{10}T_{20}$										×
$-\Delta y$	$1/T_{10}+1/T_{20}$		×	×	×	×	×	×	×	×	×
$-x_K\Delta y$	$K_0(1/T_{10}+1/T_{20})/K_1$								×	×	×
$-x_{T_1}\Delta y$	↓: $1/T_{11}$	↑: $T_{11}/T_{10}T_{20}$			↑	↑	↑	↑		↑	↑
$-x_{T_2}\Delta y$	↓: $1/T_{21}$	↑: $T_{21}/T_{10}T_{20}$				↑		↑			↑
$-x_Kx_{T_1}\Delta y$	↓: K_0/K_1T_{11}	↑: $K_0T_{11}/K_1T_{10}T_{20}$								↑	↑
$-x_Kx_{T_2}\Delta y$	↓: K_0/K_1T_{21}	↑: $K_0T_{21}/K_1T_{10}T_{20}$									↑

A=$K_0T_{10}T_{20}$ B=$K_1T_{10}T_{20}$ C=$K_0T_{11}T_{20}$ D=$K_0T_{11}T_{21}$ E=$K_1T_{11}T_{20}$ F=$K_1T_{11}T_{21}$ G=$\bar{K}_1^{-1}T_{10}T_{20}$ H=$\bar{K}_1^{-1}T_{11}T_{20}$ I=$\bar{K}_1^{-1}T_{11}T_{21}$

The variables $\nabla y(k)$, $\Delta y(k)$, $\tilde{u}(k)$ and $\tilde{y}(k)$ depend on the transformation method used. Tables 1.6.5 to 1.6.7 summarizes the difference equations when the above variables are substituted with (1.6.17) or (1.6.21), respectively, valid for the Euler approximation. Tables 1.6.8 to 1.6.10 summarizes the same cases for the bilinear approximation with (1.6.26) or (1.6.30), respectively. The difference equations of Tables 1.6.5 to 1.6.10 are rearranged to the output signal – as is usual with the parameter estimation.

For simplicity there was only one signal $x(t)$ instead of $x_K(t)$, $x_T(t)$ and $x_\xi(t)$ on which the parameters were assumed to depend.

An application that reports about the identification of processes with signal dependent parameters of the continuous time description is:

- *biological membrane* (Haber and Wernstedt, 1979).

TABLE 1.6.3b Difference equations of second-order aperiodic systems with signal dependent parameters

Component	*Parameter*		L	M	N	O	P	R	S	U	V
∇y	1		×	×	×	×	×	×	×	×	×
$-x_K \nabla y$	K_0/K_1						×	×			×
$-x_{T_1} \nabla y$	T_{11}/T_{10}			×		×		×			
$-x_{T_2} \nabla y$	T_{21}/T_{20}										
$-x_K x_{T_1} \nabla y$	$K_0 T_{11}/K_1 T_{10}$							×			
$-x_K x_{T_2} \nabla y$	$K_0 T_{21}/K_1 T_{20}$										
$-x_{T_1} x_{T_2} \nabla y$	$T_{11} T_{21}/T_{10} T_{20}$										
$-x_K x_{T_1} x_{T_2} \nabla y$	$K_0 T_{11} T_{21}/K_1 T_{10} T_{20}$										
$-\Delta y$	$1/T_{10} + 1/T_{20}$		×	×	×	×	×	×	×	×	×
$-x_K \Delta y$	$K_0(1/T_{10} + 1/T_{20})/K_1$						×	×			×
$-x_{T_1} \Delta y$	↓: $1/T_{11}$	↑: $T_{11}/T_{10} T_{20}$		↑		↑		↑	↓	↓	↓
$-x_{T_2} \Delta y$	↓: $1/T_{21}$	↑: $T_{21}/T_{10} T_{20}$	↓	↓	↓	↓	↓	↓	↓	↓	↓
$-x_K x_{T_1} \Delta y$	↓: $K_0/K_1 T_{11}$	↑: $K_0 T_{11}/K_1 T_{10} T_{20}$						↑			↓
$-x_K x_{T_2} \Delta y$	↓: $K_0/K_1 T_{21}$	↑: $K_0 T_{21}/K_1 T_{10} T_{20}$					↓	↓			↓

L=$K_0 T_{10} \bar{T}_{21}^{-1}$ M=$K_0 T_{11} \bar{T}_{21}^{-1}$ N=$K_1 T_{10} \bar{T}_{21}^{-1}$ O=$K_1 T_{11} \bar{T}_{21}^{-1}$ P=$\bar{K}_1^{-1} T_{10} \bar{T}_{21}^{-1}$ R=$\bar{K}_1^{-1} T_{11} \bar{T}_{21}^{-1}$ S=$K_0 \bar{T}_{11}^{-1} \bar{T}_{21}^{-1}$ U=$K_1 \bar{T}_{11}^{-1} \bar{T}_{21}^{-1}$ V=$\bar{K}_1^{-1} \bar{T}_{11}^{-1} \bar{T}_{21}^{-1}$

1.6.4 Discrete time models with signal dependent parameters

So far we have dealt with the signal dependence of the parameters of the continuous time process. A signal dependence can be imagined also for the parameters of the discrete time difference equation.

Example 1.6.14 *First-order system with linearly signal dependent pole of the pulse transfer function*

The pulse transfer function

$$H(z^{-1}) = b_1 z^{-1} / (1 + a_1 z^{-1})$$

has one pole at $z = a_1$. The difference equation of the system is:

$$y(k) = b_1 u(k-1) - a_1 y(k-1)$$

Assume that b_0 is constant, $b_1 = b_{10}$ and the pole depends linearly on a signal $x(k)$

$$a_1(k) = a_{10} + a_{11}x(k)$$

The input–output difference equation is linear in the parameters

$$y(k) = b_1u(k) + a_{10}y(k-1) + a_{11}x(k)y(k-1)$$ ■

TABLE 1.6.3.c Difference equations of second-order aperiodic systems with signal dependent parameters

Component	*Parameter*	A	B	C	D	E	F	G	H	I
$-x_{T_1}x_{T_2}\Delta y$	$T_{11}/T_{10}T_{21}$									
$-x_Kx_{T_1}x_{T_2}\Delta y$	$K_0T_{11}/K_1T_{10}T_{21}$									
$-\bar{y}$	$1/T_{10}T_{20}$	×	×	×	×	×	×	×	×	×
$-x_K\bar{y}$	$K_0/K_1T_{10}T_{20}$							×	×	×
$-x_{T_1}\bar{y}$	$1/T_{11}T_{20}$									
$-x_{T_2}\bar{y}$	$1/T_{21}T_{10}$									
$-x_Kx_{T_1}\bar{y}$	$K_0/K_1T_{11}T_{20}$									
$-x_Kx_{T_2}\bar{y}$	$K_0/K_1T_{10}T_{21}$									
$-x_{T_1}x_{T_2}\bar{y}$	$1/T_{11}T_{21}$									
$-x_Kx_{T_1}x_{T_2}\bar{y}$	$K_0/K_1T_{11}T_{21}$									
$\bar{u}$	$K_0/T_{10}T_{20}$	×	×	×	×	×	×	×	×	×
$x_K\bar{u}$	$K_1/T_{10}T_{20}$		×				×	×		
$x_{T_1}\bar{u}$	$K_0/T_{11}T_{20}$									
$x_{T_2}\bar{u}$	$K_0/T_{10}T_{21}$									
$x_Kx_{T_1}\bar{u}$	$K_1/T_{11}T_{20}$									
$x_Kx_{T_2}\bar{u}$	$K_1/T_{10}T_{21}$									
$x_{T_1}x_{T_2}\bar{u}$	$K_0/T_{11}T_{21}$									
$x_Kx_{T_1}x_{T_2}\bar{u}$	$K_1/T_{11}T_{21}$									

A=$K_0T_{10}T_{20}$ B=$K_1T_{10}T_{20}$ C=$K_0T_{11}T_{20}$ D=$K_0T_{11}T_{21}$ E=$K_1T_{11}T_{20}$ F=$K_1T_{11}T_{21}$ G=$\bar{K}_1^{-1}T_{10}T_{20}$
H=$\bar{K}_1^{-1}T_{11}T_{20}$ I=$\bar{K}_1^{-1}T_{11}T_{21}$

Example 1.6.15 *First-order system with inverse linearly signal dependent pole of the pulse transfer function*

The equations of the system are:

$$y(k) = b_1u(k-1) - a_1y(k-1), \quad b_1 = b_{10}$$
$$a_1^{-1}(k) = \bar{a}_{10}^{-1} + \bar{a}_{11}^{-1}x(k) \tag{1.6.36}$$

(1.6.36) can be rewritten into a form linear-in-parameters in three steps

$$a_1^{-1}y(k) = a_1^{-1}b_1u(k-1) - y(k-1)$$
$$\bar{a}_{10}^{-1}y(k) + \bar{a}_{11}^{-1}x(k)y(k) = \bar{a}_{10}^{-1}b_{10}u(k-1) + \bar{a}_{11}^{-1}b_{10}x(k)u(k-1) - y(k-1) \tag{1.6.37}$$
$$y(k) = -\bar{a}_{10}\bar{a}_{11}^{-1}x(k)y(k) + b_{10}u(k-1)$$
$$+\bar{a}_{10}\bar{a}_{11}^{-1}b_{10}x(k)u(k-1) - \bar{a}_{10}y(k-1)$$

(1.6.37) is a linear form in the parameters $\bar{a}_{10}\bar{a}_{11}^{-1}$, b_{10}, $\bar{a}_{10}\bar{a}_{11}^{-1}b_{10}$, $\bar{a}_{10}$. The number of parameters is four, one more than in (1.6.36). ■

From examples 1.6.14 and 1.6.15 the following conclusions can be drawn:

1. The setting up of the model linear-in-parameters is easier for the signal dependence of the difference equation than that of the differential equation;
2. There is no restriction for the mode of the transformation. A step response equivalent pulse transfer function model can be assumed if the parameters are constant. (As we know, in the practical case when the input signal is constant in the sampling instant that is the best transformation method.);
3. A signal dependence of any parameter of the difference equation cannot be interpreted always as clear as a change in the parameters of the differential equation. (In Examples 1.6.14 and 1.6.15 a change in the pole a_1 does not mean that only the time constant changes but also the static gain $K = b_{10}/(1+a_1)$ varies.);
4. In some cases the consideration of a signal dependence in the difference equation can be more practical than the same in the differential equation form. This is shown in the next example.

Example 1.6.16 *First-order system with a known constant gain and linearly signal dependent pole of the pulse transfer function (Haber et al., 1986)*

Assume the static gain is known either from physical considerations or from steady state measurements. The difference equations of the system are:

$$y(k) = b_1 u(k-1) - a_1 y(k-1) \tag{1.6.38a}$$

$$K = K_0 = \frac{b_1}{1+a_1} \tag{1.6.38b}$$

$$a_1(k) = a_{10} + a_{11}x(k) \tag{1.6.38c}$$

Substitute b_1 from (1.6.38b) and a_1 from (1.6.38c) into (1.6.38a)

$$y(k) = K_0[1 + a_{10} + a_{11}x(k)]u(k-1) - [a_{10} + a_{11}x(k)]y(k-1)$$

and after having rearranged it we obtain

$$\begin{aligned} y(k) - K_0 u(k-1) = a_{10}[K_0 u(k-1) - y(k-1)] \\ + a_{11}x(k)[K_0 u(k-1) - y(k-1)] \end{aligned} \tag{1.6.39}$$

(1.6.39) is linear in the parameters and is not redundant, because the two unknown parameters a_{10} and a_{11} can be estimated unambiguously. The time constant of the process is then

$$T(t) = \frac{-\Delta T}{\ln(a_1(t))} = \frac{-\Delta T}{\ln(a_{10} + a_{11}x(t))}$$

A practical example for such a system is a mill in which the steady state relation between the inlet and outlet flows is $K_0 = 1$, and the time constant depends on the working point $x(t) = y(t)$. This will be investigated for a cement mill among the applications in Chapter 7. ■

Applications that report about the identification of processes with signal dependent parameters of the discrete time description are listed below:

1. *Vulcanization curves* (Vuchkov *et al.*, 1985);
2. *Concentration curves in furfural production* (Vuchkov *et al.*, 1985);
4. *Chemical reaction* (Velev, 1985);
5. *Cement mill* (Haber *et al.*, 1986).

1.6.5 The effect of changing the parameters of the continuous time models to the parameters of the discrete time equivalent description

A linear change in the parameters of the differential equation can lead to a highly nonlinear functional relation in the difference equation and *vice versa*. To decide which functional dependence could be assumed in a concrete case in the easiest way, we show how the parameters of the pulse transfer function change when the parameters of the transfer function vary linearly. Since the effects are similar with the different transformation methods, only the step response equivalent transformation will be presented. (Haber (1979b) shows the same examples using the bilinear transformation, as well.)

The transformation equations are (e.g., Isermann, 1987):

- first-order system (1.6.4)

$$b_0 = 0, \qquad b_1 = K\left[1 - \exp(-\Delta T/T)\right], \qquad a_1 = -\exp(-\Delta T/T)$$

- second-order aperiodic system (1.6.5):

$$z_1 = \exp(-\Delta T/T_1), \qquad z_2 = \exp(-\Delta T/T_2)$$

$$b_1 = \frac{K}{T_1 - T_2}\left[T_1(1 - z_1) - T_2(1 - z_2)\right]$$

$$b_2 = \frac{K}{T_1 - T_2}\left[T_2 z_1(1 - z_2) - T_1 z_2(1 - z_1)\right]$$

$$a_1 = -(z_1 + z_2), \qquad a_2 = z_1 z_2$$

- second-order under-damped system (1.6.6):

$$\delta_0 = \frac{\xi}{T}, \qquad \omega_0 = \frac{\sqrt{1-\xi^2}}{T}, \qquad z_0 = \exp(-\Delta T/T)$$

$$b_1 = K\left(1 - z_0\left[\cos\omega_0\Delta T + \frac{\delta_0}{\omega_0}\sin\omega_0\Delta T\right]\right)$$

$$b_2 = K\left(z_0 - \cos\omega_0\Delta T + \frac{\delta_0}{\omega_0}\sin\omega_0\Delta T\right)$$

$$a_1 = -2z_0\cos\omega_0\Delta T, \qquad a_2 = z_0^2$$

TABLE 1.6.3d Difference equations of second-order aperiodic systems with signal dependent parameters

Component	*Parameter*	L	M	N	O	P	R	S	U	V
$-x_{T_1}x_{T_2}\Delta y$	$T_{11}/T_{10}T_{21}$		×		×		×			
$-x_K x_{T_1}x_{T_2}\Delta y$	$K_0T_{11}/K_1T_{10}T_{21}$						×			
$-\bar{y}$	$1/T_{10}T_{20}$	×	×	×	×	×	×	×	×	×
$-x_K\bar{y}$	$K_0/K_1T_{10}T_{20}$					×	×			×
$-x_{T_1}\bar{y}$	$1/T_{11}T_{20}$							×	×	×
$-x_{T_2}\bar{y}$	$1/T_{21}T_{10}$	×	×	×	×	×	×	×	×	×
$-x_K x_{T_1}\bar{y}$	$K_0/K_1T_{11}T_{20}$									×
$-x_K x_{T_2}\bar{y}$	$K_0/K_1T_{10}T_{21}$					×	×			×
$-x_{T_1}x_{T_2}\bar{y}$	$1/T_{11}T_{21}$							×	×	×
$-x_K x_{T_1}x_{T_2}\bar{y}$	$K_0/K_1T_{11}T_{21}$									×
$\bar{u}$	$K_0/T_{10}T_{20}$	×	×	×	×	×	×	×	×	×
$x_K\bar{u}$	$K_1/T_{10}T_{20}$			×	×				×	
$x_{T_1}\bar{u}$	$K_0/T_{11}T_{20}$							×	×	×
$x_{T_2}\bar{u}$	$K_0/T_{10}T_{21}$	×	×	×	×	×	×	×	×	×
$x_K x_{T_1}\bar{u}$	$K_1/T_{11}T_{20}$								×	
$x_K x_{T_2}\bar{u}$	$K_1/T_{10}T_{21}$			×	×				×	
$x_{T_1}x_{T_2}\bar{u}$	$K_0/T_{11}T_{21}$							×	×	×
$x_K x_{T_1}x_{T_2}\bar{u}$	$K_1/T_{11}T_{21}$								×	

$L=K_0T_{10}\bar{T}_{21}^{-1}$ $M=K_0T_{11}\bar{T}_{21}^{-1}$ $N=K_1T_{10}\bar{T}_{21}^{-1}$ $O=K_1T_{11}\bar{T}_{21}^{-1}$ $P=\bar{K}_1^{-1}T_{10}\bar{T}_{21}^{-1}$ $R=\bar{K}_1^{-1}T_{11}\bar{T}_{21}^{-1}$ $S=K_0\bar{T}_{11}^{-1}\bar{T}_{21}^{-1}$
$U=K_1\bar{T}_{11}^{-1}\bar{T}_{21}^{-1}$ $V=\bar{K}_1^{-1}\bar{T}_{11}^{-1}\bar{T}_{21}^{-1}$

TABLE 1.6.4a Difference equations of second-order under-damped systems with signal dependent parameters

Component	*Parameter*	A	B	C	D	E	F	G	H	I
∇y	1	×	×	×	×	×	×	×	×	×
$-x_K\nabla y$	K_0/K_1									×
$-x_T\nabla y$	$2T_1/T_0$			×	×			×	×	
$-x_T^2\nabla y$	T_1^2/T_0^2			×	×			×	×	
$-x_\xi\nabla y$	ξ_0/ξ_1									
$-x_K x_T\nabla y$	$2K_0T_1/K_1T_0$									
$-x_K x_T^2\nabla y$	$K_0T_1^2/K_1T_0^2$									
$-x_K x_\xi\nabla y$	$K_0\xi_0/K_1\xi_1$									
$-x_T x_\xi\nabla y$	$2T_1\xi_0/T_0\xi_1$									
$-x_T^2 x_\xi\nabla y$	$T_1^2\xi_0/T_0^2\xi_1$									
$-x_K x_T x_\xi\nabla y$	$2K_0T_1\xi_0/K_1T_0\xi_1$									
$-x_K x_T^2 x_\xi\nabla y$	$K_0T_1^2\xi_0/K_1T_0^2\xi_1$									
$-\Delta y$	$2\xi_0/T_0$	×	×	×	×	×	×	×	×	×
$-x_K\Delta y$	$2K_0\xi_0/K_1T_0$									×
$-x_T\Delta y$	↓: $2\xi_0/T_1$; ↑: $2\xi_0T_1/T_0^2$			↑	↑			↑	↑	
$-x_T^2\Delta y$										
$-x_\xi\Delta y$	$2\xi_1/T_1$					×	×	×	×	
$-x_K x_T\Delta y$	↓: $2K_0\xi_0/K_1T_1$; ↑: $2K_0T_1\xi_0/K_1T_0^2$									

$A=K_0T_0\xi_0$ $B=K_1T_0\xi_0$ $C=K_0T_1\xi_0$ $D=K_1T_1\xi_0$ $E=K_0T_0\xi_1$ $F=K_1T_0\xi_1$ $G=K_0T_1\xi_1$
$H=K_1T_1\xi_1$ $I=\bar{K}_1^{-1}T_0\xi_0$

TABLE 1.6.4b Difference equations of second-order under-damped systems with signal dependent parameters

Component	*Parameter*	L	M	N	O	P	R	S	U	V
∇y	1	×	×	×	×	×	×	×	×	×
$-x_K \nabla y$	K_0 / K_1	×	×	×		×		×		
$-x_T \nabla y$	$2T_1 / T_0$	×		×			×	×		×
$-x_T^2 \nabla y$	T_1^2 / T_0^2	×		×			×	×		×
$-x_\xi \nabla y$	ξ_0 / ξ_1				×	×	×	×	×	×
$-x_K x_T \nabla y$	$2K_0 T_1 / K_1 T_0$	×		×				×		
$-x_K x_T^2 \nabla y$	$K_0 T_1^2 / K_1 T_0^2$	×		×				×		
$-x_K x_\xi \nabla y$	$K_0 \xi_0 / K_1 \xi_1$					×		×		
$-x_T x_\xi \nabla y$	$2T_1 \xi_0 / T_0 \xi_1$						×	×		×
$-x_T^2 x_\xi \nabla y$	$T_1^2 \xi_0 / T_0^2 \xi_1$						×	×		×
$-x_K x_T x_\xi \nabla y$	$2K_0 T_1 \xi_0 / K_1 T_0 \xi_1$							×		
$-x_K x_T^2 x_\xi \nabla y$	$K_0 T_1^2 \xi_0 / K_1 T_0^2 \xi_1$							×		
$-\Delta y$	$2\xi_0 / T_0$	×	×	×	×	×	×	×	×	×
$-x_K \Delta y$	$2K_0 \xi_0 / K_1 T_0$	×	×	×		×		×		
$-x_T \Delta y$	↓: $2\xi_0 / T_1$; ↑: $2\xi_0 T_1 / T_0^2$	↑		↑			↑	↑		↑
$-x_T^2 \Delta y$										
$-x_\xi \Delta y$	$2\xi_1 / T_1$		×	×						
$-x_K x_T \Delta y$	↓: $2K_0 \xi_0 / K_1 T_1$; ↑: $2K_0 T_1 \xi_0 / K_1 T_0^2$	↑		↑				↑		

L$=\bar{K}_1^{-1} T_1 \xi_0$ M$=\bar{K}_1^{-1} T_0 \xi_1$ N$=\bar{K}_1^{-1} T_1 \xi_1$ O$=K_0 T_0 \bar{\xi}_1^{-1}$ P$=\bar{K}_1^{-1} T_0 \bar{\xi}_1^{-1}$ R$=K_0 T_1 \bar{\xi}_1^{-1}$ S$=\bar{K}_1^{-1} T_1 \bar{\xi}_1^{-1}$
U$=K_1 T_0 \bar{\xi}_1^{-1}$ V$=K_1 T_1 \bar{\xi}_1^{-1}$

TABLE 1.6.4c Difference equations of second-order under-damped systems with signal dependent parameters

Component	*Parameter*	$\bar{A}$	$\bar{B}$	$\bar{C}$	$\bar{D}$	$\bar{E}$	$\bar{F}$	$\bar{G}$	$\bar{H}$	$\bar{I}$
∇y	1	×	×	×	×	×	×	×	×	×
$-x_K \nabla y$	K_0 / K_1		×		×	×	×			×
$-x_T \nabla y$	$2T_1 / T_0$									
$-x_T^2 \nabla y$	T_1^2 / T_0^2									
$-x_\xi \nabla y$	ξ_0 / ξ_1							×	×	×
$-x_K x_T \nabla y$	$2K_0 T_1 / K_1 T_0$									
$-x_K x_T^2 \nabla y$	$K_0 T_1^2 / K_1 T_0^2$									
$-x_K x_\xi \nabla y$	$K_0 \xi_0 / K_1 \xi_1$									×
$-x_T x_\xi \nabla y$	$2T_1 \xi_0 / T_0 \xi_1$									
$-x_T^2 x_\xi \nabla y$	$T_1^2 \xi_0 / T_0^2 \xi_1$									
$-x_K x_T x_\xi \nabla y$	$2K_0 T_1 \xi_0 / K_1 T_0 \xi_1$									
$-x_K x_T^2 x_\xi \nabla y$	$K_0 T_1^2 \xi_0 / K_1 T_0^2 \xi_1$									
$-\Delta y$	$2\xi_0 / T_0$	×	×	×	×	×	×	×	×	×
$-x_K \Delta y$	$2K_0 \xi_0 / K_1 T_0$		×		×					×
$-x_T \Delta y$	↓: $2\xi_0 / T_1$; ↑: $2\xi_0 T_1 / T_0^2$	↓	↓	↓	↓	↓	↓	↓	↓	↓
$-x_T^2 \Delta y$										
$-x_\xi \Delta y$	$2\xi_1 / T_1$			×	×		×			
$-x_K x_T \Delta y$	↓: $2K_0 \xi_0 / K_1 T_1$; ↑: $2K_0 T_1 \xi_0 / K_1 T_0^2$		↓		↓					↓

$\bar{A}=K_0 \bar{T}_1^{-1} \xi_0$ $\bar{B}=\bar{K}_1^{-1} \bar{T}_1^{-1} \xi_0$ $\bar{C}=K_0 \bar{T}_1^{-1} \xi_1$ $\bar{D}=\bar{K}_1^{-1} \bar{T}_1^{-1} \xi_1$ $\bar{E}=K_1 \bar{T}_1^{-1} \xi_0$ $\bar{F}=K_1 \bar{T}_1^{-1} \xi_1$ $\bar{G}=K_0 \bar{T}_1^{-1} \bar{\xi}_1^{-1}$
$\bar{H}=K_1 \bar{T}_1^{-1} \bar{\xi}_1^{-1}$ $\bar{I}=\bar{K}_1^{-1} \bar{T}_1^{-1} \bar{\xi}_1^{-1}$

TABLE 1.6.4d Difference equations of second-order under-damped systems with signal dependent parameters

Component	*Parameter*	A	B	C	D	E	F	G	H	I
$-x_K x_T^2 \Delta y$										
$-x_K x_\xi \Delta y$	$2K_0\xi_1 / K_1T_0$									
$-x_T x_\xi \Delta y$	$\downarrow: 2\xi_1 / T_1$; $\uparrow: 2\xi_1 T_1 / T_0^2$							↑	↑	
$-x_T^2 x_\xi \Delta y$										
$-x_K x_T x_\xi \Delta y$	$\downarrow: 2K_0\xi_1 / K_1T_1$; $\uparrow: 2K_0T_1\xi_1 / K_1T_0^2$									
$-x_K x_T^2 x_\xi \Delta y$										
$-\bar{y}$	$1 / T_0^2$	×	×	×	×	×	×	×	×	×
$-x_K \bar{y}$	$K_0 / K_1T_0^2$									×
$-x_T \bar{y}$	$2 / T_0T_1$									
$-x_T^2 \bar{y}$	$1 / T_1^2$									
$-x_\xi \bar{y}$	$\xi_0 / T_0^2\xi_1$									
$-x_K x_T \bar{y}$	$2K_0 / K_1T_0T_1$									
$-x_K x_T^2 \bar{y}$	$K_0 / K_1T_1^2$									
$-x_K x_\xi \bar{y}$	$K_0\xi_0 / K_1T_0^2\xi_1$									
$-x_T x_\xi \bar{y}$	$2\xi_0 / T_0T_1\xi_1$									
$-x_T^2 x_\xi \bar{y}$	$\xi_0 / T_1^2\xi_1$									
$-x_K x_T x_\xi \bar{y}$	$2K_0\xi_0 / K_1T_0T_1\xi_1$									
$-x_K x_T^2 x_\xi \bar{y}$	$K_0\xi_0 / K_1T_1^2\xi_1$									

A=$K_0T_0\xi_0$ B=$K_1T_0\xi_0$ C=$K_0T_1\xi_0$ D=$K_1T_1\xi_0$ E=$K_0T_0\xi_1$ F=$K_1T_0\xi_1$ G=$K_0T_1\xi_1$
H=$K_1T_1\xi_1$ I=$\bar{K}_1^{-1}T_0\xi_0$

TABLE 1.6.4e Difference equations of second-order under-damped systems with signal dependent parameters

Component	*Parameter*	L	M	N	O	P	R	S	U	V
$-x_K x_T^2 \Delta y$										
$-x_K x_\xi \Delta y$	$2K_0\xi_1 / K_1T_0$		×	×						
$-x_T x_\xi \Delta y$	$\downarrow: 2\xi_1 / T_1$; $\uparrow: 2\xi_1 T_1 / T_0^2$			↑						
$-x_T^2 x_\xi \Delta y$										
$-x_K x_T x_\xi \Delta y$	$\downarrow: 2K_0\xi_1 / K_1T_1$; $\uparrow: 2K_0T_1\xi_1 / K_1T_0^2$			↑						
$-x_K x_T^2 x_\xi \Delta y$										
$-\bar{y}$	$1 / T_0^2$	×	×	×	×	×	×	×	×	×
$-x_K \bar{y}$	$K_0 / K_1T_0^2$	×	×	×		×		×		
$-x_T \bar{y}$	$2 / T_0T_1$									×
$-x_T^2 \bar{y}$	$1 / T_1^2$									
$-x_\xi \bar{y}$	$\xi_0 / T_0^2\xi_1$				×	×	×	×	×	×
$-x_K x_T \bar{y}$	$2K_0 / K_1T_0T_1$									
$-x_K x_T^2 \bar{y}$	$K_0 / K_1T_1^2$									
$-x_K x_\xi \bar{y}$	$K_0\xi_0 / K_1T_0^2\xi_1$					×		×		
$-x_T x_\xi \bar{y}$	$2\xi_0 / T_0T_1\xi_1$									
$-x_T^2 x_\xi \bar{y}$	$\xi_0 / T_1^2\xi_1$									
$-x_K x_T x_\xi \bar{y}$	$2K_0\xi_0 / K_1T_0T_1\xi_1$									
$-x_K x_T^2 x_\xi \bar{y}$	$K_0\xi_0 / K_1T_1^2\xi_1$									

L=$\bar{K}_1^{-1}T_1\xi_0$ M=$\bar{K}_1^{-1}T_0\xi_1$ N=$\bar{K}_1^{-1}T_1\xi_1$ O=$K_0T_0\bar{\xi}_1^{-1}$ P=$\bar{K}_1^{-1}T_0\bar{\xi}_1^{-1}$ R=$K_0T_1\bar{\xi}_1^{-1}$ S=$\bar{K}_1^{-1}T_1\bar{\xi}_1^{-1}$
U=$K_1T_0\bar{\xi}_1^{-1}$ V=$K_1T_1\bar{\xi}_1^{-1}$

TABLE 1.6.4f Difference equations of second-order under-damped systems with signal dependent parameters

Component	Parameter	$\overline{A}$	$\overline{B}$	$\overline{C}$	$\overline{D}$	$\overline{E}$	$\overline{F}$	$\overline{G}$	$\overline{H}$	$\overline{I}$
$-x_K x_T^2 \Delta y$										
$-x_K x_\xi \Delta y$	$2K_0\xi_1 / K_1T_0$				×					
$-x_T x_\xi \Delta y$	$\downarrow: 2\xi_1 / T_1$ $\uparrow: 2\xi_1 T_1 / T_0^2$			$\downarrow$	$\downarrow$		$\downarrow$			
$-x_T^2 x_\xi \Delta y$										
$-x_K x_T x_\xi \Delta y$	$\downarrow: 2K_0\xi_1 / K_1T_1$ $\uparrow: 2K_0T_1\xi_1 / K_1T_0^2$				$\downarrow$					
$-x_K x_T^2 x_\xi \Delta y$										
$-\bar{y}$	$1 / T_0^2$	×	×	×	×	×	×	×	×	×
$-x_K \bar{y}$	$K_0 / K_1T_0^2$		×		×					×
$-x_T \bar{y}$	$2 / T_0T_1$	×	×	×	×	×	×	×	×	×
$-x_T^2 \bar{y}$	$1 / T_1^2$	×	×	×	×	×	×	×	×	×
$-x_\xi \bar{y}$	$\xi_0 / T_0^2\xi_1$							×	×	×
$-x_K x_T \bar{y}$	$2K_0 / K_1T_0T_1$		×		×					×
$-x_K x_T^2 \bar{y}$	$K_0 / K_1T_1^2$		×		×					×
$-x_K x_\xi \bar{y}$	$K_0\xi_0 / K_1T_0^2\xi_1$									×
$-x_T x_\xi \bar{y}$	$2\xi_0 / T_0T_1\xi_1$							×	×	×
$-x_T^2 x_\xi \bar{y}$	$\xi_0 / T_1^2\xi_1$							×	×	×
$-x_K x_T x_\xi \bar{y}$	$2K_0\xi_0 / K_1T_0T_1\xi_1$									×
$-x_K x_T^2 x_\xi \bar{y}$	$K_0\xi_0 / K_1T_1^2\xi_1$									×

$\overline{A}=K_0\overline{T}_1^{-1}\xi_0$ $\overline{B}=\overline{K}_1^{-1}\overline{T}_1^{-1}\xi_0$ $\overline{C}=K_0\overline{T}_1^{-1}\xi_1$ $\overline{D}=\overline{K}_1^{-1}\overline{T}_1^{-1}\xi_1$ $\overline{E}=K_1\overline{T}_1^{-1}\xi_0$ $\overline{F}=K_1\overline{T}_1^{-1}\xi_1$ $\overline{G}=K_0\overline{T}_1^{-1}\overline{\xi}_1^{-1}$
$\overline{H}=K_1\overline{T}_1^{-1}\overline{\xi}_1^{-1}$ $\overline{I}=\overline{K}_1^{-1}\overline{T}_1^{-1}\overline{\xi}_1^{-1}$

TABLE 1.6.4g Difference equations of second-order under-damped systems with signal dependent parameters

Component	Parameter	A	B	C	D	E	F	G	H	I
$\bar{u}$	K_0 / T_0^2	×	×	×	×	×	×	×	×	×
$x_K \bar{u}$	K_1 / T_0^2		×		×		×		×	
$x_T \bar{u}$	$2K_0 / T_0T_1$									
$x_T^2 \bar{u}$	K_0 / T_1^2									
$x_\xi \bar{u}$	$K_0\xi_0 / T_0^2\xi_1$									
$x_K x_T \bar{u}$	$2K_1 / T_0T_1$									
$x_K x_T^2 \bar{u}$	K_1 / T_1^2									
$x_K x_\xi \bar{u}$	$K_1\xi_0 / T_0^2\xi_1$									
$x_T x_\xi \bar{u}$	$2K_0\xi_0 / T_0 T_1 \xi_1$									
$x_T^2 x_\xi \bar{u}$	$K_0\xi_0 / T_1^2 \xi_1$									
$x_K x_T x_\xi \bar{u}$	$2K_1\xi_0 / T_0 T_1 \xi_1$									
$x_K x_T^2 x_\xi \bar{u}$	$K_1\xi_0 / T_1^2 \xi_1$									

A=$K_0T_0\xi_0$ B=$K_1T_0\xi_0$ C=$K_0T_1\xi_0$ D=$K_1T_1\xi_0$ E=$K_0T_0\xi_1$ F=$K_1T_0\xi_1$ G=$K_0T_1\xi_1$
H=$K_1T_1\xi_1$ I=$\overline{K}_1^{-1}T_0\xi_0$

TABLE 1.6.4h Difference equations of second-order under-damped systems with signal dependent parameters

Component	*Parameter*	L	M	N	O	P	R	S	U	V
$\bar{u}$	K_0 / T_0^2	×	×	×	×	×	×	×	×	×
$x_K \bar{u}$	K_1 / T_0^2								×	×
$x_T \bar{u}$	$2K_0 / T_0 T_1$									
$x_T^2 \bar{u}$	K_0 / T_1^2									
$x_\xi \bar{u}$	$K_0 \xi_0 / T_0^2 \xi_1$				×	×	×	×	×	×
$x_K x_T \bar{u}$	$2K_1 / T_0 T_1$									
$x_K x_T^2 \bar{u}$	K_1 / T_1^2									
$x_K x_\xi \bar{u}$	$K_1 \xi_0 / T_0^2 \xi_1$								×	×
$x_T x_\xi \bar{u}$	$2K_0 \xi_0 / T_0 T_1 \xi_1$									
$x_T^2 x_\xi \bar{u}$	$K_0 \xi_0 / T_1^2 \xi_1$									
$x_K x_T x_\xi \bar{u}$	$2K_1 \xi_0 / T_0 T_1 \xi_1$									
$x_K x_T^2 x_\xi \bar{u}$	$K_1 \xi_0 / T_1^2 \xi_1$									

L=$\bar{K}_1^{-1} T_1 \xi_0$ M=$\bar{K}_1^{-1} T_0 \xi_1$ N=$\bar{K}_1^{-1} T_1 \xi_1$ O=$K_0 T_0 \bar{\xi}_1^{-1}$ P=$\bar{K}_1^{-1} T_0 \bar{\xi}_1^{-1}$ R=$K_0 T_1 \bar{\xi}_1^{-1}$ S=$\bar{K}_1^{-1} T_1 \bar{\xi}_1^{-1}$
U=$K_1 T_0 \bar{\xi}_1^{-1}$ V=$K_1 T_1 \bar{\xi}_1^{-1}$

TABLE 1.6.4i Difference equations of second-order under-damped systems with signal dependent parameters

Component	*Parameter*	$\bar{A}$	$\bar{B}$	$\bar{C}$	$\bar{D}$	$\bar{E}$	$\bar{F}$	$\bar{G}$	$\bar{H}$	$\bar{I}$
$\bar{u}$	K_0 / T_0^2	×	×	×	×	×	×	×	×	×
$x_K \bar{u}$	K_1 / T_0^2					×	×		×	
$x_T \bar{u}$	$2K_0 / T_0 T_1$	×	×	×	×	×	×	×	×	×
$x_T^2 \bar{u}$	K_0 / T_1^2	×	×	×	×	×	×	×	×	×
$x_\xi \bar{u}$	$K_0 \xi_0 / T_0^2 \xi_1$							×	×	×
$x_K x_T \bar{u}$	$2K_1 / T_0 T_1$					×	×		×	
$x_K x_T^2 \bar{u}$	K_1 / T_1^2					×	×		×	
$x_K x_\xi \bar{u}$	$K_1 \xi_0 / T_0^2 \xi_1$								×	
$x_T x_\xi \bar{u}$	$2K_0 \xi_0 / T_0 T_1 \xi_1$							×	×	×
$x_T^2 x_\xi \bar{u}$	$K_0 \xi_0 / T_1^2 \xi_1$							×	×	×
$x_K x_T x_\xi \bar{u}$	$2K_1 \xi_0 / T_0 T_1 \xi_1$								×	
$x_K x_T^2 x_\xi \bar{u}$	$K_1 \xi_0 / T_1^2 \xi_1$								×	

$\bar{A}$=$K_0 \bar{T}_1^{-1} \xi_0$ $\bar{B}$=$\bar{K}_1^{-1} \bar{T}_1^{-1} \xi_0$ $\bar{C}$=$K_0 \bar{T}_1^{-1} \xi_1$ $\bar{D}$=$\bar{K}_1^{-1} \bar{T}_1^{-1} \xi_1$ $\bar{E}$=$K_1 \bar{T}_1^{-1} \xi_0$ $\bar{F}$=$K_1 \bar{T}_1^{-1} \xi_1$ $\bar{G}$=$K_0 \bar{T}_1^{-1} \bar{\xi}_1^{-1}$
$\bar{H}$=$K_1 \bar{T}_1^{-1} \bar{\xi}_1^{-1}$ $\bar{I}$=$\bar{K}_1^{-1} \bar{T}_1^{-1} \bar{\xi}_1^{-1}$

TABLE 1.6.5 Discretization of first-order systems using the Euler transformation

Model	*Difference equation*
K_0, T_0	$y(k) = -\left(\frac{\Delta T}{T_0} - 1\right)y(k-1) + K_0 \frac{\Delta T}{T_0} u(k-1)$
K_1, T_0	$y(k) = -\left(\frac{\Delta T}{T_0} - 1\right)y(k-1) + K_0 \frac{\Delta T}{T_0} u(k-1) + K_1 \frac{\Delta T}{T_0} x(k)u(k-1)$
K_0, T_1	$y(k) = -\left(\frac{\Delta T}{T_0} - 1\right)y(k-1) + K_0 \frac{\Delta T}{T_0} u(k-1) - \frac{T_1}{T_0}[x(k)(y(k) - y(k-1))]$
$\overline{K}_1^{-1}, T_0$	$y(k) = -\left(\frac{\Delta T}{T_0} - 1\right)y(k-1) + K_0 \frac{\Delta T}{T_0} u(k-1) - \frac{K_0}{\overline{K}_1} x(k)y(k) - \frac{K_0}{\overline{K}_1}\left(\frac{\Delta T}{T_0} - 1\right)x(k)y(k-1)$
$K_0, \overline{T}_1^{-1}$	$y(k) = -\left(\frac{\Delta T}{T_0} - 1\right)y(k-1) + K_0 \frac{\Delta T}{T_0} u(k-1) + K_0 \frac{\Delta T}{\overline{T}_1} x(k)u(k-1) - \frac{\Delta T}{\overline{T}_1} x(k)y(k-1)$

TABLE 1.6.6 Discretization of second-order aperiodic systems using the Euler transformation

Model	*Difference equation*
K_0, T_{10}, T_{20}	$y(k) = -a_1 y(k-1) - a_2 y(k-2) + K_0 \frac{\Delta T^2}{T_{10}T_{20}} u(k-2)$
K_1, T_{10}, T_{20}	$y(k) = -a_1 y(k-1) - a_2 y(k-2) + K_0 \frac{\Delta T^2}{T_{10}T_{20}} u(k-2) + K_1 \frac{\Delta T^2}{T_{10}T_{20}} x(k)u(k-2)$
K_0, T_{11}, T_{20}	$y(k) = -a_1 y(k-1) - a_2 y(k-2) + K_0 \frac{\Delta T^2}{T_{10}T_{20}} u(k-2) + \frac{T_{11}}{T_{10}}[x(k)(-y(k) + 2y(k-1) - y(k-2))] - \frac{T_{11}}{T_{10}} \frac{\Delta T}{T_{20}}[x(k)(y(k-1) - y(k-2))]$
K_0, T_{11}, T_{21}	$y(k) = -a_1 y(k-1) - a_2 y(k-2) + K_0 \frac{\Delta T^2}{T_{10}T_{20}} u(k-2) + \left(\frac{T_{11}}{T_{10}} + \frac{T_{21}}{T_{20}}\right)[x(k)(-y(k) + 2y(k-1) - y(k-2))] -$ $-\Delta T \frac{T_{11} + T_{21}}{T_{10}T_{20}}[x(k)(y(k-1) - y(k-2))] - \frac{T_{11}T_{21}}{T_{10}T_{20}}\left[x^2(k)(-y(k) + 2y(k-1) - y(k-2))\right]$
$\overline{K}_1^{-1}, T_{10}, T_{20}$	$y(k) = -a_1 y(k-1) - a_2 y(k-2) + K_0 \frac{\Delta T^2}{T_{10}T_{20}} u(k-2) - \frac{K_0}{\overline{K}_1} a_1 x(k)y(k-1) - \frac{K_0}{\overline{K}_1} a_2 x(k)y(k-2) - \frac{K_0}{\overline{K}_1} x(k)y(k)$
$K_0, \overline{T}_{11}^{-1}, T_{20}$	$y(k) = -a_1 y(k-1) - a_2 y(k-2) + K_0 \frac{\Delta T^2}{\overline{T}_{11}T_{20}} x(k)u(k-2) - \frac{\Delta T^2}{\overline{T}_{11}T_{20}} x(k)y(k-2) + K_0 \frac{\Delta T^2}{T_{10}T_{20}} u(k-1) - \frac{\Delta T}{\overline{T}_{11}}[x(k)(y(k-1) - y(k-2))]$
$K_0, \overline{T}_{11}^{-1}, \overline{T}_{21}^{-1}$	$y(k) = -a_1 y(k-1) - a_2 y(k-2) + K_0\left(\frac{\Delta T^2}{T_{10}\overline{T}_{21}} + \frac{\Delta T^2}{\overline{T}_{11}T_{20}}\right)x(k)u(k-2) + K_0 \frac{\Delta T^2}{\overline{T}_{11}\overline{T}_{21}} x^2(k)u(k-2) + K_0 \frac{\Delta T^2}{T_{10}T_{20}} u(k) -$ $-\left(\frac{\Delta T^2}{T_{10}\overline{T}_{21}} + \frac{\Delta T^2}{\overline{T}_{11}T_{20}}\right)x(k)y(k-2) - \left(\frac{\Delta T}{\overline{T}_{11}} + \frac{\Delta T}{\overline{T}_{21}}\right)[x(k)(y(k-1) - y(k-2))] - \frac{\Delta T^2}{\overline{T}_{11}\overline{T}_{21}} x^2(k)y(k-2)$

$$a_1 = \frac{\Delta T}{T_{10}} + \frac{\Delta T}{T_{20}} - 2 \qquad a_2 = \frac{\Delta T^2}{T_{10}T_{20}} - \left(\frac{\Delta T}{T_{10}} + \frac{\Delta T}{T_{20}}\right) + 1$$

TABLE 1.6.7 Discretization of second-order under-damped systems using the Euler transformation

Model	*Difference equation*
K_0, ξ_0, T_0	$y(k) = -a_1 y(k-1) - a_2 y(k-2) + K_0 \frac{\Delta T^2}{T_0^2} u(k-2)$
K_1, ξ_0, T_0	$y(k) = -a_1 y(k-1) - a_2 y(k-2) + K_0 \frac{\Delta T^2}{T_0^2} u(k-2) + K_1 \frac{\Delta T^2}{T_0^2} x(k) u(k-2)$
K_0, ξ_1, T_0	$y(k) = -a_1 y(k-1) - a_2 y(k-2) + K_0 \frac{\Delta T^2}{T_0^2} u(k-2) - 2\xi_1 \frac{\Delta T}{T_0} [x(k)(y(k-1) - y(k-2))]$
K_0, ξ_0, T_1	$y(k) = -a_1 y(k-1) - a_2 y(k-2) + K_0 \frac{\Delta T^2}{T_0^2} u(k-2) + 2\frac{T_1}{T_0} [x(k)(-y(k) + 2y(k-1) - y(k-2))] -$ $-2\xi_0 \frac{T_1}{T_0} \frac{\Delta T}{T_0} [x(k)(y(k-1) - y(k-2))] + \frac{T_1^2}{T_0^2} [x^2(k)(-y(k) + 2y(k-1) - y(k-2))]$
$\bar{K}_1^{-1}, \xi_0, T_0$	$y(k) = -a_1 y(k-1) - a_2 y(k-2) + K_0 \frac{\Delta T^2}{T_0^2} u(k-2) - \frac{K_0}{\bar{K}_1} x(k) y(k) - \frac{K_0}{\bar{K}_1} a_1 x(k) y(k-1) - \frac{K_0}{\bar{K}_1} a_2 x(k) y(k-2)$
$K_0, \bar{\xi}_1^{-1}, T_0$	$y(k) = -a_1 y(k-1) - a_2 y(k-2) + K_0 \frac{\Delta T^2}{T_0^2} u(k-2) + K_0 \frac{\Delta T^2}{T_0^2} \frac{\xi_0}{\bar{\xi}_1} x(k) u(k-2) + \frac{\xi_0}{\bar{\xi}_1} [x(k)(-y(k) + 2y(k-1) - y(k-2))] - \frac{\Delta T^2}{T_0^2} \frac{\xi_0}{\bar{\xi}_1} x(k) y(k-2)$
$K_0, \xi_0, \bar{T}_1^{-1}$	$y(k) = -a_1 y(k-1) - a_2 y(k-2) + K_0 \frac{\Delta T^2}{T_0^2} u(k-2) + 2K_0 \frac{\Delta T^2}{T_0 \bar{T}_1} x(k) u(k-2) + K_0 \frac{\Delta T^2}{\bar{T}_1^2} x^2(k) u(k-2) -$ $-2\frac{\Delta T^2}{T_0 \bar{T}_1} x(k) y(k-2) - \frac{\Delta T^2}{T_0^2} x^2(k) y(k-2) - 2\xi_0 \frac{\Delta T}{\bar{T}_1} [x(k)(y(k-1) - y(k-2))]$

$$a_1 = 2\xi_0 \frac{\Delta T}{T_0} - 2 \qquad a_2 = \frac{\Delta T^2}{T_0^2} - 2\xi_0 \frac{\Delta T}{T_0} + 1$$

TABLE 1.6.8 Discretization of first-order systems using the bilinear transformation

Model	*Difference equation*
K_0, T_0	$y(k) = -\frac{\Delta T - 2T_0}{\Delta T + 2T_0} y(k-1) + K_0 \frac{\Delta T}{\Delta T + 2T_0} [u(k) + u(k-1)]$
K_1, T_0	$y(k) = -\frac{\Delta T - 2T_0}{\Delta T + 2T_0} y(k-1) + K_0 \frac{\Delta T}{\Delta T + 2T_0} [u(k) + u(k-1)] +$ $+K_1 \frac{\Delta T}{\Delta T + 2T_0} x(k)[u(k) + u(k-1)]$
K_0, T_1	$y(k) = -\frac{\Delta T - 2T_0}{\Delta T + 2T_0} y(k-1) + K_0 \frac{\Delta T}{\Delta T + 2T_0} [u(k) + u(k-1)] -$ $-2\frac{T_1}{\Delta T + 2T_0} x(k)[y(k) - y(k-1)]$
$\bar{K}_1^{-1}, T_1$	$y(k) = -\frac{\Delta T - 2T_0}{\Delta T + 2T_0} y(k-1) + K_0 \frac{\Delta T}{\Delta T + 2T_0} [u(k) + u(k-1)] - \frac{K_0}{\bar{K}_1} x(k) y(k) -$ $-\frac{K_0}{\bar{K}_1} \frac{\Delta T - 2T_0}{\Delta T + 2T_0} x(k) y(k-1)$
$K_0, \bar{T}_1^{-1}$	$y(k) = -\frac{\Delta T - 2T_0}{\Delta T + 2T_0} y(k-1) + K_0 \frac{\Delta T}{\Delta T + 2T_0} [u(k) + u(k-1)] +$ $+\frac{T_0}{\bar{T}_1} K_0 \frac{\Delta T}{\Delta T + 2T_0} [x(k)(u(k) + u(k-1))] - \frac{T_0}{\bar{T}_1} \frac{\Delta T}{\Delta T + 2T_0} x(k)[y(k) + y(k-1)]$

TABLE 1.6.9 Discretization of second-order aperiodic systems using the bilinear transformation

Model	*Difference equation*
K_0, T_{10}, T_{20}	$y(k) = -\frac{2\Delta T^2 - 8T_{10}T_{20}}{D_0} y(k-1) - \frac{N_0}{D_0} y(k-2) + K_0 \frac{\Delta T^2}{D_0} u_1$
K_1, T_{10}, T_{20}	$y(k) = -\frac{2\Delta T^2 - 8T_{10}T_{20}}{D_0} y(k-1) - \frac{N_0}{D_0} y(k-2) + K_0 \frac{\Delta T^2}{D_0} u_1 + K_1 \frac{\Delta T^2}{D_0} x(k) u_1$
K_0, T_{11}, T_{20}	$y(k) = -\frac{2\Delta T^2 - 8T_{10}T_{20}}{D_0} y(k-1) - \frac{N_0}{D_0} y(k-2) + K_0 \frac{\Delta T^2}{D_0} u_1 + 4\frac{T_{11}T_{20}}{D_0} x(k) y_1 -$ $-2\Delta T \frac{T_{11}}{D_0} x(k)[y(k) - y(k-2)]$
K_0, T_{11}, T_{21}	$y(k) = -\frac{2\Delta T^2 - 8T_{10}T_{20}}{D_0} y(k-1) - \frac{N_0}{D_0} y(k-2) + K_0 \frac{\Delta T^2}{D_0} u_1 + 4\frac{T_{10}T_{21} + T_{20}T_{11}}{D_0} x(k) y_1 -$ $-2\Delta T \frac{T_{11} + T_{21}}{D_0} x(k)[y(k) - y(k-2)] + 4\frac{T_{11}T_{21}}{D_0} x(k) y_1$
$\overline{K}_1^{-1}, T_{10}, T_{20}$	$y(k) = -\frac{2\Delta T^2 - 8T_{10}T_{20}}{D_0} y(k-1) - \frac{N_0}{D_0} y(k-2) + K_0 \frac{\Delta T^2}{D_0} u_1 - \frac{K_0}{\overline{K}_1} x(k) y(k) -$ $-\frac{K_0}{\overline{K}_1} \frac{2\Delta T^2 - 8T_{10}T_{20}}{D_0} x(k) y(k-1) - \frac{K_0}{\overline{K}_1} \frac{N_0}{D_0} x(k) y(k-2)$
$K_0, \overline{T}_{11}^{-1}, T_{20}$	$y(k) = -\frac{2\Delta T^2 - 8T_{10}T_{20}}{D_0} y(k-1) - \frac{N_0}{D_0} y(k-2) + K_0 \frac{\Delta T^2}{D_0} u_1 + K_0 \frac{\Delta T^2}{D_0} \frac{T_{10}}{\overline{T}_{11}} x(k) u_1 -$ $-\frac{\Delta T^2}{D_0} \frac{T_{10}}{\overline{T}_{11}} x(k) y_2 - 2\Delta T \frac{T_{20}}{D_0} \frac{T_{10}}{\overline{T}_{11}} x(k)[y(k) - y(k-2)]$
$K_0, \overline{T}_{11}^{-1}, \overline{T}_{21}^{-1}$	$y(k) = -\frac{2\Delta T^2 - 8T_{10}T_{20}}{D_0} y(k-1) - \frac{N_0}{D_0} y(k-2) + K_0 \frac{\Delta T^2}{D_0} u_1 + K_0 \frac{\Delta T^2}{D_0} \left(\frac{T_{10}}{\overline{T}_{11}} + \frac{T_{20}}{\overline{T}_{21}} \right) x(k) u_1 +$ $+K_0 \frac{\Delta T^2}{D_0} \frac{T_{10}}{\overline{T}_{11}} \frac{T_{20}}{\overline{T}_{21}} x^2(k) u_1 - \frac{\Delta T^2}{D_0} \left(\frac{T_{10}}{\overline{T}_{11}} + \frac{T_{20}}{\overline{T}_{21}} \right) x(k) y_2 -$ $-2\frac{\Delta T^2}{D_0} \left(T_{10} \frac{T_{20}}{\overline{T}_{21}} + T_{20} \frac{T_{10}}{\overline{T}_{11}} \right) x(k)[y(k) - y(k-2)] - \frac{\Delta T^2}{D_0} \frac{T_{10}}{\overline{T}_{11}} \frac{T_{20}}{\overline{T}_{21}} x^2(k) y_2$

$$u_1 = u(k) + 2u(k-1) + u(k-2) \qquad y_1 = -y(k) + 2y(k-1) - y(k-2)$$

$$y_2 = y(k) + 2y(k-1) + y(k-2) \qquad N_0 = 4T_{10}T_{20} - 2\Delta T(T_{10} + T_{20}) + \Delta T^2$$

$$D_0 = 4T_{10}T_{20} + 2\Delta T(T_{10} + T_{20}) + \Delta T^2$$

TABLE 1.6.10 Discretization of second-order under-damped systems using the bilinear transformation

Model	*Difference equation*
K_0, ξ_0, T_0	$y(k) = -\frac{2\Delta T^2 - 8T_0^2}{D_0} y(k-1) - \frac{N_0}{D_0} y(k-2) + K_0 \frac{\Delta T^2}{D_0} u_1$
K_1, ξ_0, T_0	$y(k) = -\frac{2\Delta T^2 - 8T_0^2}{D_0} y(k-1) - \frac{N_0}{D_0} y(k-2) + K_0 \frac{\Delta T^2}{D_0} u_1 + K_1 \frac{\Delta T^2}{D_0} x(k) u_1$
K_0, ξ_1, T_0	$y(k) = -\frac{2\Delta T^2 - 8T_0^2}{D_0} y(k-1) - \frac{N_0}{D_0} y(k-2) + K_0 \frac{\Delta T^2}{D_0} u_1 -$ $-4\Delta T \frac{T_0 \xi_1}{D_0} x(k)[y(k) - y(k-2)]$
K_0, ξ_0, T_1	$y(k) = -\frac{2\Delta T^2 - 8T_0^2}{D_0} y(k-1) - \frac{N_0}{D_0} y(k-2) + K_0 \frac{\Delta T^2}{D_0} u_1 + 8 \frac{T_0 T_1}{D_0} x(k) y_1 -$ $-4\Delta T \frac{T_1 \xi_0}{D_0} x(k)[y(k) - y(k-2)] + 4 \frac{T_1^2}{D_0} x^2(k) y_1$
$\bar{K}_1^{-1}, \xi_0, T_0$	$y(k) = -\frac{2\Delta T^2 - 8T_0^2}{D_0} y(k-1) - \frac{N_0}{D_0} y(k-2) + K_0 \frac{\Delta T^2}{D_0} u_1 - \frac{K_0}{\bar{K}_1} x(k) y(k) -$ $-\frac{K_0}{\bar{K}_1} \frac{2\Delta T^2 - 8T_0^2}{D_0} x(k) y(k-1) - \frac{K_0}{\bar{K}_1} \frac{N_0}{D_0} x(k) y(k-2)$
$K_0, \bar{\xi}_1^{-1}, T_0$	$y(k) = -\frac{2\Delta T^2 - 8T_0^2}{D_0} y(k-1) - \frac{N_0}{D_0} y(k-2) + K_0 \frac{\Delta T^2}{D_0} u_1 + K_0 \frac{\Delta T^2}{D_0} \frac{\xi_0}{\bar{\xi}_1} x(k) u_1 +$ $+4 \frac{T_0^2}{D_0} \frac{\xi_0}{\bar{\xi}_1} x(k) y_1 - \frac{\Delta T^2}{D_0} \frac{\xi_0}{\bar{\xi}_1} x(k) y_2$
$K_0, \xi_0, \bar{T}_1^{-1}$	$y(k) = -\frac{2\Delta T^2 - 8T_0^2}{D_0} y(k-1) - \frac{N_0}{D_0} y(k-2) + K_0 \frac{\Delta T^2}{D_0} u_1 + 2K_0 \frac{\Delta T^2}{D_0} \frac{T_0}{\bar{T}_1} x(k) u_1 +$ $+K_0 \frac{\Delta T^2}{D_0} \left(\frac{T_0}{\bar{T}_1} \right)^2 x^2(k) u_1 - 2 \frac{\Delta T^2}{D_0} \frac{T_0}{\bar{T}_1} x(k) y_2 -$ $-4\Delta T \frac{\xi_0}{D_0} \frac{T_0^2}{\bar{T}_1} x(k)[y(k) - y(k-2)] - \frac{\Delta T^2}{D_0} \left(\frac{T_0}{\bar{T}_1} \right)^2 x^2(k) y_2$

$$u_1 = u(k) + 2u(k-1) + u(k-2) \qquad y_1 = -y(k) + 2y(k-1) - y(k-2)$$

$$y_2 = y(k) + 2y(k-1) + y(k-2) \qquad N_0 = 4T_0^2 - 4\Delta T T_0 \xi_0 + \Delta T^2$$

$$D_0 = 4T_0^2 + 4\Delta T T_0 \xi_0 + \Delta T^2$$

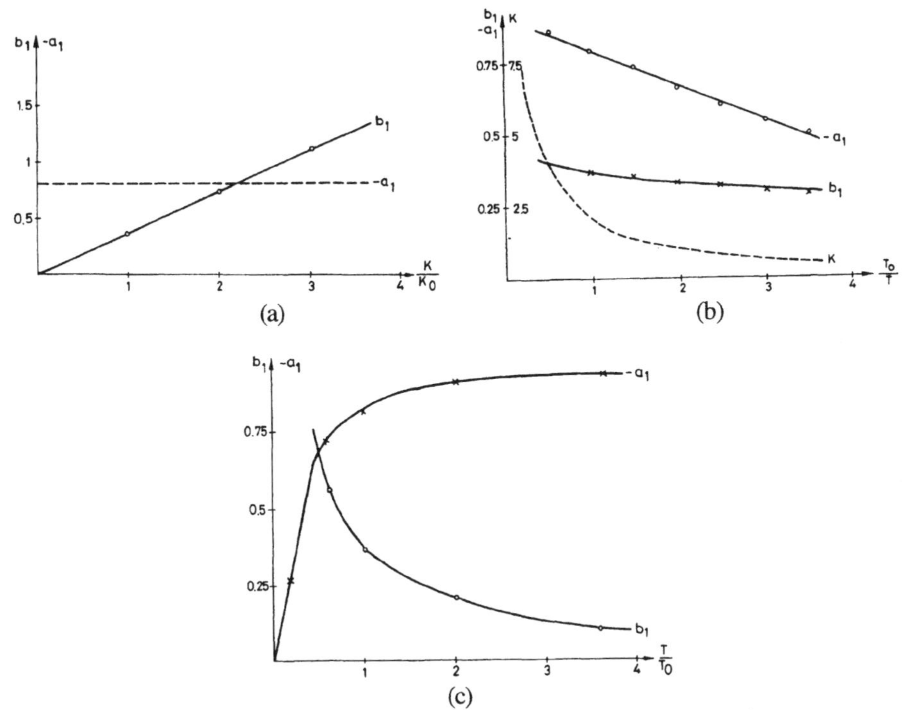

(c)

Fig. 1.6.7 Coefficients of the first-order pulse transfer function: (a) $b_1(K/K_0)$, $a_1(K/K_0)$; (b) $b_1(T_0/T)$, $a_1(T_0/T)$; (c) $b_1(T/T_0)$, $a_1(T/T_0)$

Example 1.6.17 *First-order system (Haber, 1979b)*
Figure 1.6.7a to 1.6.7c show how the parameters of the pulse transfer function

$$H(z^{-1}) = \frac{b_1 z^{-1}}{1 + a_1 z^{-1}} \tag{1.6.40}$$

change if the parameters (θ_i) of the transfer function

$$G(s) = \frac{K}{1+sT} = \frac{2}{1+10s} = \frac{0.2}{s+0.1} = \frac{\beta_1}{s+\alpha_1}$$

vary one by one in the domain $[0 < \theta_i/\theta_{i0} \le 4]$. ■

The following relations can be seen from the diagrams:

1. There is no relation between the parameters: $a_1(K)$.
2. There is an (at least approximately) linear relationship between the parameters: $b_1(K), a_1(1/T)$.
3. There is a nonlinear relation between the parameters: $b_1(1/T), b_1(T), a_1(T)$

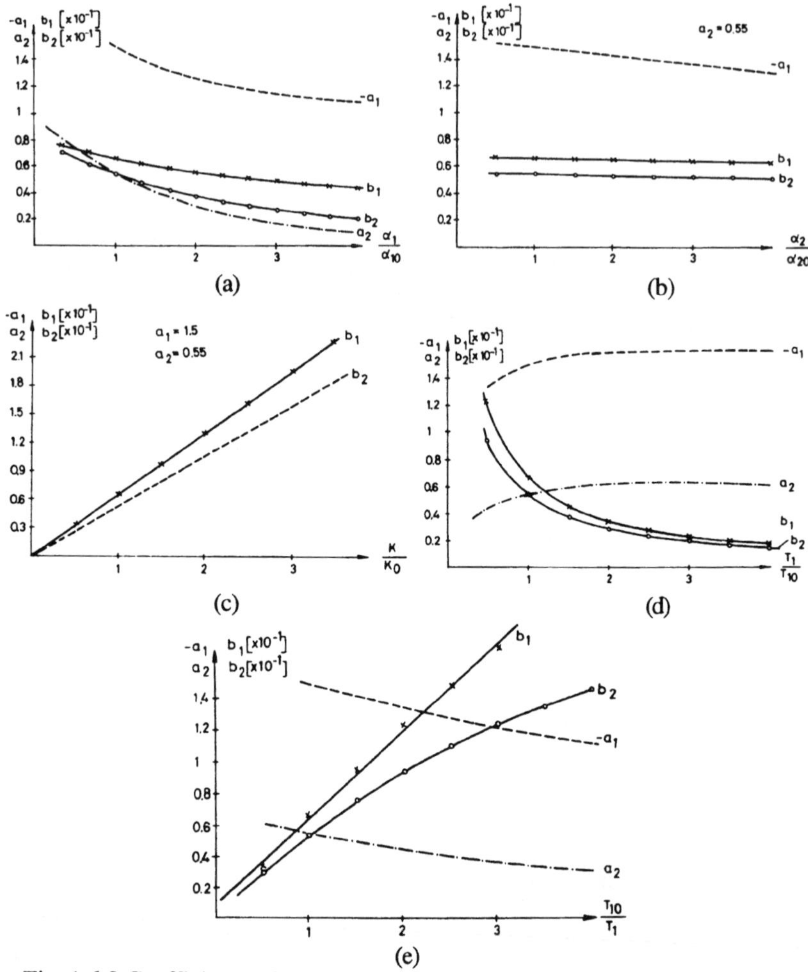

Fig. 1.6.8 Coefficients of the second-order aperiodic pulse transfer function:
(a) $b_1(\alpha_1/\alpha_0)$, $b_2(\alpha_1/\alpha_0)$, $a_1(\alpha_1/\alpha_0)$, $a_2(\alpha_1/\alpha_0)$;
(b) $b_1(\alpha_2/\alpha_0)$, $b_2(\alpha_2/\alpha_0)$, $a_1(\alpha_2/\alpha_0)$, $a_2(\alpha_2/\alpha_0)$;
(c) $b_1(K/K_0)$, $b_2(K/K_0)$, $a_1(K/K_0)$, $a_2(K/K_0)$;
(d) $b_1(T_1/T_{10})$, $b_2(T_1/T_{10})$, $a_1(T_1/T_{10})$, $a_2(T_1/T_{10})$;
(e) $b_1(T_{10}/T_1)$, $b_2(T_{10}/T_1)$, $a_1(T_{10}/T_1)$, $a_2(T_{10}/T_1)$

Example 1.6.18 *Second-order aperiodic system (Haber, 1979b)*
Figure 1.6.8a to 1.6.8e show how the parameters of the pulse transfer function

$$H(z^{-1}) = \frac{b_1 z^{-1} + b_2 z^{-2}}{1 + a_1 z^{-1} + a_2 z^{-2}} \tag{1.6.41}$$

change if the parameters (θ_i) of the transfer function

$$G(s) = \frac{K}{(1+sT_1)(1+sT_2)} = \frac{2}{(1+10s)(1+5s)} = \frac{0.04}{s^2+0.3s+0.02}$$
$$= \frac{\beta_0 s^2 + \beta_1 s + \beta_2}{s^2 + \alpha_1 s + \alpha_2}$$

vary one by one in the domain $[0 < \theta_i/\theta_{i0} \le 4]$. ■

The following relations can be seen from the diagrams:

1. There is no relation between the parameters: $a_2(\alpha_2), a_1(K), a_2(K)$;
2. There is an (at least approximately) linear relationship between the parameters: $b_1(K), b_2(K), b_1(\alpha_2), b_2(\alpha_2), a_1(\alpha_2), b_1(1/T_1), a_1(1/T_1), a_2(1/T_1)$;
3. There is a nonlinear relation between the parameters: $b_1(\alpha_1), b_2(\alpha_1), a_1(\alpha_1), a_2(\alpha_1), b_1(T_1), b_2(T_1), a_1(T_1), a_2(T_1)$.

Remember that $\alpha_1 = 1/T_1 + 1/T_2$ and $\alpha_2 = 1/(T_1T_2)$. The dependence on T_2 has the same character as the dependence on T_1.

Example 1.6.19 *Second-order under-damped system (Haber, 1979b)*
Figure 1.6.9a to 1.6.9e show how the parameters of the pulse transfer function

$$H(z^{-1}) = \frac{b_1 z^{-1} + b_2 z^{-2}}{1 + a_1 z^{-1} + a_2 z^{-2}}$$

change if the parameters (θ_i) of the transfer function

$$G(s) = \frac{K}{1+2\xi sT^2 + s^2T^2} = \frac{2}{1+2\cdot 0.5\cdot 10s + 10^2 s^2} = \frac{0.02}{s^2+0.1s+0.01} = \frac{\beta_0 s^2 + \beta_1 s + \beta_2}{s^2 + \alpha_1 s + \alpha_2}$$

vary one by one in the domain $[0 < \theta_i/\theta_{i0} \le 4]$. ■

The following relations can be seen from the diagrams:

1. There is no relation between the parameters: $a_1(K), a_2(K)$;
2. There is an (at least approximately) linear relationship between the parameters: $b_1(\alpha_1), b_2(\alpha_1), a_1(\alpha_1), a_2(\alpha_1), b_1(\alpha_2), b_2(\alpha_2), a_1(\alpha_2), a_2(\alpha_2), b_1(K), b_2(K), a_1(1/T), a_2(1/T)$;
3. There is a nonlinear relation between the parameters: $b_1(T), b_2(T), a_1(T), a_2(T), b_1(1/T), b_2(1/T)$

The dependence on ξ is similar to that of α_1 and the dependence on $1/T^2$ is equal to that of α_2 because $\alpha_1 = 2\xi$ and $\alpha_2 = 1/T^2$.

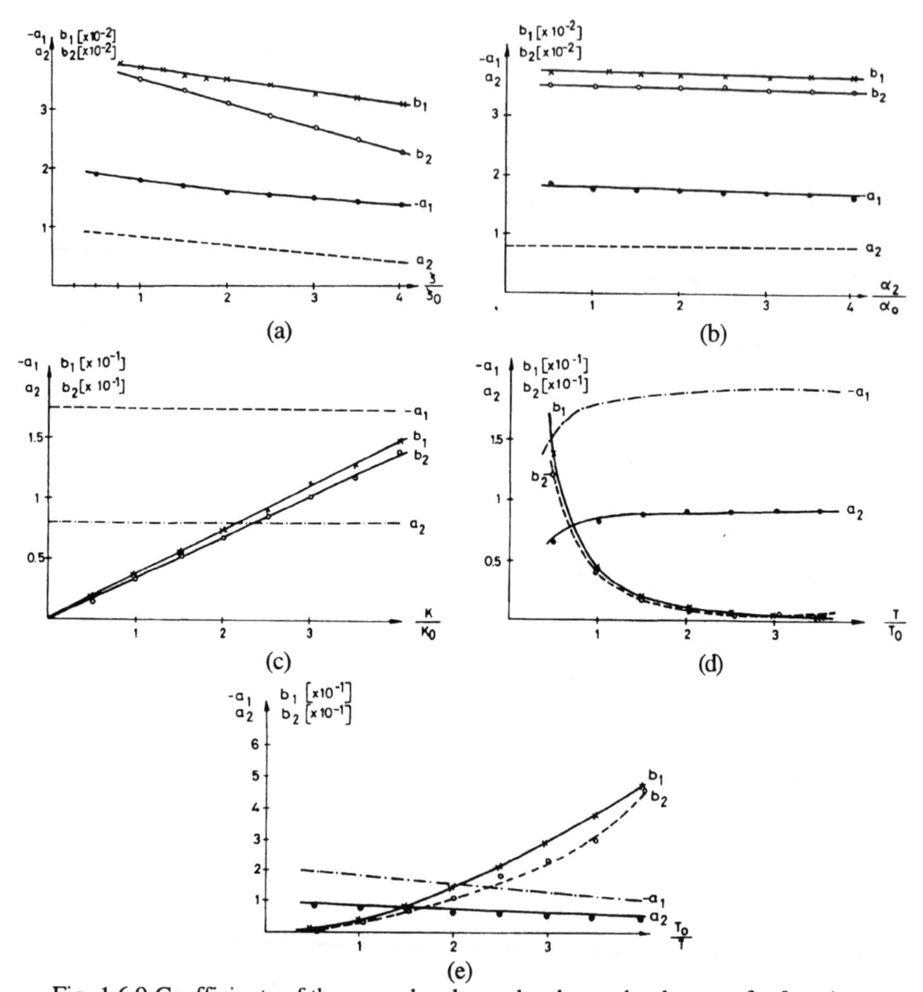

Fig. 1.6.9 Coefficients of the second-order under-damped pulse transfer function:

(a) $b_1(\xi_1/\xi_0)$, $b_2(\xi_1/\xi_0)$, $a_1(\xi_1/\xi_0)$, $a_2(\xi_1/\xi_0)$;

(b) $b_1(\alpha_2/\alpha_0)$, $b_2(\alpha_2/\alpha_0)$, $a_1(\alpha_2/\alpha_0)$, $a_2(\alpha_2/\alpha_0)$;

(c) $b_1(K/K_0)$, $b_2(K/K_0)$, $a_1(K/K_0)$, $a_2(K/K_0)$;

(d) $b_1(T/T_0)$, $b_2(T/T_0)$, $a_1(T/T_0)$, $a_2(T/T_0)$;

(e) $b_1(T_0/T)$, $b_2(T_0/T)$, $a_1(T_0/T)$, $a_2(T_0/T)$

1.6.6 The effect of changing the parameters of the discrete time models to the parameters of the continuous time equivalent description

Sometimes a highly nonlinear signal dependence in the parameters of the differential equation of the process can be described by a much simpler (e.g., linear) signal dependence in the parameters of the pulse transfer function. The following examples demonstrate this. As the relations are similar for the different transformation methods, the step response equivalent transformation is considered here. (The same examples using the bilinear transformation are in Haber (1979b)). The transformation equations between the discrete and the continuous time descriptions are as follows (e.g., Keviczky, 1977):

- first-order system (1.6.40):

$$\alpha_1 = \frac{1}{T} = -\frac{\Delta T}{\ln(a_1)}, \qquad \beta_0 = 0, \qquad \beta_1 = \frac{K}{T} = \frac{b_1}{1+a_1}\alpha_1$$

- second-order system :

$$\alpha_1 = -\frac{1}{\Delta T}\ln(a_2), \quad \alpha_2 = \frac{1}{2\Delta T}\left[\ln(a_2)\right]^2 + \left[\frac{1}{\Delta T}\arccos\left(\frac{-a_1}{2\sqrt{a_2}}\right)\right]^2$$

$$\beta_1 = -\frac{1}{2\Delta T}\ln(a_2)\frac{b_1+b_2}{1+a_1+a_2} + \frac{1}{\Delta T}\arccos\left(\frac{-a_1}{2\sqrt{a_2}}\right)\frac{(a_1+2a_2)(b_1-b_2)}{(1+a_1+a_2)\sqrt{4a_2-a_1^2}}$$

$$\beta_2 = \frac{\alpha_2(b_1+b_2)}{1+a_1+a_2}$$

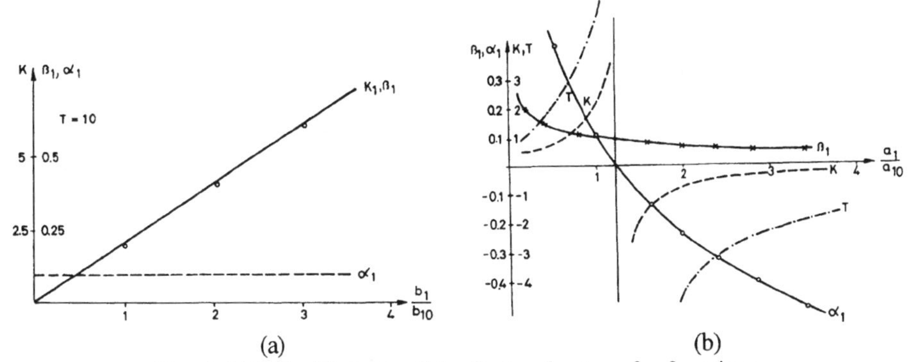

(a) (b)

Fig. 1.6.10 Coefficients of the first-order transfer function:
(a) $\beta_1(b_1/b_{10})$, $\alpha_1(b_1/b_{10})$; (b) $\beta_1(a_1/a_{10})$, $\alpha_1(a_1/a_{10})$

Example 1.6.20 *First-order transfer system (Haber, 1979b)*

Figure 1.6.10a and Figure 1.6.10b show how the parameters of the transfer function

$$G(s) = \frac{\beta_1}{s+\alpha_1}$$

change if the parameters (θ_i) of the pulse transfer function

$$H(z^{-1}) = \frac{b_1}{1+a_1 z^{-1}} = \frac{0.3626 z^{-1}}{1+0.8187 z^{-1}}$$

vary one by one in the domain $[0 < \theta_i/\theta_{i0} \le 4]$. The following relations can be seen from the diagrams:

1. There is no relation between the parameters: $\alpha_1(b_1)$.;
2. There is a linear relationship between the parameters: $\beta_1(b_1)$;
3. There is a nonlinear relation between the parameters:
   $\beta_1(a_1), \alpha_1(a_1), K(a_1), T(a_1)$.

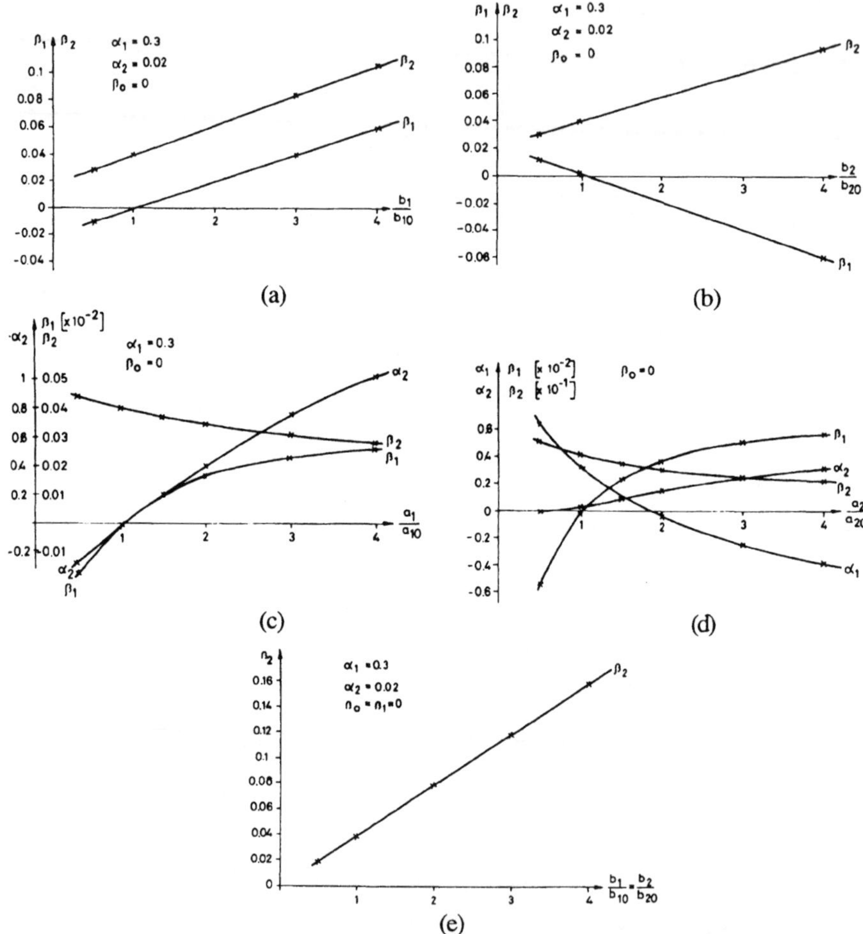

Fig. 1.6.11 Coefficients of the second-order aperiodic transfer function:
(a) $\beta_1(b_1/b_{10})$, $\beta_2(b_1/b_{10})$, $\alpha_1(b_1/b_{10})$, $\alpha_2(b_1/b_{10})$;
(b) $\beta_1(b_2/b_{20})$, $\beta_2(b_2/b_{20})$, $\alpha_1(b_2/b_{20})$, $\alpha_2(b_2/b_{20})$;
(c) $\beta_1(a_1/a_{10})$, $\beta_2(a_1/a_{10})$, $\alpha_1(a_1/a_{10})$, $\alpha_2(a_1/a_{10})$;
(d) $\beta_1(a_2/a_{20})$, $\beta_2(a_2/a_{20})$, $\alpha_1(a_2/a_{20})$, $\alpha_2(a_2/a_{20})$;
(e) $\beta_1(b_1/b_{10} = b_2/b_{20})$, $\beta_2(b_1/b_{10} = b_2/b_{20})$, $\alpha_1(b_1/b_{10} = b_2/b_{20})$, $\alpha_2(b_1/b_{10} = b_2/b_{20})$

Example 1.6.21 *Second-order aperiodic system (Haber, 1979b)*
Figure 1.6.11a to Figure 1.6.11e show how the parameters of the transfer function

$$G(s)=\frac{\beta_0 s^2+\beta_1 s+\beta_2}{s^2+\alpha_1 s+\alpha_2} \tag{1.6.42}$$

change if the parameters (θ_i) of the pulse transfer function

$$H\left(z^{-1}\right)=\frac{b_1 z^{-1}+b_2 z^{-2}}{1+a_1 z^{-1}+a_2 z^{-2}}=\frac{0.06572 z^{-1}+0.0538 z^{-2}}{1-1.489 z^{-1}+0.5488 z^{-2}}$$

vary one by one in the domain $\left[0<\theta_i/\theta_{i0}\leq 4\right]$. ■

The following relations can be seen from the diagrams:

1. There is no relation between the parameters:
   $\alpha_1(b_1), \alpha_2(b_1), \beta_0(b_1), \alpha_1(b_2), \alpha_2(b_2), \beta_0(b_2), \alpha_1(a_1), \beta_0(a_1), \beta_0(a_2)$;
2. There is a linear relationship between the parameters:
   $\beta_2(b_1), \beta_1(b_1), \beta_2(b_2), \beta_1(b_2)$;
3. There is a nonlinear relation between the parameters:
   $\alpha_2(a_1), \beta_1(a_1), \beta_2(a_1), \alpha_1(a_2), \alpha_2(a_2), \beta_1(a_2), \beta_2(a_2)$;
4. If the parameters b_1 and b_2 change simultaneously then α_1, α_2, β_0 and β_1 remain constants and β_2 changes linearly.

Example 1.6.22 *Second-order under-damped system (Haber, 1979b)*
Figure 1.6.12a to Figure 1.6.12e show how the parameters of the transfer function (1.6.42) change if the parameters (θ_i) of the pulse transfer function

$$H\left(z^{-1}\right)=\frac{b_1 z^{-1}+b_2 z^{-2}}{1+a_1 z^{-1}+a_2 z^{-2}}=\frac{0.03734 z^{-1}+0.03493 z^{-2}}{1-1.7826 z^{-1}+0.81873 z^{-2}}$$

vary one by one in the domain $\left[0<\theta_i/\theta_{i0}\leq 4\right]$. ■

The following relations can be seen from the diagrams:

1. There is no relation between the parameters:
   $\alpha_1(b_1), \alpha_2(b_1), \beta_0(b_1), \alpha_1(b_2), \alpha_2(b_2), \beta_0(b_2), \alpha_1(a_1), \beta_0(a_1), \beta_1(a_1), \beta_0(a_2)$;
2. There is a linear relationship between the parameters:
   $\beta_2(b_1), \beta_1(b_1), \beta_2(b_2), \beta_1(b_2)$;
3. There is a nonlinear relation between the parameters:
   $\alpha_2(a_1), \beta_2(a_1), \alpha_1(a_2), \alpha_2(a_2), \beta_1(a_2), \beta_2(a_2)$;
4. If the parameters b_1 and b_2 change simultaneously then α_1, α_2, β_0 and β_1 remain constants and β_2 changes linearly.

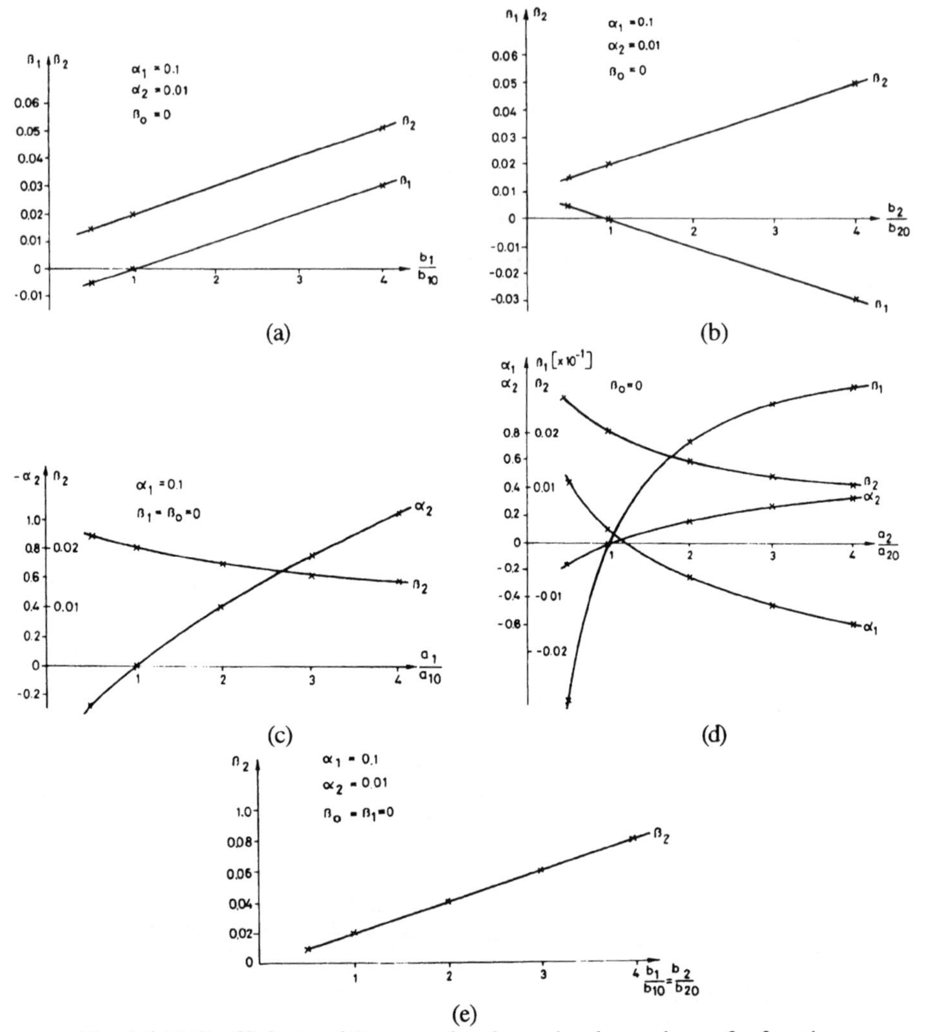

Fig. 1.6.12 Coefficients of the second-order under-damped transfer function:

(a) $\beta_1(b_1/b_{10})$, $\beta_2(b_1/b_{10})$, $\alpha_1(b_1/b_{10})$, $\alpha_2(b_1/b_{10})$

(b) $\beta_1(b_2/b_{20})$, $\beta_2(b_2/b_{20})$, $\alpha_1(b_2/b_{20})$, $\alpha_2(b_2/b_{20})$

(c) $\beta_1(a_1/a_{10})$, $\beta_2(a_1/a_{10})$, $\alpha_1(a_1/a_{10})$, $\alpha_2(a_1/a_{10})$

(d) $\beta_1(a_2/a_{20})$, $\beta_2(a_2/a_{20})$, $\alpha_1(a_2/a_{20})$, $\alpha_2(a_2/a_{20})$

(e) $\beta_1(b_1/b_{10} = b_2/b_{20})$, $\beta_2(b_1/b_{10} = b_2/b_{20})$, $\alpha_1(b_1/b_{10} = b_2/b_{20})$ $\alpha_2(b_1/b_{10} = b_2/b_{20})$

1.7 GATE FUNCTION MODELS

There are different gate function type of models. Common features of the models are the followings:

1. They describe the input–output relation by different models in different states, e.g., in different working points.
2. The estimation of the parameters of the different models is independent from each other.

The linear multi-models are special cases of these model types.

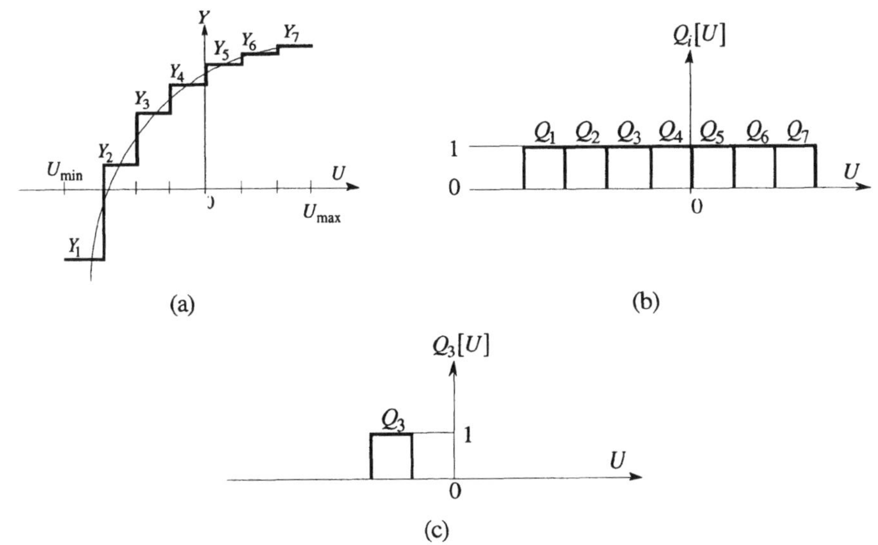

Fig. 1.7.1 Stepwise approximation of a function:
(a) mean values in the intervals; (b) gate functions; (c) single gate

1.7.1 Static models

1. Single-input single-output model

Consider a single-input single-output static model first. If the static relation cannot be described easily by an analytic function the stepwise approximation is a practical alternative. The range between the minimum and maximum values of the input signal is divided into M intervals, where the width of each interval is $\left(U_{\max}-U_{\min}\right)/M$. Figure 1.7.1a shows this for $M=7$. In each interval the function is approximated by its expected (or mean) value:

$$Y_i = \mathrm{E}\left\{y(k) \middle| U_i \le u(k) < U_{i+1}\right\}$$

Another formulation of the same model is the so-called gate function model introduced by Bose (1959) and discussed by Schetzen (1965b). The gate function Q_i is defined unit if the input signal is within the ith interval and it is zero elsewhere (Figure 1.7.1b)

$$Q_i\left[u(k)\right] = \begin{cases} 1 & U_i \le u(k) < U_{i+1} \\ 0 & \text{otherwise} \end{cases}$$

Then the stepwise approximation is

$$Y_i = \mathrm{E}\{y(k)Q_i[u(k)]\}$$

Figure 1.7.1c shows a single gate function (for $i = 3$).

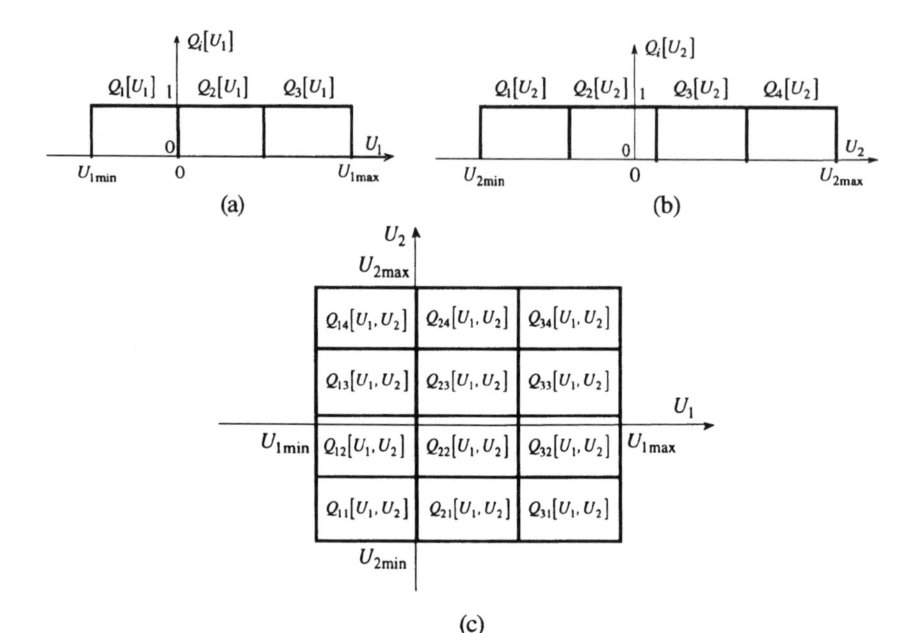

(c)

Fig. 1.7.2 One- and two-dimensional gate functions:
(a) gate functions for the first variable; (b) gate functions for the second variable;
(c) two-variable gate functions

2. Multi-input single-output model

It is easy to extend the method for multi-input single-output systems. Denote now the number of inputs by m. In this case all input signals have to be quantized and the gate function $Q_i[u_1(k), u_2(k), \ldots, u_m(k)]$ is the product of the one-dimensional gate functions

$$Q_{11\ldots1}[u_1(k), u_2(k), \ldots, u_m(k)] = Q_1[u_1(k)]Q_1[u_2(k)]\ldots Q_1[u_m(k)]$$

$$Q_{11\ldots2}[u_1(k), u_2(k), \ldots, u_m(k)] = Q_1[u_1(k)]Q_1[u_2(k)]\ldots Q_2[u_m(k)]$$

$$Q_{M_1M_2\ldots M_m}[u_1(k), u_2(k), \ldots, u_m(k)] = Q_{M_1}[u_1(k)]Q_{M_2}[u_2(k)]\ldots Q_{M_m}[u_m(k)]$$

Figure 1.7.2 shows the gate functions for two input signals.

So far we have discussed only a uniform division of the range of the input signal. There is no object again an other type of quantization. An example can be the case when the input signal is not bounded. Then the width of the first and of the last intervals is infinite because (in one dimensional case) $U_1 = -\infty$ and $U_M = \infty$. The shape of the nonlinear function can require that in a certain range the intervals are chosen more frequently than otherwise. Such so-called tapered quantization is recommended in Roy and DeRusso (1962) for dynamic systems.

1.7.2 Nonparametric dynamic models

1. Nonparametric dynamic spatial model

The output signal of a nonparametric SISO nonlinear dynamic model is a function of the actual and of the delayed input signals: $u(k),\ldots,u(k-m)$. Here m is the maximum memory of the system. With the Volterra series only polynomial steady state relation could be described. The Zadeh series is an extension to other analytical functional relationships. If, however, the relation is not analytical, then the gate function approach can be applied.

The MISO static gate function model can be extended to the case of nonparametric spatial model if the input signals are replaced by the actual and delayed input signals of the SISO dynamic model

$$u(k) \to u_1(k),\ u(k-1) \to u_2(k),\ u(k-2) \to u_3(k),\ \ldots,\ u(k-m) \to u_{m+1}(k)$$

The gate function is defined as $Q[u(k),\ldots,u(k-m)]$, and differs from zero if all shifted signals are in their intervals. The number of intervals is M^{m+1} because M intervals can be assumed for all components. The scheme of the model is seen in Figure 1.7.3. The following papers report about the successful modeling by this gate function model:

1. *Simple Wiener-Hammerstein model with saturation nonlinear characteristic* (Roy and DeRusso, 1962; Harris and Lapidus, 1967a);
2. *Continuous stirred tank reactor with a single exothermic reaction* (Harris and Lapidus, 1967a, 1967b);
3. *Closed loop nonlinear process* (Miller and Roy, 1964).

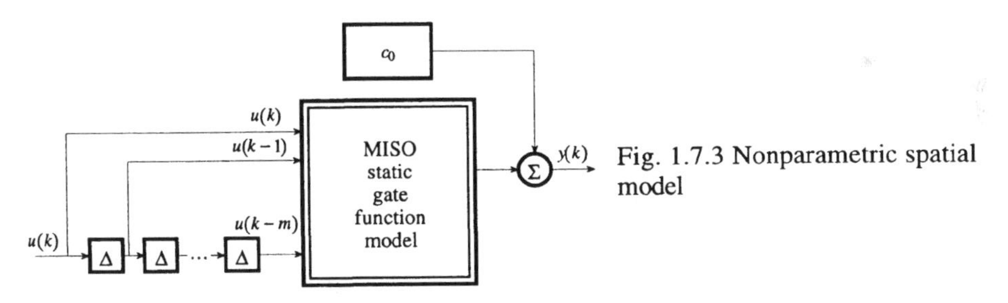

Fig. 1.7.3 Nonparametric spatial model

2. Wiener's gate function model

The Wiener series model consists of two parts

- parallel linear dynamic channels, whose outputs are orthogonal to each other, and
- a static multi-input single-output part,

where the inputs of the MISO nonlinear part are the outputs of the linear orthogonal submodels. By measuring the input signal and by selecting the linear dynamic orthogonal functions, the outputs of the dynamical part are known, which means that they are computable. The task of the identification is to estimate the nonlinear part. Wiener suggested applying the Hermite functionals as an orthogonal base for the nonlinear part. The unknown parameters can be calculated independently from each other only if the input signal is a Gaussian white noise. Applying, however, the gate functions instead of the Hermite functionals allows an independent estimation of the parameters even if the input signal is not Gaussian. This technique was suggested by Schetzen (1965) and was described in detail in Schetzen (1980).

Denote the outputs of the parallel linear dynamic channels by $v_i(k), i = 0, 1, 2, \ldots, m$ then the gate function is defined as $Q[v_0(k), \ldots, v_m(k)]$, and the MISO model can be applied if the input signals are replaced by $v_i(k), i = 0, 1, 2, \ldots, m$,

$$v_0(k) \to u_1(k),\ v_1(k) \to u_2(k),\ v_2(k) \to u_3(k), \ldots, v_m(k) \to u_{m+1}(k)$$

Figure 1.7.4 shows the scheme of the model.

The following paper reports about the successful modeling by this gate function model:

two-phase induction motor with speed control (Pincock and Atherton, 1973).

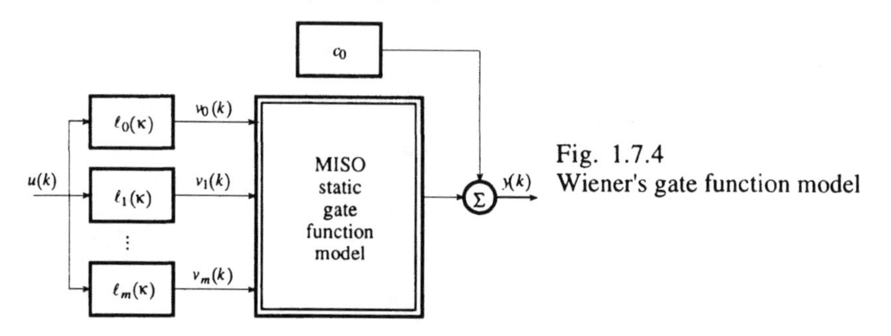

Fig. 1.7.4
Wiener's gate function model

1.7.3 Parametric spatial model

A further extension of the MISO static gate function model is if the input signals are the actual and delayed output signals of the SISO dynamic process

$$u(k) \to u_1(k),\ u(k-1) \to u_2(k), \ldots, u(k-n) \to u_{n+1}(k)$$
$$y(k-1) \to u_{n+2}(k), \ldots, y(k-n) \to u_{2n+1}(k)$$

The gate function is defined as $Q[u(k), \ldots, u(k-n), y(k-1), \ldots, y(k-n)]$ and it differs from zero if all signal terms are in their intervals. The number of intervals is M^{2n+1} if the range of both the input and the output signal is quantized into M intervals. The scheme of the parametric dynamic model is given in Figure 1.7.5.

Billings and Voon (1987) report about the approximation of the model to a simple Wiener model.

1.7.4 The elementary gate function model and the linear multi-model

1. Elementary gate function model

So far the gate model method has been applied with the following assumptions:

1. The Q-functions were functions only of the input signal(s)
2. The elementary models belonging to the single intervals were static.

Both assumptions are not essential, and they can be lifted. We define the elementary gate model (EGM) as follows. Choose a signal $x(t)$ the models depend on. Quantize this signal similarly to the technique dealt with up to now. Define the $Q_i[x(t)]$ function as

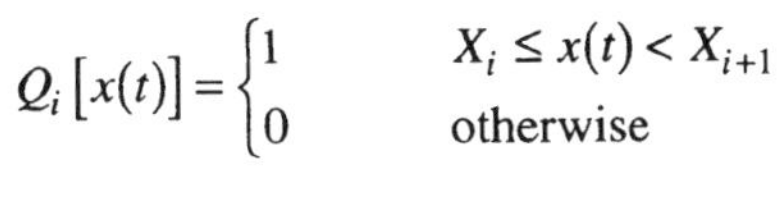

$$Q_i\left[x(t)\right]=\begin{cases}1 & X_i \le x(t) < X_{i+1}\\ 0 & \text{otherwise}\end{cases}$$

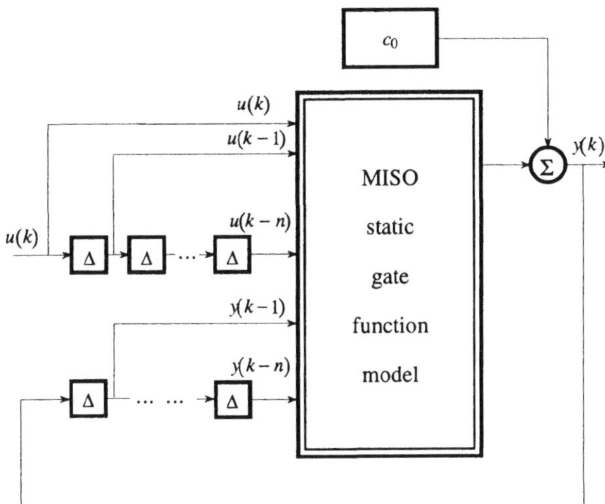

Fig. 1.7.5
Parametric spatial model

Cut from the measured input and output signals the gated ones

$$u_i^*(t)=Q_i\left[x(t)\right]u(t)=\begin{cases}u(t) & X_i \le x(t) < X_{i+1}\\ 0 & \text{otherwise}\end{cases}$$

$$y_i^*(t)=Q_i\left[x(t)\right]y(t)=\begin{cases}y(t) & X_i \le x(t) < X_{i+1}\\ 0 & \text{otherwise}\end{cases}$$

Of course the sum of the gated input and output signals is the total input and output signal:

$$u(t)=\sum_{i=1}^{M} u_i^*(t)$$

and

$$y(t)=\sum_{i=1}^{M} y_i^*(t)$$

An elementary gate model can be identified between the input and output signals cut by the same gate function. The whole model of the process is the *sum* of the elementary gate models as seen in Figure 1.7.6. The *sum* means that at any time only one of the elementary models is valid, and then the process output is equal to the output of the corresponding elementary gate model.

An alternative way to build the output of the globally valid model is to calculate the weighted sum of the outputs of the locally valid submodels (Johansen and Foss, 1993; Nelles *et al.*, 1997).

2. Linear multi-model

If the elementary gate models are linear then they are called linear multi-models. The equation of the submodels is

$$y_i^*(k) = \frac{B_i(q^{-1})}{A_i(q^{-1})} u_i^*(k)$$

The corresponding scheme is seen in Figure 1.7.7.

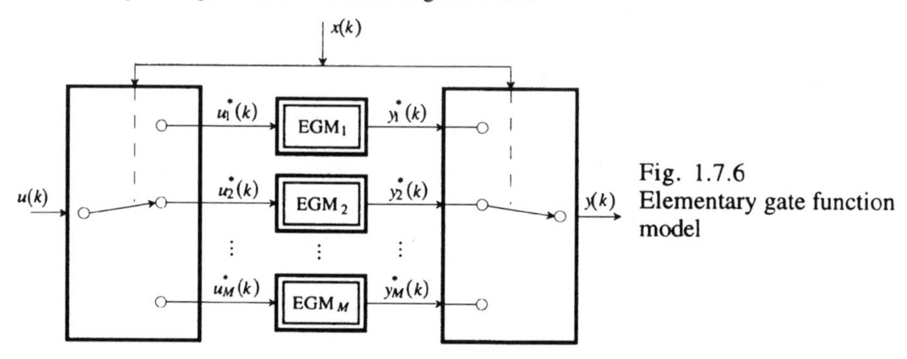

Fig. 1.7.6
Elementary gate function model

The signal by which the models are selected is usually the value of the working point (output signal). Some processes are not symmetrical to their input signals. They are direction dependent, as well.

Linear multi-models are frequently used in the practice. The reasons therefore are as follows:

1. The submodels can be described by few parameters, which is a great advantage mainly with adaptive control when the parameters have to be estimated recursively;
2. Control methods elaborated for linear systems can be applied.

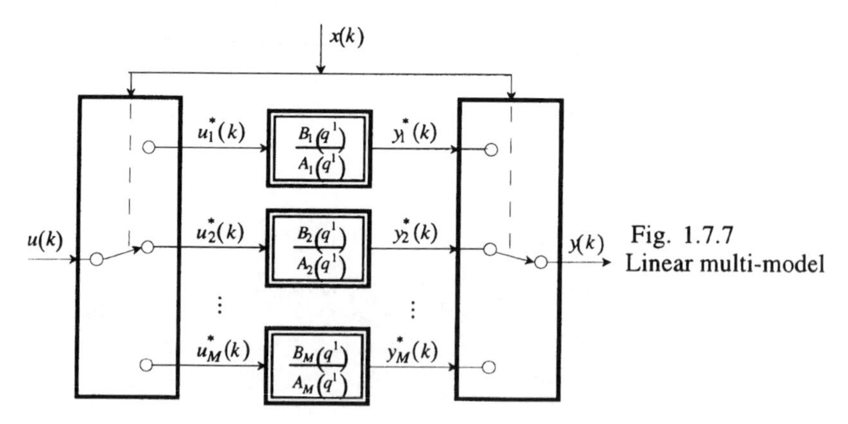

Fig. 1.7.7
Linear multi-model

The linear multi-models and the quasi-linear models with signal dependent parameters are related to each other. With quasi-linear models the parameters of the same structure depend analytically on a signal, say on the working point. With the multi-models, however, there are some diverse working points and in each of them there are linear(ized) models valid with perhaps different structures.

As it has already been mentioned with the elementary gate model, an alternative way to build the output of the globally valid model is to calculate the weighted sum of the outputs of the locally valid submodels (Johansen and Foss, 1993; Nelles *et al.*, 1997).

A typical case for the difference of the two classes of models is when the parameters are direction dependent. In this case we have two submodels, and the parameters are not functions of a continuously changing signal. On the other hand, many multi-models with the same structure are possibly modelable by a quasi-linear model with signal dependent parameters. This can be checked by regressing the parameters of the model against the signal the dependence on which is assumed.

Some applications from the references are listed now:

1. *railway vehicle* (Broersen, 1973);
2. *electrical motor–generator system* (Diekmann and Unbehauen, 1985);
3. *air transport* (Diekmann and Unbehauen, 1985);
4. *air pressure turbine driving a synchronous generator* (Jedner and Unbehauen, 1986);
5. *closed circuit ball grinding cement mill* (Keviczky, 1976);
6. *superheater in thermal power plant* (Mien and Normand–Cyrot, 1984);
7. *distillation column* (Moczek *et al.*, 1965; Weigand *et al.*, 1972; Karim and Stainthorp, 1979);
8. *model of the human pupil system* (Maletinsky, 1979);
9. *fermentation process* (Johansen and Foss, 1995);
10. *exothermic chemical reaction in a continuous stirred tank reactor* (Johansen and Foss, 1993);
11. *pH-process* (Pottmann *et al.*, 1993);
12. *transportation system for bulk goods* (Pickhardt,1997);
13. *heat exchanger* (Nelles *et al.*, 1997).

TABLE 1.8.1 Continuous and discontinuous nonlinearities

No	*Type*	*Characteristic*	*Derivative*
1	Continuous function	Y, U	continuous
2	Continuous function with discontinuous derivative	Y, U, *	discontinuous (* breakpoint)
3	Discontinuous function with discontinuous derivative	Y, U	discontinuous

TABLE 1.8.2 Stepwise approximation of curves

No	*Type*	*Graphical representation*	*Equation*
1	Original curve	Y, U, U_i U_{i+1}	$Y = f(U)$
2	Stepwise linear	Y, U, U_i U_{i+1}	$Y = f(U_i) + (U - U_i) \times$ $\times \frac{f(U_{i+1}) - f(U_i)}{U_{i+1} - U_i}$ if $U_i \le U < U_{i+1}$
3	Stepwise constant	Y, U, U_i U_{i+1}	$Y = f(U_i)$ if $U_i \le U < U_{i+1}$

1.8 TYPES OF NONLINEAR STATIC TERMS

In physical systems nonlinear static terms can be frequently separated from the other dynamic parts. Such systems were treated in this book as block oriented systems. Table 1.8.1 reviews the single-valued nonlinear continuous and discontinuous characteristics. A static nonlinearity is called continuous, if both the function itself and its derivative are continuous. To the discontinuous nonlinearities belong those characteristics that have either breakpoints or their derivative is a discontinuous function.

Nonlinear functions are often approximated by linear curves. This is the case if either too many parameters are needed to describe the nonlinear function or the parameter estimation methods cannot handle the nonlinear functional relationship. Two common approximations are the stepwise linear approximation and stepwise constant approximation (Table 1.8.2).

A nonlinear relation is called single-valued if there is an unambiguous relationship between the input and the output signals. From the fact that the output is an unambiguous function of the input it does not follow that the input is also an unambiguous function of the output signal. We can distinguish the following characteristics:

- single-valued function with single-valued inverse function;
- single-valued function with double-valued or multi-valued inverse function;
- multi-valued function with single-valued inverse function;
- multi-valued function with multi-valued inverse function.

They are summarized in Table 1.8.3. The terminology single- or multi-valued depends also on the interval of the possible working points. A function can be single-valued in a given range but multi-valued globally.

TABLE 1.8.3 Single- and not single-valued nonlinearities

No	*Name*	*Function*	*Inverse function*
1	Single-valued with single-valued inverse	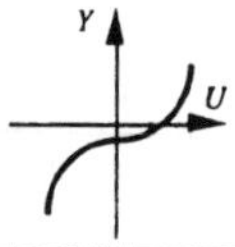 Y U	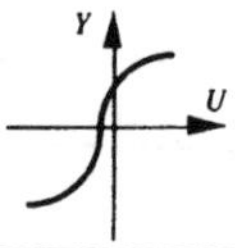 Y U
2	Single-valued with double-valued inverse	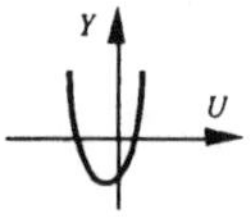 Y U	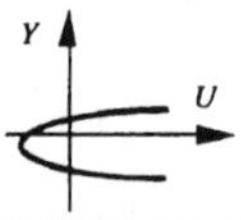 Y U
3	Double-valued with single-valued inverse	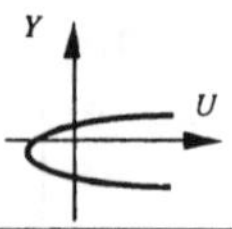 Y U	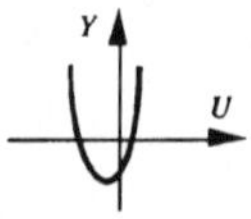 Y U
4	Double-valued with double-valued inverse	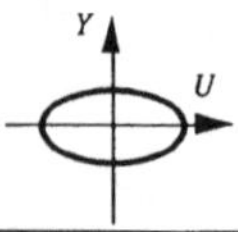 Y U	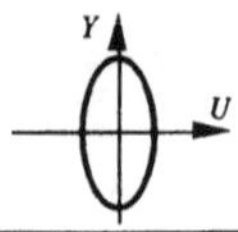 Y U

TABLE 1.8.4 Characteristics with and without hysteresis

No	Name	Characteristics	Application
1	Linear	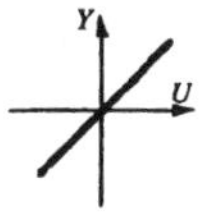	
2	Backlash (hysteresis)	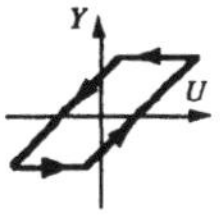	mechanical gear U: position of the driving gear Y: position of the driven gear
3	Saturation	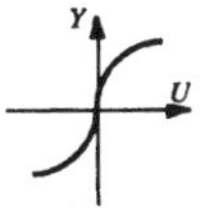	
4	Hysteresis	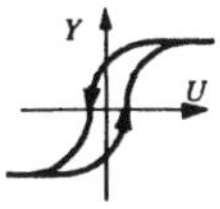	magnetization curve U: current Y: flux

Example 1.8.1 *Simple Hammerstein model with quadratic steady state characteristics*

Assume that the parameters of the static nonlinear and the linear dynamic parts are estimated iteratively, which means that the static part is estimated from the input signal $u(k)$ and the computed inner signal $v(k)$, and the linear dynamic term is estimated from the computed inner variable $v(k)$ and the process output signal $y(k)$. The inner variable can be computed from both the input and the output signals, and the identification is executable.

Example 1.8.2 *Simple Wiener model with quadratic steady state characteristics*

Assume that the parameters of the static nonlinear and the linear dynamic parts are estimated iteratively, which means that the linear dynamic term is estimated from the input signal $u(k)$ and the computed inner signal $v(k)$, and the static nonlinear part is estimated from the computed inner variable $v(k)$ and the process output signal $y(k)$. Assume that the excitation is such that the inverse of the static nonlinear function is two-valued. The inner variable can be computed from the input signal but cannot be computed unambiguously from the output signals and the identification is not executable in this way. ■

A typical double-valued nonlinearity is hysteresis (Table 1.8.4). The relation between the current and the magnetic flux of an electromagnet depends on the direction of the change in the current. In mechanical gears the position of the driven gear is behind the position of the driving gear because of the backlash between the cogs of the wheels. The only difference between the two characteristics apart from the hysteresis is that the magnetic curve is nonlinear (with saturation) and the backlash in gears has a linear function (with saturation).

Some typical discontinuous nonlinearities are presented in Table 1.8.5. Common to the nonlinearities is that the characteristics consist of straight lines and have breakpoints:

- *Variable gain*: The slope changes at each breakpoint;
- *Saturation*: The output is bounded between a low and a high limit;
- *Threshold*: The output is zero till the input is greater than a threshold value;
- *Dead zone*: The output is zero if the input is in a range;
- *Preload*: If the input is positive then the output signal is equal to the input signal plus a constant bias. If the input is negative then the output signal is equal to the input signal minus a constant bias;
- *Absolute value*: The output is equal for the same positive and negative input values.

TABLE 1.8.5
Some typical discontinuous static nonlinearities

No	*Name*	*Characteristics*	*Occurance*
1	Variable gain	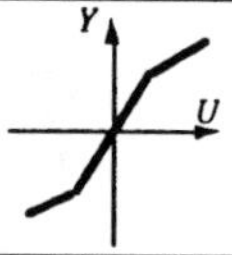 Y, U	PID controller
2	Saturation	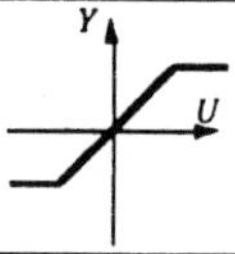 Y, U	actuator
3	Threshold	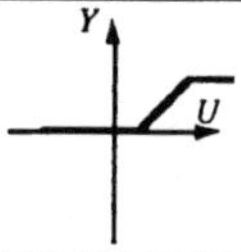 Y, U	physiological system
4	Dead zone	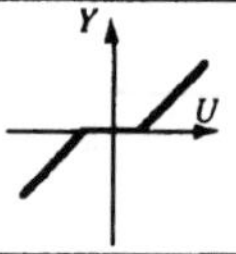 Y, U	actuator
5	Preload	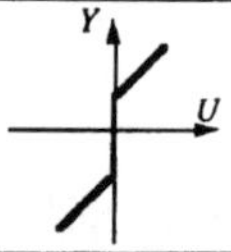 Y, U	compensation of dead zone
6	Absolute value	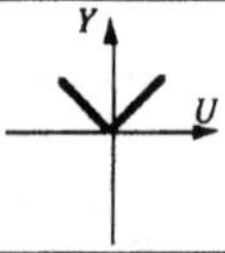 Y, U	

On–off controls have relay characteristics. Table 1.8.6 (*No* 1) and 1.8.6 (*No* 2) show the two-valued relay characteristics without and with hysteresis. Such terms are used, e.g., in temperature or in level controls. Relays with three-values are used in extruders where both heating and cooling are needed. The third state is no intervention. Such a relationship is also called on–off relay characteristics with a dead zone (Table 1.8.6 (*No* 3)). To spare the actuator the same relay characteristic is supplied with a hysteresis (Table 1.8.6 (*No* 4)).

No	*Name*	*Characteristics*	*Application*
1	On-off	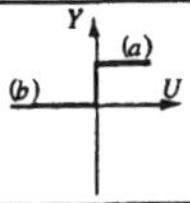 Y, U, (a), (b)	temperature control (a) heating (b) no action
2	On-off with hysteresis	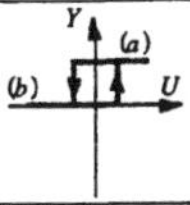 Y, U, (a), (b)	temperature control (a) heating (b) no action
3	On-off with dead zone	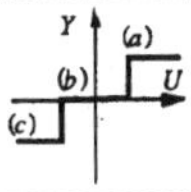 Y, U, (a), (b), (c)	temperature control (a) heating (b) no action (c) cooling
4	On-off with dead zone and hysteresis	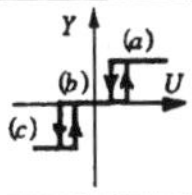 Y, U, (a), (b), (c)	temperature control (a) heating (b) no action (c) cooling

TABLE 1.8.6 Relay characteristics

No	*Name*	*Characteristics*	*Application*
1	Coulomb	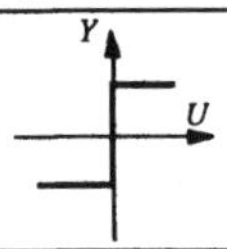 Y, U	between two solid masses
2	Viscous	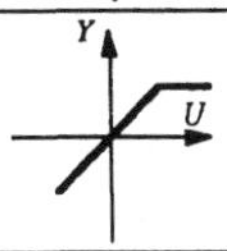 Y, U	laminar flow
3	Velocity squared	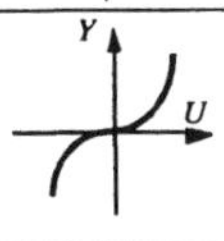 Y, U	turbulent flow
4	Sticking	 Y, U	to get a mass into motion
5	A possible combination	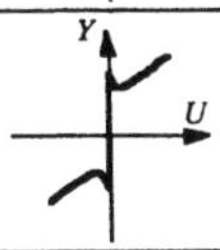 Y, U	

TABLE 1.8.7
Different friction relationships
(U: velocity, Y: frictional force)

Friction forces have different functions. The Coulomb force occurs between two solid masses, acts against the moving force and is independent of the velocity (Table 1.8.7 (*No* 1)). Viscous force occurs with laminar flow in gases and fluids and is linearly proportional to the speed (Table 1.8.7 (*No* 2)). The force is proportional to the square of the speed in a turbulent flow (Table 1.8.7 (*No* 3)). A sticking is a force that appears only at zero speed and tries to hinder the start of a movement between two solid masses (Table 1.8.7 (*No* 4)). A possible combination of the above forces is seen in Table 1.8.7 (*No* 5). All friction relationships are odd functions.

The following books give good survey on the static nonlinearities: Atherton (1982), Dransfield (1968), Laspe and Stout.

1.9 REFERENCES

Agarwal, M. and D.E. Seborg (1985). Self-tuning controllers for nonlinear systems. *Prepr. IFAC Workshop on Adaptive Control in Chemical Engineering*, (Frankfurt: F.R.G).

Agarwal, M. and D.E. Seborg (1986). A self-tuning controller for MIMO nonlinear systems. *Prepr. 2nd IFAC Workshop on Adaptive Systems in Control and Signal Processing*, (Lund:Sweden), pp. 393–398.

Ahmed, N.U. (1970). Closure and completeness of Wiener's orthogonal set $\{G_n\}$ in the class $L^2(\Omega, B, \mu)$ and its application to stochastic hereditary differential systems, *Information and Control*, Vol. 17, pp. 161–174.

Alper, P. (1965). A consideration of the discrete Volterra series. *IEEE Trans. on Automatic Control*, AC-10, pp. 322–327.

Amorocho, J. and W.E. Hart (1964). A critique of current methods in hydrologic systems investigation. *Trans. American Geophysical Union*, Vol. 45, 2, pp. 307-321.

Atherton, D.P. (1982). *Nonlinear Control Engineering*. Van Nostrand Reinhold Company, (London: UK).

Bamberger, W. (1978). Methods for on-line optimization of the steady state behavior of nonlinear dynamical processes (in German). *PDV-Bericht, KfK-PDV 159*, 173 p.

Bamberger, W. R. and R. Isermann (1976). On-line optimization of the steady state behavior of nonlinear dynamical processes (in German). *Prepr. 21. Int. Wissenschaftliches Kolloquium*, (Ilmenau: GDR), pp. 67–70.

Bamberger, W. and R. Isermann (1977). Adaptive on-line optimization of the steady state behavior of slow dynamic processes with process computers. *Prepr. 5th IFAC Symposium on Digital Computer Applications to Process Control*, pp. 543–550.

Bamberger, W. and R. Isermann (1978). Adaptive on-line steady state optimization of slow dynamic processes. *Automatica*, Vol. 14, pp. 223–230.

Bard, Y. and L. Lapidus (1970). Nonlinear system identification. *Ind. Eng. Chem. Fundam.*, Vol. 9, 4, pp. 628–633.

Barker, H.A. (1969). Synchronous sampling theorem for nonlinear systems. *Electronics Letters*, Vol. 5, 25, pp. 657.

Barrett, J.F. (1963). The use of functionals in the analysis of nonlinear physical systems, *Journal of Electronics and Control*, Vol. 15, pp. 567–615.

Baumgarten, S.L. and W.S. Rugh (1975). Complete identification of a class of nonlinear systems from steady state frequency response, *IEEE Trans. on Circuits and Systems*, Vol. 22, pp. 753–759.

Beghelli, S. and R. Guidorzi (1976). Bilinear systems identification from input–output sequences. *Prepr. 4th IFAC Symp. on Identification and System Parameter Estimation*, (Tbilisi: USSR), pp. 360–370.

Billings, S.A. (1980). Identification of nonlinear systems – A survey. *Proc. IEE*, Part D, Vol. 127, 6, pp. 272–285.

Billings, S.A. and M.B. Fadzil (1985).The practical identification of systems with nonlinearities. *Prepr. 7th IFAC/IFORS Symp. on Identification and System Parameter Estimation*, (York: UK), pp. 155–160.

Billings, S.A. and S.Y. Fakhouri, (1979a). Identification of S_m systems. *Int. Journal of Systems Science*, Vol. 10, 12, pp. 1401–1408.

Billings, S.A. and S.Y. Fakhouri (1979b). Identification of factorable Volterra systems, *Proc. IEE*, Part D, Vol. 126, 10, pp. 1018–1024.

Billings, S.A. and W.S.F. Voon (1987). Piecewise linear identification of nonlinear systems. *Int. Journal of Control*, Vol. 46, 1, pp. 215–235.

Bose, A.G. (1959). Nonlinear system characterization and optimization. *IRE Trans. on Circuit Theory*, Vol. 6 (spec. Suppl.), pp. 30–40.

Broersen, P.M.T. (1973). Estimation of multivariable railway vehicle dynamics from normal operating data. *Prepr. 3rd IFAC Symp. on Identification and System Parameter Estimation*, (Hague: The Netherlands), pp. 425–433.

Chantre, P. (1987). Identification using a fast to compute Volterra type model applied to a nonlinear industrial process. *Prepr. Int. Conf. on Industrial and Applied Mathematics*, (Paris: France).

Chantre, P. (1989). Nonlinear predictive control by means of the Volterra model (in French). *Prepr. IASTED Symp. on Modeling, Identification and Control*, (Grindewald: Switzerland), pp. 136–139.

Chen, S. and S.A. Billings (1989). Representations of nonlinear systems: the NARMAX model. *Int.. Journal of Control*, Vol. 49, 3, pp. 1013–1032.

Chen, S., S.A. Billings and W. Luo (1989). Orthogonal least squares methods and their applications to nonlinear system identification. *Int. Journal of Control*, Vol. 50, 5, pp. 1873–1896.

Corlis, R.G. and R. Luus (1969). Use of residuals in the identification and control of two-input, single-output

systems. *I & EC Fundamentals*, Vol. 8, 5, pp. 246–253.
DeLorenzo, F., G. Guardabassi, A. Locatelli, and S. Rinaldi (1972). On the asymmetric behavior of distillation systems. *Chemical Engineering Science*, Vol. 27, pp. 1211–1221.
Diaz, H. (1986). Modeling of nonlinear systems from input–output data. *Ph.D. Thesis*, Rensselaer Polytechnic Institute, (Troy: N.Y., USA).
Diaz, H. and A.A. Desrochers (1988). Modeling of nonlinear discrete time systems from input–output data. *Automatica*, Vol. 24, 5, pp. 629–641.
Diekmann, K. and H., Unbehauen (1985). On-line parameter estimation in a class of nonlinear systems via modified least squares and instrumental variable algorithms. *Prepr. 7th IFAC Symp. on Identification and System Parameter Estimation*, (York: UK), pp. 149–153.
Diskin, M.H. and A. Boneh (1973). Determination of optimal kernels for second-order stationary surface run–off systems. *Water Resources Research*, Vol. 9, pp. 311-325.
Dochain, D. and G. Bastin (1984). Adaptive identification and control algorithms for nonlinear bacterial growth system, *Automatica*, Vol. 20, 5, pp. 621–634.
Dransfield, P. (1968). *Engineering Systems and Automatic Control*, Prentice–Hall, (Englewood Cliffs: New Jersey, USA).
Economakos, E. (1971). Identification of groups of internal signals of zero memory nonlinear systems. *Electronics Letters*, Vol. 7, 4, pp. 99–100.
Eykhoff, P. (1974). *System Identification – Parameter and State Estimation*. John Wiley and Sons, (London: UK).
Fliess, M. and D. Normand–Cyrot (1982). On the approximation of nonlinear systems by some simple state space models. *Prepr. 6th IFAC Symp. on Identification and System Parameter Estimation*, (Washington D.C.:USA).
Franck, G. and H. Rake (1985). Identification of large water heated cross flow heat exchanger with binary multifrequency signals. *Prepr. 7th IFAC/IFORS Symp. on Identification and System Parameter Estimation,* (York: UK), pp. 1859–1863.
Frechet, M. (1910). About continuous functionals (in French). *Ann. Ec. Norm. Sup.*, Vol. 27, pp. 193–219.
Fronza, G., G. Guardabassi, A. Locatelli and S. Rinaldi (1969). A simple nonlinear model for binary distillation columns. *Report, Laboratorio Controlli Automatici, Instituto di Elettrotecnica ed Elettronica*, Polytecnico di Milano.
Gallman, P.G. (1971). Identification of nonlinear dynamical systems using the Uryson operator . *Ph.D. Thesis, Yale University*, (New Haven: Conn., USA).
Gallman, P.G. and K.S. Narendra (1976). Representations of nonlinear systems via the Stone–Weierstrass theorem. *Automatica,* Vol. 12, pp. 619–622.
Gardiner, A.B. (1973). Identification of processes containing single-valued nonlinearities. *Int. Journal of Control,* Vol. 18, 5, pp. 1029–1039.
Gilles, G. and N. Laggoune (1985a). Digital control of bilinear continuous processes. *Prepr. 7th IFAC Conf. on Digital Computer Applications to Process Control,* (Vienna: Austria) pp. 119–124.
Gilles, G. and N. Laggoune (1985b). Linearizing control of a class of nonlinear continuous processes. *Proc. IFAC/IFORS Conference on Control Science and Technology for Development,* (Beijing: China), pp. 981–992.
Godfrey, K.R. and P.A.N. Briggs (1973). The identification of processes with direction-dependent dynamic responses, *Prepr. 3rd IFAC Symp. on Identification and System Parameter Estimation,* (Hague: The Netherlands), pp. 809–818.
Godfrey, K.R. and D.J. Moore (1974). Identification of processes having direction dependent response with gas turbine engine applications. *Automatica*, Vol. 10, pp. 469–481.
Goodwin, G.C. and K.S. Sin (1984). *Adaptive Filtering, Prediction and Control.* Prentice Hall Inc., (Englewood Cliffs: UK).
Goodwin, G.C., B. McInnis and R.S. Long (1982). Adaptive control algorithms for waste water treatment and pH neutralization. *Optimal Control Applications and Methods,* Vol. 3, pp. 443–459.
Grabbe, E.M., S. Ramo and D.E. Woddridge (1958). *Handbook of Automation, Computation and Control,* Wiley, (London: UK).
Haber, R. (1979a). Parametric identification of nonlinear dynamic systems. *Prepr. 5th IFAC Symp. on Identification and System Parameter Estimation,* (Darmstadt: FRG), pp. 515–522.
Haber, R. (1979b). Identification of nonlinear dynamic systems having signal dependent parameters. *Report, Center for Control Sciences, University of Minnesota,* (Minneapolis: USA), 226 p.
Haber, R. (1982). Comparison of different nonlinear dynamic models at the identification of a flood process. *Int. Journal of Systems Science,* Vol. 8, 4, pp. 95–101.
Haber, R. and L. Keviczky (1974). Nonlinear structures for system identification. *Periodica Polytechnica*, Electrical Engineering, Vol. 18, 4, pp. 393–414.
Haber, R. and L. Keviczky (1976). Identification of nonlinear dynamic systems – Survey paper. *Prepr. 4th IFAC Symposium on Identification and System Parameter Estimation.* (Tbilisi: USSR), pp. 62–112.
Haber, R. and L. Keviczky (1979). Parametric description of dynamic systems having signal dependent parameters. *Prepr. Joint American Control Conference*, (Denver: USA), pp. 681– 686.
Haber, R. and L. Keviczky (1985). Identification of "linear" systems having signal dependent parameters. *Int. Journal of Systems Science*, Vol. 16, 7, pp. 869–884.
Haber, R. and H. Unbehauen (1990). Structure identification of nonlinear dynamic systems – A survey on input/output approaches. *Automatica,* Vol. 26, 4, pp. 651-677.
Haber, R. and I. Wernstedt (1979). New nonlinear dynamic models for the simulation of electrically excited

biological membrane. *Int. Juornal of Systems Science*, Vol. 5, 2, pp. 227–233.
Haber, R. and R. Zierfuss (1991). Identification of nonlinear models between the reflux flow and the top temperature of a distillation column, *Prepr. 9th IFAC/IFORS Symp. on Identification and System Parameter Estimation*, (Budapest: Hungary), pp. 486–491.
Haber, R., L. Keviczky and M. Hilger (1986). Modeling and identification of a cement mill by a "linear" model with signal dependent parameters. *Prepr. IMACS/IFAC Symposium on Modeling and Simulation for Control of Lumped and Distributed Parameter Systems*, (Lille: France), pp. 273–276.
Hammerstein, A. (1930). Nonlinear integral equation and its application (in German). *Acta Math.*, Vol . 54, pp. 117–176.
Harper, T.R. and W.J. Rugh (1976). Structural features of factorable Volterra systems. *IEEE Trans. on Automatic Control*, Vol. AC-21, 6, pp. 822–832.
Harris, G.H. and L. Lapidus (1967a). The identification of nonlinear reaction systems with two-level inputs. *A.I.Ch.E. Journal*, Vol. 13, 2, pp. 291–302.
Harris, G.H. and L. Lapidus (1967b). The identification of nonlinear systems. *Industrial and Engineering Chemistry*, Vol. 59, 6, 2, pp. 67–81.
Hoffmeyer–Zlotnik, H.-J., P. Otto, A. Seifert, H. Heym and H. Luckert (1979). Application of experimental models for automatic control and prediction of water quantities in the central range of mountains. (in German). *Preprints 24th Int. Scientific Symposum*, (Ilmenau: GDR), pp. 173–177.
Hu, K.P. and Z.D. Yuan (1988). Identification of streamflow processes. *Prepr. 8th IFAC/IFORS Symp. on Identification and System Parameter Estimation*, (Beijing: China), pp. 1777–1781.
Hubbell, P.G. (1968). Identification of a class of nonlinear systems. *Prepr. 2nd Asilomar Conf. Circuits & Systems*, Pacific Grove, (California: USA), pp. 511–514.
Hunter, I.W. (1985a). Nonlinearities in isolated frog muscle fiber mechanical dynamics. *Soc. Neurosci. Abstr.*, Vol. 11, pp. 406.
Hunter, I.W. (1985b). Frog muscle fiber dynamic stiffness determined using nonlinear system identification techniques. *Biophys. Journal*, Vol. 47, pp. 287.
Isermann, R. (1987). *Digital Control Systems* (in German), Springer Verlag, (Berlin: FRG).
Jedner, U. and H. Unbehauen (1986). Identification of a class of nonlinear systems by parameter estimation of a linear multi-model. *Prepr. IMACS/IFAC Symp. on Modeling and Simulation for Control of Lumped and Distributed Parameter Systems*, (Lille: France), pp. 287–290.
Johansen, T.A. and B.A. Foss (1993). Constructing NARMAX models using ARMAX models. *Int. Journal of Control*, Vol. 58, 5, pp. 1125–1153.
Johansen, T.A. and B.A. Foss (1995). Semi-empirical modeling of nonlinear dynamic systems through identification of operating regimes and local models. *Modeling, Identification and Control*, Vol. 16, 4, pp. 213–232.
Kaminskas, V., K. Sidlauskas and C. Tallat–Kelpsa (1988). Self-tuning minimum variance control of nonlinear Wiener–Hammerstein type systems. *Prepr. 8th IFAC/IFORS Symp. on Identification and System Parameter Estimation*, (Beijing: China), pp. 384–389.
Karim, N.N. and F.P. Stainthorp (1979). Asymmetric dynamics and feed forward control of a fractionating column. *Proc. Joint Automatic Control Conference*, (Denver: USA), pp. 825– 828.
Keviczky, L. (1976). Nonlinear dynamic identification of a cement mill to be optimized. *Prepr.4th IFAC Symp. on Identification and System Parameter Estimation*, (Tbilisi: USSR), pp. 388–396.
Keviczky, L. (1977). On the equivalence of discrete and continuous transfer functions. *Problems of Control and Information Theory*, Vol. 6, 2, pp. 111–128.
King, R.E. and P.N. Paraskevopoulos (1979). Parametric identification of discrete time SISO systems. *Int. Journal of Control*, Vol. 30, 6, pp. 1023–1029.
Kitajima, T. and K. Hara (1987). Bilinear model of the nerve cell. *Prepr. 10th IFAC World Congress*, (Munich: FRG), Vol. 5, pp. 103–108.
Klein, S. and S. Yasui (1979). Nonlinear system analysis with non-Gaussian white stimuli: general basis functional and kernels. *IEEE Trans. on Information Theory*, IT-25, 4, pp. 495–501.
Kortmann, M. and H. Unbehauen (1987). New algorithms for automatic selection of optimal model structure in the identification of nonlinear systems (in German). *Automatisierungstechnik*, Vol. 35, 12, pp. 491–498.
Kortmann, M. and H. Unbehauen (1988a). Structure detection in the identification of nonlinear systems. *Automatique Productique Informatique Indiustrielle*, Vol. 22, pp. 5-25.
Kortmann, M. and H. Unbehauen (1988b). Two algorithms for model structure determination of nonlinear dynamic systems with applications to industrial processes, *Prepr. 8th IFAC/IFORS Symp. on Identification and System Parameter Estimation*, (Beijing: China), pp. 939–946.
Kortmann, M., K. Janiszowski and H. Unbehauen (1988). Application and comparison of different identification schemes under industrial conditions. *Int. Journal of Control*, Vol. 48, 6, pp. 2275–2296.
Lachmann, K.H. (1983). Parameter adaptive control algorithms for a certain class of nonlinear processes with single-valued nonlinearities (in German). *VDI-Bericht*, Series 8, N° 66.
Lachmann, K.H. (1985a). Self-tuning nonlinear control algorithms for a certain class of nonlinear processes (in German). *Automatisierungstechnik*, Vol. 33, 7, pp. 210-218.
Lachmann, K.H. (1985b). Control of different nonlinear processes with nonlinear parameter adaptive control methods (in German). Part 1 and 2. *Automatisierungstechnik*, 1: Vol. 33, 9, pp. 280–284; 2: Vol. 33, 10, pp. 318–321.
Lachmann, K.H. and W. Goedecke (1982). A Parameter adaptive controller for nonlinear processes (in German). *Regelungstechnik*, Vol. 30, 6, pp. 197–206.

\Lammers, H.C., H.B. Verbruggen and E. de Boer (1979).An identification method for a combined Wiener–Hammerstein filter describing the encoding part of the Cochlear system. *Prepr. 5th IFAC Symp. on Identification and System Parameter Estimation,* (Darmstadt: FRG), pp. 485–491.
Laspe, C.G. and T. M. Stout (manuscript). Nonlinearities. In: *Chemical Process Control Systems.*
Li, Y. and S. Chen (1988). Identification of model for a cement rotary kiln. *Prepr. 8th IFAC/IFORS Symp. on Identification and System Parameter Estimation,* (Beijing: China), pp. 711–716.
Leontaritis, I.J. and S.A. Billings (1985a). Input–output parametric models for non-linear systems. Part I: Deterministic non-linear systems. *Int. Journal of Control,* Vol. 41, 2, pp. 303–328.
Leontaritis, I.J. and S.A. Billings (1985b).Input-output parametric models for non-linear systems. Part II: Stochastic non-linear systems. *Int. Journal of Control,* Vol. 41, 2, pp. 329–341.
Luyben, W.L. (1973). *Process Modeling, Simulation and Control for Chemical Engineers,* McGraw–Hill, (New York: USA).
Maletinsky, V. (1979). Modeling an identification of the pupillary light reflex system. *Proc. IEEE Conf. on Decision and Control,* (Florida: USA), pp. 526– 527.
Marmarelis, P.Z. and K.-I. Naka (1972). White noise analysis of neuron chain: An application of the Wiener theory, *Science*, Vol. 175, 3, pp. 1276–1278.
Marmarelis, P.Z. and K.-I. Naka (1974.).Identification of multi-input biological systems, *IEEE Trans. on Biomedical Engineering*, BME-21, 2, pp. 88–101.
Marmarelis, V.Z. (1979). Error analysis and optimal estimation procedures in identification of nonlinear Volterra systems. *Automatica*, Vol. 15, pp. 161–174.
Marmarelis, P.Z. and V.Z. Marmarelis (1978). *Analysis of Physiological Systems. The White Noise Approach* . Plenum Press, (New York: USA).
McInnis, G.C., L.-z Deng and R. Vogt (1985). Adaptive pole assignment control of blood pressures using bilinear modes. *Prepr. 7th IFAC Symp. on Identification and System Parameter Estimation,* (York: UK), pp. 1209–1211.
Melton, R.B. and J.P. Kroeker (1982) Wiener functionals for an N-level uniformly distributed discrete random process. *Proc. of 6th IFAC Symp. on Identification and System Parameter Estimation,* (Washington D.C.: USA), pp. 1307–1312.
Mien, H.D.V. and D. Normand–Cyrot (1984). Nonlinear state affine identification methods: Applications to electric power plants. *Automatica,* Vol. 20, 2, pp. 175-188.
Miller, R. W. and R. Roy (1964). Nonlinear process identification using decision theory. *IEEE Trans . on Automatic Control,* Vol. 9, 10, pp. 538–540.
Mizuno, H., Y. Watanabe, Y. Nishimura and M. Matsubara (1972). Asymmetric properties of continuous distillation column dynamics. *Chemical Engineering Science*, Vol. 27, pp. 129–136.
Moczek, J.S., R.E. Otto and T.J. Williams (1965). Approximation models for the dynamic response of large distillation columns, *Chem. Eng. Progr. Symp.Series,* Vol. 61, pp. 432/1–432/9.
Moss, G.C. (manuscript). Application of statistical correlation techniques to gas analysis by chromatography.
Nelles, O., O. Hecker and R. Isermann (1997). Automatic model selection in local linear model trees (LOLIMOT) for nonlinear system identification of a transport delay process. *Prepr. 12th IFAC Symposium on System Identification,* (Fukuoka: Japan), Vol. 2, pp. 727–732.
Pajunen, G.A. (1985a). Identification and adaptive control of Wiener type nonlinear processes. *Prepr. 7th IFAC Conf. on Digital Computer Applications to Process Control,* (Vienna: Austria), pp. 559–606.
Pajunen, G.A. (1985b). Recursive identification of Wiener type nonlinear systems. *Prepr. American Control Conference,* (USA), pp. 1365–1370.
Parker, G.A. and E.L. Moore (1982). Practical nonlinear system identification using a modified Volterra series approach. *Automatica*, Vol. 18, 1, pp. 85–91.
Pickhardt, R. (1997). Adaptive control using a multi-model approach to a transportation system for bulk goods (in German), *Automatisierungstechnik*, Vol. 45, 3, pp. 113–120.
Pincock D.G. and D.P. Atherton (1973) The identification of a two phase induction machine with SCR speed control. *Prepr. 3rd IFAC Symp. on Identification and System Parameter Estimation,* (Hague: The Netherlands), pp. 505–512.
Pottmann, M., H. Unbehauen and D.E. Seborg (1993). Application of a general multi-model approach for identification of highly nonlinear processes. – A case study. *Int. Journal of Control*, Vol. 57, 1, pp. 97–120.
Proudfoot, C.G., D.W. Clarke, O.L.R. Jacobs and P.S. Tuffs (1985). Comparative study of self-tuning controllers regulating pH in an industrial process. *CONTROL'85 Conference,* (Cambridge: UK), pp. 359–363.
Proudfoot, C.G., P.J. Gawthrop, D. Phil and O.L.R. Jacobs (1983). Self-tuning PI control of a pH neutralization process. *IEEE Proc.*, Part D., Vol. 130, 5, pp. 267-272.
Rault, A., J. Richalet, A. Barbot and J.P. Sergenton (manuscript). Identification and modeling of a jet engine.
Roy, R. J. and P. M. DeRusso (1962). A digital orthogonal model for nonlinear processes with two level inputs. *IRA Transactions on Automatic Control*, October, pp. 93–101.
Rugh, W.J. (1981). *Nonlinear System Theory. The Volterra/Wiener Approach*. The John Hopkins University Press, (Baltimore: USA).
Schetzen, M. (1965a). Synthesis of a class of nonlinear systems. *Int. Journal of Control,* 1, pp. 251–263.
Schetzen, M. (1965b). Determination of optimum nonlinear systems for generalized criteria based on the use of gate functions. *IEEE Trans. on Information Theory,* IT-11, pp. 117–125.
Schetzen, M. (1980). *The Volterra and Wiener Theories of Nonlinear Systems.* John Wiley & Sons, (New

York: USA).
Simpson, H.R. (1965). A sampled data nonlinear filter. *Proc. IEE*, Vol. 112, 6, pp. 1187–1196.
Sontag, E.D. (1979a). Realization theory of discrete time nonlinear systems: Part I – The bounded case. *IEEE Trans. on Circuits and Systems*, CAS-26, 4, pp. 342–356.
Sontag, E.D. (1979b). *Polynomial Response Maps*. Springer, (Berlin: FRG).
Spekreijse, H. (1969). Rectification in the Goldfish retina: Analysis by sinusoidal and auxiliary stimulation. *Vision Res.*, Vol. 9, pp. 1461–1472.
Spekreijse, H. and J. Oosting (1970). Linearizing: a method for analyzing and synthesizing nonlinear systems. *Kybernetik*, Vol. 7, pp. 23–31.
Stark, L. (1969). The pupillary control system: Its nonlinear adaptive and stochastic engineering design characteristics. *Automatica*, Vol. 5, pp. 655–676.
Taylor, L.W. and A.V. Balakrishnan (1967). Identification of human response models in manual control systems. *Prepr. 1st IFAC Symp. on Identification and Parameter Estimation*, (Prague: Czechoslovakia), 1.7/1–8.
Thathachar, M.A.L. and S. Ramaswamy (1973). Identification of a class of nonlinear systems. *Int. Journal of Control*, Vol. 18, 4, pp. 741–752.
Tuis, L. (1976). Identification of nonlinear systems by means of multilevel pseudo-random signals applied to a water–turbine unit. *Prepr. 4th IFAC Symp. on Identification and Parameter Estimation*, (Tbilisi: USSR), pp. 569–579.
Tuschák, R., I. Bézi, G. Tevesz, J. Hetthéssy and R. Haber (1982). Practical experiences on the setup and identification of a distillation pilot plant. *Prepr. 6th IFAC Symp.on Identification and System Parameter Estimation*, (Washington D.C.: USA), pp. 663–668.
Velev, K. (1985). Interactive software package for identification of parametrically dependent processes. *Prepr. 30th Int. Scientific Sysmposium*, Technical University, (Ilmenau: GDR), pp. 351– 354.
Volterra, V. (1959). *Theory of Functionals and of Integral and Integro Differential Equations*. Dover, (New York: USA).
Vuchkov, I.N., K.D. Velev and V.K. Tsochev. (1985). Identification of parametrically dependent processes with applications to chemical technology. *Prepr. 7th IFAC/IFORS Symp. on Identification and System Parameter Estimation*, (York: UK), pp. 1089–1093.
Weigand, W.A., A.K. Jhawar and T.J. Williams (1972). Calculation method for the response time to step inputs for approximate dynamic models of distillation columns. *AIChE Journal*, Vol. 18, 6, pp. 1243–1252.
Weiss, T.F. (1966). A model of the peripheral auditory system. *Kybernetik*, Vol. 3, pp. 153–175.
Wiener, N. (1958). *Nonlinear Problems in Random Theory*. MIT Press, (Cambridge: Mass., USA).
Wysocki, E.M. and W.J. Rugh (1976). Further results on the identification problem for the class of nonlinear systems S_M. *IEEE Trans. on Circuits Systems*, Vol. 23, 11, pp. 664–670.
Yi, G., Y.B. Hwang, H.N. Chang and K.S. Lee (1989). Computer control of cell mass concentration in continuous culture. *Automatica*, Vol. 25, 2, pp. 243–249.
Zadeh, L.A. (1953a). Optimum nonlinear filters. *Journal of Applied Physics*, Vol. 24, 4, pp. 396–404.
Zadeh, L.A. (1953b). A contribution theory of nonlinear systems. *Journal of the Franklin Institute*, Vol. 255, pp. 387–408.

2. Test Signals for Identification

2.1 INTRODUCTION

With linear system identification the only assumption about the test signal is that it has to excite the process properly. This assumption relates to the time shape of the signal but not to its amplitude distribution.

In identifying linear systems the test signal has to have at least two amplitude levels. This is the case with a step function or with any binary signals. The first example shows that the application of a single unit step is not sufficient if we do not know whether the system is linear or not.

Example 2.1.1 *Identification by a unit step*
A system is excited by a unit step

$$u(t) = 1(t)$$

The system's response (Figure 2.1.1a) can be described by the formula

$$y(t) = 1 - 2\exp\left(-\frac{t}{T}\right) + \exp\left(-\frac{2t}{T}\right)$$

The linear transfer function can be obtained by dividing the Laplace transformation of the output signal by those of the input signal. The transformation of the input signal is

$$u(s) = \mathscr{L}\{u(t)\} = \frac{1}{s} \tag{2.1.1}$$

and that of the output signal is

$$y(s) = \mathscr{L}\{y(t)\} = \frac{1}{s} - \frac{2}{s + \frac{1}{T}} + \frac{1}{s + \frac{2}{T}} = \frac{1}{s(1+sT)(1+sT/2)} \tag{2.1.2}$$

The transfer function can be obtained by dividing (2.1.2) by (2.1.1)

$$G(s) = \frac{y(s)}{u(s)} = \frac{1}{(1+sT)(1+sT/2)}$$

This would lead to the conclusion that the process is a linear second-order one (Figure 2.1.1b). However, the first-order linear system followed by a square element (Figure 2.1.1c) would result in the same step response. Namely, the response of the first-order lag is

$$v(t) = 1 - \exp\left(-\frac{t}{T}\right)$$

and the square of it is

$$y(t) = v^2(t) = 1 - \exp\left(-\frac{t}{T}\right) + \exp\left(-\frac{2t}{T}\right)$$ ■

Generally a nonlinear system cannot be identified by a two-level signal. In the second example we show that there are nonlinear processes that can be identified by such a signal.

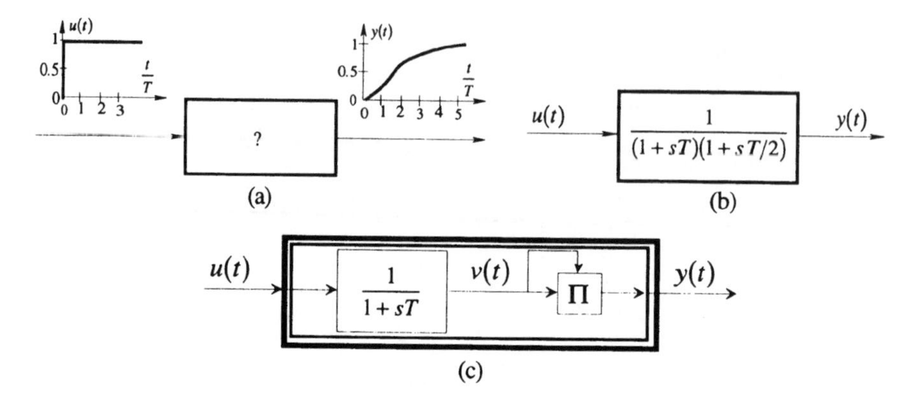

Fig. 2.1.1 Identification by a unit step in Example 2.1.1
(a) input and output signals; (b) estimated linear second-order process;
(c) estimated first-order quadratic simple Wiener model

Example 2.1.2 *Identification of the simple Wiener model by a two-level signal with the levels -2 and 2*

The scheme of the simple Wiener model is seen in Figure 2.1.2a. The transfer function of the linear part is

$$G(s) = \frac{1}{1+10s}$$

and the equation of the static polynomial is

$$y(t) = 2 + v(t) + 0.5v^2(t)$$

The sampling time of the simulation is $\Delta T = 2[\mathrm{s}]$, thus the pulse transfer function used in the discrete-time simulation is

$$G\left(z^{-1}\right) = \frac{0.1813z^{-1}}{1-0.8187z^{-1}}$$

The test signal was a pseudo-random binary signal (PRBS) with maximum length 31, levels -2 and 2. The minimum switching time of the PRBS was 5 times the sampling time, $10[\mathrm{s}]$. $N = 31 \cdot 5 = 155$ data pairs were used for the identification. A bilinear model was fitted to the sampled data by the LS method. The model equation with the estimated parameters and with their standard deviations is as follows:

$$\begin{aligned}\hat{y}(k) = {} & (0.7473 \pm 0.0073)y(k-1) + (0.5602 \pm 0.02) \\ & -(0.08237 \pm 0.01)u(k-1) + (0.1051 \pm 0.0036)u(k-1)y(k-1)\end{aligned}$$

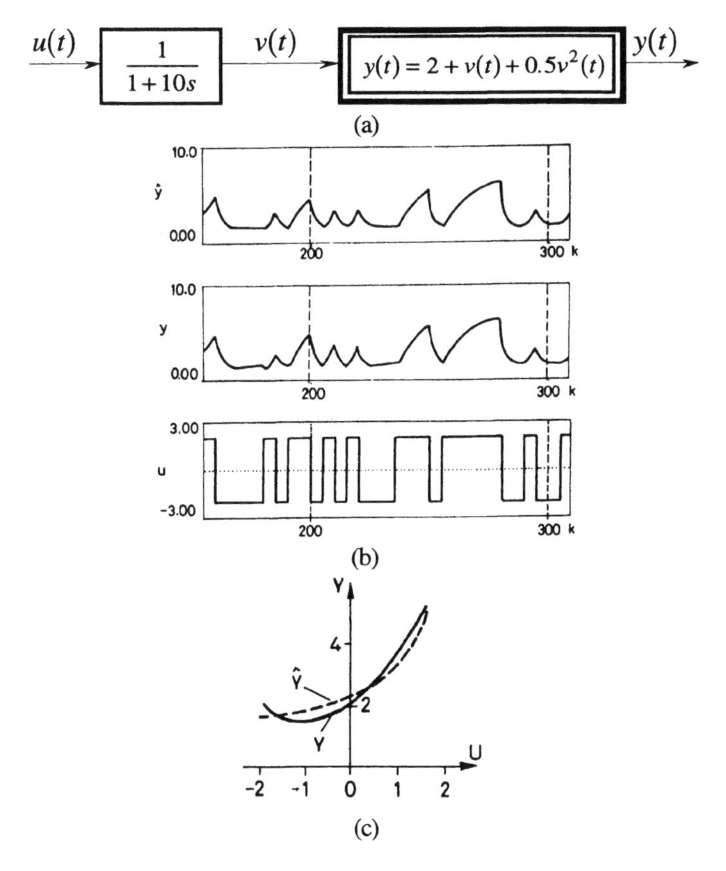

Fig. 2.1.2 Identification of the quadratic simple Wiener model using a PRBS with levels -2 and 2 in Example 2.1.2: (a) scheme of the process; (b) input (u), measured (y) and computed output $(\hat{y})$ based on the estimated model signals; (c) steady state characteristics of the simulated (Y) and of the estimated process $(\hat{Y})$

The identification was successful, as is seen from the relatively good fit between the measured output signal $y(k)$ and the computed output signal $\hat{y}(k)$ based on the estimated model (Figure 2.1.2b). The steady state characteristics of the simulated and the estimated models are also near to each other in the domain $(-2 \le U \le 2)$, as is seen in Figure 2.1.2c. The static relation of the bilinear model was computed from

$$(1-0.7473)Y = 0.5602 - 0.08237U + 0.1051UY$$

that is

$$Y = \frac{0.5602 - 0.08237U}{0.2527 - 0.1051U}$$

■

The next example shows a nonlinear dynamic process that cannot be identified by a two-level test signal.

Example 2.1.3 *Identification of the simple Hammerstein model by a two-level signal with the levels -2 and 2*

The simple Hammerstein model (Figure 2.1.3a) has the same linear dynamic and static nonlinear terms as the simple Wiener model in Example 2.1.2. Both the test signal and the sampling times were the same as in Example 2.1.2. The input and output signals are seen in Figure 2.1.3b. A perfect fit between the measured input and output signals can be achieved by a linear first-order model:

$$\hat{y}(t) = 0.8187y(k-1) + 0.7273 + 0.1813u(k-1)$$

The standard deviations of the parameters are zero. A good fit can be seen on the plot of the computed output signal $\hat{y}(k)$ based on the linear model (Figure 2.1.3b). The scheme of the estimated continuous time linear model is given in Figure 2.1.3c. The reason for that the nonlinear process cannot be identified is that the two-level test signal excites the parabola described by three parameters in only two points. With these experimental conditions the parabola can be replaced by a line crossing the parabola $V = V(U)$ in the intersections V_1 and V_2, determined by the two levels of the test signal

$$V_1 = 2 + U + 0.5U^2\Big|_{U=-2} = 2$$

$$V_2 = 2 + U + 0.5U^2\Big|_{U=2} = 6$$

The static parabola and the replacement line of the estimated linear model are also shown in Figure 2.1.3d.

Example 2.1.4 *Identification of the simple Wiener model by a two-level signal with the levels -2 and 0*

The simple Wiener model (Figure 2.1.4a) is the same as in Example 2.1.2. The levels of the test signal are -2 and 0, all other parameters of the test signal and the sampling times are the same as in Example 2.1.2. The input and output signals are seen in Figure 2.1.4b. A parametric Volterra model was fitted to the input–output measurements. The model equation with the estimated parameters and with their standard deviations is as follows:

$$\begin{aligned}\hat{y}(k) = {} & (1.479 \pm 0.019)y(k-1) - (0.5901 \pm 0.019)y(k-1) \\ & +(0.2069 \pm 0.014) + (0.09556 \pm 0.0031)u(k-1) \\ & +(0.09363 \pm 0.0031)u(k-2) + (0.09413 \pm 0.0022)u(k-1)u(k-2)\end{aligned}$$

The identification was successful as seen from the relatively good fit between the measured output signal $y(k)$ and the computed output signal $\hat{y}(k)$ based on the estimated model (Figure 2.1.4b). The steady state characteristics of the simulated and the estimated models are also near to each other in the domain $(-2 \le U \le 2)$, as seen in Figure 2.1.4c. The static relation of the parametric Volterra model was computed from

$$(1-1.479+0.5901)Y=0.2069+(0.09566+0.09363)U+0.09413U^2$$

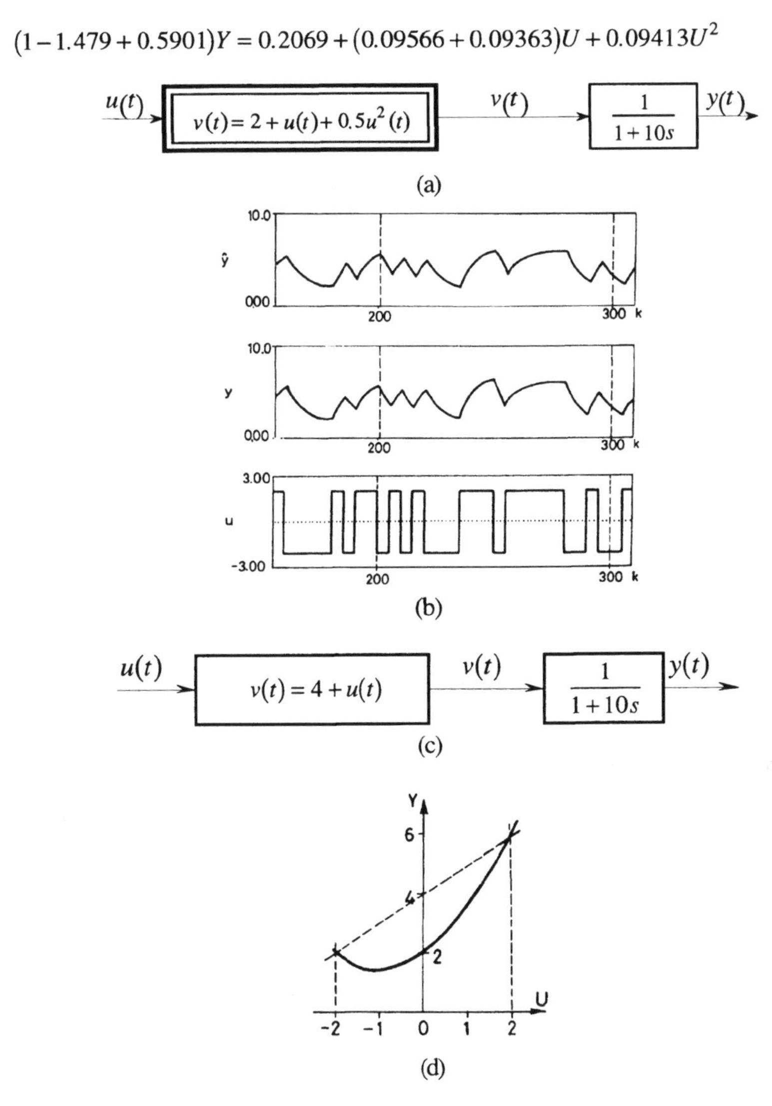

Fig. 2.1.3 Identification of the quadratic simple Hammerstein model using a PRBS with levels -2 and 2 in Example 2.1.3
(a) scheme of the process; (b) input (u), measured (y) and computed output $(\hat{y})$ based on the estimated model signals; (c) scheme of the estimated linear model with the linear static characteristic; (d) steady state characteristics of the simulated (Y) and of the estimated process $(\hat{Y})$

that is

$$Y=1.8623+1.637U+0.84725U^2$$

■

The next example again shows a nonlinear dynamic process that cannot be identified by a two-level test signal.

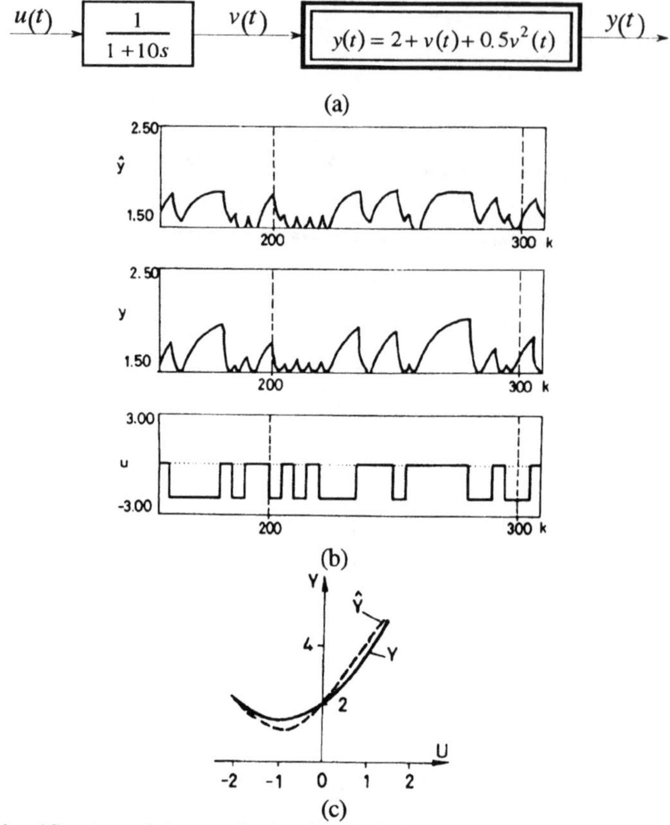

(c)

Fig. 2.1.4 Identification of the quadratic simple Wiener model using a PRBS with levels -2 and 0 in Example 2.1.4

(a) scheme of the process; (b) input (u), measured (y) and computed output $(\hat{y})$ based on the estimated model signals; (c) steady state characteristics of the simulated (Y) and of the estimated process $(\hat{Y})$

Example 2.1.5 *Identification of the simple Hammerstein model by a two-level signal with the levels -2 and 0*

The simple Hammerstein model (Figure 2.1.5a) is the same as in Example 2.1.3. Both the test signal and the sampling times were the same as of Example 2.1.4. The input and output signals are seen in Figure 2.1.5b. The output signal is constant, only a constant model can be fitted to the input–output data pairs:

$$\hat{y}(k) = 2$$

The reason for that is that the two-level test signal excites the static parabola $V = V(U)$ in such two points to which the same ordinate values belong

$$V_1 = 2 + U + 0.5U^2\Big|_{U=-2} = 2$$

$$V_2 = 2 + U + 0.5U^2\Big|_{U=0} = 2$$

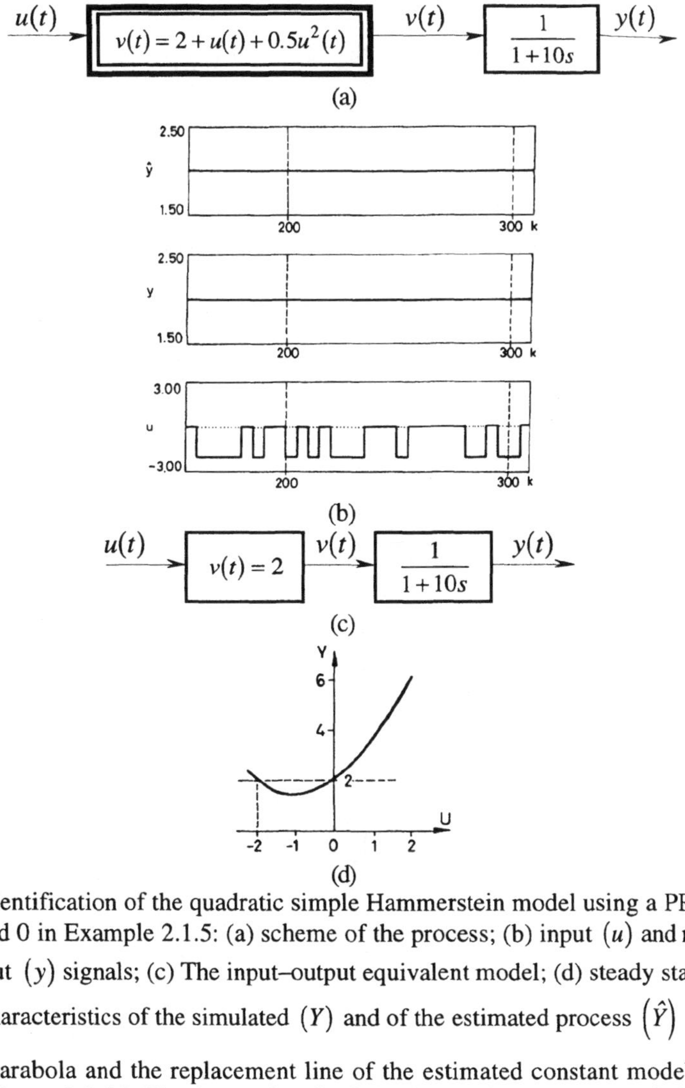

Fig. 2.1.5 Identification of the quadratic simple Hammerstein model using a PRBS with levels -2 and 0 in Example 2.1.5: (a) scheme of the process; (b) input (u) and measured output (y) signals; (c) The input–output equivalent model; (d) steady state characteristics of the simulated (Y) and of the estimated process $(\hat{Y})$

The static parabola and the replacement line of the estimated constant model are also shown in Figure 2.1.5d. The input signal of the linear dynamic part of the simple Wiener model, however, is a constant signal, which makes an identification of neither of the nonlinear nor of the dynamic part of the process possible. ■

The following conclusions can be drawn from the above examples:

1. Two-level test signals (both steps and pseudo-random binary signals) are suited

for identification of nonlinear dynamic systems if the experiment is repeated in several working points;

2. If a two-level test signal is used, then the number of the working points should be greater than or equal to the degree of the highest power of the polynomial steady state characteristic of the process;
3. A nonlinear system should be investigated by a multi-level test signal with more levels than the degree of the highest power of the polynomial steady state characteristic of the process. The steady state output values belonging to the excited working points should differ from each other;
4. If the inner structure of the process is known, and a single static nonlinear term can be separated from the linear dynamic parts, then Paragraphs 2 and 3 should be fulfilled for the static nonlinear term. That means in special cases that a lower level test signal than the degree of the highest power of the polynomial can be sufficient. This is the case with the simple Wiener model, where the inner variable, which is the input to the static nonlinear term, has a wide range amplitude distribution even for a binary excitation, because the two-level test signal is passed through a linear dynamic filter.

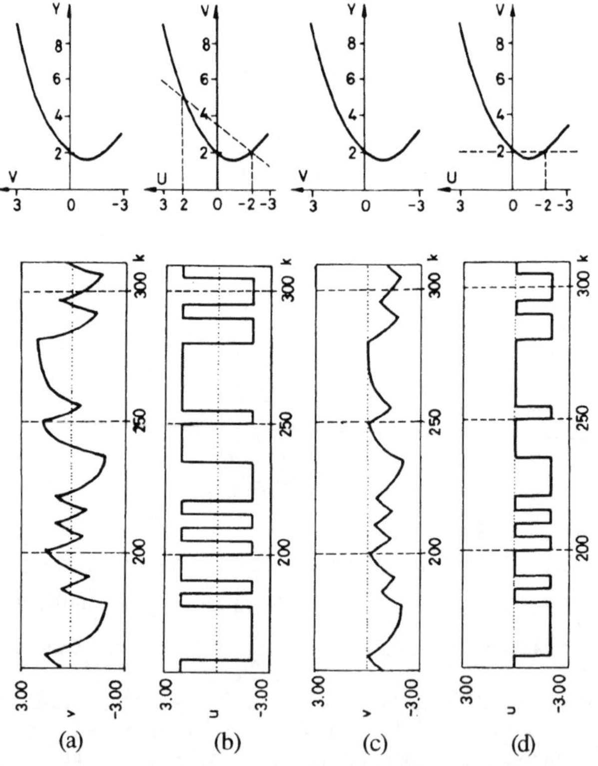

Fig. 2.1.6 The input signals of the static parabolics in Examples 2.1.2 to 2.1.5
(a) identification of the simple Wiener model in Example 2.1.2;
(b) identification of the simple Hammerstein model in Example 2.1.3;
(c) identification of the simple Wiener model in Example 2.1.4;
(d) identification of the simple Hammerstein model in Example 2.1.5

Figure 2.1.6 shows the input signals of the nonlinear static terms in Examples 2.1.2 to 2.1.5. These signals are the input signals in the case of a Hammerstein model and the signals passed the linear dynamic filter in the case of the Wiener model. We can see why the Wiener models can be identified by means of a binary excitation. However, it should be mentioned that the PRBS signals used are certainly not the optimal test signals for the Wiener model.

Concerning the need of a proper amplitude distribution of the test signal, the dynamic behavior of the process has to be excited sufficiently, as well. As is known in the linear system identification, white noise would be the best choice in this respect. In practice this is approximated by a pseudo-random test signal that has a bounded frequency spectrum. The advantage of the usage of the pseudo-random signals is their easy generation and that they are reproducible.

The ideal test signal for the investigation of nonlinear systems is white noise with a Gaussian amplitude distribution. Several identification methods need relatively few computations if this test signal is used. The disadvantage of this method is that very long samples are needed to push the advantages of reducing the computations. A practical compromise is the usage of the so-called pseudo-random multi-level test signals. They have a wide range but district amplitude distribution and their auto-correlation functions are similar to those of white noise even for short periods.

Leontaritis and Billings (1987) performed a theoretical and experimental study in which they compared white noise test signals with uniform and normal distributions in nonlinear system identification. Their investigations can be summed up in the following theorems.

Theorem 2.1.1 The optimal test signal for a nonlinear dynamic system is white noise and has a Gaussian distribution if the power of the signal is constrained (Leontaritis and Billings, 1987). ■

Theorem 2.1.2 The optimal test signal for a nonlinear dynamic system is white noise and has a uniformly distribution if the amplitude of the signal is constrained (Leontaritis and Billings, 1987). ■

The theorems lead to the following conclusion:

Corollary 2.1.1 In nonlinear system identification a Gaussian white noise is before a white noise of uniformly distributed and with the same standard deviation in respect of the relation parameter to its standard deviation and of the standard deviation of the residuals (Leontaritis and Billings, 1987). ■

Leontaritis and Billings (1987) compared the parameter estimation of the process

$$\begin{aligned} y(k) &= \theta_1 y(k-1) + \theta_2 u(k-1) + \theta_3 y^3(k-1) + \theta_4 u^2(k-1) y(k-1) \\ &\quad + \theta_5 u(k-1) y(k-1) e(k-1) + \theta_6 e(k-1) + e(k) \end{aligned}$$

where $e(k)$ is Gaussian white noise. The simulation results supported their statement.

According to the authors' experience the difference in the goodness of the identification between the two test signals is very little. If the structure of the process and the model used in the identification were the same then the Gaussian signal was a bit before the uniformly distributed test signal, but if the structures were different the uniformly distributed signal was little better than the Gaussian one.

Example 2.1.6 *Identification of the simple Wiener model by white noise test signal with Gaussian and uniform distributions (Haber, 1990)*

The simulated simple Wiener model was the same as in Examples 2.1.2 to 2.1.5. The sampling time was $\Delta T = 2$ [s]. $N = 1000$ data pairs were used for the identification. The mean value of the test signals was zero and the standard deviation was $\sigma_u = 1$. No noise was simulated. As the Wiener model is nonlinear in the parameters parametric Volterra and bilinear models were fitted by the LS method to the input–output data. The model equations with the estimated parameters and with their standard deviations are as follows.

(a) parametric Volterra model and Gaussian test signal:

$$\begin{aligned}\hat{y}(k) = &(0.7033 \pm 0.02605)\hat{y}(k-1) + (0.08673 \pm 0.02127)\hat{y}(k-2) \\ &+(0.4162 \pm 0.01264) + (0.1796 \pm 0.001298)u(k-1) \\ &+(0.02035 \pm 0.004883)u(k-2) + (0.01495 \pm 0.0009974)u^2(k-1) \\ &-(0.0008429 \pm 0.001076)u^2(k-2) + (0.02855 \pm 0.00139)u(k-1)u(k-2)\end{aligned}$$

The estimated standard deviation of the residuals was $\sigma_\varepsilon = 0.0407$. The estimated steady state characteristic is

$$\hat{Y} = 1.9822 + 0.9522U + 0.2032U^2$$

(b) parametric Volterra model and uniformly distributed test signal:

$$\begin{aligned}\hat{y}(k) = &(0.7544 \pm 0.02491)\hat{y}(k-1) + (0.04189 \pm 0.02012)\hat{y}(k-2) \\ &+(0.4001 \pm 0.01256) + (0.184 \pm 0.001254)u(k-1) \\ &+(0.01189 \pm 0.004771)u(k-2) + (0.01852 \pm 0.001363)u^2(k-1) \\ &-(0.000388 \pm 0.001435)u^2(k-2) + (0.02552 \pm 0.001256)u(k-1)u(k-2)\end{aligned}$$

The estimated standard deviation of the residuals was $\sigma_\varepsilon = 0.03953$. The estimated steady state characteristic is

$$\hat{Y} = 1.964 + 0.9616U + 0.2143U^2$$

(c) bilinear model and Gaussian test signal:

$$\begin{aligned}\hat{y}(k) = &(0.7972 \pm 0.002464)\hat{y}(k-1) + (0.4151 \pm 0.005186) \\ &-(0.1003 \pm 0.005186)u(k-1) + (0.1368 \pm 0.002711)u(k-1)\hat{y}(k-1)\end{aligned}$$

The estimated standard deviation of the residuals was $\sigma_\varepsilon = 0.02827$. The estimated steady state characteristic is

$$\hat{Y} = \frac{0.4151 - 0.1003U}{0.2028 - 0.1368U}$$

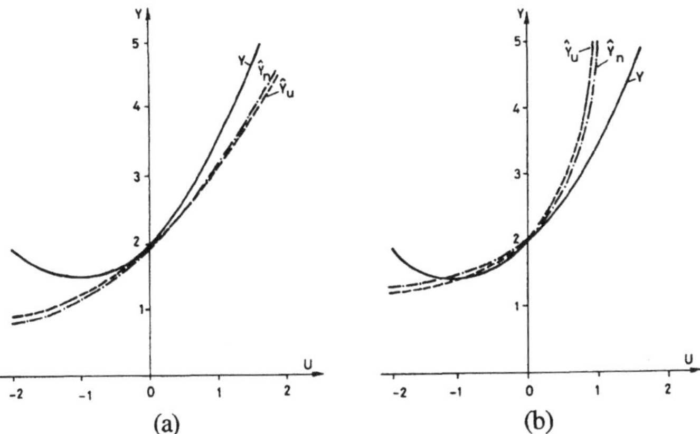

Fig. 2.1.7 Identification of the quadratic simple Wiener model
(a) fitting a second-order quadratic parametric Volterra model; (b) fitting a first-order bilinear Volterra model (Y: simulated process, $\hat{Y}_u$: identification with uniformly distributed test signal, $\hat{Y}_n$: identification with Gaussian test signal)

(d) bilinear model and uniformly distributed test signal:

$$\hat{y}(k) = (0.8111 \pm 0.001897)\hat{y}(k-1) + (0.3869 \pm 0.003969)$$
$$-(0.09112 \pm 0.003849)u(k-1) + (0.1329 \pm 0.001828)u(k-1)\hat{y}(k-1)$$

The estimated standard deviation of the residuals was $\sigma_\varepsilon = 0.02112$. The estimated steady state characteristic is

$$\hat{Y} = \frac{0.3869 - 0.09112U}{0.1889 - 0.1329U}$$

The standard deviations of both the parameters and of the residuals are less in the case of a uniformly distributed signal. The difference is very small, as is seen on the plots of the curves of the estimated steady state characteristics (Figures 2.1.7a and 2.1.7b).

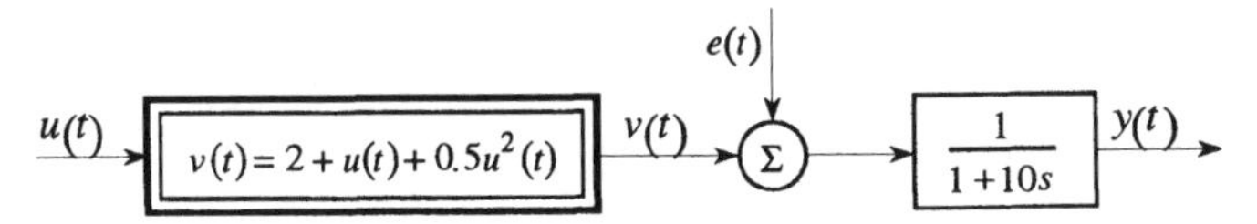

Fig. 2.1.8 Noisy simple Hammerstein model

Example 2.1.7 *Identification of the simple Hammerstein model by a white noise test signal with Gaussian and uniform distributions (Haber, 1990)*

The scheme of the noisy simple Hammerstein model is seen in Figure 2.1.8. The transfer functions of the static quadratic and of the linear dynamic parts are the same as of the simple Wiener model in Examples 2.1.2 to 2.1.5. The sampling time was $\Delta T = 2$ [s] $N = 1000$ data pairs were used for the identification. The mean value of the test signals was zero and the standard deviation was $\sigma_u = 1$. A Gaussian noise was superposed on the inner signal between the static nonlinear and linear dynamic terms. Its standard deviation was chosen in such a way that the ratio of the noise at the output to the noise-free output signal was 10% and 30%. To these values belonged a standard deviation of the source noise $\sigma_e = 0.11$ and $\sigma_e = 0.33$, respectively. As the noise passes the dynamic part of the process the LS estimation can be applied. The model equations with the estimated parameters and with their standard deviations are as follows:

(a) Gaussian test signal and noise to signal ratio 10%:

$$\hat{y}(k) = (0.8205 \pm 0.00162)\hat{y}(k-1) + (0.3581 \pm 0.004116)$$
$$+(0.1805 \pm 0.0006309)u(k-1) + (0.09115 \pm 0.0004839)u^2(k-1)$$

The estimated standard deviation of the residuals was $\sigma_\varepsilon = 0.01982$. The estimated steady state characteristic is

$$\hat{Y} = 1.9995 + 1.00067U + 0.508U^2$$

(b) uniformly distributed test signal and noise to signal ratio 10%:

$$\hat{y}(k) = (0.8194 \pm 0.001634)\hat{y}(k-1) + (0.3607 \pm 0.00418)$$
$$+(0.1809 \pm 0.0006285)u(k-1) + (0.09113 \pm 0.000683)u^2(k-1)$$

The estimated standard deviation of the residuals was $\sigma_\varepsilon = 0.01984$. The estimated steady state characteristic is

$$\hat{Y} = 1.9997 + 1.0U + 0.504U^2$$

(c) Gaussian test signal and noise to signal ratio 30%:

$$\hat{y}(k) = (0.8254 \pm 0.004579)\hat{y}(k-1) + (0.3454 \pm 0.01173)$$
$$+(0.1789 \pm 0.001892)u(k-1) + (0.09215 \pm 0.00145)u^2(k-1)$$

The estimated standard deviation of the residuals was $\sigma_\varepsilon = 0.05842$. The estimated steady state characteristic is

$$\hat{Y} = 1.978 + 1.025U + 0.528U^2$$

(d) uniformly distributed test signal and noise to signal ratio 30%:

$$\hat{y}(k) = (0.8228 \pm 0.004717)\hat{y}(k-1) + (0.3520 \pm 0.01212)$$
$$+(0.1802 \pm 0.001885)u(k-1) + (0.0921 \pm 0.002049)u^2(k-1)$$

The estimated standard deviation of the residuals was $\sigma_\varepsilon = 0.05952$. The estimated steady state characteristic is

$$\hat{Y} = 1.986 + 1.017U + 0.520U^2$$

There is no big difference among the applications of the different test signals. However, the Gaussian test signal causes a bit smaller standard deviation of both the parameters and of the residuals. ■

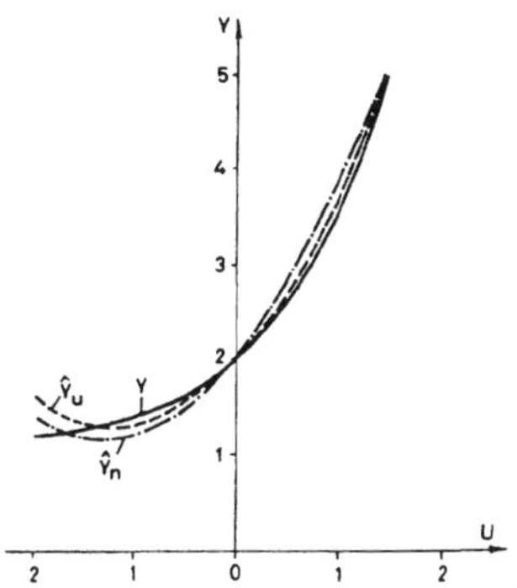

Fig. 2.1.9 Steady state characteristics of the cubic static process
(Y: simulated process, $\hat{Y}_u$: identification with uniformly distributed test signal, $\hat{Y}_n$: identification with Gaussian test signal)

Example 2.1.8 *Identification of a static cubic system by a quadratic model with a white noise test signal with Gaussian and uniform distribution (Haber, 1990)*

The equation of the simulated process was

$$y(k) = 2 + u(k) + 0.5u^2(k) + 0.1u^3(k)$$

and a quadratic model was fitted to the noiseless input–output data. $N = 1000$ data pairs were used for the identification. The mean value of the test signals was zero and the standard deviation was $\sigma_u = 1$.

LS parameter estimation can be applied. The model equations with the estimated parameters and with their standard deviations are as follows:

(a) Gaussian test signal:

$$\hat{y}(k) = (2.009 \pm 0.008131) + (1.269 \pm 0.006507)u(k) + (0.485 \pm 0.004985)u^2(k-1)$$

The estimated standard deviation of the residuals was $\sigma_\varepsilon = 0.2045$.

(b) uniformly distributed test signal:

$$\hat{y}(k) = (1.999 \pm 0.003767) + (1.184 \pm 0.002552)u(k) + (0.4991 \pm 0.002775)u^2(k-1)$$

The estimated standard deviation of the residuals was $\sigma_\varepsilon = 0.08065$.

The identification seems to be better with the uniformly distributed test signal. The difference in the estimated static curves is minimal (Figure 2.1.9). ■

It should be remarked that some identification methods need a Gaussian test signal, and do not work with a uniformly distributed test signal. The reverse case is not relevant.

Pearson *et al.*, (1996) shows examples how the choice of the test signal influences the parameter estimation of the Volterra series model of different nonlinear dynamic models.

In the sequel the features of the following test signals will be summarized:

- Gaussian white noise;
- pseudo-random two-level signals with maximum length;
- pseudo-random three-level signals with maximum length;
- pseudo-random five-level signals with maximum length.

Further on, a subroutine for the generation of pseudo-random multi-level signals will be presented.

Chapter 2 is based on the report Haber (1990). The examples illustrating the generation of the two-, three- and five-level signals originate from there.

2.2 RANDOM SIGNAL WITH NORMAL (GAUSSIAN) DISTRIBUTION

There are two reasons why we deal with white noise with a Gaussian distribution. The first is that measurement errors are usually of this type. The second is that it serves as a theoretically ideal test signal for the investigation of nonlinear processes.

Assumption 2.2.1 The random signal $x(t)$ has a normal distribution with mean value $\bar{x}$ and standard deviation σ_x. ■

Definition 2.2.1 The time function of the normalized Gaussian white noise signal is defined as

$$x'(t) = \frac{x(t) - \bar{x}}{\sigma_x}$$ ■

With Assumption 2.2.1 and with Definition 2.2.1 $x(t)$ and $x'(t)$ have the following properties.

Property 2.2.1 The probability density function of $x(t)$ is

$$p(x) = \frac{1}{\sqrt{2\pi}\sigma_x} \exp\left(-\frac{(x-\bar{x})^2}{2\sigma_x^2} \right)$$

and the cumulative probability density function is

$$P(x) = \frac{1}{\sqrt{2\pi}\sigma_x} \int_{-\infty}^{x} \exp\left(-\frac{(x_1-\bar{x})^2}{2\sigma_x^2} \right) dx_1$$

The probability density function of $x'(t)$ is

$$p(x') = \frac{1}{\sqrt{2\pi}} \exp\left(-\frac{x^2}{2} \right)$$

and the cumulative probability density function is

$$P(x') = \frac{1}{\sqrt{2\pi}} \int_{-\infty}^{x} \exp\left(-\frac{x_1^2}{2}\right) dx_1$$ ■

The theoretical curves $p(x')$ and $P(x')$ are drawn in Figures 2.2.1a and 2.2.1b. ■

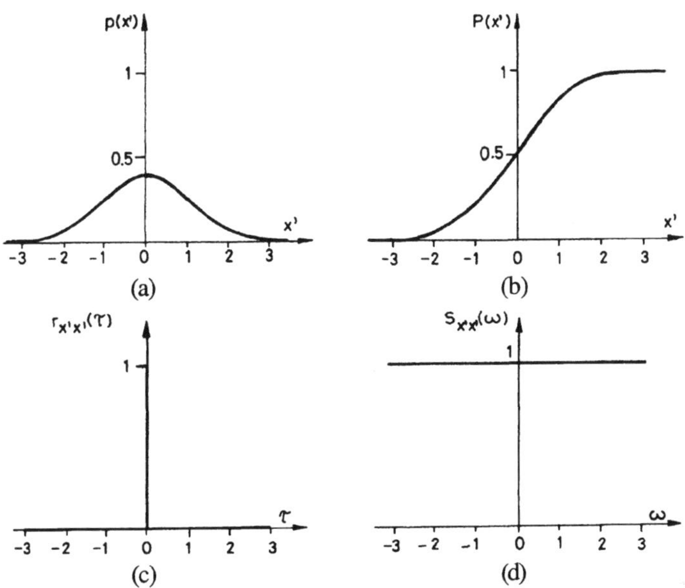

Fig. 2.2.1 Some features of the normalized Gaussian signal $x'(t)$ $(\bar{x} = 0,\ \sigma = 1)$
(a) probability density function; (b) cumulative probability density function;
(c) auto-correlation function; (d) power density spectrum

Property 2.2.2 The auto-correlation function of the normalized signal $x'(t)$ is

$$r_{x'x'}(\tau) = \delta(\tau)$$ ■

A signal with a Dirac delta like auto-correlation function is called white noise.

Property 2.2.3 The power density spectrum of the normalized signal is

$$S_{x'x'}(j\omega) = 1$$

for every frequency. ■

Property 2.2.3 has the consequence that a white noise excites the dynamics of a system persistently. The shapes of the auto-correlation and of the power density spectra of a normalized Gaussian white noise are drawn in Figures 2.2.1c and 2.2.1d.

Property 2.2.4 The even-order auto-correlation functions of a normalized Gaussian white noise are zero for every shifting time. In other words, all odd degree moments are zero.

$$r_{x'}(\tau_1,\ldots,\tau_n)=\mathrm{E}\{x'(t)x'(t+\tau_1)\ldots x'(t+\tau_n)\}$$
$$=\mathrm{E}\left\{x'(t)\prod_{i=1}^{n} x'(t+\tau_i)\right\}=0 \qquad n=2,4,6,\ldots$$

(The proof is given in Schetzen (1980).)

Property 2.2.5 The odd-order auto-correlation functions of the normalized Gaussian white noise can be calculated by

$$r_{x'}(\tau_1,\ldots,\tau_n)=\mathrm{E}\{x'(t)x'(t+\tau_1)\ldots x'(t+\tau_n)\}$$
$$=\mathrm{E}\left\{x'(t)\prod_{i=1}^{n} x'(t+\tau_i)\right\}=\sum\prod \mathrm{E}\{x'(t+\tau_i)x'(t+\tau_j)\} \tag{2.2.1}$$
$$i\neq j$$

where $\sum\prod$ means 'the sum of all completely distinct ways of partitioning $x'(t)x'(t+\tau_1)\ldots x'(t+\tau_n)$ into pairs'. (Cited from Schetzen, (1980)). The expression (2.2.1) is also called the even degree moment of the signal $x'(t)$. ■

(The proof is given in Schetzen (1980).)

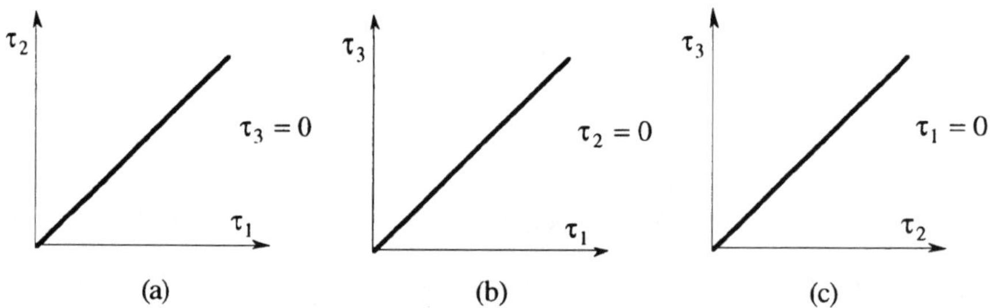

Fig. 2.2.2 Locations where the third-order auto-correlation function of the Gaussian white-noise differs from zero: (a) plane $\tau_1=\tau_2$, $\tau_3=0$; (b) plane $\tau_1=\tau_3$, $\tau_2=0$; (c) plane $\tau_2=\tau_3$, $\tau_1=0$

Example 2.2.1 *Third-order auto-correlation function of the normalized Gaussian white noise.*

$$r_{x'}(\tau_1,\tau_2,\tau_3)=\mathrm{E}\{x'(t)x'(t+\tau_1)x'(t+\tau_2)x'(t+\tau_3)\}$$
$$=\mathrm{E}\{x'(t)x'(t+\tau_1)\}\mathrm{E}\{x'(t+\tau_2)x'(t+\tau_3)\}$$
$$+\mathrm{E}\{x'(t)x'(t+\tau_2)\}\mathrm{E}\{x'(t+\tau_1)x'(t+\tau_3)\}$$
$$+\mathrm{E}\{x'(t)x'(t+\tau_3)\}\mathrm{E}\{x'(t+\tau_1)x'(t+\tau_2)\}$$
$$=r_{x'x'}(\tau_1)r_{x'x'}(\tau_3-\tau_2)+r_{x'x'}(\tau_2)r_{x'x'}(\tau_3-\tau_1)+r_{x'x'}(\tau_3)r_{x'x'}(\tau_2-\tau_1)$$

which has the value

$$r_{x'}(\tau_1,\tau_2,\tau_3)=\begin{cases}0 & \text{if} \quad \tau_1\neq\tau_2\neq\tau_3\neq 0\\ 1 & \text{if} \quad \tau_1=0 \quad \text{and} \quad \tau_2=\tau_3\neq 0\\ 1 & \text{if} \quad \tau_2=0 \quad \text{and} \quad \tau_1=\tau_3\neq 0\\ 1 & \text{if} \quad \tau_3=0 \quad \text{and} \quad \tau_1=\tau_2\neq 0\\ 3 & \text{if} \quad \tau_1=\tau_2=\tau_3=0\end{cases}$$

The locations where $r_{x'}(\tau_1,\tau_2,\tau_3)$ differs from zero are drawn in Figure 2.2.2. ■

Example 2.2.2 *Fifth-order auto-correlation function of the normalized Gaussian white noise.*

$$\begin{aligned}
&r_{x'}(\tau_1,\tau_2,\tau_3,\tau_4,\tau_5)\\
&=\mathrm{E}\{x'(t)x'(t+\tau_1)x'(t+\tau_2)x'(t+\tau_3)x'(t+\tau_4)x'(t+\tau_5)\}\\
&=\mathrm{E}\{x'(t)x'(t+\tau_1)\}\mathrm{E}\{x'(t+\tau_2)x'(t+\tau_3)\}\mathrm{E}\{x'(t+\tau_4)x'(t+\tau_5)\}\\
&\quad+\mathrm{E}\{x'(t)x'(t+\tau_1)\}\mathrm{E}\{x'(t+\tau_2)x'(t+\tau_4)\}\mathrm{E}\{x'(t+\tau_3)x'(t+\tau_5)\}\\
&\quad+\mathrm{E}\{x'(t)x'(t+\tau_1)\}\mathrm{E}\{x'(t+\tau_2)x'(t+\tau_5)\}\mathrm{E}\{x'(t+\tau_3)x'(t+\tau_4)\}\\
&\quad+\mathrm{E}\{x'(t)x'(t+\tau_2)\}\mathrm{E}\{x'(t+\tau_1)x'(t+\tau_3)\}\mathrm{E}\{x'(t+\tau_4)x'(t+\tau_5)\}\\
&\quad+\mathrm{E}\{x'(t)x'(t+\tau_2)\}\mathrm{E}\{x'(t+\tau_1)x'(t+\tau_4)\}\mathrm{E}\{x'(t+\tau_3)x'(t+\tau_5)\}\\
&\quad+\mathrm{E}\{x'(t)x'(t+\tau_2)\}\mathrm{E}\{x'(t+\tau_1)x'(t+\tau_5)\}\mathrm{E}\{x'(t+\tau_3)x'(t+\tau_4)\}\\
&\quad+\mathrm{E}\{x'(t)x'(t+\tau_3)\}\mathrm{E}\{x'(t+\tau_1)x'(t+\tau_2)\}\mathrm{E}\{x'(t+\tau_4)x'(t+\tau_5)\}\\
&\quad+\mathrm{E}\{x'(t)x'(t+\tau_3)\}\mathrm{E}\{x'(t+\tau_1)x'(t+\tau_4)\}\mathrm{E}\{x'(t+\tau_2)x'(t+\tau_5)\}\\
&\quad+\mathrm{E}\{x'(t)x'(t+\tau_3)\}\mathrm{E}\{x'(t+\tau_1)x'(t+\tau_5)\}\mathrm{E}\{x'(t+\tau_2)x'(t+\tau_4)\}\\
&\quad+\mathrm{E}\{x'(t)x'(t+\tau_4)\}\mathrm{E}\{x'(t+\tau_1)x'(t+\tau_2)\}\mathrm{E}\{x'(t+\tau_3)x'(t+\tau_5)\}\\
&\quad+\mathrm{E}\{x'(t)x'(t+\tau_4)\}\mathrm{E}\{x'(t+\tau_1)x'(t+\tau_3)\}\mathrm{E}\{x'(t+\tau_2)x'(t+\tau_5)\}\\
&\quad+\mathrm{E}\{x'(t)x'(t+\tau_4)\}\mathrm{E}\{x'(t+\tau_1)x'(t+\tau_5)\}\mathrm{E}\{x'(t+\tau_2)x'(t+\tau_3)\}\\
&\quad+\mathrm{E}\{x'(t)x'(t+\tau_5)\}\mathrm{E}\{x'(t+\tau_1)x'(t+\tau_2)\}\mathrm{E}\{x'(t+\tau_3)x'(t+\tau_4)\}\\
&\quad+\mathrm{E}\{x'(t)x'(t+\tau_5)\}\mathrm{E}\{x'(t+\tau_1)x'(t+\tau_3)\}\mathrm{E}\{x'(t+\tau_2)x'(t+\tau_4)\}\\
&\quad+\mathrm{E}\{x'(t)x'(t+\tau_5)\}\mathrm{E}\{x'(t+\tau_1)x'(t+\tau_4)\}\mathrm{E}\{x'(t+\tau_2)x'(t+\tau_3)\}\\
&=r_{x'x'}(\tau_1)r_{x'x'}(\tau_3-\tau_2)r_{x'x'}(\tau_5-\tau_4)+r_{x'x'}(\tau_1)r_{x'x'}(\tau_4-\tau_2)r_{x'x'}(\tau_5-\tau_3)\\
&\quad+r_{x'x'}(\tau_1)r_{x'x'}(\tau_5-\tau_2)r_{x'x'}(\tau_4-\tau_3)+r_{x'x'}(\tau_2)r_{x'x'}(\tau_3-\tau_1)r_{x'x'}(\tau_5-\tau_4)\\
&\quad+r_{x'x'}(\tau_2)r_{x'x'}(\tau_4-\tau_1)r_{x'x'}(\tau_5-\tau_3)+r_{x'x'}(\tau_2)r_{x'x'}(\tau_5-\tau_1)r_{x'x'}(\tau_4-\tau_3)\\
&\quad+r_{x'x'}(\tau_3)r_{x'x'}(\tau_2-\tau_1)r_{x'x'}(\tau_5-\tau_4)+r_{x'x'}(\tau_3)r_{x'x'}(\tau_4-\tau_1)r_{x'x'}(\tau_5-\tau_2)\\
&\quad+r_{x'x'}(\tau_3)r_{x'x'}(\tau_5-\tau_1)r_{x'x'}(\tau_4-\tau_2)+r_{x'x'}(\tau_4)r_{x'x'}(\tau_2-\tau_1)r_{x'x'}(\tau_5-\tau_3)\\
&\quad+r_{x'x'}(\tau_4)r_{x'x'}(\tau_3-\tau_1)r_{x'x'}(\tau_5-\tau_2)+r_{x'x'}(\tau_4)r_{x'x'}(\tau_5-\tau_1)r_{x'x'}(\tau_3-\tau_2)
\end{aligned}$$

$$+r_{x'x'}(\tau_5)r_{x'x'}(\tau_2-\tau_1)r_{x'x'}(\tau_4-\tau_3)+r_{x'x'}(\tau_5)r_{x'x'}(\tau_3-\tau_1)r_{x'x'}(\tau_4-\tau_2)$$
$$+r_{x'x'}(\tau_5)r_{x'x'}(\tau_4-\tau_1)r_{x'x'}(\tau_3-\tau_2)$$

which differs from zero only if some shifting time pairs are equal to each other and at least one shifting time is zero. The possible maximum value is 15 when $\tau_1=\tau_2=\tau_3=\tau_4=\tau_5=0$. ■

2.3 PSEUDO-RANDOM MULTI-LEVEL SIGNALS (PRMS-S) WITH MAXIMAL LENGTH

Pseudo-random multi-level signals with maximal length are artificially generated signals with features that resemble to those of the Gaussian white noise:

- They have some, free choosable district amplitude values;
- The sequence repeats itself after a period, which can be chosen by the user;
- The whole sequence can be determined by some parameters. This fact allows to reproduce the sequence and the whole identification experiment any time;
- The mean value of a sequence in a period is zero;
- The first-order auto-correlation function is similar in a given domain of the shifting time to that of the white noise;
- The even order auto-correlation function of the normalized signal – except the binary one – is zero as with the Gaussian white noise.

Above properties made the PRMS with maximal length a popular and useful test signal in the identification of processes.

In practice the following PRMS-s are used:

- with two levels: pseudo-random binary signal (PRBS);
- with three levels: pseudo-random ternary signal (PRTS);
- with five levels: pseudo-random quinary signal (PRQS).

2.3.1 Generating PRMS-s by multi-level shift registers

A shift register (Figure 2.3.1a) is built up from some n_r registers which store the information until new information is fed in. Every register except the first one receives the information from the previous one that is on their left. The transfer of the information happens periodically and synchronously. The signal that enables the transfer is called the clock signal. The minimum switching time ΔT_e is the interval between two shifting procedures.

Each register may have r levels like $0,1,\ldots,r-1$. Then r^{n_r} different states are imaginable. A *state* means the vector of the actual levels of the n_r registers. Since the information is always shifted to the right, the states depend on the information fed into the first register on the left. Assume that this is done in such a way that all possible states occur.

The basic idea of generating pseudo-random signals is that the state of one register is considered as the test signal. This signal has r levels with the most randomness if the shift register has all possible states in sequel. (The integer values have to be transformed to real ones depending on the actual application.) The degree of randomness increases very fast with n_r because r^{n_r} increases exponentially. The question arises of how all states of the shift register can be generated.

It was shown (e.g., Davies, 1970) that a suitable feedback from the last, and perhaps from other stages, to the first stage results in a cyclic behavior of the shift register with a period $r^{n_r}-1$. The first register gets the linear combinations of the states of the other (so-called feedback) registers (Figure 2.3.1b). (The feedback registers are those registers where c_i differs from zero.) The number of all different states of the shift register is therefore by one less than r^{n_r}, because the case that all registers have zero values has to be excluded. Otherwise the information fed to the first register would be always zero, and the shift register would be frozen by full zero state. The number of levels may be the prime numbers 2, 3, 5, 7, etc.

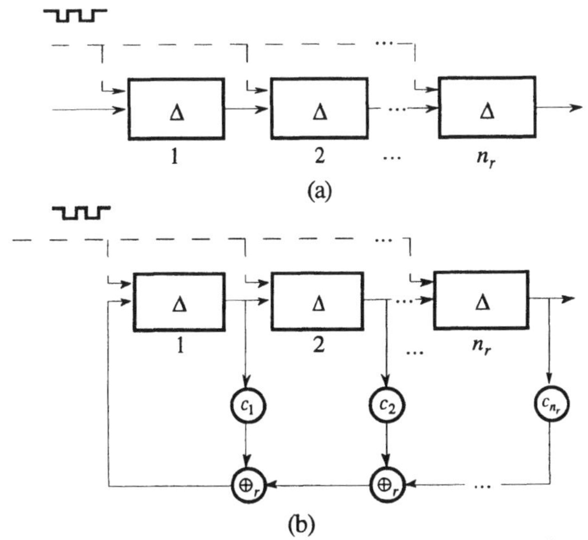

Fig. 2.3.1 Multilevel shift registers; (All operations in modulo-r, where r is the number of the levels): (a) open loop mode; (b) closed loop mode

Both the additions and multiplications should be done in modulo-r. The rules of the operations multiplication, addition and subtraction are given in Tables 2.3.1 to 2.3.3 for modulo-2, 3 and 5 and they are denoted by $\otimes_n, \oplus_n, \ominus_n$ Furthermore, the modulo division $\oslash_n$ calculates the remainder of an integer division, rather than the quotient.

The suitable feedback coefficients for generating PRMS with maximal length are tabulated in several publications (Zierler, 1959; Davies, 1970; Barker, 1969b, 1993; Godfrey, 1993). Tables 2.3.4 to 2.3.6 summarize some possible choices of the feedback coefficients for generating two-, three- and five-level pseudo-random signals with maximal length, respectively. The values are taken over from (Barker, 1969b).

Example 2.3.1 *Generation of a PRBS with maximum length of 7 (Haber, 1990).*
A PRBS with maximal length of 7 can be generated if the registers 2 and 3 are fed back and both multipliers are, e.g., $c_1 = c_2 = -1$ (Barker, 1969b). The scheme of the shift register with the feedbacks is drawn in Figure 2.3.2. Any initial condition except $[0,0,0]^T$ can be chosen. Let it, e.g., $[0,0,1]^T$. Denote the values in the scheme by x with a subscript showing their places. Then the components before the summation are

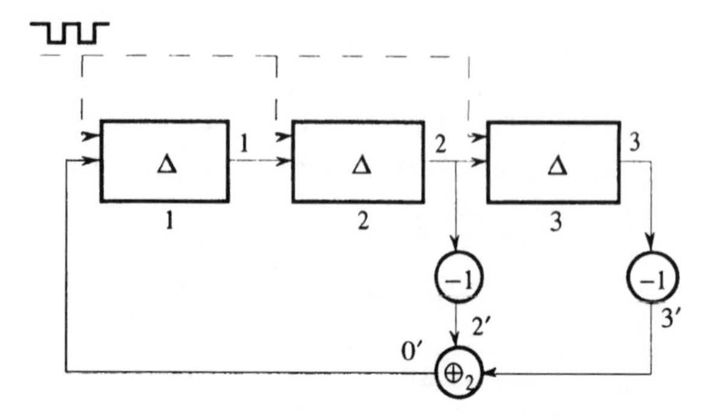

Fig. 2.3.2 The two-level shift register of Example 2.3.1

TABLE 2.3.1 Operations in modulo-2

$x_1 \oplus_2 x_2$		$x_2 = 0$	$x_2 = 1$
x_1	0	0	1
	1	1	0

(a) addition

$x_1 \ominus_2 x_2$		$x_2 = 0$	$x_2 = 1$
x_1	0	0	1
	1	1	0

(b) subtraction

$x_1 \otimes_2 x_2$		$x_2 = 0$	$x_2 = 1$
x_1	0	0	0
	1	0	1

(c) multiplication

$$x_{2'} = c_1 \otimes_2 x_2 = (-1) \otimes_2 0 = 0$$

and

$$x_{3'} = c_2 \otimes_2 x_3 = (-1) \otimes_2 1 = 1$$

where $\otimes_2$ stands for modulo-2.

The sum of the values of $2'$ and $3'$ is

$$x_{0'} = x_{2'} \oplus_2 x_{3'} = 0 \oplus_2 1 = 1$$

which is fed back to the first register in the next step. In the next step the values of the registers are shifted to the right by one

$$x_2 \equiv x_2(k) = x_1(k-1) = 0$$

$$x_3 \equiv x_3(k) = x_2(k-1) = 0$$

and, as calculated,

$$x_1 \equiv x_1(k) = x_{0'}(k-1) = 1$$

The computations have to be continued in the same manner and the results are given in Table 2.3.7. The values of the registers repeat themselves after 7 steps, the period is 7 that is the maximum that can be achieved by 3 registers $N_p = 2^3 - 1 = 7$. The states of the three registers are drawn in Figure 2.3.3 as a function of the time. As is seen, the states of the registers are delayed by one to their left neighbor. The output of any register can be considered as a PRBS signal. In practice the integer values 0 and 1 are transformed to real ones depending on the actual application. ■

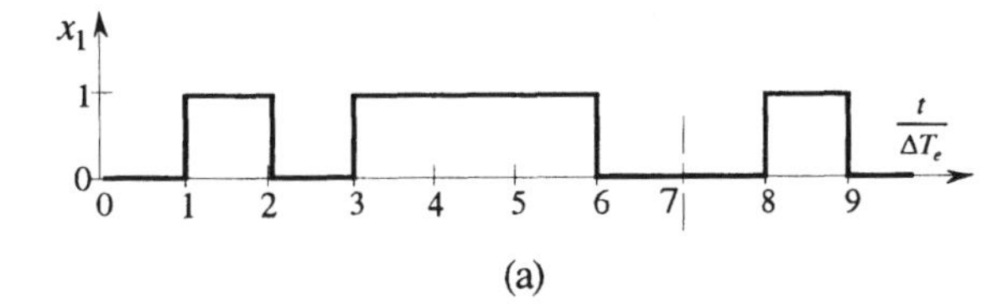

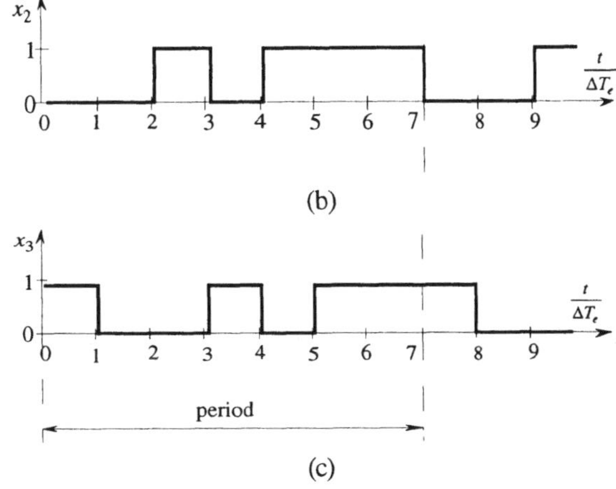

Fig. 2.3.3 Time plots of the states of the registers of Figure 2.3.2
(a) register 1; (b) register 2; (c) register 3;
The period is 7.

Example 2.3.2 *Generation of a PRTS with maximum length of 8 (Haber, 1990).*
A PRTS with a maximum length of 8 can be generated if the registers 2 and 3 are fed back and the multipliers are, e.g., $c_1 = -1$ and $c_2 = +1$ (Barker, 1969b). The scheme of the shift register with the feedbacks is drawn in Figure 2.3.4. Any initial condition except $[0, 0]^{\mathrm{T}}$ can be chosen. Let it, e.g., be $[0, 1]^{\mathrm{T}}$. Denote the values in the scheme by x with a subscript showing their places. Then the components before the summation are

$$x_{1'} = c_1 \otimes_3 x_1 = (-1) \otimes_2 0 = 0$$

TABLE 2.3.2 Operations in modulo-3

$x_1 \oplus_3 x_2$		x_2		
		0	1	2
x_1	0	0	1	2
	1	1	2	0
	2	2	0	1

(a) addition

$x_1 \ominus_3 x_2$		x_2		
		0	1	2
x_1	0	0	2	1
	1	1	0	2
	2	2	1	0

(b) subtraction

$x_1 \otimes_3 x_2$		x_2		
		0	1	2
x_1	0	0	0	0
	1	0	1	2
	2	0	2	1

(c) multiplication

and

$$x_{2'} = c_2 \otimes_3 x_2 = (+1) \otimes_2 1 = 1$$

The sum of the values of $1'$ and $2'$ is

$$x_{0'} = c_{1'} \oplus_3 x_{2'} = 0 \oplus_3 1 = 1$$

which is fed back to the first register in the next step. In the next step the values of the registers are shifted to the right by one

$$x_2 \equiv x_2(k) = x_1(k-1) = 0$$

and, as calculated,

$$x_1 \equiv x_1(k) = x_{0'}(k-1) = 1$$

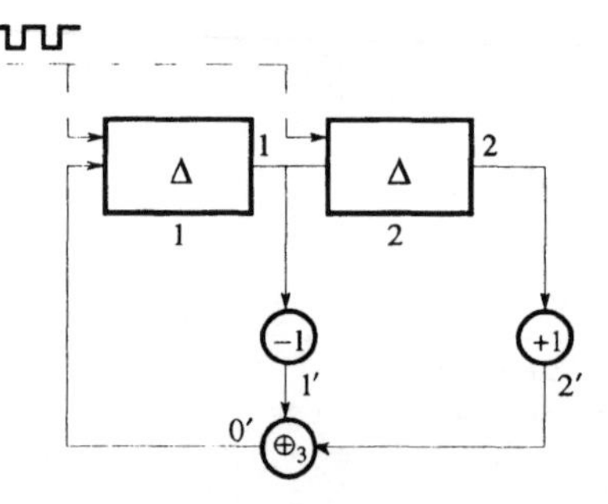

Fig. 2.3.4 The three-level shift register of Example 2.3.2

The computations have to be continued in the same manner and the results are given in Table 2.3.8. The values of the registers repeat themselves after 8 steps, the period is 8. This value is the maximal that can be achieved by 2 registers $N_p = 3^2 - 1 = 8$. The states of the three registers are drawn in Figure 2.3.5 as a function of the time. As is seen, the states of the registers are delayed by one to their left neighbor. The output of both registers can be considered as a PRTS signal. In practice the integer values 0, 1 and

2 are transformed to real ones, depending on the actual application. ■

TABLE 2.3.3 Operations in modulo-5

$x_1 \oplus_5 x_2$		x_2 = 0	1	2	3	4
x_1	0	0	1	2	3	4
	1	1	2	3	4	0
	2	2	3	4	0	1
	3	3	4	0	1	2
	4	4	0	1	2	3

(a) addition

$x_1 \ominus_5 x_2$		x_2 = 0	1	2	3	4
x_1	0	0	4	3	2	1
	1	1	0	4	3	2
	2	2	1	0	4	3
	3	3	2	1	0	4
	4	4	3	2	1	0

(b) subtraction

$x_1 \otimes_5 x_2$		x_2 = 0	1	2	3	4
x_1	0	0	0	0	0	0
	1	0	1	2	3	4
	2	0	2	4	1	3
	3	0	3	1	4	2
	4	0	4	3	2	1

(c) multiplication

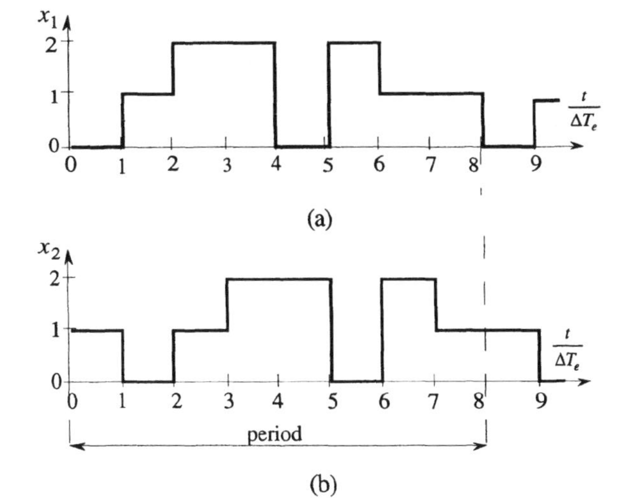

Fig. 2.3.5 Time plots of the states of the registers of Figure 2.3.4
(a) register 1;
(b) register 2;
The period is 8.

TABLE 2.3.4 Possible feedback coefficients of the two-level shift register generators generating PRBS-s with maximal length (based on Table 1 from (Barker, 1969b))

No	*Number of registers*	*Feedback coefficients*											*Period*
		c_1	c_2	c_3	c_4	c_5	c_6	c_7	c_8	c_9	c_{10}	c_{11}	
1	2	-1	-1										3
2	3	0	-1	-1									7
3		-1	0	-1									7
4	4	0	0	-1	-1								15
5		-1	0	0	-1								15
6	5	0	-1	-1	-1	-1							31
7		-1	-1	-1	0	-1							31
8	6	0	0	0	0	-1	-1						63
9		-1	0	0	0	0	-1						63
10	7	0	0	-1	-1	-1	0	-1					127
11		0	-1	-1	0	0	0	-1					127
12	8	0	-1	-1	0	0	0	-1	-1				255
13		-1	0	0	0	0	-1	-1	-1				255
14	9	0	0	-1	0	-1	-1	0	0	-1			511
15		0	0	-1	-1	0	-1	0	0	-1			511
16	10	-1	-1	0	0	0	-1	0	-1	-1	-1		1023
17		-1	-1	0	-1	0	0	0	-1	-1	-1		1023

No	*Number of registers*	*Feedback coefficients*							*Period*
		c_1	c_2	c_3	c_4	c_5	c_6	c_7	
1	2	-1	1						8
2		1	1						8
3	3	1	-1	1					26
4		-1	1	-1					26
5	4	-1	0	0	-1				80
6		0	0	1	1				80
8	5	1	1	0	1	-1			242
9		1	0	1	1	-1			242
10	6	-1	1	-1	-1	0	1		728
11		0	1	1	-1	1	1		728

TABLE 2.3.5 Possible feedback coefficients of the three-level shift register generators generating PRTS-s with maximal length (based on Table 2 from (Barker, 1969b))

2.3.2 Generating PRMS-s by solving difference equations

The shift register is a chain of delay operators. If the input to the first register is $x(k)$ then the output of the first register is $x(k-1)$, that of the second $x(k-2)$, etc. (Figure 2.3.6.). $x(k)$ can be determined by the following difference equation

$$x(k) \equiv c_1 \otimes_r x(k-1) \oplus_r c_2 \otimes_r x(k-2) \oplus_r \dots c_{n_r} \otimes_r x(k-n_r) \qquad (2.3.1)$$

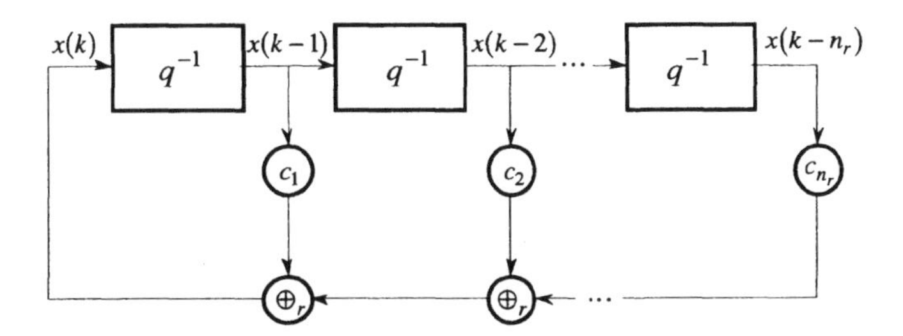

Fig. 2.3.6 The difference equation in modulo-r algebra presented by means of a shift register generator

The polynomial

$$\begin{aligned} C^{(r)}\left(q^{-1}\right) &= 1\ominus_r c_1 \otimes_r q^{-1} \ominus_r c_2 \otimes_r q^{-2} \ominus_r \ldots \ominus_r c_{n_r} \otimes_r q^{-n_r} \\ &= 1\oplus_r (-c_1)\otimes_r q^{-1} \oplus_r (-c_2)\otimes_r q^{-2} \oplus_r \ldots \oplus_r \left(-c_{n_r}\right)\otimes_r q^{-n_r} \end{aligned} \tag{2.3.2}$$

is called the characteristic polynomial defined in the modulo-r algebra (Galois field).

The parameters given in Tables 2.3.4 to 2.3.6 can be interpreted as the coefficients of the characteristic polynomials generating maximal length PRBS, PRTS and PRQS, respectively.

No	Number of registers	Feedback coefficients				Period
		c_1	c_2	c_3	c_4	
1	2	-1	-2			24
2		2	2			24
3	3	-2	1	-2		124
4		-2	-1	2		124
5	4	2	0	-1	2	724
6		1	-1	1	-2	724

TABLE 2.3.6 Possible feedback coefficients of the five-level shift register generators generating PRQS-s with maximal length (based on Table 3 from (Barker, 1969b))

No	States of the registers			Feedback components		Feedback value
	1	2	3	2'	3'	0'
0	0	0	1	0	1	1
1	1	0	0	0	0	0
2	0	1	0	1	0	1
3	1	0	1	0	1	1
4	1	1	0	1	0	1
5	1	1	1	1	1	0
6	0	1	1	1	1	0
7	0	0	1	0	1	1

TABLE 2.3.7 Calculation of the states of the two-level shift register generator in Example 2.3.1.

No	States of the registers		Feedback components		Feedback value
	1	2	2'	3'	0
0	0	1	0	1	1
1	1	0	2	0	2
2	2	1	1	1	2
3	2	2	1	2	0
4	0	2	0	2	2
5	2	0	1	0	1
6	1	2	2	2	1
7	1	1	2	1	0

TABLE 2.3.8 Calculation of the states of the three-level shift register generator in Example 2.3.2.

Example 2.3.3 *Generation of a PRBS with maximum length of 7 (Haber, 1990).* The shift register of Example 2.3.1 corresponds to the following difference equation

$$x(k) = (-1) \otimes_2 x(k-2) \oplus_2 (-1) \otimes_2 x(k-3) \tag{2.3.3}$$

Any initial condition, except that all registers have zero values, results in a PRBS with a period of 7. Let

$$x(-1) = 0 \qquad x(-2) = 0 \qquad x(-3) = 1$$

which correspond to the case of Example 2.3.1 as $x(k-1)$ correspond to x_1, $x(k-2)$ to x_2 and $x(k-3)$ to x_3. The first term of $x(k)$ is

$$x(0) = (-1) \otimes_2 x(0-2) \oplus_2 (-1) \otimes_2 x(0-3) = (-1) \otimes_2 0 \oplus_2 (-1) \otimes_2 1 = 1$$

the second one is

$$x(1) = (-1) \otimes_2 x(1-2) \oplus_2 (-1) \otimes_2 x(1-3) = (-1) \otimes_2 0 \oplus_2 (-1) \otimes_2 0 = 0$$

and the third one becomes

$$x(2) = (-1) \otimes_2 x(2-2) \oplus_2 (-1) \otimes_2 x(2-3) = (-1) \otimes_2 1 \oplus_2 (-1) \otimes_2 0 = 1$$

etc. The recursive solution of the difference equation (2.3.3) leads to the column $0'$ of the feedback values in Table 2.3.7. The values of this column are shifted to the first register in every step. ■

Example 2.3.4 *Generation of a PRTS with maximum length of 8 (Haber, 1990).* The shift register of Example 2.3.2 corresponds to the following difference equation

$$x(k) = (-1) \otimes_3 x(k-1) \oplus_3 (+1) \otimes_3 x(k-2) \tag{2.3.4}$$

Any initial condition, except that both registers have zero values, results in a PRTS with a period of 8. Let

$$x(-1) = 0 \qquad x(-2) \doteq 1$$

which correspond to the case of Example 2.3.2 as $x(k-1)$ correspond to x_1 and $x(k-2)$ to x_2. The first term of $x(k)$ is

$$x(0) = (-1) \otimes_3 x(0-1) \oplus_3 (+1) \otimes_3 x(0-2) = (-1) \otimes_3 0 \oplus_3 (+1) \otimes_3 1 = 1$$

the second one is

$$x(1) = (-1) \otimes_3 x(1-1) \oplus_3 (+1) \otimes_3 x(1-2) = (-1) \otimes_3 1 \oplus_3 (-1) \otimes_3 0 = 2$$

and the third one becomes

$$x(2) = (-1) \otimes_3 x(2-1) \oplus_3 (+1) \otimes_3 x(2-2) = (-1) \otimes_3 2 \oplus_3 (-1) \otimes_3 1 = 2$$

etc. The recursive solution of the difference equation (2.3.4) leads to the column 0′ of the feedback values in Table 2.3.8. The values of this column are shifted to the first register in every step. ■

2.3.3 Generating PRMS-s by polynomial division

The difference equation (2.3.2) can be solved by polynomial division, as well (Davies, 1970). The ratio

$$\frac{D^{(r)}\left(q^{-1}\right)}{C^{(r)}\left(q^{-1}\right)} = X\left(q^{-1}\right) = x(k) + x(k-1)q^{-1} + x(k-2)q^{-2} + \ldots \tag{2.3.5}$$

in modulo- r algebra is the discrete time Laplace transform of the sequence. That means that the values $x(k-i)$ of the sequence are the coefficients of the ith power of the backward delay operator. In (2.3.5) $D^{(r)}\left(q^{-1}\right)$ is a nonzero polynomial with coefficients depending on the initial conditions.

Example 2.3.5 *Generation of a PRBS with maximum length of 7 (Haber, 1990).* From (2.3.3) the characteristic polynomial is

$$C^{(2)}\left(q^{-1}\right) = 1 \ominus_2 (-1) \otimes_2 q^{-1} \ominus_2 (-1) \otimes_2 q^{-2} = 1 \oplus_2 q^{-1} \oplus_2 q^{-2}$$

and let

$$D^{(2)}\left(q^{-1}\right) = 1$$

The modulo-2 polynomial division is done in the following steps:

$$1 \odot_2 \left(1 + q^{-2} + q^{-3}\right) = 1 + q^{-2} + q^{-3} + q^{-4} + q^{-7} + q^{-9} + q^{-10} + \ldots$$

$$1 + q^{-2} + q^{-3}$$

$$q^{-2} + q^{-3}$$

$$q^{-2} + q^{-4} + q^{-5}$$

$$q^{-3}+q^{-4}+q^{-5}$$
$$q^{-3}+q^{-5}+q^{-6}$$
$$\overline{\qquad q^{-4}+q^{-6}}$$
$$q^{-4}+q^{-6}+q^{-7}$$
$$\overline{\qquad q^{-7}}$$
$$q^{-7}+q^{-9}+q^{-10}$$
$$\overline{\qquad q^{-9}+q^{-10}}$$
$$q^{-9}+q^{-11}+q^{-12}$$
$$\overline{\qquad q^{-10}+q^{-11}+q^{-12}}$$

The coefficients of the increasing powers of the backward shift operator are:

1, 0, 1, 1, 1, 0, 0, 1, 0, 1, 1, ...

which are the elements in the column $0'$ of the feedback values in Table 2.3.7. The values of this column are shifted to the first register in every step. ■

Example 2.3.6 *Generation of a PRTS with maximum length of 8 (Haber, 1990).* From (2.3.4) the characteristic polynomial is

$$C^{(3)}\left(q^{-1}\right)=1\ominus_3(-1)\otimes_3 q^{-1}\ominus_3(+1)\otimes_3 q^{-2}=1\oplus_3 q^{-1}\ominus_3 q^{-2}$$

and let

$$D^{(3)}\left(q^{-1}\right)=1$$

The modulo-3 polynomial division is done in the following steps:

$$1\odot_3\left(1+q^{-1}+q^{-2}\right)=1+2q^{-1}+2q^{-2}+2q^{-4}+q^{-5}+q^{-6}+q^{-8}+\ldots$$
$$1+q^{-1}+q^{-2}$$
$$\overline{\qquad 2q^{-2}+2q^{-3}}$$
$$2q^{-2}+2q^{-3}-2q^{-4}$$
$$\overline{\qquad 2q^{-4}}$$
$$2q^{-4}+2q^{-5}-2q^{-6}$$
$$\overline{\qquad q^{-5}+2q^{-6}}$$
$$q^{-5}+q^{-6}-q^{-7}$$
$$\overline{\qquad}$$

$$
\begin{array}{r}
q^{-6}+q^{-7} \\
q^{-6}+q^{-7}-q^{-8} \\
\hline
q^{-8}
\end{array}
$$

The coefficients of the increasing powers of the backward shift operator are:

1, 2, 2, 0, 2, 1, 1, 0, 1, ...

which are the elements in the column 0′ of the feedback values in Table 2.3.8. The values of this column are shifted to the first register in every step. ■

2.3.4 Mapping PRMS-s to centered signals

Consider the PRTS of Example 2.3.2. It has the mean value $(1+0+1+2+2+0+2+1)/8=9/8=1.125$ which differs from the middle value 1 of the levels 0, 1 and 2. Consequently the subtraction of the middle value of the levels from all values does not result in a mean value of zero, which would be advantageous in several applications. A zero mean can be achieved by the following transformation.

Lemma 2.3.1

(Barker, 1969a) A PRMS with the number of levels greater than 2 can be centered by the transformation

$$
\begin{array}{lll}
0 & \rightarrow & 0 \\
1 & \rightarrow & U/[(r-1)/2]=2U/(r-1) \\
2 & \rightarrow & 2U/[(r-1)/2]=4U/(r-1) \\
\vdots & & \vdots \\
(r-1)/2-1 & \rightarrow & (r-3)U/(r-1) \\
(r-1)/2 & \rightarrow & U \\
1+(r-1)/2 & \rightarrow & -U \\
2+(r-1)/2 & \rightarrow & -(r-3)U/(r-1) \\
\vdots & & \vdots \\
(r-2) & \rightarrow & -4U/(r-1) \\
(r-1) & \rightarrow & -2U/(r-1)
\end{array}
$$

where U is the amplitude of the signal (the half of the peak-to-peak value), which is equal to the upper value $(U_{\max})$ and the negative lower value $(-U_{\min})$. ■

Example 2.3.7 *Mapping a PRTS with maximum length of 8 to a centered signal with amplitude 2 (Haber, 1990).*

The transformation equations are with $r=3$:

$$
\begin{array}{lll}
0 & \rightarrow & 0.0 \\
1 & \rightarrow & U=\ 2.0 \\
2 & \rightarrow & -U=-2.0
\end{array}
$$

The transformation of the uncentered sequence of Figure 2.3.5b, repeated in Figure 2.3.7a, results in the signal drawn in Figure 2.3.7b. ■

Now consider the PRBS of Example 2.3.1. It has the mean value $(1+0+0+1+0+1+1)/7=4/7=0.571$ that differs from the mean value of 0 and 1; i.e., (0+1)/2=0.5. There are two possibilities to achieve a mean value of zero.

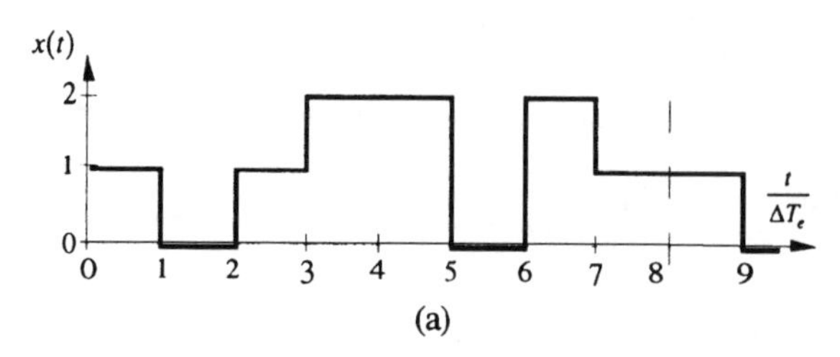

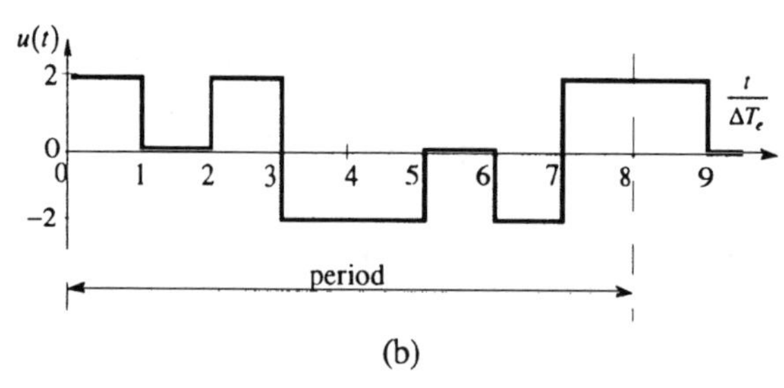

Fig. 2.3.7 Centerization of the PRTS in Example 2.3.7
(a) uncentered sequence (Figure 2.3.5b);
(b) centered sequence

Lemma 2.3.2
A PRBS can be centered by the transformation

$$0 \quad \rightarrow \quad -2U(N_p+1)/(2N_p)$$
$$1 \quad \rightarrow \quad 2U(N_p-1)/(2N_p)$$

where U is the amplitude of the signal (the half of the peak-to-peak value). ■

Example 2.3.8 *Mapping a PRBS with maximum length of 7 to a centered signal with peak-to-peak value 4 (Haber, 1990).*
The transformation equations are with $N_p=7$ and $U=2.0$:

$$0 \quad \rightarrow \quad -2\cdot 2.0(7+1)/(2\cdot 7)=-2.286$$
$$1 \quad \rightarrow \quad 2\cdot 2.0(7-1)/(2\cdot 7)=1.714$$

The transformation of the uncentered sequence of Figure 2.3.3c, repeated in Figure 2.3.8a, results in the signal drawn in Figure 2.3.8b. ■

As is seen, the absolute values of the upper and lower values differ from each other. A PRBS with *symmetrical amplitude* can be generated if the period of the signal would be the double of the *normal* PRBS and the number of upper and lower values would be equal. That is done by the so-called antisymmetric PRBS.

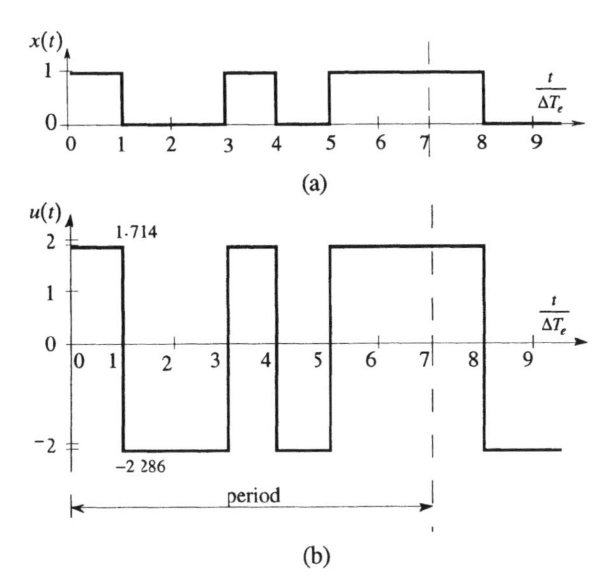

Fig. 2.3.8 Centerization of the PRBS in Example 2.3.8 (a) uncentered sequence (Figure 2.3.3c); (b) centered sequence

Lemma 2.3.3
(Barker and Obidegwu, 1973) The antisymmetric PRBS can be obtained from two periods of the non-centered PRBS by the following transformation

$$0 \rightarrow -U$$
$$1 \rightarrow U$$
$$u(k) \rightarrow (-1)^k u(k)$$

where U is the amplitude of the signal (the half of the peak-to-peak value) and the formula means that the sign of every second value has to be changed. The new signal has a period that is the double of that of the original PRBS. ■

Example 2.3.9 *Mapping a PRBS with maximum length of 7 to an antisymmetric, centered signal with peak-to-peak value 4 (Haber, 1990).*

The sequences of two periods of the output of the last register in Example 2.3.1 (Figure 2.3.3c) are

1, 0, 0, 1, 0, 1, 1, 1, 0, 0, 1, 0, 1, 1,...

The transformation equations of the absolute values of the levels are with $U = 2.0$:

$$0 \rightarrow -2.0$$
$$1 \rightarrow 2.0$$

Above transformation leads to

2.0, −2.0, −2.0, 2.0, −2.0, 2.0, 2.0,
2.0, −2.0, −2.0, 2.0, −2.0, 2.0, 2.0,...

Finally the sign of every second element has to be changed

$$2.0, \quad 2.0, \quad -2.0, \quad -2.0, \quad -2.0, \quad -2.0, \quad 2.0,$$
$$-2.0, \quad -2.0, \quad 2.0, \quad 2.0, \quad 2.0, \quad 2.0, \quad -2.0, \ldots$$

Figure 2.3.9 shows the two periods of the original PRBS (Figure 2.3.9a), the transformation of the amplitudes (Figure 2.3.9b) and the change of the signs of every second value (Figure 2.3.9c). ■

The word antisymmetric originates from the fact that the second half of the period is the opposite of the first half. The antisymmetric PRBS resembles white noise. This can be seen from the auto-correlation function of the signal, which is periodic, but in a given domain has a Dirac delta-like shape.

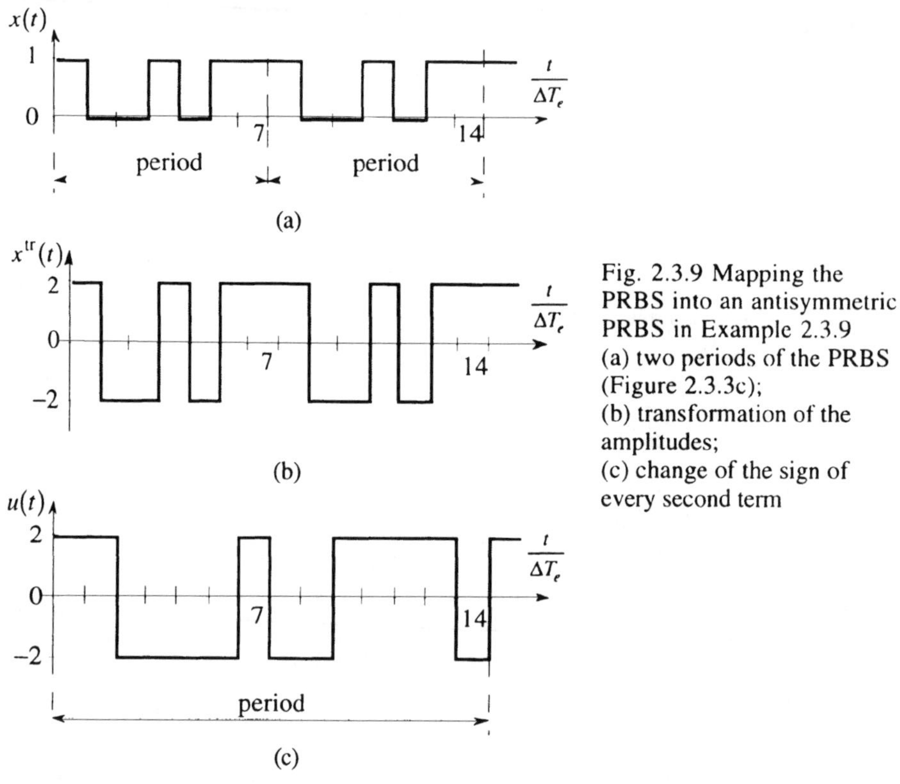

Fig. 2.3.9 Mapping the PRBS into an antisymmetric PRBS in Example 2.3.9 (a) two periods of the PRBS (Figure 2.3.3c); (b) transformation of the amplitudes; (c) change of the sign of every second term

2.3.5 A FORTRAN subroutine for generating PRMS-s

Kónya (1981) wrote a software module in FORTRAN that generates PRMS-s. In the case of a PRBS the signal is not a centered one, in other cases it is centered around the middle value of its uppermost and lowest values. The source code of the module is listed in Table 2.3.9.

The relation of the input parameters of this function to the feedback coefficients is the following:

$$C^{(r)}\left(q^{-1}\right) = 1 \ominus_r IB(1) \otimes_r q^{-IE(1)} \ominus_r IB(2) \otimes_r q^{-IE(2)} \ominus_r IB(3) \otimes_r q^{-IE(3)} \ominus_r$$
$$\ldots \ominus_r IB(M) \otimes_r q^{-IE(M)}$$

TABLE 2.3.9 List of the PRMS generator in FORTRAN

```
      FUNCTION PRMS(IA, LEV,N,M,IE,IB,UMIN,UMAX)
      IMPLICIT INTEGER*2 (I-N)
      DIMENSION IA(1),IB(1),IE(1)
C
C     Pseudo-random multi-level signal (PRMS) generator
C
C  Variable  Description                                   I/O  Type  Dimension
C
C  PRMS      Actual value of the sequence                  O    R
C  IA        Vector of the shift register                  I/O  I     N
C  LEV       Number of levels                              I    I
C  N         Number of registers                           I    I
C  M         Number of registers used for feedback         I    I
C  IE        Serial numbers of the feedback registers      I    I     N
C  IB        Feedback coefficients                         I    I     M
C  UMIN      Lowest level of the signal                    I    R
C  UMAX      Uppermost level of the signal                 I    R
C
C  Subroutines required: None
C
C  Author: Dr. László Kónya
C  Department of Automation, Technical University of Budapest
C  1981
C
      I=LEV/2
      J=IA(N)
      Z=0.5*FLOAT(LEV)
      S=FLOAT(I)
      IF(ABS(Z-S).GT.0.1) GOTO 1
      K=J-I
      IF(K.EQ.0) IY=J
      IF(K.EQ.0) GOTO 2
      IY=K
      GOTO 2
1     IF(J.GT.I) IY=LEV-J
      IF(J.GT.I) GOTO 2
      IY=-J
2     JJ=0
      DO 3 I=1,M
      K=IE(I)
3     JJ=JJ+IA(K)*IB(I)
      K=N-1
      DO 4 J=1,K
      I=N-J
      I1=I+1
4     IA(I1)=IA(I)
5     CONTINUE
      IF(JJ.GE.0.AND.JJ.LT.LEV) GOTO 6
      IF(JJ.GE.LEV) JJ=JJ-LEV
      IF(JJ.LT.0) JJ=JJ+LEV
      GOTO 5
6     IA(1)=JJ
      Z=2.0*FLOAT(LEV/2)
      S=(UMAX-UMIN)*FLOAT(IY)/Z
      Z=0.5*(UMAX+UMIN)
      PRMS=S+Z
      RETURN
      END
```

The vector ***IE*** points out the registers fed back, i.e. the nonzero powers of the backward shift operator in the characteristic polynomial (2.3.2) and the vector ***IB*** contains the feedback coefficients, i.e., the coefficients of the characteristic polynomial. The vector ***IA*** is the vector of the initial conditions of the registers:

$$IA(1) = x(n_r - 1),\ IA(2) = x(n_r - 2), \ldots,\ IA(N-1) = x(1),\ IA(N) = x(0)$$

with $N = n_r$. The generated sequence is the output of the last register.

To make the usage of the function easy, the input parameters of the function are tabulated in Tables 2.3.10 to 2.3.12 for two-, three- and five-level pseudo-random signals with maximal length. The tables are the counterparts of the Tables 2.3.4 to 2.3.6. The vector of the initial conditions of the shift registers is a possible choice, $r^{n_r}-2$ others could be still given. Three examples of calling the function PRMS are presented hereafter.

TABLE 2.3.10 Possible parameters of the FORTRAN module PRMS for generating PRBS-s with maximal length (based on Table 1 from (Barker, 1969b))

No	*Period*	*LEV*	*N*	*M*	*IA*	*IE*	*IB*
1	3	2	2	2	0, 1	1, 2	-1,-1
2	7	2	3	2	0, 0, 1	2, 3	-1,-1
3	7	2	3	2	0, 0, 1	1, 3	-1,-1
4	15	2	4	2	0, 0, 0, 1	3, 4	-1,-1
5	15	2	4	2	0, 0, 0, 1	1, 4	-1,-1
6	31	2	5	4	0, 0, 0, 0, 1	2, 3, 4, 5	-1,-1,-1,-1
7	31	2	5	4	0, 0, 0, 0, 1	1, 2, 3, 5	-1,-1,-1,-1
8	63	2	6	2	0, 0, 0, 0, 0, 1	5, 6	-1,-1
9	63	2	6	2	0, 0, 0, 0, 0, 1	1, 6	-1,-1
10	127	2	7	4	0, 0, 0, 0, 0, 0, 1	3, 4, 5, 7	-1,-1,-1,-1
11	127	2	7	4	0, 0, 0, 0, 0, 0, 1	2, 3, 4, 7	-1,-1,-1,-1
12	255	2	8	4	0, 0, 0, 0, 0, 0, 0, 1	2, 3, 7, 8	-1,-1,-1,-1
13	255	2	8	4	0, 0, 0, 0, 0, 0, 0, 1	1, 6, 7, 8	-1,-1,-1,-1
14	511	2	9	4	0, 0, 0, 0, 0, 0, 0, 0, 1	3, 5, 6, 9	-1,-1,-1,-1
15	511	2	9	4	0, 0, 0, 0, 0, 0, 0, 0, 1	3, 4, 6, 9	-1,-1,-1,-1
16	1023	2	10	6	0, 0, 0, 0, 0, 0, 0, 0, 0, 1	1, 2, 6, 8, 9, 10	-1,-1,-1,-1,-1,-1
17	1023	2	10	6	0, 0, 0, 0, 0, 0, 0, 0, 0, 1	1, 2, 4, 8, 9, 10	-1,-1,-1,-1,-1,-1

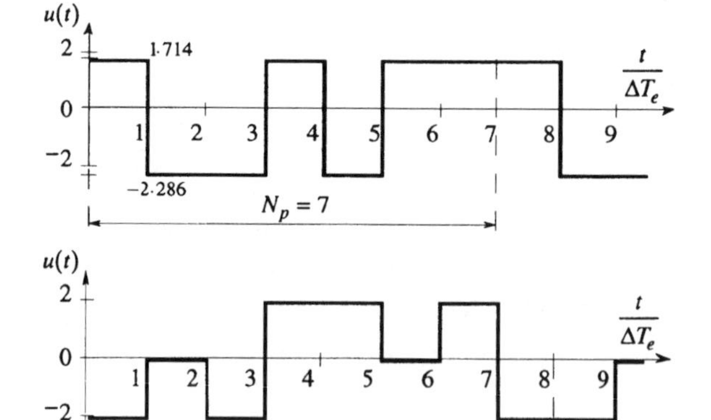

Fig. 2.3.10 The PRBS generated by the FORTRAN module PRMS in Example 2.3.10

Fig. 2.3.11 The PRTS generated by the FORTRAN module PRMS in Example 2.3.11

Example 2.3.10 *Generation of a PRBS with maximum length of 7 (Haber, 1990).* The sequence should be the same as generated by Examples 2.3.1, 2.3.3, and 2.3.5. The input parameters of the function PRMS are tabulated in row 2 of Table 2.3.10: $LEV=2$, $N=3$, $M=2$, $\boldsymbol{IA}=[0,0,1]^{\mathrm{T}}$, $\boldsymbol{IE}=[2,3]^{\mathrm{T}}$, $\boldsymbol{IB}=[-1,-1]^{\mathrm{T}}$. To have a centered signal the lower level of the signal is chosen as $-2\cdot 2(7+1)/(2\cdot 7)=-2.286$ and the upper level as $2\cdot 2(7-1)/(2\cdot 7)=1.714$. The generated signal (Figure 2.3.10) is identical with the sequence given in Figure 2.3.8b. ■

TABLE 2.3.11 Possible parameters of the FORTRAN module PRMS for generating PRTS-s with maximal length (based on Table 2 from (Barker, 1969b))

No	*Period*	*LEV*	*N*	*M*	***IA***	***IE***	***IB***
1	8	3	2	2	0, 1	1, 2	-1, 1
2	8	3	2	2	0, 1	1, 2	1, 1
3	26	3	3	3	0, 0, 1	1, 2, 3	1,-1,-1
4	26	3	3	3	0, 0, 1	1, 2, 3	-1, 1,-1
5	80	3	4	2	0, 0, 0, 1	1, 4	-1, 1
6	80	3	4	2	0, 0, 0, 1	3, 4	1, 1
7	242	3	5	4	0, 0, 0, 0, 1	1, 2, 4, 5	1, 1, 1,-1
8	242	3	5	4	0, 0, 0, 0, 1	1, 3, 4, 5	1, 1, 1,-1
9	728	3	6	5	0, 0, 0, 0, 0, 1	1, 2, 3, 4, 6	-1, 1,-1,-1, 1
10	728	3	6	5	0, 0, 0, 0, 0, 1	1, 2, 3, 4, 6	1, 1,-1, 1, 1

No	*Period*	*LEV*	*N*	*M*	***IA***	***IE***	***IB***
1	24	5	2	2	0, 1	1, 2	-1,-2
2	24	5	2	2	0, 1	1, 2	2, 2
3	124	5	3	3	0, 0, 1	1, 2, 3	-2, 1,-2
4	124	5	3	3	0, 0, 1	1, 2, 3	-2,-1, 2
5	724	5	4	3	0, 0, 0, 1	1, 3, 4	2,-2, 2
6	724	5	4	3	0, 0, 0, 1	1, 3, 4	1,-1,-2

TABLE 2.3.12 Possible parameters of the FORTRAN module PRMS for generating PRQS-s with maximal length (based on Table 3 from (Barker, 1969b))

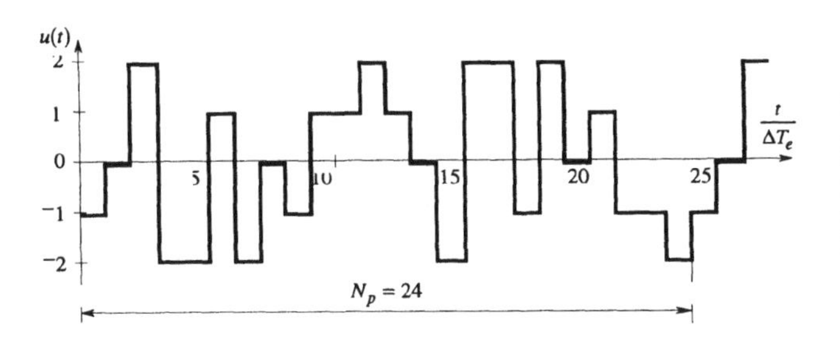

Fig. 2.3.12 The PRQS generated by the FORTRAN module PRMS in Example 2.3.12

Example 2.3.11 *Generation of a PRTS with maximum length of 8 (Haber, 1990).* The sequence should be the same as generated by Examples 2.3.2, 2.3.4, and 2.3.6. The input parameters of the function PRMS are tabulated in row 1 of Table 2.3.11: $LEV = 3$, $N = 2$, $M = 2$, $\boldsymbol{IA} = [0,1]^T$, $\boldsymbol{IE} = [1,2]^T$, $\boldsymbol{IB} = [-1,1]^T$. The lowest level of the signal is chosen as -2.0 and the uppermost as 2.0. The generated signal is plotted in Figure 2.3.11. The values are equal to the those of Figure 2.3.7 if they are multiplied by -1. The signal of Figure 2.3.7 was the result of the manual calculations of Examples 2.3.2, 2.3.4 or 2.3.6 and the centering transformation in Example 2.3.7. ■

Example 2.3.12 *Generation of a PRQS with maximum length of 24 (Haber, 1990).* A possible set of the input parameters of the function PRMS are tabulated in row 1 of Table 2.3.12: $LEV = 5$, $N = 2$, $M = 2$, $\boldsymbol{IA} = [0,1]^T$, $\boldsymbol{IE} = [1,2]^T$, $\boldsymbol{IB} = [-1,-2]^T$. The amplitude of the signal is chosen as 2: $UMIN = -2.0$, $UMAX = 2.0$. One period of the sequence is seen in Figure 2.3.12. ■

2.4 TIME FUNCTIONS OF PRMS-S WITH MAXIMAL LENGTH

2.4.1 Two-level signals
Some not centered PRBS-s were generated with the parameters of Table 2.3.10. For each period only one signal, the first one in the Table, was generated. Signals with too long periods do not give a clear picture and are omitted. The lower and upper levels are -2.0 and 2.0, respectively. A zero-order holding device is assumed between the generated points of the sequences. Figure 2.4.1 shows the plots of the time functions.

2.4.2 Three-level signals
The centered PRTS-s were generated with the parameters of Table 2.3.11. For each period only one signal, the first one in the Table, was generated. Signals with too long periods do not give a clear picture and are omitted. The lowest and uppermost levels are -2.0 and 2.0, respectively. A zero-order holding device is assumed between the generated points of the sequences. Figure 2.4.2 shows the plots of the time functions.

2.4.3 Five-level signals
The centered PRQS-s were generated with the parameters of Table 2.3.12. For each period only one signal, the first one in the Table, was generated. Signals with too long periods do not give a clear picture and are omitted. The lowest and uppermost levels are -2.0 and 2.0, respectively. A zero-order holding device is assumed between the generated points of the sequences. Figure 2.4.3 shows the plots of the time functions.

2.4.4 Properties of the signals
The properties are demonstrated on the following signals:

- PRBS with maximal length 31 (Figure 2.4.1d);
- PRTS with maximal length 26 (Figure 2.4.2b);
- PRQS with maximal length 24 (Figure 2.4.3a).

They are redrawn in Figures 2.4.4 to 2.4.6, respectively. This figures show the occurrence of the same levels in turn by two numbers in parentheses (value, number of repeats).

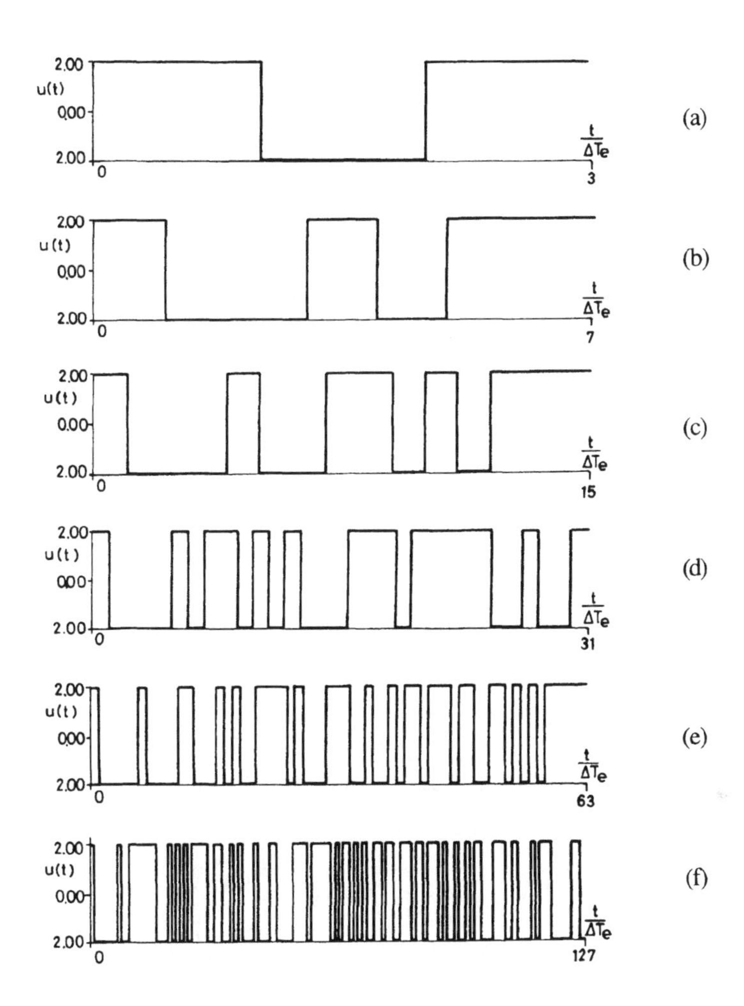

Fig. 2.4.1 Plots of the PRBS-s generated with the parameters of Table 2.3.10 (In the case of two sequences with the same period only the first one is given.)
(a) $N_p = 3$; (b) $N_p = 7$; (c) $N_p = 15$; (d) $N_p = 31$; (e) $N_p = 63$; (f) $N_p = 127$

Property 2.4.1 The period of the sequence is $N_p = r^{n_r} - 1$.

Proof. n_r shift registers with r levels each may have r^{n_r} different states. The state that all registers are zero is excluded because then all succeeding states would be zero. ■

Example 2.4.1 *PRBS with maximum length of 31 (Figure 2.4.4).*

In this case $N_p = 2^5 - 1 = 31$, as seen in Figure 2.4.4. ■

Example 2.4.2 *PRTS with maximum length of 26 (Figure 2.4.5).*

In this case $N_p = 3^3 - 1 = 26$, as seen in Figure 2.4.5. ■

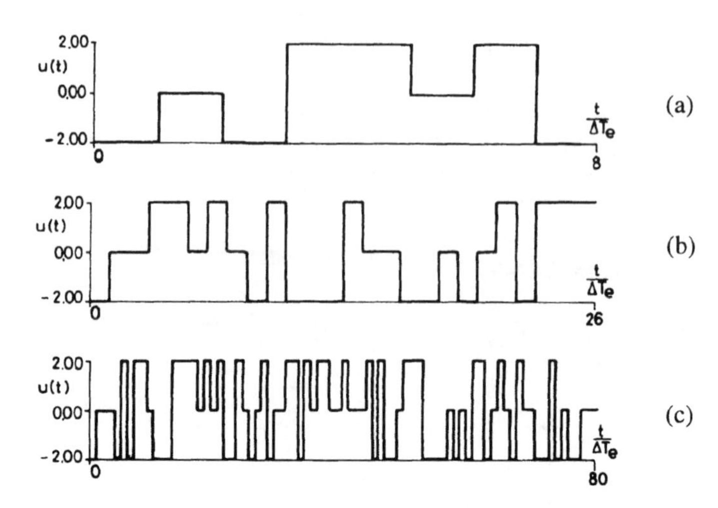

Fig. 2.4.2 Plots of the PRTS-s generated with the parameters of Table 2.3.11
(In case of two sequences with the same period only the first one is given.)
(a) $N_p = 8$; (b) $N_p = 26$; (c) $N_p = 80$

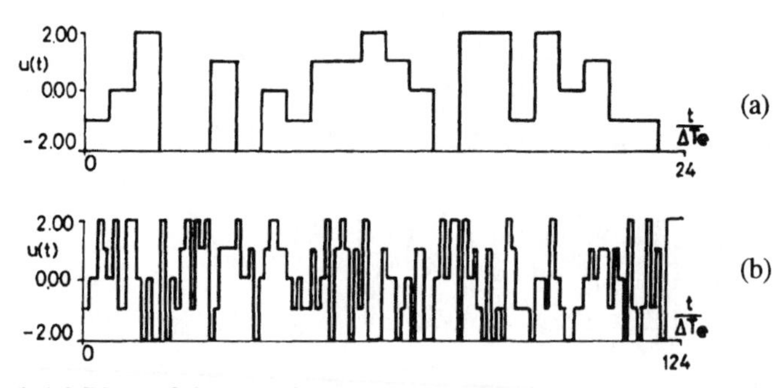

Fig. 2.4.3 Plots of the PRQS-s generated with the parameters of Table 2.3.12
(In case of two sequences with the same period only the first one is given.)
(a) $N_p = 24$; (b) $N_p = 124$

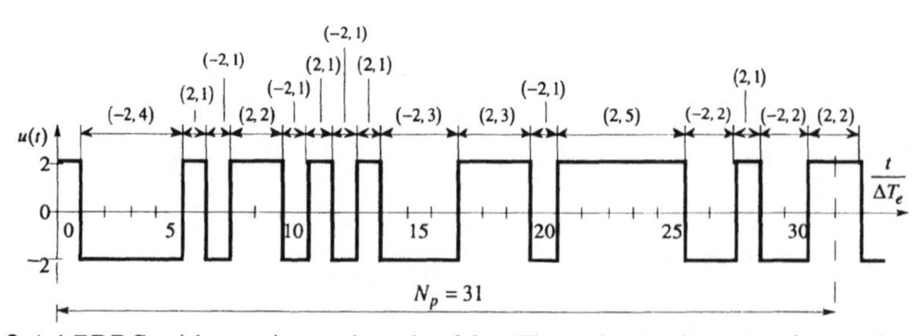

Fig. 2.4.4 PRBS with maximum length of 31. The pairs in the parentheses show how often (second number) the same value (first number) occurs in turn in the sequence.

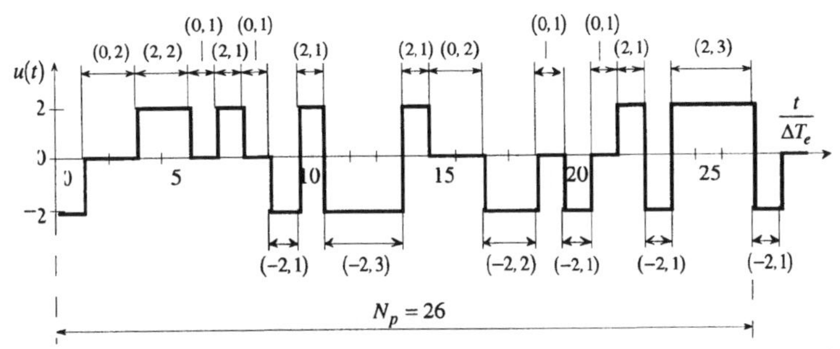

Fig. 2.4.5 PRTS with maximum length of 26. The pairs in the parentheses show how often (second number) the same value (first number) occurs in turn in the sequence.

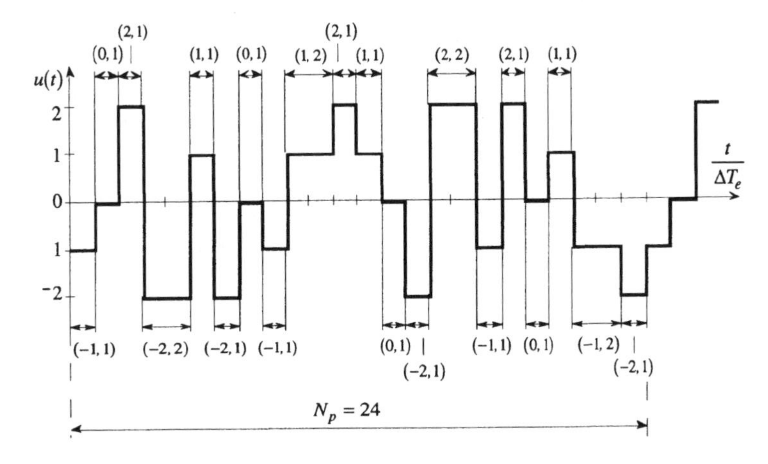

Fig. 2.4.6 PRQS with maximum length of 24. The pairs in the parentheses show how often (second number) the same value (first number) occurs in turn in the sequence.

Example 2.4.3 *PRQS with maximum length of 24 (Figure 2.4.6).*
In this case $N_p = 5^2 - 1 = 24$, as seen in Figure 2.4.6. ■

Definition 2.4.1 Denote the lower value of the PRBS and the medium value of the other sequences by u^0. (The index 0 means that these values arise from the zero value in the Galois field.) ■

Property 2.4.2 (Davies, 1970) The values occur r^{n_r-1}-times except the u^0 value that occurs $\left(r^{n_r-1}-1\right)$-times in one period.

Proof. This is a corollary of the proof of Property 2.4.1.

The period is $N_p = r^{n_r} - 1 = r \cdot r^{n_r-1} - 1$. r levels occur $\left(N_p + 1\right)/r = r^{n_r-1}$-times except the value u^0, which occurs by once less. ■

Example 2.4.4 *PRBS with maximum length of 31 (Figure 2.4.4).*
The value 2.0 occurs $2^{5-1} = 16$-times and the value -2.0 16-1=15-times, as is seen in Figure 2.4.4. ■

Example 2.4.5 *PRTS with maximum length of 26 (Figure 2.4.5).*
The values -2.0 and 2.0 occur $3^{3-1} = 9$-times and the value 0.0 9-1=8-times, as is seen in Figure 2.4.5. ■

Example 2.4.6 *PRQS with maximum length of 24 (Figure 2.4.6).*
The values -2.0, -1.0, 1.0 and 2.0 occur $5^{2-1} = 5$-times and the value 0.0 5-1=4-times, as is seen in Figure 2.4.6. ■

Property 2.4.3 The amplitude distribution of a PRMS is discontinuous. The relative frequency distribution of a centered PRMS with more than two levels is symmetrical to its mean value. The relative frequency values are:

- at the lower value of the PRBS and in the mean value of a centered higher-level PRMS

$$\frac{r^{n_r-1}-1}{r^{n_r}-1} = \frac{1}{r}\left(1+\frac{1}{N_p}\right)-\frac{1}{N_p}$$

- at the upper value of the PRBS and at every value of a centered higher-level PRMS except the mean value

$$\frac{r^{n_r-1}}{r^{n_r}-1} = \frac{1}{r}\left(1+\frac{1}{N_p}\right)$$

Proof. Property 2.4.3 is a corollary of Property 2.4.2. ■

Example 2.4.7 *PRBS with maximum length of 31 (Figure 2.4.4).*
The relative frequency values are as follows:

1. at -2.0: $(1/2)(1+1/31)-1/31 = 0.484$, which is equal to 15/31,
2. at 2.0: $(1/2)(1+1/31) = 0.516$, which is equal to 16/31.

Their sum is $0.484+0.516 = 1$. The histogram is drawn in Figure 2.4.7. ■

Example 2.4.8 *PRTS with maximum length of 26 (Figure 2.4.5).*
The relative frequency values are as follows:

1. at -2.0: $(1/3)(1+1/26) = 0.346$, which is equal to 9/26,
2. at 0.0: $(1/3)(1+1/26)-1/26 = 0.308$, which is equal to 8/26,
3. at 2.0: $(1/3)(1+1/26) = 0.346$, which is equal to 9/26.

Their sum is $0.346+0.308+0.346 = 1$. The histogram is drawn in Figure 2.4.8. ■

Example 2.4.9 *PRQS with maximum length of 24 (Figure 2.4.6).*
The relative frequency values are as follows:

1. at -2.0: $(1/5)(1+1/24) = 0.208$, which is equal to 5/24,

2. at -1.0: $(1/5)(1+1/24) = 0.208$, which is equal to 5/24,
3. at 0.0: $(1/5)(1+1/24) - 1/24 = 0.167$, which is equal to 4/24,
4. at 1.0: $(1/5)(1+1/24) = 0.208$, which is equal to 5/24,
5. at 2.0: $(1/5)(1+1/24) = 0.208$, which is equal to 5/24.

Their sum is $4 \cdot 0.208 + 0.167 = 1$. The histogram is drawn in Figure 2.4.9. ■

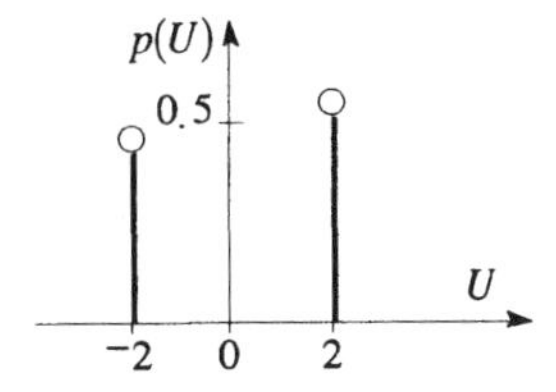

Fig. 2.4.7 Histogram of the relative frequency values for the PRBS of maximal length 31

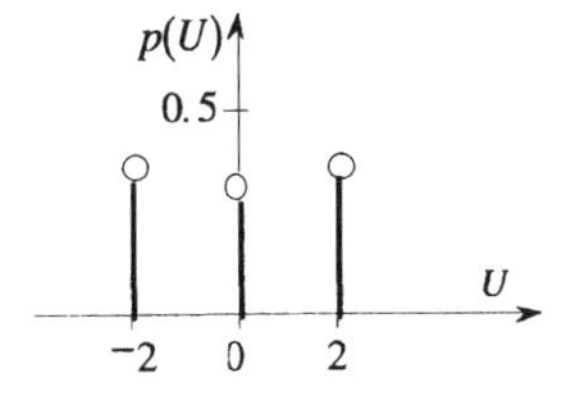

Fig. 2.4.8 Histogram of the relative frequency values for the PRTS of maximal length 26

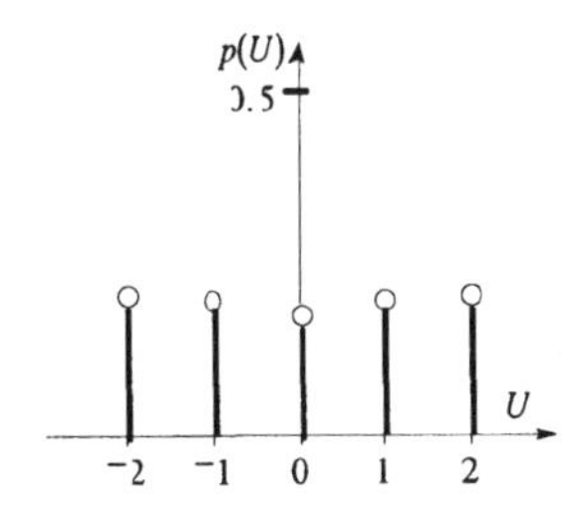

Fig. 2.4.9 Histogram of the relative frequency values for the PRQS of maximal length 24

Property 2.4.4 Let $U_{\max}$ the uppermost and $U_{\min}$ the lowest value of a PRMS. Then the mean value of the PRBS is

$$\bar{u} = \frac{U_{\max} + U_{\min}}{2} + \frac{U_{\max} - U_{\min}}{2N_p}$$

and that of the other, centered, PRMS-s is

$$\bar{u} = \frac{U_{\max} + U_{\min}}{2}$$

Proof. We use the results of Property 2.4.2.

(a) case PRBS:

The upper value occurs $r^{n_r - 1} = r^{n_r}/2 = \left(N_p + 1\right)/2$-times and the lower value $r^{n_r - 1} - 1 = r^{n_r}/2 - 1 = \left(N_p + 1\right)/2 - 1 = \left(N_p - 1\right)/2$-times. Thus the mean value is

$$\bar{u} = \frac{U_{\max}\frac{N_p+1}{2} + U_{\min}\frac{N_p-1}{2}}{N_p} = \frac{U_{\max}+U_{\min}}{2} + \frac{U_{\max}-U_{\min}}{2N_p}$$

(b) case PRMS except PRBS:

All values except the middle one occur the same times, therefore the mean value is equal to the middle value, which is $(U_{\max}+U_{\min})/2$. ■

Example 2.4.10 *PRBS with maximum length of 31 (Figure 2.4.4).*
The lower and upper values are 2.0, and -2.0, respectively. The mean value is

$$\bar{u} = \frac{2.0+(-2.0)}{2} + \frac{2.0-(-2.0)}{2 \cdot 31} = 0.0645$$ ■

Example 2.4.11 *PRTS with maximum length of 26 (Figure 2.4.5).*
Since the uppermost and the lowest values have the same absolute values but opposite signs, both the middle and the mean values are zero. ■

Example 2.4.12 *PRQS with maximum length of 24 (Figure 2.4.6).*
Since the uppermost and the lowest values have the same absolute values but opposite signs, both the middle and the mean values are zero. ■

Property 2.4.5 (Barker, 1969a) The second half of a centered PRMS with more than two levels is the negative of the first half

$$u(k) = -u\left(k + \frac{N_p}{2}\right)$$ ■

This feature of the signal is called also antisymmetric or inverse-repeat property. A PRBS is not an inverse-repeat one, but an antisymmetric PRBS with the twice of the period of the normal PRBS can be generated (See Lemma 2.3.3.). ■

Example 2.4.13 *PRTS with maximum length of 26 (Figure 2.4.5).*
The first half of the sequence is

$$-2, 0, 0, 2, 2, 0, 2, 0, -2, 2, -2, -2, -2$$

which is the negative of the second half

$$2, 0, 0, -2, -2, 0, -2, 0, 2, -2, 2, 2, 2$$ ■

Example 2.4.14 *PRQS with maximum length of 24 (Figure 2.4.6).*
The first half of the sequence is

$$-1, 0, 2, -2, -2, 1, -2, 0, -1, 1, 1, 2$$

which is the negative of the second half

$$1, 0, -2, 2, 2, -1, 2, 0, 1, -1, -1, -2$$ ■

Property 2.4.6 (Dotsenko *et al.*, 1971) The following relation to the delayed parts of the not centered sequence exists

$$x(k) = \alpha^i \otimes_r x\left(k - \frac{iN_p}{r-1}\right), \qquad i = 0, 1, 2, 3, \ldots \tag{2.4.1}$$

where $0 \le \alpha \le r-1$. ■

Remarks:

1. For $r = 2$ (2.4.1) says only that a PRBS has the period N_p.
2. For $i = 0$ (2.4.1) reduces itself to an identity $x(k) = x(k)$.
3. For $i = j(r-1); j = 1, 2, 3, \ldots$ (2.4.1) shows only that the sequence has the period N_p because then

   $$x(k) = \alpha^{j(r-1)} \otimes_r x(k - jN_p), \qquad j = 1, 2, 3, \ldots$$

   and

   $$\alpha^{j(r-1)} = 1 \qquad j = 1, 2, 3, \ldots$$

   in modulo-r algebra.
4. Only the values $i = 1, \ldots, (r-1)$ are of interest because $\alpha^i = \alpha^{i-jr}; j = 1, 2, 3, \ldots$ in modulo-r.

Example 2.4.15 *PRTS with maximum length of 26 (Figure 2.4.5).*
The not centered signal can be obtained from the centered one by the transformation

$$-2 \to 2 \qquad 0 \to 0 \qquad 2 \to 1$$

The only case $i = 1$ has to be investigated because $i = 2$ shows only that the sequence has the period $N_p = 26$. The first half of the not centered sequence is

$$x_1(k): \qquad 2, 0, 0, 1, 1, 0, 1, 0, 2, 1, 2, 2, 2$$

and the second half is

$$x_2(k): \qquad 1, 0, 0, 2, 2, 0, 2, 0, 1, 2, 1, 1, 1$$

It can be checked that with α=2

$$x_1(k) = 2 \otimes_3 x_2(k) \qquad \text{for} \qquad k = 1, 2, \ldots, 13$$

■

Example 2.4.16 *PRQS with maximum length of 24 (Figure 2.4.6).*
The not centered signal can be obtained from the centered one by the transformation

$$-2 \to 3 \qquad -1 \to 4 \qquad 0 \to 0 \qquad 1 \to 1 \qquad 2 \to 2$$

Now the cases $i = 1, 2, 3$ have to be investigated. The sequence is

4, 0, 2, 3, 3, 1, 3, 0, 4, 1, 1, 2,
1, 0, 3, 2, 2, 4, 2, 0, 1, 4, 4, 3,...

It can be seen that

$$x(k) = 3x(k-6) = 3x(k-1\cdot 24/4) \quad \text{for} \quad \alpha = 3 \quad i = 1$$
$$x(k) = 4x(k-12) = 4x(k-2\cdot 24/4) \quad \text{for} \quad \alpha = 4 \quad i = 2$$
$$x(k) = 2x(k-18) = 2x(k-3\cdot 24/4) \quad \text{for} \quad \alpha = 2 \quad i = 3$$

■

Definition 2.4.2 *Run* is the successive occurrence of the same state in the sequence (Davies, 1970). ■

Definition 2.4.3 *Number of runs* is the number of runs during one period (Davies, 1970). Denote them by $N_r(u,m)$, where m is the number in turn of the same value u. ■

Property 2.4.7 (Dotsenko *et al.*, 1971) The number of runs with the lengths $\ell_r = 1, 2, \ldots, n_r - 2$ is

$$N_r(u, \ell_r) = r^{n_r - 2 - \ell_r}(r-1)^2 \tag{2.4.2}$$

The number of the runs is the same for the different values u. ■

Example 2.4.17 *PRBS with maximum length of 31 (Figure 2.4.4).*
Since the number of registers is $n_r = 5$, the number of runs can be calculated by Property 2.4.7 only for the lengths of the runs $\ell_r = 1, 2, 3$.

$$N_r(-2.0,\ 1) = N_r(2.0,\ 1) = 2^{5-2-1}(2-1)^2 = 4$$
$$N_r(-2.0,\ 2) = N_r(2.0,\ 1) = 2^{5-2-2}(2-1)^2 = 2$$
$$N_r(-2.0,\ 3) = N_r(2.0,\ 1) = 2^{5-2-3}(2-1)^2 = 1$$

The correctness of the above values is easy to check in Figure 2.4.4. ■

Example 2.4.18 *PRTS with maximum length of 26 (Figure 2.4.5).*
Since the number of registers is $n_r = 3$, the number of runs can be calculated by Property 2.4.7 only for the length of the runs $\ell_r = 1$.

$$N_r(-2.0,\ 1) = N_r(0.0,\ 1) = N_r(2.0,\ 1) = 2^{3-2-1}(3-1)^2 = 4$$

The correctness of the above values is easy to check in Figure 2.4.5. ■

Example 2.4.19 *PRQS with maximum length of 24 (Figure 2.4.6).*
Since the number of registers is $n_r = 2$, the number of runs cannot be calculated by Property 2.4.7. ■

Property 2.4.8 (Dotsenko *et al.*, 1971) There is no run of length n_r of u^0, and its longest run has $(r-1)$ elements:

$$N_r(u^0, n_r) = 0 \tag{2.4.3}$$

$$N_r(u^0, n_r - 1) = r - 1 \tag{2.4.4}$$

All other values occur with the length equal to the number of registers once and $r-2$-times with a length shorter by one:

$$N_r(u \neq u^0, n_r) = 1 \tag{2.4.5}$$

$$N_r(u \neq u^0, n_r - 1) = r - 2 \tag{2.4.6}$$

■

Example 2.4.20 *PRBS with maximum length of 31 (Figure 2.4.4).*
Now $r = 2$ and $n_r = 5$. From (2.4.3) and (2.4.4) we obtain

$$N_r(-2.0, 5) = 0 \qquad N_r(-2.0, 4) = 1$$

and from (2.4.5) and (2.4.6)

$$N_r(2.0, 5) = 1 \qquad N_r(2.0, 4) = 0$$

The correctness of the above values is easy to check in Figure 2.4.4. ■

Example 2.4.21 *PRTS with maximum length of 26 (Figure 2.4.5).*
Now $r = 3$ and $n_r = 3$. From (2.4.3) and (2.4.4) we obtain

$$N_r(0, 3) = 0 \qquad N_r(0, 2) = 2$$

and from (2.4.5) and (2.4.6)

$$N_r(-2.0, 3) = 1, \quad N_r(-2.0, 2) = 1, \quad N_r(2.0, 3) = 1, \quad N_r(2.0, 2) = 1$$

The correctness of the above values is easy to check in Figure 2.4.5. ■

Example 2.4.22 *PRQS with maximum length of 24 (Figure 2.4.6).*
Now $r = 5$ and $n_r = 2$. From (2.4.3) and (2.4.4) we obtain

$$N_r(0, 2) = 0 \qquad N_r(0, 1) = 4$$

and from (2.4.5) and (2.4.6)

$$N_r(-2.0, 2) = 1, \qquad N_r(-2.0, 1) = 3, \qquad N_r(-1.0, 2) = 1$$

$$N_r(-1.0, 1) = 3, \qquad N_r(1.0, 2) = 1, \qquad N_r(1.0, 1) = 3$$

$$N_r(2.0, 2) = 1, \qquad N_r(2.0, 1) = 3$$

The correctness of the above values is easy to check in Figure 2.4.6. ■

Property 2.4.9 (Dotsenko *et al.*, 1971) The sum of all runs of each value is $N_r = r^{n_r-2}(r-1)$. ■

Example 2.4.23 *PRBS with maximum length of 31 (Figure 2.4.4).*
As $r=2$ and $n_r=5$ the number of all runs of each value is

$$N_r = 2^{5-2}(2-1) = 8$$

This can be checked by means of the runs of different lengths calculated already both for the value -2.0

$$N_r = \sum_{i=1}^{5} N_r(-2.0, i) = 4+2+1+1+0 = 8$$

and for the value 2.0

$$N_r = \sum_{i=1}^{5} N_r(2.0, i) = 4+2+1+0+1 = 8$$

Example 2.4.24 *PRTS with maximum length of 26 (Figure 2.4.5).*
As $r=3$ and $n_r=3$ the number of all runs of each value is

$$N_r = 3^{3-2}(3-1) = 6$$

This can be checked by means of the runs of different lengths calculated already both for the value 0.0

$$N_r = \sum_{i=1}^{3} N_r(0.0, i) = 4+2+0 = 6$$

and for the values -2.0 and 2.0

$$N_r = \sum_{i=1}^{3} N_r(-2.0, i) = \sum_{i=1}^{3} N_r(2.0, i) = 4+1+1 = 6$$ ■

Example 2.4.25 *PRQS with maximum length of 24 (Figure 2.4.6).*
As $r=5$ and $n_r=2$ the number of all runs of each value is

$$N_r = 5^{2-2}(5-1) = 4$$

This can be checked by means of the runs of different lengths calculated already both for the value 0.0

$$N_r = \sum_{i=1}^{2} N_r(0.0, i) = 4+0 = 4$$

and for the values -2.0, -1.0, 1.0 and 2.0

$$N_r = \sum_{i=1}^{2} N_r(-2.0, i) = \sum_{i=1}^{2} N_r(-1.0, i) = \sum_{i=1}^{2} N_r(1.0, i) = \sum_{i=1}^{2} N_r(2.0, i) = 1 + 3 = 4$$

■

2.5 AUTO-CORRELATION FUNCTIONS OF PRMS-S WITH MAXIMAL LENGTH

2.5.1 Relation between the continuous and the discrete time auto-correlation functions

Definition 2.5.1 The discrete time nth-order auto-correlation function of the test signal sequence $u(k)$ with periodicity N_p is

$$r_u(\kappa_1, \ldots, \kappa_n) = \frac{1}{N_p} \sum_{k=0}^{N_p - 1} u(k) \ldots u(k + \kappa_n)$$

$u(k+\kappa)$ means $u(k+\kappa \Delta T_e)$ where ΔT_e is the minimum switching time of the exciting pseudo-random signal. ■

Definition 2.5.2 The continuous time nth-order auto-correlation function of the test signal $u(t)$ with periodicity $N_p \Delta T_e$ is

$$r_u(\tau_1, \ldots, \tau_n) = \frac{1}{N_p \Delta T_e} \int_{t=0}^{N_p \Delta T_e} u(t) \ldots u(t + \tau_n) dt$$ ■

Property 2.5.1 (Haber, 1990) Define the following:

- $\kappa_i = \mathrm{entier}(\tau_i / \Delta T_e) \qquad i = 1, \ldots, n$
- $\tau_i^* = \tau_i - \kappa_i \qquad i = 1, \ldots, n$
- $\tau_1^* \geq \tau_2^* \geq \ldots \geq \tau_n^*$

That means that the τ_i^*-s are in decreasing order. The continuous time nth order auto-correlation function can be calculated from the corresponding discrete time nth order auto-correlation functions by the formula

$$\begin{aligned} r_u(\tau_1, \ldots, \tau_n) = & \left(1 - \frac{\tau_1^*}{\Delta T_e}\right) r_u(\kappa_1, \ldots, \kappa_{n-1}, \kappa_n) \\ & + \frac{1}{\Delta T_e}(\tau_1^* - \tau_2^*) r_u(\kappa_1 + 1, \kappa_2, \ldots, \kappa_{n-1}, \kappa_n) \\ & + \frac{1}{\Delta T_e}(\tau_2^* - \tau_3^*) r_u(\kappa_1 + 1, \kappa_2 + 1, \ldots, \kappa_{n-1}, \kappa_n) \\ & \ldots + \frac{1}{\Delta T_e}(\tau_{n-1}^* - \tau_n^*) r_u(\kappa_1 + 1, \kappa_2 + 1, \ldots, \kappa_{n-1} + 1, \kappa_n) \\ & + \frac{1}{\Delta T_e} \tau_n^* r_u(\kappa_1 + 1, \kappa_2 + 1, \ldots, \kappa_{n-1} + 1, \kappa_n + 1) \end{aligned}$$

$$= \sum_{i=1}^{n+1} \frac{1}{\Delta T_e}\left(\tau_{i-1}^* - \tau_i^*\right) r_u\left(\kappa_1 + 1, \kappa_2 + 1, \ldots, \kappa_{i-1} + 1, \kappa_i, \kappa_{i+1}, \ldots, \kappa_{n-1}, \kappa_n\right)$$

where

$$\tau_0^* \equiv \Delta T_e$$

and

$$\tau_{n+1}^* \equiv 0$$

Proof. In order to illustrate the different shifted signals $u(t+\tau_i)$ Figure 2.5.1a shows the PRTS of Example 2.3.7 (Figure 2.3.7). The relation between the discrete and continuous times is $t = k\Delta T_e$. The continuous time $t = 0$ correspond to the discrete time $k = 0$. Two shifts are seen with τ_1 and τ_2 thus that $\tau_1^* \geq \tau_2^*$ Figure 2.5.1b, 2.5.1c and 2.5.1d show the same signal shifted by $\kappa_1\Delta T_e$, τ_1 and $(\kappa_1+1)\Delta T_e$ and Figure 2.5.1e, 2.5.1f and 2.5.1g show the same signal shifted by $\kappa_2\Delta T_e$, τ_2 and $(\kappa_2+1)\Delta T_e$ respectively. The continuous time higher-order auto-correlation function can be calculated as

$$r_u(\tau_1, \ldots, \tau_n) = \frac{1}{N_p\Delta T_e} \int_{t=0}^{N_p\Delta T_e} u(t)u(t+\tau_1)\ldots u(t+\tau_n)dt$$

$$= \frac{1}{N_p\Delta T_e}\left[\int_{t=0}^{\Delta T_e - \tau_1^*} u(t)u(t+\kappa_1\Delta T_e)\ldots u(t+\kappa_n\Delta T_e)dt\right.$$

$$+ \int_{t=\Delta T_e - \tau_1^*}^{\Delta T_e - \tau_2^*} u(t)u(t+(\kappa_1+1)\Delta T_e)u(t+\kappa_2\Delta T_e)\ldots u(t+\kappa_n\Delta T_e)dt + \ldots$$

$$+ \int_{t=\Delta T_e}^{2\Delta T_e - \tau_1^*} u(t)u(t+\kappa_1\Delta T_e)u(t+\kappa_2\Delta T_e)\ldots u(t+\kappa_n\Delta T_e)dt$$

$$\left. + \int_{t=2\Delta T_e - \tau_1^*}^{2\Delta T_e - \tau_2^*} u(t)u(t+(\kappa_1+1)\Delta T_e)u(t+\kappa_2\Delta T_e)\ldots u(t+\kappa_n\Delta T_e)dt + \ldots\right]$$

$$= \frac{1}{N_p\Delta T_e}\left[\left(\Delta T_e - \tau_1^*\right)u(0\Delta T_e)u(\kappa_1\Delta T_e)u(\kappa_2\Delta T_e)\ldots u(\kappa_n\Delta T_e)\right.$$

$$+\left(\tau_1^* - \tau_2^*\right)u(0\Delta T_e)u((\kappa_1+1)\Delta T_e)u(\kappa_2\Delta T_e)\ldots u(\kappa_n\Delta T_e) + \ldots$$

$$+\left(\Delta T_e - \tau_1^*\right)u(\Delta T_e)u((1+\kappa_1)\Delta T_e)u((1+\kappa_2)\Delta T_e)\ldots u((1+\kappa_n)\Delta T_e)$$

$$\left.+\left(\tau_1^* - \tau_2^*\right)u(\Delta T_e)u([1+(\kappa_1+1)]\Delta T_e)u((1+\kappa_2)\Delta T_e)\ldots u((1+\kappa_n)\Delta T_e) + \ldots\right]$$

$$= \frac{1}{N_p\Delta T_e}\left[\left(\Delta T_e - \tau_1^*\right)\sum_{k=0}^{N_p-1} u(k\Delta T_e)u((k+\kappa_1)\Delta T_e)u((k+\kappa_2)\Delta T_e)\ldots u((k+\kappa_n)\Delta T_e)\right.$$

$$+\left(\tau_1^* - \tau_2^*\right) \sum_{k=0}^{N_p - 1} u(k\Delta T_e) u\left((k+\kappa_1+1)\Delta T_e\right) u\left((k+\kappa_2)\Delta T_e\right) \ldots u\left((k+\kappa_n)\Delta T_e\right) + \ldots \Bigg]$$

$$= \left(1 - \frac{\tau_1^*}{\Delta T_e}\right) r_u(\kappa_1, \ldots, \kappa_n) + \frac{1}{\Delta T_e}\left(\tau_1^* - \tau_2^*\right) r_u(\kappa_1 + 1, \kappa_2, \ldots, \kappa_n) + \ldots \qquad \text{q.e.d.} \blacksquare$$

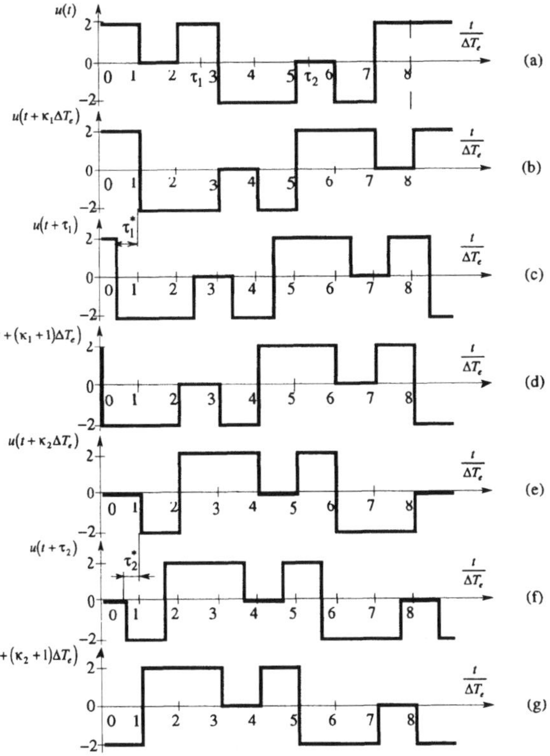

Fig. 2.5.1 PRTS with period 8 from Example 2.3.2
(a) the original sequence from Figure 2.3.5a;
(b) as (a) but shifted left by $\kappa_1 \Delta T_e$, $\kappa_1 = 2$;
(c) as (a) but shifted left by τ_1^*, $2\Delta T_e \le \tau_1 \le 3\Delta T_e$;
(d) as (a) but shifted left by $(\kappa_1 + 1)\Delta T_e$, $\kappa_1 + 1 = 3$;
(e) as (a) but shifted left by $\kappa_2 \Delta T_e$, $\kappa_2 = 5$;
(f) as (a) but shifted left by τ_2^*, $5\Delta T_e \le \tau_2 \le 6\Delta T_e$;
(g) as (a) but shifted left by $(\kappa_2 + 1)\Delta T_e$, $\kappa_2 + 1 = 6$

Corollary 2.5.1 The first-order continuous time auto-correlation function $r_{uu}(\tau)$ can be calculated from the corresponding discrete time auto-correlation function $r_{uu}(\kappa)$ by

$$r_{uu}(\tau) = \left(1 - \frac{\tau^*}{\Delta T_e}\right) r_{uu}(\kappa) + \frac{\tau^*}{\Delta T_e} r_{uu}(\kappa + 1)$$

with

$$\kappa = \text{entier}(\tau / \Delta T_e) \qquad \tau^* = \tau - \kappa$$

Proof. Apply Property 2.5.1 for $n = 1$ and replace τ_1 and κ_1 by τ and κ, respectively. A detailed proof for the first-order auto-correlation function is presented in (Krempl, 1973). $\blacksquare$

Consequently it is enough to compute the discrete time auto-correlation function of the sampled PRMS if the continuous time auto-correlation function is wanted. In the sampling points the continuous and the discrete time auto-correlation functions coincide. The values between the sampled auto-correlation function values can be calculated by interpolating linearly between the neighboring discrete time values.

Henceforth, if the order of the auto-correlation function is not given explicitly then it means the first-order one. The (first-order) auto-correlation function of the discrete time signal $x(k)$ is denoted by $r_{xx}(\kappa)$ and that of the continuous time signal $x(t)$ by $r_{xx}(\tau)$.

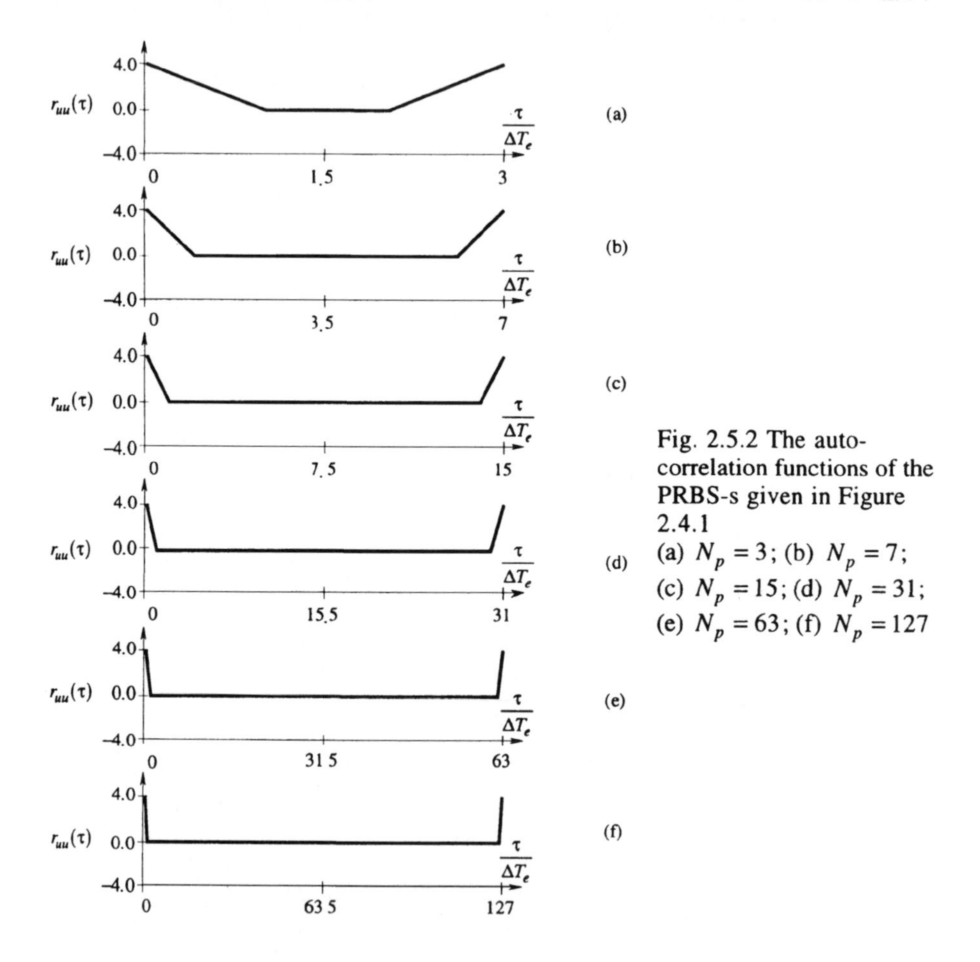

Fig. 2.5.2 The auto-correlation functions of the PRBS-s given in Figure 2.4.1
(a) $N_p = 3$; (b) $N_p = 7$; (c) $N_p = 15$; (d) $N_p = 31$; (e) $N_p = 63$; (f) $N_p = 127$

2.5.2 Two-level signals

Figure 2.5.2 shows the (first-order) auto-correlation functions of the PRBS-s of Figure 2.4.1 for the positive shifting times. The auto-correlation functions have the same period as the signal itself. Inside the period there is only one positive peak at the zero shifting time. Constant negative values belong to the shifting times greater than the minimum switching time (clock period) and less than the period minus the minimum switching time. These values are nearer and nearer to zero with increasing period. The auto-correlation function of a PRBS resembles the auto-correlation function of the white noise the better the longer the period of the PRBS is.

Property 2.5.2 (Davies, 1970) The (first-order) auto-correlation function of the not centered pseudo-random binary sequence $u(k)$ with amplitude U around zero (peak-to-peak value $2U$) is

$$r_{uu}(\kappa)=\begin{cases} U^2 & \text{if} \quad \kappa = iN_p \qquad i=0,\pm 1,\pm 2,\dots \\ \dfrac{U^2}{N_p} & \text{if} \quad \kappa \neq iN_p \qquad i=0,\pm 1,\pm 2,\dots \end{cases} \tag{2.5.1}$$

■

Example 2.5.1 *PRBS with period 31 and amplitude 2 (Figure 2.4.4).*
From (2.5.1) we get

$$r_{uu}(\kappa)=\begin{cases} 2^2=4 & \text{if} \quad \kappa=0, 31, 62, 93,\dots \\ \dfrac{2^2}{31}=0.129 & \text{for all other values} \end{cases}$$

The calculated values coincide with the computed ones in Figure 2.5.2d. ■

Property 2.5.3 (Davies, 1970) The (first-order) auto-correlation function of the antisymmetric PRBS with amplitude U is

$$r_{uu}(\kappa)=\begin{cases} U^2 & \text{if} \quad \kappa=0, \pm 2N_p, \pm 4N_p,\dots \\ -U^2 & \text{if} \quad \kappa=0, \pm N_p, \pm 3N_p,\dots \\ \dfrac{U^2}{N_p} & \text{if} \quad \kappa=\pm 1, \pm 3, \pm 5,\dots \quad \text{and} \quad \kappa \neq \pm N_p, \pm 3N_p, \pm 5N_p,\dots \\ -\dfrac{U^2}{N_p} & \text{if} \quad \kappa=\pm 2, \pm 4, \pm 6,\dots \quad \text{and} \quad \kappa \neq \pm 2N_p, \pm 4N_p, \pm 6N_p,\dots \end{cases}$$

Here N_p is the period of the original PRBS, i.e., the period of the antisymmetric PRBS is $2N_p$. ■

Example 2.5.2 *Auto-correlation function of the antisymmetric PRBS with period of 2·7=14 and peak-to-peak value 4.*
The auto-correlation function of the signal given in Example 2.3.9 (Figure 2.3.9c) was computed and is plotted between the shifting times 0 and 14 in Figure 2.5.3. The computed values coincide with those calculated from Property 2.5.3:

$$r_{uu}(0)=r_{uu}(14)=2^2=4$$
$$r_{uu}(7)=-2^2=-4$$
$$r_{uu}(1)=r_{uu}(3)=r_{uu}(5)=r_{uu}(9)=r_{uu}(11)=r_{uu}(13)=2^2/7=0.5714$$
$$r_{uu}(2)=r_{uu}(4)=r_{uu}(6)=r_{uu}(8)=r_{uu}(10)=r_{uu}(12)=-2^2/7=-0.5714$$

■

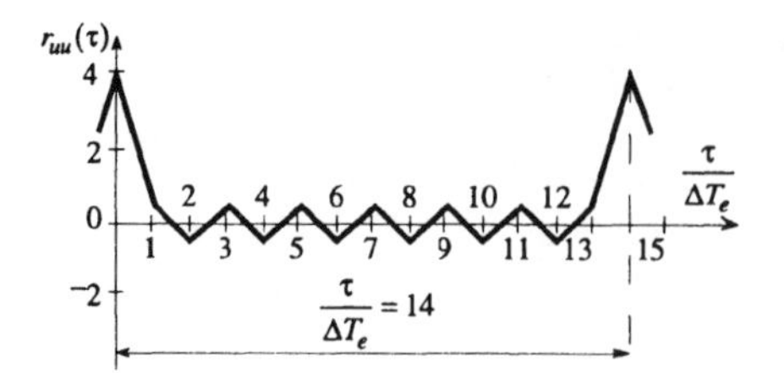

Fig. 2.5.3 Auto-correlation function of the antisymmetric PRBS with period 14 and peak-to-peak value 4

2.5.3 Three-level signals

Figure 2.5.4 shows the (first-order) auto-correlation functions of the PRTS-s of Figure 2.4.2 for the positive shifting times. The auto-correlation functions have the same period as the signal itself. Inside the period there is only one positive peak at the zero shifting time and two negative peaks at the shifting times equal to the plus/minus half of the period. The values between all other discrete shifting times are zero. The auto-correlation function of a PRTS resembles the auto-correlation function of the white noise in the shifting time domain where the absolute value of the shifting time is less than the half period.

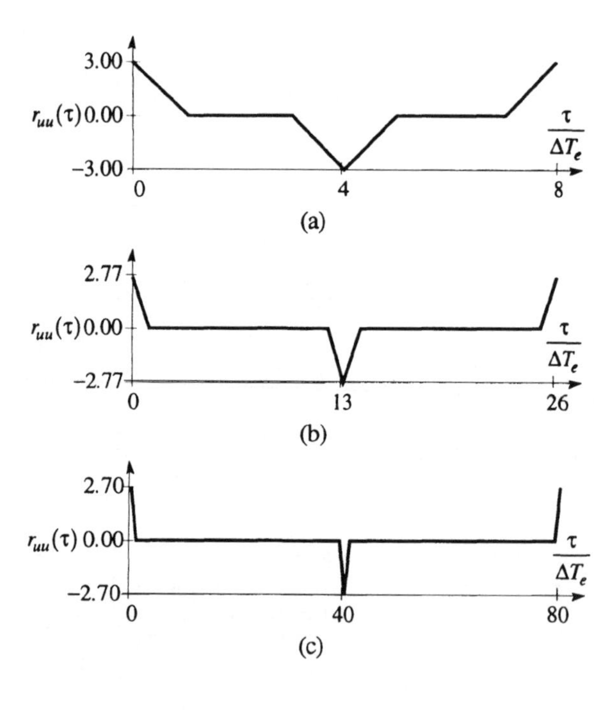

Fig. 2.5.4 The auto-correlation functions of the PRTS-s given in Figure 2.4.2.
(a) $N_p = 8$; (b) $N_p = 26$; (c) $N_p = 80$

Property 2.5.4 (Godfrey, 1966; Davies, 1970) The (first-order) auto-correlation function of the centered pseudo-random ternary sequence $u(k)$ with amplitude steps $\pm U$ (peak-to-peak value $2U$) is

$$r_{uu}(\kappa)=\begin{cases}\frac{2}{3}U^2\left(1+\frac{1}{N_p}\right) & \text{if} \quad \kappa=iN_p \quad i=0,\pm1,\pm2,\dots\\ -\frac{2}{3}U^2\left(1+\frac{1}{N_p}\right) & \text{if} \quad \kappa=\frac{iN_p}{2} \quad i=\pm1,\pm3,\dots\\ 0 & \text{if} \quad \kappa\neq\frac{iN_p}{2} \quad i=0,\pm1,\pm2,\dots\end{cases} \tag{2.5.2}$$

Proof. It is presented in Corollary 2.5.3. ■

Example 2.5.3 *PRTS with period 26 and amplitude 2 (Figure 2.4.5).*
From (2.5.2) we get

$$r_{uu}(\kappa)=\begin{cases}\frac{2}{3}U^2\left(1+\frac{1}{26}\right)=2.77 & \text{if} \quad \kappa=0,\pm26,\pm52,\dots\\ -\frac{2}{3}U^2\left(1+\frac{1}{26}\right)=-2.77 & \text{if} \quad \kappa=\pm13,\pm39,\pm65,\dots\\ 0 & \text{for all other values}\end{cases}$$

The calculated values coincide with the computed ones in Figure 2.5.4b.

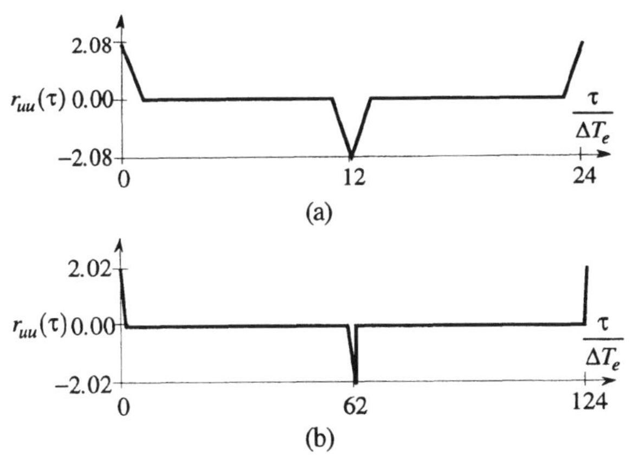

Fig. 2.5.5 The auto-correlation functions of the PRQS-s given in Figure 2.4.3
(a) $N_p=24$; (b) $N_p=124$

2.5.4 Five-level signals

Figure 2.5.5 shows the (first-order) auto-correlation functions of the PRQS-s of Figure 2.4.3 for the positive shifting times. The auto-correlation functions have the same period as the signal itself. Inside the period there is only one positive peak at the zero shifting time and two negative peaks at the shifting times equal to the plus/minus half of the period. The values between all other discrete shifting times are zero. The auto-correlation function of a PRTS resembles the auto-correlation function of the white noise in the shifting time domain where the absolute value of the shifting time is less than the half period. Moreover, the auto-correlation functions of a PRTS and a PRQS have very similar shapes.

Property 2.5.5 (Chang, 1966; Davies, 1970) The (first-order) auto-correlation function of the centered pseudo-random quinary sequence $u(k)$ with amplitude steps $\pm U$ (peak-to-peak value $4U$) is

$$r_{uu}(\kappa)=\begin{cases}2U^2\left(1+\frac{1}{N_p}\right) & \text{if} \quad \kappa=iN_p \quad i=0,\pm1,\pm2,\dots\\ -2U^2\left(1+\frac{1}{N_p}\right) & \text{if} \quad \kappa=\frac{iN_p}{2} \quad i=\pm1,\pm3,\dots\\ 0 & \text{if} \quad \kappa\neq\frac{iN_p}{2} \quad i=0,\pm1,\pm2,\dots\end{cases} \tag{2.5.3}$$

Proof. The proof is presented in Corollary 2.5.4. ■

Example 2.5.4 *PRQS with period 24 and peak-to-peak value 4 (Figure 2.4.6).* The amplitude steps are $U=4/4=1$. Form (2.5.3) we obtain

$$r_{uu}(\kappa)=\begin{cases}2\cdot1^2\left(1+\frac{1}{24}\right)=2.083 & \text{if} \quad \kappa=0,\pm24,\pm48,\dots\\ -2\cdot1^2\left(1+\frac{1}{24}\right)=-2.083 & \text{if} \quad \kappa=\pm12,\pm36,\pm60,\dots\\ 0 & \text{for all other values}\end{cases}$$

The calculated values coincide with the computed ones in Figure 2.5.5a. ■

2.5.5 Properties of the first-order auto-correlation functions

Property 2.5.6 (Zierler, 1959; Yuen, 1973) The (first-order) auto-correlation function of the sampled pseudo-random multi-level sequence with maximal length and more than two levels is

$$r_{uu}(\kappa)=\begin{cases}\dfrac{r^{n_r-1}}{r^{n_r}-1}\sum\limits_{j=0}^{r-1}u_ju_{j\beta^i}-u_0^2 & \text{if} \quad \kappa=i\dfrac{r^{n_r}-1}{r-1} \quad i=0,1,2,\dots\\ \dfrac{r^{n_r-2}}{r^{n_r}-1}\sum\limits_{j_1=0}^{r-1}\sum\limits_{j_2=0}^{r-1}u_{j_1}u_{j_2}-u_0^2 & \text{for all other } \kappa-\text{s}\end{cases}$$

where β is the highest prime element less than the number of levels r and the u_j-s are the levels of the signal corresponding to the generating numbers of the Galois field, i.e.

$$0\to u_0 \qquad 1\to u_1 \qquad 2\to u_2 \qquad \dots \qquad (r-1)\to u_{r-1}$$

If the index j is greater than r, then j has to be replaced by the rest of the modulo-r division. ■

Corollary 2.5.2 The auto-correlation function of the sampled centered pseudo-random multi-level sequence with more than two levels, zero mean and amplitude steps $\pm U$ (that means peak-to-peak value $(r-1)U$) is:

1. $$r_{uu}\left(iN_p\right)=U^2\frac{r^{n_r}}{r^{n_r}-1}\frac{r^2-1}{12} \qquad i=0,1,2,\dots$$

2. $$r_{uu}\left(\frac{iN_p}{r-1}\right)=\frac{r^{n_r-1}}{r^{n_r}-1}\sum_{j=0}^{r-1} u_j u_{j\beta^i}$$

$$i=1,2,\ldots,r-2,r,r+1,\ldots,2(r-1)-1,2(r-1)+1,\ldots$$

3. $$r_{uu}(\kappa)=0 \qquad \text{if} \qquad \kappa\neq\frac{iN_p}{r-1} \qquad i=0,1,2,3,\ldots$$

Proof. (Haber, 1990) The levels of the PRMS are

$$\begin{aligned}&u_0=0,\ u_1=U,\ u_2=2U,\ \ldots\ ,\ u_{(r-1)/2}=\frac{r-1}{2U},\\&u_{(r+1)/2}=-\frac{r-1}{2U},\ \ldots\ ,\ u_{r-2}=-2U,\ u_{r-1}=-U\end{aligned} \tag{2.5.4}$$

Observe, furthermore, that

$$u_0^2=0$$

and

$$\frac{r^{n_r}-1}{r-1}=\frac{N_p}{r-1}$$

1. *Case* $\kappa=i=0$

$$r_{uu}(\kappa)=\frac{r^{n_r-1}}{r^{n_r}-1}\sum_{j=0}^{r-1}u_j^2$$

$$=\frac{r^{n_r-1}}{r^{n_r}-1}\left[0^2+U^2+(2U)^2+\ldots+\left(\frac{r-1}{2}U\right)^2+\left(-\frac{r-1}{2}U\right)^2+\ldots+(-2U)^2+(-U)^2\right]$$

$$=U^2\frac{r^{n_r-1}}{r^{n_r}-1}\left[0^2+1^2+2^2+\ldots+\left(\frac{r-1}{2}\right)^2+\left(-\frac{r-1}{2}\right)^2+\ldots+(-2)^2+(-1)^2\right]$$

$$=U^2\frac{r^{n_r-1}}{r^{n_r}-1}2\frac{1}{6}\frac{r-1}{2}\left(\frac{r-1}{2}+1\right)\left(2\frac{r-1}{2}+1\right)=U^2\frac{r^{n_r}}{r^{n_r}-1}\frac{r^2-1}{12}$$

2. *Case* $\kappa=i\left[N_p/(r-1)\right]$; $i=1,2,\ldots,r-2,r,r+1,\ldots,2(r-1)-1,2(r-1)+1,\ldots$
This is the unaltered form of Property 2.5.6 for

$$\kappa=i\frac{r^{n_r}-1}{r-1}=\frac{iN_p}{r-1} \qquad i\neq 0,\ r-1,\ 2(r-1),\ldots$$

3. *Case* $\kappa\neq i\left[N_p/(r-1)\right]$; $i=0,1,2,3,\ldots$

Now

$$\sum_{j_1=0}^{r-1}\sum_{j_2=0}^{r-1} u_{j_1}u_{j_2}=0$$

thus the auto-correlation function is zero. ■

Remark: The constant factor in Property 2.5.6 can be simplified like

$$\frac{r^{n_r-1}}{r^{n_r}-1}=\frac{\frac{N_p+1}{r}}{N_p}=\frac{1}{r}\left(1+\frac{1}{N_p}\right) \tag{2.5.5}$$

Corollary 2.5.3 The formula for the auto-correlation function of a centered PRTS in Property 2.5.4 can be derived from Corollary 2.5.2.

Proof. (Haber, 1990)
1. From Corollary 2.5.2/1

$$r_{uu}\left(iN_p\right)=U^2\frac{r^{n_r}}{r^{n_r}-1}\frac{r^2-1}{12}=U^2\frac{N_p+1}{N_p}\frac{3^2-1}{12}=\frac{2}{3}U^2\left(1+\frac{1}{N_p}\right)$$
$$i=0,1,2,\ldots$$

2. From Corollary 2.5.2/2 with $\beta=2$ and $u_0=0, u_1=U, u_2=-U$ for $i=1$, $1+2=3$, $3+2=5$, ...

$$r_{uu}(\kappa)=\frac{r^{n_r-1}}{r^{n_r}-1}\sum_{j=0}^{r-1}u_j u_{j\beta^i}=\frac{\frac{N_p+1}{3}}{N_p}\sum_{j=0}^{2}u_j u_{2j}$$

$$=\frac{\frac{N_p+1}{3}}{N_p}\left[0\cdot 0+U\cdot(-U)+(-U)\cdot U\right]=-\frac{2}{3}U^2\left(1+\frac{1}{N_p}\right)$$

$$\text{if } \kappa=i\frac{r^{n_r}-1}{r-1}=i\frac{N_p}{2} \qquad i=1,3,5,\ldots$$

3. From Corollary 2.5.2/3

$$r_{uu}(k)=0 \qquad \text{if} \qquad \kappa\neq\frac{iN_p}{r-1}=\frac{iN_p}{2} \qquad i=0,1,2,3,\ldots$$ ■

Corollary 2.5.4 The formula for the auto-correlation function of a centered PRQS in Property 2.5.5 can be derived from Corollary 2.5.2.

Proof. (Haber, 1990)
1. From Corollary 2.5.2/1

$$r_{uu}\left(iN_p\right)=U^2\frac{r^{n_r}}{r^{n_r}-1}\frac{r^2-1}{12}=U^2\frac{N_p+1}{N_p}\frac{5^2-1}{12}=2U^2\left(1+\frac{1}{N_p}\right)$$

$$i=0,1,2,\ldots$$

2. From Corollary 2.5.2/2 with $\beta=3$ and $u_0=0$, $u_1=U$, $u_2=2U$, $u_3=-2U$, $u_4=-U$ for $i=1$, $1+4=5, 5+4=9$:

$$r_{uu}(\kappa)=\frac{r^{n_r-1}}{r^{n_r}-1}\sum_{j=0}^{r-1}u_j u_{j\beta^i}=\frac{\frac{N_p+1}{5}}{N_p}\sum_{j=0}^{4}u_j u_{3j}$$

$$=\frac{\frac{N_p+1}{5}}{N_p}\left[0\cdot 0+U\cdot(-2U)+2U\cdot U+(-2U)\cdot(-U)+(-U)\cdot 2U\right]=0$$

$$\text{if}\qquad \kappa=i\frac{r^{n_r}-1}{r-1}=i\frac{N_p}{4}\qquad i=1,5,9,\ldots$$

for $i=2$, $2+4=6$, $6+4=10,\ldots$

$$r_{uu}(\kappa)=\frac{r^{n_r-1}}{r^{n_r}-1}\sum_{j=0}^{r-1}u_j u_{j\beta^i}=\frac{\frac{N_p+1}{5}}{N_p}\sum_{j=0}^{4}u_j u_{9j}=$$

$$=\frac{\frac{N_p+1}{5}}{N_p}\left[0\cdot 0+U\cdot(-U)+2U\cdot(-2U)+(-2U)\cdot 2U+(-U)\cdot U\right]=$$

$$=-2U^2\left(1+\frac{1}{N_p}\right)$$

$$\text{if}\qquad \kappa=i\frac{r^{n_r}-1}{r-1}=i\frac{N_p}{4}\qquad i=2,6,10,\ldots$$

for $i=3$, $3+4=7$, $7+4=11,\ldots$

$$r_{uu}(\kappa)=\frac{r^{n_r-1}}{r^{n_r}-1}\sum_{j=0}^{r-1}u_j u_{j\beta^i}=\frac{\frac{N_p+1}{5}}{N_p}\sum_{j=0}^{4}u_j u_{27j}$$

$$=\frac{\frac{N_p+1}{5}}{N_p}\left[0\cdot 0+U\cdot 2U+2U\cdot(-U)+(-2U)\cdot U+(-U)\cdot(-2U)\right]=0$$

$$\text{if} \quad \kappa = i\frac{r^{n_r}-1}{r-1} = i\frac{N_p}{4} \qquad i = 3, 7, 11, \ldots$$

3. From Corollary 2.5.2/3

$$r_{uu}(k) = 0 \qquad \text{if} \quad \kappa \neq \frac{iN_p}{r-1} = \frac{iN_p}{4} \qquad i = 0, 1, 2, 3, \ldots$$

■

Corollary 2.5.5 The (first-order) auto-correlation function of a centered seven-level pseudo-random signal is

$$r_{uu}(\kappa) = \begin{cases} 4U^2\left(1+\frac{1}{N_p}\right) & \text{if} \quad \kappa = iN_p \qquad i = 0, \pm 1, \pm 2, \ldots \\ -4U^2\left(1+\frac{1}{N_p}\right) & \text{if} \quad \kappa = \frac{iN_p}{2} \qquad i = \pm 1, \pm 3, \pm 5, \ldots \\ 2U^2\left(1+\frac{1}{N_p}\right) & \text{if} \quad \kappa = \frac{iN_p}{6} \qquad i = \pm 1, \pm 5, \pm 7, \ldots \\ -2U^2\left(1+\frac{1}{N_p}\right) & \text{if} \quad \kappa = \frac{iN_p}{6} \qquad i = \pm 2, \pm 4, \pm 8, \ldots \\ 0 & \text{if} \quad \kappa \neq \frac{iN_p}{6} \qquad i = 0, \pm 1, \pm 2, \ldots \end{cases} \tag{2.5.6}$$

Proof. (Haber, 1990)

1. From Corollary 2.5.2/1:

$$r_{uu}(iN_p) = U^2 \frac{r^{n_r}}{r^{n_r}-1}\frac{r^2-1}{12} = U^2\frac{N_p+1}{N_p}\frac{7^2-1}{12} = 4U^2\left(1+\frac{1}{N_p}\right)$$
$$i = 0, 1, 2, \ldots$$

2. From Corollary 2.5.2/2 with $\beta = 5$ and $u_0 = 0$, $u_1 = U$, $u_2 = 2U$, $u_3 = 3U$, $u_4 = -3U$, $u_5 = -2U$, $u_6 = -U$ for $i = 1$, $1+6=7$, $7+6=13, \ldots$:

$$r_{uu}(\kappa) = \frac{r^{n_r-1}}{r^{n_r}-1}\sum_{j=0}^{r-1} u_j u_{j\beta^i} = \frac{\frac{N_p+1}{7}}{N_p}\sum_{j=0}^{6} u_j u_{5j} = \frac{\frac{N_p+1}{7}}{N_p}$$
$$\times\left[0\cdot 0 + U\cdot(-2U) + 2U\cdot 3U + 3U\cdot U + (-3U)\cdot(-U) + (-2U)\cdot(-3U) + (-U)\cdot 2U\right]$$
$$= 2U^2\left(1+\frac{1}{N_p}\right) \qquad \text{if} \quad \kappa = i\frac{r^{n_r}-1}{r-1} = i\frac{N_p}{6} \qquad i = 1, 7, 13, \ldots$$

for $i = 2$, $2+6=8$, $8+6=14, \ldots$:

$$r_{uu}(\kappa)=\frac{r^{n_r-1}}{r^{n_r}-1}\sum_{j=0}^{r-1}u_ju_{j\beta^i}=\frac{\frac{N_p+1}{7}}{N_p}\sum_{j=0}^{6}u_ju_{25j}=\frac{\frac{N_p+1}{7}}{N_p}$$

$$\times[0\cdot 0+U\cdot(-3U)+2U\cdot U+3U\cdot(-2U)+(-3U)\cdot 2U+(-2U)\cdot(-U)+(-U)\cdot 3U]$$

$$=-2U^2\left(1+\frac{1}{N_p}\right)\qquad\text{if}\qquad \kappa=i\frac{r^{n_r}-1}{r-1}=i\frac{N_p}{6}\qquad i=2,8,14,\ldots$$

for $i=3,\ 3+6=9,\ 9+6=15,\ldots$:

$$r_{uu}(\kappa)=\frac{r^{n_r-1}}{r^{n_r}-1}\sum_{j=0}^{r-1}u_ju_{j\beta^i}=\frac{\frac{N_p+1}{7}}{N_p}\sum_{j=0}^{6}u_ju_{125j}=\frac{\frac{N_p+1}{7}}{N_p}$$

$$\times[0\cdot 0+U\cdot(-U)+2U\cdot(-2U)+3U\cdot(-3U)+(-3U)\cdot 3U+(-2U)\cdot 2U+(-U)\cdot U]$$

$$=-4U^2\left(1+\frac{1}{N_p}\right)\qquad\text{if}\qquad \kappa=i\frac{r^{n_r}-1}{r-1}=i\frac{N_p}{6}\qquad i=3,9,15,\ldots$$

for $i=4,\ 4+6=10,\ 10+6=16,\ldots$:

$$r_{uu}(\kappa)=\frac{r^{n_r-1}}{r^{n_r}-1}\sum_{j=0}^{r-1}u_ju_{j\beta^i}=\frac{\frac{N_p+1}{7}}{N_p}\sum_{j=0}^{6}u_ju_{625j}=\frac{\frac{N_p+1}{7}}{N_p}$$

$$\times[0\cdot 0+U\cdot 2U+2U\cdot(-3U)+3U\cdot(-U)+(-3U)\cdot U+(-2U)\cdot 3U+(-U)\cdot(-2U)]$$

$$=-2U^2\left(1+\frac{1}{N_p}\right)\qquad\text{if}\qquad \kappa=i\frac{r^{n_r}-1}{r-1}=i\frac{N_p}{6}\qquad i=4,10,16,\ldots$$

for $i=5,\ 5+6=11,\ 11+6=17,\ldots$:

$$r_{uu}(\kappa)=\frac{r^{n_r-1}}{r^{n_r}-1}\sum_{j=0}^{r-1}u_ju_{j\beta^i}=\frac{\frac{N_p+1}{7}}{N_p}\sum_{j=0}^{6}u_ju_{3125j}=\frac{\frac{N_p+1}{7}}{N_p}$$

$$\times[0\cdot 0+U\cdot 3U+2U\cdot(-U)+3U\cdot 2U+(-3U)\cdot(-2U)+(-2U)\cdot U+(-U)\cdot(-3U)]$$

$$=2U^2\left(1+\frac{1}{N_p}\right)\qquad\text{if}\qquad \kappa=i\frac{r^{n_r}-1}{r-1}=i\frac{N_p}{6}\qquad i=5,11,17,\ldots$$

3. From Corollary 2.5.2/3:

$$r_{uu}(k)=0\qquad\text{if}\qquad \kappa\neq\frac{iN_p}{r-1}=\frac{iN_p}{6}\qquad i=0,1,2,3,\ldots$$

■

Example 2.5.5 *Pseudo-random seven-level signal with period 48 and peak-to-peak value 2.*

Figure 2.5.6 shows the auto-correlation function of a seven-level pseudo-random signal with period 48 and peak-to-peak value 4. Now $U = 4/6 = 0.667$. The non-zero values of the auto-correlation function are in coincidence with (2.5.6), i.e.

$$r_{uu}(0) = 4 \cdot 0.667^2(1+1/48) = 1.817$$

$$r_{uu}(48/6) = r(8) = 2 \cdot 0.667^2(1+1/48) = 0.908$$

$$r_{uu}(2 \cdot 48/6) = r(16) = -2 \cdot 0.667^2(1+1/48) = -0.908$$

$$r_{uu}(48/2) = r(24) = -4 \cdot 0.667^2(1+1/48) = -1.817$$

$$r_{uu}(4 \cdot 48/6) = r(32) = -2 \cdot 0.667^2(1+1/48) = -0.908$$

$$r_{uu}(5 \cdot 48/6) = r(40) = 2 \cdot 0.667^2(1+1/48) = 0.908$$

$$r_{uu}(\kappa) = 0 \quad \text{if} \quad \kappa = 1, 2, \ldots, 6, 7, 9, 10, \ldots, 14, 15, 17, 18, \ldots, 22, 23, 25, 26, \ldots$$

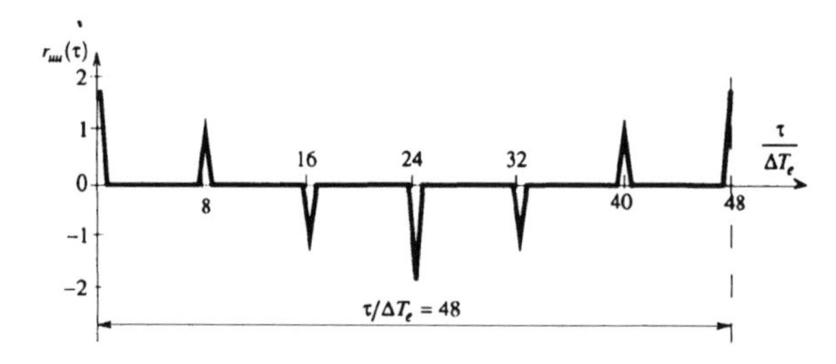

Fig. 2.5.6 Auto-correlation function of a seven-level PRMS with period 48 and peak-to-peak value 4

The corresponding time function can be generated by two registers. The first and second registers are fed back by the coefficients -3 and 2, respectively. The plot is seen in Figure 2.5.7. ■

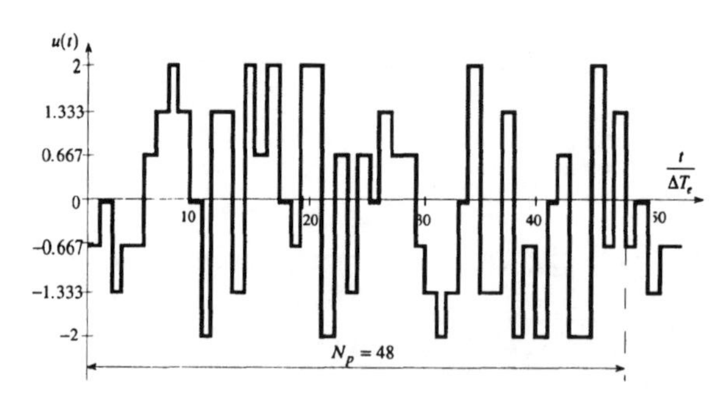

Fig. 2.5.7 Plot of a seven-level PRMS with period 48 and peak-to-peak value 4

Corollary 2.5.6 (Dotsenko *et al.*, 1971) The (first-order) auto-correlation function of the sampled centered pseudo-random multi-level sequence with more than two levels, zero mean and amplitude steps $\pm U$ (i.e. peak-to-peak value $(r-1)U$) is

$$
r_{uu}(\kappa)=\begin{cases} U^2 \dfrac{r^{n_r-1}}{r^{n_r}-1}\left[\sum\limits_{j=1}^{(r-1)/2} j\left[j\otimes_r \gamma^i\right]-\sum\limits_{j=1}^{(r-1)/2} j\left[(-j)\otimes_r \gamma^i\right]\right] \\ \qquad\qquad \text{if} \quad \kappa = i\dfrac{N_p}{r-1} \quad i=0,1,2,\dots \\ 0 \qquad \text{for all other } \kappa \end{cases} \tag{2.5.7}
$$

where $\gamma=\beta-r$ and β is the highest prime element less than the number of levels r. The elements of the modulo-r algebra have to lie in the domain $\left[-(r-1)/2,\,(r-1)/2\right]$.

Proof. (Haber, 1990)

1. *Case* $i=\kappa=0$*:*

Replace $\gamma^i=\gamma^0=1$ into (2.5.7), then we obtain

$$
r_{uu}(\kappa)=U^2\frac{r^{n_r-1}}{r^{n_r}-1}\left[1^2+2^2+\dots+\left(\frac{r-1}{2}\right)^2+\left(-\frac{r-1}{2}\right)^2+\dots+(-2)^2+(-1)^2\right]
$$

$$
=U^2\frac{r^{n_r}}{r^{n_r}-1}\,\frac{r^2-1}{12}
$$

as shown in the proof of Corollary 2.5.2.

2. *Case* $\kappa=i\left[N_p/(r-1)\right]$; $i=1,2,\dots,r-2,r,r+1,\dots,2(r-1)-1,2(r-1)+1,\dots$

Divide the amplitude levels of the test signal u_i by the amplitude step U and denote them by x_i

$$
x_i=\frac{u_i}{U} \qquad i=0,1,\dots,r-1
$$

Using (2.5.4) the normalized amplitude values x_i are as follows

$$
x_0=0 \qquad x_1=1 \qquad x_2=2 \quad \dots \qquad x_{(r-1)/2}=\frac{r-1}{2}
$$

$$
x_{(r+1)/2}=-\frac{r-1}{2} \quad \dots \qquad x_{r-2}=-2 \qquad x_{r-1}=-1
$$

From Corollary 2.5.2/2

$$
r_{uu}(\kappa)=\frac{r^{n_r-1}}{r^{n_r}-1}\sum_{j=0}^{r-1} u_j u_{j\beta^i}=U^2\frac{r^{n_r-1}}{r^{n_r}-1}\sum_{j=0}^{r-1} x_j x_{j\beta^i}
$$

$$= U^2 \frac{r^{n_r-1}}{r^{n_r}-1}\left[\sum_{j=1}^{(r-1)/2} jx_{j\beta^i} + \sum_{j=(r+1)/2}^{(r-1)} jx_{j\beta^i}\right] \tag{2.5.8}$$

because $x_0 = 0$ is zero with a centered signal with zero mean. (2.5.8) is defined in the domain $[0, r-1]$. As Property 2.5.7 is defined in the domain $[-(r-1)/2, (r-1)/2]$, the second sum in (2.5.8) has to be proceed not from $j = (r+1)/2$ till $(r-1)$ but from $j' = (r+1)/2 - r = -(r-1)/2$ till $(r-1) - r = -1$. Thus (2.5.8) can be rewritten to

$$r_{uu}(\kappa) = U^2 \frac{r^{n_r-1}}{r^{n_r}-1}\left[\sum_{j=1}^{(r-1)/2} jx_{j\beta^i} + \sum_{j=-(r-1)/2}^{-1} jx_{j\beta^i}\right]$$

$$= U^2 \frac{r^{n_r-1}}{r^{n_r}-1}\left[\sum_{j=1}^{(r-1)/2} jx_{j\beta^i} - \sum_{j=1}^{(r-1)/2} jx_{j\beta^i}\right] \tag{2.5.9}$$

The proof is complete if the following equalities are understood

$$x_{j\beta^i} = j \otimes_r (\beta - r)^i = j \otimes_r \gamma^i \qquad \text{for} \qquad j \le \frac{r-1}{2} \tag{2.5.10}$$

and

$$x_{j\beta^i} = (-j) \otimes_r (\beta - r)^i = (-j) \otimes_r \gamma^i \qquad \text{for} \qquad j > \frac{r-1}{2} \tag{2.5.11}$$

Both are true as a consequence of the natures of the modulo-r algebra. Remember, that the modulo-r algebra has to be applied on the left sides of (2.5.10) and (2.5.11) in the domains $[0, r-1]$ and on the right sides in the domain $[-(r-1)/2, (r-1)/2]$. As an example see the modulo-5 algebra

$$x_0 = 0 \qquad x_1 = 1 \qquad x_2 = 2 \qquad x_3 = -2 \qquad x_4 = -1$$

with $\beta = 3$ and $i = 2$. The validity of (2.5.10) can be demonstrated for $j = 2 \le 2 = (5-1)/2 = (r-1)/2$ by

$$x_{2\beta^i} = x_{2\cdot 3^2} = x_{2\cdot 9} = x_{2\cdot(9-5)} = x_{2\cdot 4} = x_8 = x_{8-5} = x_3 = -2$$

$$j \otimes_r (\beta - r)^i = 2 \otimes_5 (3-5)^2 = 2 \otimes_5 (-2)^2 = 2 \otimes_5 4 = 2 \otimes_5 (4-5) = 2 \otimes_5 (-1) = -2$$

The validity of (2.5.11) can be demonstrated for $j = 4 > 2 = (5-1)/2 = (r-1)/2$ by

$$x_{4\beta^i} = x_{4\cdot 3^2} = x_{4\cdot 9} = x_{4\cdot(9-5)} = x_{4\cdot 4} = x_{16} = x_{16-15} = x_1 = 1$$

$$j \otimes_r (\beta - r)^i = 4 \otimes_5 (3-5)^2 = 4 \otimes_5 (-2)^2 = 4 \otimes_5 4 = 4 \otimes_5 (4-5) =$$
$$= 4 \otimes_5 (-1) = -4 = -4 + 5 = 1$$

3. *Case* $\kappa \neq i[N_p/(r-1)],\ i = 1, 2, 3, \ldots$

This is the case of Corollary 2.5.2/3. ∎

Corollary 2.5.7 (Haber, 1990) The (first-order) auto-correlation function of the sampled centered pseudo-random multi-level sequence with more than two levels, zero mean and amplitude steps $\pm U$ (i.e., peak-to-peak value $(r-1)U$) is

$$r_{uu}(\kappa) = \begin{cases} \frac{2}{r} U^2 \left(1 + \frac{1}{N_p}\right)^{(r-1)/2} \sum_{j=1} j[j \otimes_r \gamma^i] & \text{if} \quad \kappa = \frac{iN_p}{r-1} \quad i = 0, 1, \ldots \\ 0 & \text{for all other } \kappa \end{cases} \tag{2.5.12}$$

where $\gamma = \beta - r$ and β is the highest prime element less than the number of levels r. The elements of the modulo-r algebra have to be lain in the domain $[-(r-1)/2,\ (r-1)/2]$.

Proof. The auto-correlation function was given by (2.5.7) in Corollary 2.5.6. By means of (2.5.5) it can be seen that (2.5.12) is twice the first part of (2.5.7). In the proof of Corollary 2.5.6 it was shown that (2.5.8) is equal to (2.5.7). Consequently it has to be proven that the second half (2.5.8) is equal to the half of (2.5.12)

$$\sum_{j=1}^{(r-1)/2} x_j x_{j\beta^i} = \sum_{j=(r+1)/2}^{(r-1)} j x_{j\beta^i}$$

Instead of presenting a generally valid proof, the cases of the three-, five- and seven-level pseudo-random signals in the derivations of the Corollaries 2.5.3 to 2.5.5, respectively, are cited, where the correctness of the above equality could be seen. ∎

Corollary 2.5.8 (Haber, 1990) The formula for the auto-correlation function of a centered PRTS given by Property 2.5.4 can be derived from Corollary 2.5.7, as well.

Proof.

1. *Case* $\kappa = iN_p/2,\ i = 0,\ 0+2 = 2,\ 2+2 = 4, \ldots$

(2.5.12) leads to the same proof as with Corollary 2.5.2.

2. *Case* $\kappa = iN_p/2,\ i = 1,\ 1+2 = 3,\ 3+2 = 5, \ldots$

From (2.5.12) with $\gamma = \beta - r = 2 - 3 = -1$

$$r_{uu}(\kappa) = \frac{2}{3}U^2\left(1+\frac{1}{N_p}\right)\left[1\left[1 \otimes_3 (-1)^1\right]\right] = \frac{2}{3}U^2\left(1+\frac{1}{N_p}\right)(-1) = -\frac{2}{3}U^2\left(1+\frac{1}{N_p}\right)$$

3. *Case* $\kappa \neq iN_p/2;\ i = 0, 1, 2, 3, \ldots$

$$r_{uu}(\kappa) = 0$$

■

Corollary 2.5.9 (Haber, 1990) The formula for the auto-correlation function of a centered PRQS given by Property 2.5.5 can be derived from Corollary 2.5.7, as well.

Proof.

1. *Case* $\kappa = iN_p,\ i = 0, 1, 2, \ldots$

(2.5.12) leads to the same proof as with Corollary 2.5.2.

2. *Case* $\kappa = iN_p/4,\ i \neq 0, 4, 8, \ldots$

Apply (2.5.12) with $\gamma = \beta - r = 3 - 5 = -2$. For $i = 1,\ 1+4=5,\ 5+4=9, \ldots$:

$$r_{uu}(\kappa) = \frac{2}{5}U^2\left(1+\frac{1}{N_p}\right)\left[1\left[1 \otimes_5 (-2)^1\right] + 2\left[2 \otimes_5 (-2)^1\right]\right]$$

$$= \frac{2}{5}U^2\left(1+\frac{1}{N_p}\right)\left[(-2)+2\right] = 0$$

for $i = 2,\ 2+4=6,\ 6+4=10,\ \ldots$:

$$r_{uu}(\kappa) = \frac{2}{5}U^2\left(1+\frac{1}{N_p}\right)\left[1\left[1 \otimes_5 (-2)^2\right] + 2\left[2 \otimes_5 (-2)^2\right]\right]$$

$$= \frac{2}{5}U^2\left(1+\frac{1}{N_p}\right)\left[(-1)+(-4)\right] = -2U^2\left(1+\frac{1}{N_p}\right)$$

for $i = 3,\ 3+4=7,\ 7+4=11,\ \ldots$:

$$r_{uu}(\kappa) = \frac{2}{5}U^2\left(1+\frac{1}{N_p}\right)\left[1\left[1 \otimes_5 (-2)^3\right] + 2\left[2 \otimes_5 (-2)^3\right]\right]$$

$$= \frac{2}{5}U^2\left(1+\frac{1}{N_p}\right)\left[2+(-2)\right] = 0$$

3. *Case* $\kappa \neq iN_p/4,\ i = 0, 1, 2, \ldots$

$$r_{uu}(\kappa) = 0$$

■

Corollary 2.5.10 (Haber, 1990) The formula for the auto-correlation function of a centered seven-level PRMS given by Corollary 2.5.5 can be derived from Corollary 2.5.7, as well.

Proof.

1. *Case* $\kappa = N_p,\ i = 0, 1, 2, \ldots$

(2.5.12) leads to the same proof as with Corollary 2.5.2.

2. *Case* $\kappa = iN_p/6,\ i \neq 0, 6, 12, \ldots$

Apply (2.5.12) with $\gamma = \beta - r = 5 - 7 = -2$. For $i = 1,\ 1+6=7,\ 7+6=13, \ldots$:

$$r_{uu}(\kappa) = \frac{2}{7}U^2\left(1+\frac{1}{N_p}\right)\left[1\left[1\otimes_7(-2)^1\right]+2\left[2\otimes_7(-2)^1\right]+3\left[3\otimes_7(-2)^1\right]\right]$$

$$= \frac{2}{7}U^2\left(1+\frac{1}{N_p}\right)\left[(-2)+6+3\right] = 2U^2\left(1+\frac{1}{N_p}\right)$$

for $i = 2, 2+6=8, 8+6=14, \ldots$:

$$r_{uu}(\kappa) = \frac{2}{7}U^2\left(1+\frac{1}{N_p}\right)\left[1\left[1\otimes_7(-2)^2\right]+2\left[2\otimes_7(-2)^2\right]+3\left[3\otimes_7(-2)^2\right]\right]$$

$$= \frac{2}{7}U^2\left(1+\frac{1}{N_p}\right)\left[(-3)+2+(-6)\right] = -2U^2\left(1+\frac{1}{N_p}\right)$$

for $i = 3, 3+6=9, 9+6=15, \ldots$:

$$r_{uu}(\kappa) = \frac{2}{7}U^2\left(1+\frac{1}{N_p}\right)\left[1\left[1\otimes_7(-2)^3\right]+2\left[2\otimes_7(-2)^3\right]+3\left[3\otimes_7(-2)^3\right]\right]$$

$$= \frac{2}{7}U^2\left(1+\frac{1}{N_p}\right)\left[(-1)+(-4)+(-9)\right] = -4U^2\left(1+\frac{1}{N_p}\right)$$

for $i = 4, 4+6=10, 10+6=16, \ldots$:

$$r_{uu}(\kappa) = \frac{2}{7}U^2\left(1+\frac{1}{N_p}\right)\left[1\left[1\otimes_7(-2)^4\right]+2\left[2\otimes_7(-2)^4\right]+3\left[3\otimes_7(-2)^4\right]\right]$$

$$= \frac{2}{7}U^2\left(1+\frac{1}{N_p}\right)\left[2+(-6)+(-3)\right] = -2U^2\left(1+\frac{1}{N_p}\right)$$

for $i = 5, 5+6=11, 11+6=17, \ldots$:

$$r_{uu}(\kappa) = \frac{2}{7}U^2\left(1+\frac{1}{N_p}\right)\left[1\left[1\otimes_7(-2)^5\right]+2\left[2\otimes_7(-2)^5\right]+3\left[3\otimes_7(-2)^5\right]\right]$$

$$= \frac{2}{7}U^2\left(1+\frac{1}{N_p}\right)\left[3+(-2)+6\right] = 2U^2\left(1+\frac{1}{N_p}\right)$$

3. *Case* $\kappa \neq iN_p/6,\ i \neq 0,1,2,\ldots$

$$r_{uu}(\kappa) = 0$$

■

Property 2.5.7 (Dotsenko *et al.*, 1971) The auto-correlation function of the sampled centered pseudo-random multi-level sequence with more than two levels and with zero mean has the following symmetry properties

1. $r_{uu}\left(\dfrac{N_p}{2}\right) = -r_{uu}(0)$

2. $r_{uu}\left(\dfrac{iN_p}{r-1}\right) = r_{uu}\left(\dfrac{jN_p}{r-1}\right) \qquad i,j = 0,1,2,3,\ldots,(r-1) \qquad i+j = r-1$

3. $r_{uu}\left(\dfrac{iN_p}{r-1}\right) = -r_{uu}\left(\dfrac{jN_p}{r-1}\right)$ if $i+j = \dfrac{r-1}{2}$

4. $r_{uu}\left(\dfrac{N_p}{4}\right) = 0$ if $\dfrac{r-1}{4}$ is an integer ■

Property 2.5.8 (Barker, 1969a) The auto-correlation function of the sampled centered pseudo-random multi-level sequence with more than two levels and with zero mean has the following symmetry properties:

1. periodic with the period of the signal (N_p)

$$r_{uu}(\kappa) = r_{uu}(\kappa + N_p)$$

2. an even function

$$r_{uu}(\kappa) = r_{uu}(-\kappa)$$

3. antisymmetric to the half period if the signal is a centered one and has more than two-levels

$$r_{uu}(\kappa) = -r_{uu}\left(\frac{N_p}{2} - \kappa\right)$$

Proof. (Only for Properties 2.5.8/1 and 2.5.8/2.)
1. Property 2.5.8/1 is a feature of any auto-correlation function.
2. Property 2.5.8/2 follows from the fact that a PRMS is a periodic signal. ■

Corollary 2.5.11 (Barker, 1969a) With a centered PRMS with more than two levels the even, periodic and antisymmetric features of the auto-correlation function can be summarized in the following relations

$$r_{uu}(\kappa)=-r_{uu}\left(\frac{N_p}{2}-\kappa\right)=-r_{uu}\left(\frac{N_p}{2}+\kappa\right)=-r_{uu}\left(N_p-\kappa\right)=r_{uu}\left(N_p+\kappa\right) \qquad (2.5.13)$$

Proof. Corollary 2.5.11 is a consequence of Property 2.5.8. ■
Corollary 2.5.12 A PRTS has the following symmetry properties:

1. from Property 2.5.7/1: $r_{uu}\left(N_p/2\right)=-r_{uu}(0)$;
2. from Property 2.5.7/2: $r_{uu}(0)=r_{uu}\left(N_p\right)$;
3. from Property 2.5.7/3: $r_{uu}(0)=-r_{uu}\left(N_p/2\right)$.

Proof. Substitute $r=3$ into the formulas of Property 2.5.7. ■

Example 2.5.6 *PRTS with period 26 and amplitude 2 (Figure 2.5.4b).*
From Example 2.5.3 we know that $r_{uu}(0)=2.77$ and $r_{uu}(26/2)=r_{uu}(13)=-2.77$, thus Corollary 2.5.12 is fulfilled. The validity of the relations (2.5.13) can be also checked. ■

Corollary 2.5.13 A PRQS has the following symmetry properties:
1. from Property 2.5.7/1: $r_{uu}\left(N_p/2\right)=-r_{uu}(0)$
2. from Property 2.5.7/2:
$r_{uu}(0)=r_{uu}\left(N_p\right)$ $\qquad$ $r_{uu}\left(N_p/4\right)=r_{uu}\left(3N_p/4\right)$
3. from Property 2.5.7/3:
$r_{uu}(0)=-r_{uu}\left(N_p/2\right)$ $\qquad$ $r_{uu}\left(N_p/4\right)=-r_{uu}\left(N_p/4\right)=0$
4. from Property 2.5.7/4: $r_{uu}\left(N_p/4\right)=0$

Proof. Substitute $r=5$ into the formulas of Property 2.5.7. ■

Example 2.5.7 *PRQS with period 24 and amplitude 2 (Figure 2.5.5a).*
From Example 2.5.4 we know that $r_{uu}(0)=2.083$, $r_{uu}(24/2)=r_{uu}(12)=-2.083$ and $r_{uu}(6)=r_{uu}(18)=0$, thus Corollary 2.5.13 is fulfilled. The validity of the relations (2.5.13) can be also checked. ■

Corollary 2.5.14 A seven-level pseudo-random signal has the following symmetry properties:
1. from Property 2.5.7/1:

$$r_{uu}(N_p/2) = -r_{uu}(0)$$

2. from Property 2.5.7/2:

$$r_{uu}(0) = r_{uu}(N_p) \qquad r_{uu}(N_p/6) = r_{uu}(5N_p/6)$$

$$r_{uu}(2N_p/6) = r_{uu}(4N_p/6)$$

3. from Property 2.5.7/3:

$$r_{uu}(0) = -r_{uu}(N_p/2) \qquad r_{uu}(N_p/6) = -r_{uu}(2N_p/6)$$

Proof. Substitute $r = 7$ into the formulas of Property 2.5.7. ■

Example 2.5.8 *Seven-level pseudo-random signal with period 48 and amplitude 2 (Figure 2.5.6).*

From Example 2.5.5 we know that $r_{uu}(0) = 1.817$, $r_{uu}(48/2) = r_{uu}(24) = -1.817$ and $r_{uu}(8) = r_{uu}(40) = 0.908$ and $r_{uu}(16) = r_{uu}(32) = -0.908$, thus Corollary 2.5.14 is fulfilled. The validity of the relations (2.5.13) can also be checked. ■

2.5.6 Properties of the higher-order auto-correlation functions

The first-order auto-correlation function of a PRMS resembles that of the Gaussian white noise for small shifting times. The question arises of whether this resemblance is valid for higher-order auto-correlation functions.

Property 2.5.9 (Barker and Pradisthayon, 1970) The even order auto-correlation functions of the centered PRMS-s with more levels than two are zero

$$r_{uu}(\kappa_1, \ldots, \kappa_n) = 0 \qquad \text{if} \quad n = 2, 4, 6, \ldots$$ ■

Property 2.5.10 (Barker and Pradisthayon, 1970) The odd order auto-correlation functions of the centered PRMS-s with more levels than two are zero for several combinations of the shifting times, but they differ from zero in distinct points. ■

The points, where the odd order auto-correlation functions differ from zero are called anomalies (Barker and Pradisthayon, 1970). The locations can be determined for a given PRMS analytically. Procedures are given by (Barker and Pradisthayon, 1970; Kichatov *et al.*, 1970; Dotsenko *et al.*, 1971).

2.6 POWER DENSITY SPECTRA OF PRMS-S WITH MAXIMAL LENGTH

The power density spectrum of a random signal with auto-correlation function $r_{uu}(\tau)$ is

$$S_{uu}(\omega) = \int_{\tau=-\infty}^{\infty} r_{uu}(\tau) e^{-j\omega\tau} d\tau$$

which can be simplified to the formula

$$S_{uu}(\omega) = \int_{\tau=-\infty}^{\infty} r_{uu}(\tau) \cos(\omega\tau) d\tau \qquad (2.6.1)$$

because the auto-correlation function is symmetric to the ordinate $\left[r_{uu}(\tau)=r_{uu}(-\tau)\right]$.

The pseudorandom signals are periodic signals, therefore, they can be expanded into Fourier series

$$r_{uu}(\tau)=\sum_{k=0}^{\infty}\alpha_k\cos(k\omega_0\tau)+\sum_{k=1}^{\infty}\beta_k\sin(k\omega_0\tau) \tag{2.6.2}$$

where the basic angular frequency is

$$\omega_0=\frac{2\pi}{T_p}=\frac{2\pi}{N_p\Delta T_e} \tag{2.6.3}$$

Further on the symbol T_p is used for the time period $T_p=N_p\Delta T_e$, as the auto-correlation function is even,

$$\beta_k=0 \qquad k=1,2,3,\dots$$

(2.6.2) can be reduced to

$$r_{uu}(\tau)=\sum_{k=0}^{\infty}\alpha_k\cos(k\omega_0\tau) \tag{2.6.4}$$

The Fourier coefficients of the auto-correlation function can be determined by the formula (e.g., Bronstein and Semenjajew, 1969)

$$\alpha_k=\frac{4}{T_p}\int_{\tau=0}^{T_p/2}r_{uu}(\tau)\cos(k\omega_0\tau)\,d\tau \qquad k=0,1,2,\dots \tag{2.6.5}$$

In the special case, when the auto-correlation function has a third-order symmetry $\left[r_{uu}(\tau+T_p/2)=-r_{uu}(\tau)\right]$ the Fourier coefficients are as follows (e.g., Bronstein and Semenjajew, 1969):

$$\alpha_{2k+1}=\frac{8}{T_p}\int_{\tau=0}^{T_p/4}r_{uu}(\tau)\cos((2k+1)\omega_0\tau)\,d\tau \qquad k=0,1,2,\dots \tag{2.6.6}$$

Lemma 2.6.1
The power density spectrum of a single cosine function

$$x(\tau)=U\cos(\omega_0\tau)$$

is

$$S_{uu}(\omega)=\frac{U}{2}\left[\delta(\omega-\omega_0)+\delta(\omega+\omega_0)\right]$$

Proof. (e.g., Isermann, 1988)

$$S_{uu}(\omega)=\int_{\tau=-\infty}^{\infty}U\cos(\omega_0\tau)\cos(\omega\tau)d\tau$$
$$=U\int_{\tau=-\infty}^{\infty}\frac{1}{2}\left[\cos((\omega-\omega_0)\tau)+\cos((\omega+\omega_0)\tau)\right]d\tau$$
$$=\frac{U}{2}\left[\delta(\omega-\omega_0)+\delta(\omega+\omega_0)\right]$$

■

Lemma 2.6.2
The power density spectrum of the auto-correlation function expanded in Fourier-series (2.6.4) is

$$S_{uu}(\omega)=\frac{1}{2}\sum_{k=-\infty}^{\infty}\alpha_k\delta(\omega-k\omega_0) \tag{2.6.7}$$

Proof. Apply (2.6.1) to (2.6.4)

$$S_{uu}(\omega)=\int_{\tau=-\infty}^{\infty}\left[\sum_{k=0}^{\infty}\alpha_k\cos(k\omega_0\tau)\right]\cos(\omega\tau)d\tau=\frac{1}{2}\sum_{k=-\infty}^{\infty}\alpha_k\delta(\omega-k\omega_0) \tag{2.6.8}$$

■

2.6.1 Two-level signals
Property 2.6.1 (e.g., Davies, 1970; Isermann, 1988) The power density spectrum of a PRBS is

$$S_{uu}(\omega)=\frac{U^2}{N_p}\left(1+\frac{1}{N_p}\right)\sum_{k=-\infty}^{\infty}\left[\frac{\sin(k\omega_0\Delta T_e/2)}{k\omega_0\Delta T_e/2}\right]^2\delta(\omega-k\omega_0)$$
$$=\frac{U^2}{N_p}\left(1+\frac{1}{N_p}\right)\sum_{k=-\infty}^{\infty}\left[\frac{\sin\left(\frac{k\pi}{N_p}\right)}{\frac{k\pi}{N_p}}\right]^2\delta\left(\omega-k\frac{2\pi}{N_p\Delta T_e}\right) \tag{2.6.9}$$

Proof. (Haber, 1990) From (2.5.1) the continuous time auto-correlation function is

$$r_{uu}(\tau)=\begin{cases}U^2\left[1-\left(1+\frac{1}{N_p}\right)\frac{\tau}{\Delta T_e}\right] & \text{if} \quad |\tau|\le\Delta T_e\\ -U^2\frac{1}{N_p} & \text{if} \quad \Delta T_e\le|\tau|\le T_p-\Delta T_e\end{cases}$$

Its Fourier coefficients can be determined by (2.6.5)

$$\alpha_k = \frac{4}{T_p}\int_{\tau=0}^{T_p/2} r_{uu}(\tau)\cos(k\omega_0\tau)d\tau = \frac{4}{N_p\Delta T_e}\int_{\tau=0}^{\Delta T_e} U^2\left[1-\left(1+\frac{1}{N_p}\right)\frac{\tau}{\Delta T_e}\right]\cos(k\omega_0\tau)d\tau$$

$$+\frac{4}{N_p\Delta T_e}\int_{\tau=\Delta T_e}^{N_p\Delta T_e/2}\left(-\frac{U^2}{N_p}\right)\cos(k\omega_0\tau)d\tau = \frac{4U^2}{N_p\Delta T_e}\left\{\left[\frac{\sin(k\omega_0\tau)}{\omega_0}\right]_0^{\Delta T_e}\right.$$

$$\left.-\left(1+\frac{1}{N_p}\right)\frac{1}{\Delta T_e}\left[\frac{\cos(k\omega_0\tau)}{k^2\omega_0^2}+\frac{\tau\sin(k\omega_0\tau)}{k\omega_0}\right]_0^{\Delta T_e} - \frac{1}{N_p}\left[\frac{\sin(k\omega_0\tau)}{k\omega_0}\right]_{\Delta T_e}^{N_p\Delta T_e/2}\right\}$$

$$= \frac{4U^2}{N_p\Delta T_e}\left\{\frac{\sin(k\omega_0\Delta T_e)}{k\omega_0}\left(1+\frac{1}{N_p}\right)\frac{1}{\Delta T_e}\left[\frac{\cos(k\omega_0\Delta T_e)}{k\omega_0^2}+\frac{\Delta T_e\sin(k\omega_0\tau)}{k\omega_0}-\frac{1}{k^2\omega_0^2}\right]\right.$$

$$\left.-\frac{1}{N_p}\left[\frac{\sin(k\omega_0 N_p\Delta T_e/2)}{k\omega_0}-\frac{\sin(k\omega_0\Delta T_e)}{k\omega_0}\right]\right\}$$

$$= \frac{4U^2}{N_p\Delta T_e^2}\left(1+\frac{1}{N_p}\right)\frac{1}{k^2\omega_0^2}\left[1-\cos(k\omega_0\Delta T_e)\right] = \frac{2U^2}{N_p}\left(1+\frac{1}{N_p}\right)\left[\frac{\sin\left(\frac{k\omega_0\Delta T_e}{2}\right)}{\frac{k\omega_0\Delta T_e}{2}}\right]^2$$

which is equal to the first form of (2.6.9). The second form can be obtained by using (2.6.3). ■

Similar derivations are in Davies (1970) and Isermann (1988).

Figure 2.6.1 shows the power density powers of the PRBS-s generated with Table 2.3.10 and seen in Figure 2.4.1. The lower and upper levels were -2 and 2, respectively.

2.6.2 Three-level signals

Property 2.6.2 (e.g., Krempl, 1973) The power density spectrum of a centered PRTS with maximal length N_p, duration of period $T_p = N_p\Delta T_e$, zero mean and amplitude steps $\pm U$ (peak-to-peak value $2U$) is

$$S_{uu}(\omega) = \frac{2}{3}\frac{U^2}{N_p}\left(1+\frac{1}{N_p}\right)\sum_{k=-\infty}^{\infty}\left[\frac{\sin\left(\frac{(2k+1)\omega_0\Delta T_e}{2}\right)}{\frac{(2k+1)\omega_0\Delta T_e}{2}}\right]^2\delta(\omega-(2k+1)\omega_0)$$

$$= \frac{2}{3}\frac{U^2}{N_p}\left(1+\frac{1}{N_p}\right)\sum_{k=-\infty}^{\infty}\left[\frac{\sin\left(\frac{(2k+1)\pi}{N_p}\right)}{\frac{(2k+1)\pi}{N_p}}\right]^2\delta\left(\omega-(2k+1)\frac{2\pi}{N_p\Delta T_e}\right) \qquad (2.6.10)$$

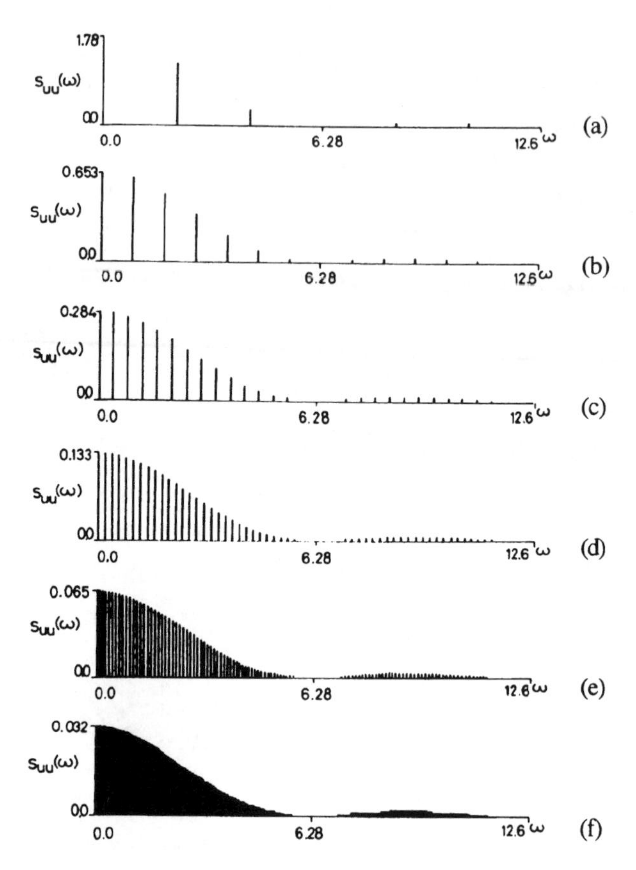

Fig. 2.6.1 The power density spectra of the PRBS-s given in Figure 2.4.1
(a) $N_p = 3$; (b) $N_p = 7$; (c) $N_p = 15$; (d) $N_p = 31$; (e) $N_p = 63$; (f) $N_p = 127$

Proof. (Haber, 1990) From (2.5.2) the continuous time auto-correlation function is in the domain $\left[-T_p/4,\ T_p/4\right]$

$$r_{uu}(\tau) = \begin{cases} r_{uu}(0)\left(1 - \dfrac{\tau}{\Delta T_e}\right) & \text{if} \quad |\tau| \le \Delta T_e \\ 0 & \text{if} \quad \Delta T_e < |\tau| \le \dfrac{N_p \Delta T_e}{4} \end{cases}$$

Its Fourier coefficients can be determined by (2.6.6). For simplicity let $i = 2k+1$, then

$$\alpha_{2k+1} = \alpha_i = \frac{8}{T_p} \int_{\tau=0}^{T_p/4} r_{uu}(\tau)\cos(i\omega_0\tau)d\tau = \frac{8}{N_p \Delta T_e} \int_{\tau=0}^{\Delta T_e} r_{uu}(0)\left(1 - \frac{\tau}{\Delta T_e}\right)\cos(i\omega_0\tau)\,d\tau$$

$$= \frac{8r_{uu}(0)}{N_p\Delta T_e}\left\{\left[\frac{\sin(i\omega_0\tau)}{i\omega_0}\right]_0^{\Delta T_e} - \frac{1}{\Delta T_e}\left[\frac{\cos(i\omega_0\tau)}{i^2\omega_0^2} + \frac{\tau\sin(i\omega_0\tau)}{i\omega_0}\right]_0^{\Delta T_e}\right\}$$

$$= \frac{8r_{uu}(0)}{N_p\Delta T_e^2 i^2\omega_0^2}\left\{\frac{\sin(i\omega_0\Delta T_e)}{i\omega_0} - \frac{1}{\Delta T_e}\left[\frac{\cos(i\omega_0\Delta T_e)}{i^2\omega_0^2} + \frac{\Delta T_e\sin(i\omega_0\Delta T_e)}{i\omega_0} - \frac{1}{i^2\omega_0^2}\right]\right\}$$

$$= \frac{8r_{uu}(0)}{N_p\Delta T_e^2 i^2\omega_0^2}\left[1-\cos(i\omega_0\Delta T_e)\right] = \frac{2r_{uu}(0)}{N_p}\left[\frac{\sin\left(\frac{i\omega_0\Delta T_e}{2}\right)}{\frac{i\omega_0\Delta T_e}{2}}\right]^2 \tag{2.6.11}$$

Substituting $r_{uu}(0)$ from (2.5.2)

$$r_{uu}(0) = \frac{2}{3}U^2\left(1+\frac{1}{N_p}\right)$$

and i by $(2k+1)$ into (2.6.11) results in (2.6.10). ■

A similar derivation is in Krempl (1973). He, however, used the additional multiplier $1/\pi$ in the definition of the spectrum density function (2.6.1).

Figure 2.6.2 shows the power density spectra of the PRTS-s of Figure 2.4.2 generated with the parameters of Table 2.3.11. The lowest and uppermost levels of the signals were -2 and 2, respectively.

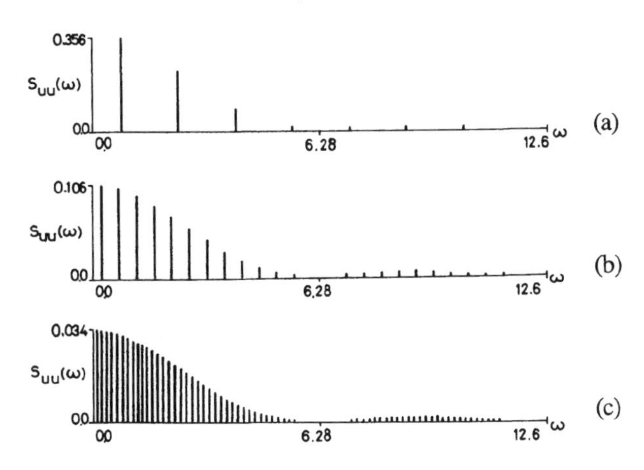

Fig. 2.6.2 The power density spectra of the PRTS-s given in Figure 2.4.2
(a) $N_p = 8$; (b) $N_p = 26$; (c) $N_p = 80$

2.6.3 Five-level signals

Property 2.6.3 (e.g., Tuis, 1975) The power density spectrum of a centered PRQS with maximal length N_p, duration of period $T_p = N_p\Delta T_e$, zero mean and amplitude steps $\pm U$ (peak-to-peak value $4U$) is

$$S_{uu}(\omega)=2\frac{U^2}{N_p}\left(1+\frac{1}{N_p}\right)\sum_{k=-\infty}^{\infty}\left[\frac{\sin\left(\frac{(2k+1)\omega_0\Delta T_e}{2}\right)}{\frac{(2k+1)\omega_0\Delta T_e}{2}}\right]^2\delta(\omega-(2k+1)\omega_0)$$

$$=2\frac{U^2}{N_p}\left(1+\frac{1}{N_p}\right)\sum_{k=-\infty}^{\infty}\left[\frac{\sin\left(\frac{(2k+1)\pi}{N_p}\right)}{\frac{(2k+1)\pi}{N_p}}\right]^2\delta\left(\omega-(2k+1)\frac{2\pi}{N_p\Delta T_e}\right) \qquad (2.6.12)$$

Proof. (Haber, 1990) As the auto-correlation functions of the centered PRTS and PRQS have the same shapes and only the values $r_{uu}(0)$ differ. (2.6.11) is also valid in this case. Substituting i by $(2k+1)$ and $r_{uu}(0)$ from (2.5.3)

$$r_{uu}(0)=2U^2\left(1+\frac{1}{N_p}\right)$$

into (2.6.11) results in (2.6.12). ■

Tuis (1975) presented a similar form. He, like Krempl (1973), used the additional multiplier $1/\pi$ in the definition of the spectrum density function (2.6.1).

Figure 2.6.3 shows the power density spectra of the PRQS-s of Figure 2.4.3 generated with the parameters of Table 2.3.12. The lowest and uppermost levels of the signals were -2.0 and 2.0, respectively.

2.6.4 *Properties of the power density spectra*

Let us summarize the formulas of the power density spectra:

- PRBS with maximal length N_p clock period ΔT_e and levels $\pm U$ (see (2.6.9))

$$S_{uu}(\omega)=\frac{U^2}{N_p}\left(1+\frac{1}{N_p}\right)\sum_{k=-\infty}^{\infty}\left[\frac{\sin(k\omega_0\Delta T_e/2)}{k\omega_0\Delta T_e/2}\right]^2\delta(\omega-k\omega_0)$$

- centered PRTS with maximal length N_p, clock period ΔT_e, zero mean and amplitude steps $\pm U$ (2.6.10)

$$S_{uu}(\omega)=\frac{2}{3}\frac{U^2}{N_p}\left(1+\frac{1}{N_p}\right)\sum_{k=-\infty}^{\infty}\left[\frac{\sin((2k+1)\omega_0\Delta T_e/2)}{(2k+1)\omega_0\Delta T_e/2}\right]^2\delta(\omega-(2k+1)\omega_0)$$

- centered PRQS with maximal length N_p, clock period ΔT_e, zero mean and amplitude steps $\pm U$ (2.6.12)

$$S_{uu}(\omega)=2\frac{U^2}{N_p}\left(1+\frac{1}{N_p}\right)\sum_{k=-\infty}^{\infty}\left[\frac{\sin((2k+1)\omega_0\Delta T_e/2)}{(2k+1)\omega_0\Delta T_e/2}\right]^2\delta(\omega-(2k+1)\omega_0)$$

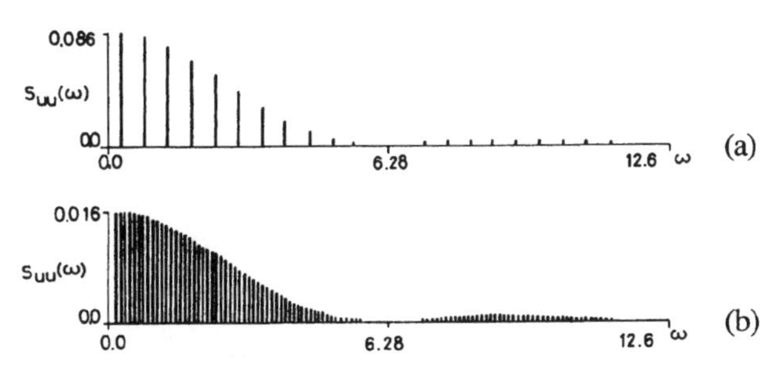

Fig. 2.6.3 The power density spectra of the PRQS-s given in Figure 2.4.3
(a) $N_p = 24$; (b) $N_p = 124$

Property 2.6.4 The PRBS and the centered PRTS and PRQS have the following properties:

1. All PRMS-s have discontinuous, so-called line spectra.
2. The distance between the neighboring lines is :
   (a) with a PRBS: $\omega_0 = 2\pi/N_p\Delta T_e$;
   (b) with a PRTS: $2\omega_0 = 4\pi/N_p\Delta T_e$;
   (c) with a PRQS: $2\omega_0 = 4\pi/N_p\Delta T_e$.
3. At the zero frequency:
   (a) a PRBS: has a non zero value;
   (b) a PRTS: has zero value;
   (c) a PRQS: has zero value.
4. The maximal value of the spectrum is at the smallest frequency. The smallest non-zero angular frequency that has a spectrum line is $\omega_0 = 2\pi/N_p\Delta T_e$.
5. The components of the spectrum decrease with increasing angular frequency up to $\omega = 2\pi/\Delta T_e$.
6. The components of the spectrum are very small after the angular frequency up to $\omega = 2\pi/\Delta T_e$.
7. As the significant angular frequency components are till $\omega = 2\pi/\Delta T_e$, this value is called the bandwidth of the signal.
8. The components belonging to the different angular frequencies are called form factors and have the common form

$$c\left[\frac{\sin(\ell\omega_0\Delta T_e/2)}{\ell\omega_0\Delta T_e/2}\right]^2$$

   where the factor c is:
   (a) with a PRBS:

$$c = \frac{U^2}{N_p}\left(1+\frac{1}{N_p}\right)$$

   (b) with a PRTS:

$$c = \frac{2}{3}\frac{U^2}{N_p}\left(1 + \frac{1}{N_p}\right)$$

(c) with a PRQS:

$$c = 2\frac{U^2}{N_p}\left(1 + \frac{1}{N_p}\right)$$

9. The angular frequency up to which the components do not decrease more than 3dB is of interest and is denoted by $\omega_{3\text{dB}}$. Its value is $\omega_{3\text{dB}} = 2\pi / 3\Delta T_e$.

Proof. Properties 2.6.4/1 to 2.6.4/8 can be easily derived from the forms of the power density spectra (2.6.9), (2.6.10) and (2.6.12).

Property 2.6.4/9 is a consequence of the inequality valid in all cases

$$20\lg\left(\left[\frac{\sin(\omega\Delta T_e/2)}{\omega\Delta T_e/2}\right]^2\right) \geq -3$$

At the boundary $\omega = \omega_{3\text{dB}}$ the equation

$$\left[\frac{\sin(\omega_{3\text{dB}}\Delta T_e/2)}{\omega_{3\text{dB}}\Delta T_e/2}\right]^2 = 0.078$$

has to be satisfied. This happens approximately for

$$\frac{\omega_{3\text{dB}}\Delta T_e}{2} = \frac{\pi}{3} \rightarrow \quad \omega_{3\text{dB}} = \frac{2\pi}{3\Delta T_e}$$ ■

Thus only a PRBS has a spectrum line at zero frequency is in coincidence with the fact that only this signal has a non zero mean value.

2.7 CHOICE OF THE PARAMETERS OF THE PRMS GENERATORS

If choosing a PRMS as a test signal for the identification the following parameters have to be selected:

1. number of the levels r;
2. lowest and uppermost values $U_{\min}$ and $U_{\max}$;
3. length of the period N_p;
4. sampling time ΔT;
5. minimum switching time (clock period) ΔT_e.

1. *Number of the levels*

   In general the number of levels should be more than the degree of the highest power of the polynomial steady state characteristic of the process.

   Strictly speaking, the only criterion which should be fulfilled is that the static nonlinear terms of the process are excited by a signal having more values than the degree of the highest power of the nonlinearity. This can also happen with a test signal that has fewer levels than the degree of the highest power of the process. For example, a quadratic simple Wiener model can be identified by means of a PRBS.

2. *Lowest and uppermost values*
   They should be selected in such a way that the range of the output signal covers the range around the working point, where the system's behavior is identified. To avoid numerical problems the peak-to-peak value of the test signal should be as large as possible. Then the signal to noise ratio is also large and the parameter estimation is more accurate.
   Signals with more levels than two have a zero mean if the amplitude values in both directions (plus and minus) are equal to each other. A symmetrical PRBS has a non-zero mean value. If the test signal has to have a zero mean value then the positive and the negative levels have to be chosen differently.

3. *Length of the period*
   The frequency components of test signal cover the range between

   - $f_{\min} = 1/T_p = 1/N_p \Delta T_e$ and
   - $f_{\max} = 1/\Delta T_e$.

   The length of period depends on the desired ratio of the highest and lowest frequency components of the test signal:

$$\frac{f_{\max}}{f_{\min}} \geq \frac{\dfrac{1}{\Delta T_e}}{\dfrac{1}{N_p \Delta T_e}} = N_p \qquad \rightarrow \qquad N_p \geq \frac{f_{\max}}{f_{\min}} \tag{2.7.1}$$

   Usual values are between:
   - 1 decade: $N_p \approx 10$ and
   - 2 decades: $N_p \approx 100$.

   One of the advantages of using PRMS-s is that the cross-correlation or the cross-dispersion function shows the dynamic behavior of the system in a form like the weighting function of the linear systems. To see the whole transients the time period of the signal should be greater than the settling time of the process. Denote the time duration within the transients settle till 5% by T_{95}. The transients are settled in about $1.5T_{95}$. The shifting time domains between two peaks with PRMS-s are as follows:
   - PRBS: $N_p \Delta T_e$;
   - PRTS: $N_p \Delta T_e / 2$;
   - PRQS: $N_p \Delta T_e / 2$.

   The auto-correlation function is constant and zero (or near zero with PRBS) within the neighboring peaks. The transients can be seen as the process having been excited by a Dirac delta function if

- PRBS: $N_p \Delta T_e \geq 1.5T_{95}$, $\rightarrow$ $N_p \geq 1.5T_{95}/\Delta T_e$; (2.7.2a)
- PRTS: $N_p \Delta T_e/2 \geq 1.5T_{95}$, $\rightarrow$ $N_p \geq 3.0T_{95}/\Delta T_e$; (2.7.2b)
- PRQS: $N_p \Delta T_e/2 \geq 1.5T_{95}$, $\rightarrow$ $N_p \geq 3.0T_{95}/\Delta T_e$. (2.7.2c)

4. *Sampling time*

   The sampling time is determined by the dynamic behavior of the process:
   1. A too small sampling time leads to an inaccurate estimation of the steady-state characteristics of the process;
   2. A too large sampling time does not allow the estimation of the small time constants of the process.

   The choice of the sampling time depends on the purpose of the identification. Isermann (1988) recommends the following rule of thumb:

$$0.05 \leq \frac{\Delta T_i}{T_{95}} \leq 0.2 \qquad (2.7.3)$$

   (ΔT_i is the sampling time of the identification.) The choice in the above range depends on the purpose of the identification:
   - for modeling purpose $\Delta T_i/T_{95} \approx 0.05$;
   - for control purpose $\Delta T_i/T_{95} \approx 0.2$.

   If the smallest time constant $(T_{\min})$ has to be exactly determined then the rule becomes $\Delta T_i \leq 0.2T_{\min}$.

5. *Minimum switching time (clock period)*

   As a rule of thumb we can say:
   - for control purpose and for modeling purpose with a test signal having a large period ($N_p > 60$): $\Delta T_e = \Delta T_i$;
   - for modeling purpose with a test signal having a short period ($N_p < 30$):

$$1 \leq \frac{\Delta T_e}{\Delta T_i} \leq 4$$

   These facts can be seen from (2.7.2b and 2.7.2c), which are valid for PRTS and PRQS, and from (2.7.3)
   - for control purposes:

$$\Delta T_e \geq 3\frac{T_{95}}{N_p} = \frac{3}{N_p} 5\Delta T_i = \frac{15}{N_p} \Delta T_i$$

   - for modeling purpose:

$$\Delta T_e \geq 3\frac{T_{95}}{N_p} = \frac{3}{N_p} 20\Delta T_i = \frac{60}{N_p} \Delta T_i$$

PRMS can be designed in order to estimate the parameters of nonlinear dynamic systems easier than with other test signals:

1. Barker and Zhuang (1997) recommended a test signal for separating in the output signal
   - the linear and odd degree nonlinear components from the even degree components in the output signal and
   - the linear components from the third-degree nonlinear components.
2. Reed and Hawksford (1996) used PRMS to generate an orthogonal basis for the identification of the Volterra series model.
3. Kashiwagi (1995), Nishiyama and Kashiwagi (1997) selected PRMS-s for estimating the Volterra kernels from the cross correlation function of the input and the output signal.

2.8 REFERENCES

Barker, H.A. (1969a). p-level sequences with primitive auto-correlation functions. *Electronics Letters*, Vol. 5, 21, pp. 531–532.

Barker, H.A. (1969b). Reference phase of pseudorandom signals. *Proc. IEE*, Vol. 116, 3, pp. 429–435.

Barker, H.A. (1993). Design of multi-level pseudo-random signals for system identification. In Godfrey, K. (Editor): *Perturbation signals for system identification.* Prentice Hall International, (Hemel Hempstad: UK), pp. 321–347.

Barker, H.A. and S.N. Obidegwu (1973). Effects of nonlinearities on the measurement of weighting functions by cross-correlation using pseudorandom signals. *Proc. IEE*, Vol. 120, 10, pp. 1293–1300.

Barker, H.A. and T. Pradisthayon (1970). High-order auto-correlation functions of pseudorandom signals based on m sequences. *Proc. IEE*, Vol. 117, 9, pp. 1857–1863.

Barker, H.A. and M. Zhuang (1997). Design of pseudo-random perturbation signals for frequency domain identification of nonlinear systems. *Prepr. 12th IFAC Symposium on Systems Identification*, (Fukuoka: Japan), Vol. 4, pp. 1753–1758.

Bronstein, L.A. and K.A. Semenjajew (1969). *Handbook of the Mathematics* (in German). (H. Deutsch Verlag, Frankfurt/Main: Germany).

Chang, J.A. (1966). Generation of 5-level maximal-length sequences. *Electronics Letters*, Vol. 2, 7, pp. 258.

Davies, W.D.T. (1970). *System Identification for Self-Adaptive Control.* Wiley-Interscience, (London: UK).

Dotsenko, V.I., R.G. Faradzhev and G.S. Chkhartishvili (1971). Properties of maximal length sequences with p-levels. *Automation and Remote Control*, Vol. 32, 8, pp. 189–194.

Godfrey, K.R. (1966). Three-level m sequences. *Electronics Letters*, Vol. 2, 7, pp. 241–243.

Godfrey, K. (1993). Introduction to perturbation signals for time domain identification. In Godfrey, K. (Editor): *Perturbation signals for system identification.* Prentice Hall International, (Hemel Hempstad: UK), pp. 1–59.

Haber, R. (1990). Generation and properties of pseudorandom multilevel signals. *Report FHK-AV-PLT-90/1, Department of Process Engineering*, Cologne Institute of Technology (Fachhochschule Köln), (Köln: FRG).

Isermann, R. (1988). *Identification of Dynamic Systems* (in German). Springer Verlag, (Berlin: FRG).

Kashiwagi, H. (1995). M-sequence and its applications to measurement and control. *Computer Science and Informatics*, Vol. 25, 1, pp. 4–12.

Kichatov, Yu.F., G.M. Tenengolts and V.N. Dynkin (1970). Nonlinear systems identification with pseudorandom multilevel sequences. *Prepr. 2nd IFAC Symposium on Identification and System Parameter Estimation*, (Prague: Czechoslovakia), Section 7.3.

Kónya L. (1981). FORTRAN subroutine for generating pseudorandom multilevel signals. *Department of Automation, Technical University of Budapest*, (Budapest: Hungary).

Krempl, R. (1973). Application of ternary pseudorandom signals for the identification of nonlinear control systems (in German). *Dissertation, Department of Measurement and Control, Ruhr University Bochum*, (Bochum: FRG).

Leontaritis, I.J. and S.A. Billings (1987). Experimental design and identifiability for non-linear systems. *Int. Journal of Control*, Vol. 18, 1, pp. 189–202.

Nishiyama, E. and H. Kashiwagi (1997). Identification of Volterra kernels of nonlinear systems by use of M-sequences correlation. *Prepr. 12th IFAC Symposium on Systems Identification*, (Fukuoka: Japan), Vol. 2, pp. 739–743.

Pearson, R.K., B.A. Ogunnaike and F.J. Doyle III (1996). Identification of structurally constrained second order Volterra models. *IEEE Trans. on Signal Processing*, Vol. 44, 11, pp. 2837–2846.

Reed, M.J. and M.O.J. Hawksford (1996). Identification of discrete Volterra series using maximum length sequences. *IEE Proc. Circuits, Devices, Systems*, Vol. 143, 5, pp. 341–348.

Schetzen, M. (1980). *The Volterra and Wiener Theories of Nonlinear Systems.* Wiley-Interscience, (New York: USA).

Söderström, T. and P. Stoica (1989). *System Identification* Prentice Hall, (London: UK).
Tuis, L. (1975). Application of multi-level pseudorandom signals for the identification of nonlinear control systems (in German). *Dissertation, Department of Measurement and Control, Ruhr University Bochum*, (Bochum: FRG).
Yuen, W.L. (1973). A mapping property of pseudorandom sequence. *Int. Journal of Control*, Vol. 17, 6, pp. 1217–1223.
Zierler, N. (1959). Linear recurring sequences. *Journal of Society Industrial Applications of Mathematics*, Vol. 7, 1, pp. 31–47.

3. Parameter Estimation Methods

3.1 INTRODUCTION

In this chapter we deal with the determination of the parameters of unknown models. Since usually both the process and the measurements can be disturbed by noise we can apply only estimation methods instead of exact calculations.

The structure of the model will be assumed known *a priori*. This is very important. In linear systems there are only two parameters that determine the structure: the dead time and the order of the (pulse) transfer function. With nonlinear systems, however, the inner structure of the process can take various forms. For example a linear dynamic and a nonlinear static term can be connected in series in two different ways. Although both the steady state characteristics and the linearized models for small excitations can be the same in both cases, the globally valid input–output relations are different. How the structure of a nonlinear dynamic process can be determined will be dealt with in the Chapter *Structure Identification.*

Different parameter estimation methods can be applied, depending on the way in which the model parameters are updated if new measurements are available. The so-called off-line methods work with all measurement data at the same time. Their counterparts, the on-line, or recursive, methods update the parameter values after each measurement. They are applied, e.g., in adaptive controls. The formal derivation of the on-line algorithms from the off-line ones is a known and not difficult procedure, at least theoretically, therefore only off-line methods will be presented here.

Different off-line parameter estimation methods will be applied:

- grapho-analytical;
- correlation;
- frequency;
- least squares and different modified ones;
- prediction error.

These methods were originally elaborated for linear dynamic systems, and their application in nonlinear systems depends on the structure of the process.

The methods will be classified according to the models to be identified:

- nonparametric model (Volterra series);
- nonparametric orthogonal models (e.g., Wiener series);
- cascade models (Wiener, Hammerstein type);
- block oriented models others than cascade;
- models linear-in-parameters;
- models nonlinear-in-parameters, etc.

As has been seen, the manifold of the methods arises from the variety of the process models and from the diversity of the parameter estimation methods.

Because of the widespread usage of digital computers and the possible application of the estimated models in control algorithms, discrete time methods are preferred. The methods presented are, therefore, all mostly of this type.

It will be assumed that the input signal fulfills the requirements of the identification. This means with the application of statistical parameter estimation algorithms that the input signal is persistently exciting. With nonlinear systems usually care has also to be taken that the input signal has more different levels, otherwise the nonlinear steady state characteristics cannot be determined.

Processes are usually corrupted by noise. We assume in the sequel that the measured input signal is not noisy and only the measured output signal contains the effects of noise. The noise-free output signal will be denoted by $w(k)$ and the one measured by $y(k)$.

A process model may have a pure dead time besides the lag terms, as well. The exact knowledge of the dead time is not critical if a nonparameric model is fitted to the measured input–output data. Therefore the dead time will not be given with nonparametric models. However, with parametric model identification the dead time will be separated from the other parts of the dynamics.

The systems considered will not have any integrators or differentiators. The unknown parameters will be summarized in the vector $\boldsymbol{\theta}$. The number of parameters is denoted by $n_{\boldsymbol{\theta}}$.

3.2 STOCHASTIC MODELS

We will assume that the measurement of the input signal is perfect and only the measured output signal involves the effect of stochastic noises. That means, either the process and/or the measurement of the output is disturbed. The disturbing noise is assumed to originate from a source noise. This source noise is assumed to be white, have a normal distribution, zero mean, and a standard deviation σ_e. Denote the source noise by $e(t)$.

Definition 3.2.1 A stochastic nonparametric model is defined by

$$y(k) = f(u(k-d),\ldots,u(k-d-nu), e(k),\ldots,e(k-ne)) \qquad \blacksquare$$

Definition 3.2.2 A stochastic parametric model is defined by

$$y(k) = f(u(k-d),\ldots,u(k-d-nu), y(k),\ldots,y(k-ny), e(k),\ldots,e(k-ne)) \qquad \blacksquare$$

Remarks:

1. A nonparametric model can be formally obtained from a parametric model by omitting the output signal terms
2. The values *nu* and *ne* are about 10 to 20 for a nonparametric model
3. The values *nu, ny* and *ne* are usually less than 5 for a parametric model

The basic idea of the parameter estimation is to compute the source noise from the measured input and output values. The computed source noise based on the estimated parameters is called the residual $\varepsilon(t)$

$$\varepsilon(k) = \hat{e}(k)$$

According to the principle of the maximum likelihood parameter estimation those parameters are expected as correct by which the sum of square of the residuals is minimum assuming Gaussian noise. To perform the minimization the residuals have to be expressed and computed from the measured input and output signals.

Definition 3.2.3 The general form of a nonparametric residual model is

$$\varepsilon(k) = f_{\varepsilon}(y(k), u(k-d),\ldots,u(k-d-nu), \varepsilon(k-1),\ldots,\varepsilon(k-ne)) \qquad \blacksquare$$

Definition 3.2.4 The general form of a parametric residual model is

$$\varepsilon(k) = f_{\varepsilon}\big(y(k), u(k-d), \ldots, u(k-d-nu), y(k-1), \ldots, y(k-ny),$$
$$\varepsilon(k-1), \ldots, \varepsilon(k-ne)\big)$$ ■

Similar remarks as listed after Definition 3.2.2 are valid again.

Definition 3.2.5 A residual model is linear in the parameters if the actual value of the residual can be expressed by a scalar product of a memory and a parameter vector

$$\varepsilon(k) = \boldsymbol{\phi}^{\mathrm{T}}(k)\boldsymbol{\theta} = \boldsymbol{\phi}^{\mathrm{T}}\big(y(k), u(k-d), \ldots, u(k-d-nu), y(k-1), \ldots, y(k-ny),$$
$$\varepsilon(k-1), \ldots, \varepsilon(k-ne)\big)\boldsymbol{\theta}$$ ■

In the sequel we study which forms the residual models of the different process models have.

3.2.1 Recursive polynomial difference equation

Definition 3.2.6 (Leontaritis and Billings, 1985) The stochastic recursive polynomial difference equation model is defined as

$$y(k) = f\big(y(k-1), \ldots, y(k-ny), u(k-d), \ldots, u(k-d-nu), \ldots,$$
$$e(k), e(k-1), \ldots, e(k-ne)\big) + e(k) \tag{3.2.1}$$

where $f(\ldots)$ is a polynomial function. ■

Example 3.2.1 *Noisy nonlinear polynomial model*

Let the noise-free process model be a first-order model including quadratic terms as

$$w(k) = -a_1 w(k-1) + b_1 u(k-1) + c_1 u(k-1)w(k-1) + d_1 w^2(k-1) \tag{3.2.2}$$

The output is disturbed by a white noise $e(k)$

$$y(k) = w(k) + e(k) \tag{3.2.3}$$

Substitute (3.2.3) into (3.2.2)

$$y(k) - e(k) = a_1 y(k-1) - a_1 e(k-1) + b_1 u(k-1) + c_1 u(k-1)y(k-1)$$
$$-c_1 u(k-1)e(k-1) + d_1 y^2(k-1) - 2d_1 y(k-1)e(k-1) + d_1 e^2(k-1)$$

which has the form of (3.2.1). The output signal can be written in a form linear-in-parameters with the white noise as an additive term

$$y(k) = \boldsymbol{\phi}^{\mathrm{T}}(k)\boldsymbol{\theta} + e(k)$$

with

$$\boldsymbol{\phi}^{\mathrm{T}}(k)=\left[y(k-1),-e(k-1),u(k-1),u(k-1)y(k-1),-u(k-1)e(k-1),y^{2}(k-1),\right.$$
$$\left.-2y(k-1)e(k-1),e^{2}(k-1)\right]$$

$$\boldsymbol{\theta}^{\mathrm{T}}=\left[a_{1},a_{1},b_{1},c_{1},c_{1},d_{1},d_{1},d_{1}\right]$$

In this special case the parameter vector can be reduced to

$$\boldsymbol{\theta}^{\mathrm{T}}=\left[a_{1},b_{1},c_{1},d_{1}\right]$$

and then the memory vector has the form

$$\boldsymbol{\phi}(k)=\left[y(k-1)-e(k-1),u(k-1),u(k-1)[y(k-1)-e(k-1)],[y(k-1)-e(k-1)]^{2}\right]^{\mathrm{T}}$$ ■

In Example 3.2.1 the stochastic model has the general form

$$y(k)=f\big(y(k-1),\ldots,y(k-ny),u(k-d),\ldots,u(k-d-nu),e(k-1),\ldots,$$
$$e(k-ne)\big)+e(k) \tag{3.2.4}$$

i.e., the source noise is additive to a model $f(\ldots)$ linear-in-parameters. Generally we cannot state that (3.2.1) can always be transformed to the form (3.2.4). On the other hand (3.2.4) seems to be a form general enough to approximate stochastic input–output models. In the parameter estimation algorithms (3.2.4) is used for computing the residuals by means of the estimated parameters

$$\varepsilon(k)=y(k)-f\big(y(k-1),\ldots,y(k-ny),u(k),\ldots,u(k-nu),\varepsilon(k-1),\ldots,\varepsilon(k-ne)\big) \tag{3.2.5}$$

In practical cases we do not know what are the real structure of the process and the noise model. Therefore the usage of the form (3.2.5) is realistic.

3.2.2 Recursive rational difference equation

Definition 3.2.7 (Billings and Chen 1989) The stochastic recursive rational difference equation model has the form

$$a\big(y(k-1),\ldots,y(k-ny),u(k-d),\ldots,u(k-d-nu),e(k-1),\ldots,e(k-ne)\big)y(k)$$
$$=b\big(y(k-1),\ldots,y(k-ny),u(k-d),\ldots,u(k-d-nu),e(k-1),\ldots,e(k-ne)\big) \tag{3.2.6}$$
$$+a\big(y(k-1),\ldots,y(k-ny),u(k-d),\ldots,u(k-d-nu),e(k-1),\ldots,e(k-ne)\big)e(k)$$

or

$$y(k)=\frac{b\big(y(k-1),\ldots,y(k-ny),u(k-d),\ldots,u(k-d-nu),e(k-1),\ldots,e(k-ne)\big)}{a\big(y(k-1),\ldots,y(k-ny),u(k-d),\ldots,u(k-d-nu),e(k-1),\ldots,e(k-ne)\big)}+e(k)$$

$a(\ldots)$ and $b(\ldots)$ are polynomials of the actual and shifted input signals. ■

The residual can be expressed by

$$\varepsilon(k)=y(k)-\frac{b\big(y(k-1),\ldots,y(k-ny),u(k-d),\ldots,u(k-d-nu),\varepsilon(k-1),\ldots,\varepsilon(k-ne)\big)}{a\big(y(k-1),\ldots,y(k-ny),u(k-d),\ldots,u(k-d-nu),\varepsilon(k-1),\ldots,\varepsilon(k-ne)\big)}$$

Example 3.2.2 *Quasi-linear first-order continuous time noisy model with output signal dependent reciprocal gain*

It was shown in Chapter 1 that the noise-free process

$$T\dot{w}(t)+w(t)=\overline{K}u(t)$$
$$\overline{K}^{-1}(t)=\overline{K}_0^{-1}+\overline{K}_1^{-1}w(t)$$
$$T(t)=T_0$$

can be described by the deterministic recursive rational difference equation (1.5.10)

$$a\big(w(k-1)\big)w(k)=b\big(w(k-1),u(k-1)\big)$$

with

$$a(\ldots)=a\big(w(k-1)\big)=\overline{K}_0^{-1}+\overline{K}_1^{-1}w(k-1)$$

$$b(\ldots)=b\big(w(k-1),u(k-1)\big)$$
$$=-\big[\overline{K}_0^{-1}+\overline{K}_1^{-1}w(k-1)\big]\big[\Delta T/T_0-1\big]w(k-1)+\big[\Delta T/T_0\big]u(k-1)$$

The assumption that the output measurements are corrupted by a white noise

$$y(k)=w(k)+e(k)$$

leads to

$$a\big(y(k-1)-e(k-1)\big)\big[y(k)-e(k)\big]=b\big(y(k-1)-e(k-1),u(k-1)\big)$$

i.e.,

$$\big\{\overline{K}_0^{-1}+\overline{K}_1^{-1}[y(k-1)-e(k-1)]\big\}y(k)=-\big\{\overline{K}_0^{-1}+\overline{K}_1^{-1}[y(k-1)-e(k-1)]\big\}$$
$$\times\big[\Delta T/T_0-1\big]\big[y(k-1)-e(k-1)\big]+\big[\Delta T/T_0\big]u(k-1)$$
$$+\big\{\overline{K}_0^{-1}+\overline{K}_1^{-1}[y(k-1)-e(k-1)]\big\}e(k)$$

Zhu and Billings (1994) presented a LS parameter estimation algorithm for stochastic rational equation models.

3.2.3 Output–affine difference equation

Definition 3.2.8 (Chen and Billings, 1988) The stochastic form of the output–affine polynomial difference equation model is

$$\sum_{i-0}^{na}a_i\big(u(k-d),u(k-d-1),\ldots,u(k-d-nu),\varepsilon(k-1),\ldots,\varepsilon(k-ne)\big)y(k-i)$$

$$= b_0\big(u(k-d), u(k-d-1), \ldots, u(k-d-nu), \varepsilon(k-1), \ldots, \varepsilon(k-ne)\big)$$
$$+\sum_{i-0}^{na} a_i\big(u(k-d), u(k-d-1), \ldots, u(k-d-nu), \varepsilon(k-1), \ldots, \varepsilon(k-ne)\big)\varepsilon(k-i) \tag{3.2.7}$$

where $a_i(\ldots)$ and $b_0(\ldots)$ are polynomials of the actual and shifted input signals and of the residuals and $a_0(\ldots) \neq 0$. ■

The residual can be expressed by

$$\varepsilon(k) = \frac{1}{a_0\big(u(k-d), \ldots, u(k-d-nu), \varepsilon(k-1), \ldots, \varepsilon(k-ne)\big)}$$
$$\times\Bigg[\sum_{i=0}^{na} a_i\big(u(k-d), \ldots, u(k-d-nu), \varepsilon(k-1), \ldots, \varepsilon(k-ne)\big)y(k-i)$$
$$-b_0\big(u(k-d), \ldots, u(k-d-nu), \varepsilon(k-1), \ldots, \varepsilon(k-ne)\big)$$
$$-\sum_{i=1}^{na} a_i\big(u(k-d), \ldots, u(k-d-nu), \varepsilon(k-1), \ldots, \varepsilon(k-ne)\big)\varepsilon(k-i)\Bigg]$$

Example 3.2.3 *Quasi-linear first-order continuous time noisy model with input signal dependent time constant*

It was shown in Chapter 1 that the noise-free process

$$T\dot{w}(t) + w(t) = Ku(t)$$
$$T(t) = T_0 + T_1 u(t)$$

can be described by the deterministic recursive rational difference equation (1.5.16)

$$a_0\big(u(k-1)\big)w(k) + a_1\big(u(k-1)\big)w(k-1) = b_0\big(u(k-1)\big)$$

with

$$a_0(\ldots) = a_0\big(u(k-1)\big) = T_0 + T_1 u(k-1)$$
$$a_1(\ldots) = a_1\big(u(k-1)\big) = \Delta T - T_0 - T_1 u(k-1)$$
$$b_0(\ldots) = b_0\big(u(k-1)\big) = K_0 \Delta T u(k-1)$$

The assumption that the output measurements are corrupted by an additive white noise

$$y(k) = w(k) + e(k)$$

leads to

$$a_0\big(u(k-1)\big)\big[y(k) - e(k)\big] + a_1\big(u(k-1)\big)\big[y(k-1) - e(k-1)\big] = b_0\big(u(k-1)\big)$$

i.e.,

$$\big[T_0 + T_1 u(k-1)\big]\big[y(k) - e(k)\big] + \big[\Delta T - T_0 - T_1 u(k-1)\big]\big[y(k-1) - e(k-1)\big]$$
$$= K_0 \Delta T u(k-1)$$

or

$$[T_0 + T_1u(k-1)]y(k) + [\Delta T - T_0 - T_1u(k-1)]y(k-1)$$
$$= K_0\Delta Tu(k-1) + [T_0 + T_1u(k-1)]e(k) + [\Delta T - T_0 - T_1u(k-1)]e(k-1)$$

■

3.3 METHODS FOR RESIDUAL MODELS LINEAR-IN-PARAMETERS

Definition 3.3.1 The residual model has the general form

$$\varepsilon(k) = y(k) - \hat{y}(k, \hat{\boldsymbol{\theta}}) \tag{3.3.1}$$

where $\hat{y}(k, \hat{\boldsymbol{\theta}})$ is the predicted output signal computed based on the estimated parameter vector $\hat{\boldsymbol{\theta}}$. ■

Definition 3.3.2 A residual model linear-in-parameters means that the predicted output signal can be expressed by a scalar product of the memory vector and of a parameter vector

$$\hat{y}(k) = \boldsymbol{\varphi}^{\mathrm{T}}\boldsymbol{\theta} \tag{3.3.2}$$

■

Generally the memory vector contains the actual and past terms of the input signal, the past terms of the output signal and the earlier residual terms:

$$\boldsymbol{\varphi}_{uy\varepsilon}(k) = f_{uy\varepsilon}\big(u(k), \ldots, u(k-nu), y(k-1), \ldots, y(k-ny), \varepsilon(k-1), \ldots, \varepsilon(k-ne)\big) \tag{3.3.3}$$

The predicted output signal is then denoted by

$$\hat{y}(k) = \boldsymbol{\varphi}_{uy\varepsilon}^{\mathrm{T}}(k)\hat{\boldsymbol{\theta}}_{uy\varepsilon} \tag{3.3.4}$$

The subscript $_{uy\varepsilon}$ shows that the memory vector contains terms of the input, output and the residual. Sometimes the residual terms fail in the memory vector. Then the predicted output signal has the form

$$\hat{y}(k) = \boldsymbol{\varphi}_{uy}^{\mathrm{T}}(k)\hat{\boldsymbol{\theta}}_{uy} \tag{3.3.5}$$

This is the case if only the parameters of the deterministic process model are estimated. If the aim of the identification is to obtain the parameters of the stochastic model we do not have a better *a priori* estimate of the residuals than that they are all zero, and (3.3.4) is reduced to (3.3.5).

Define $\boldsymbol{\varepsilon}$ as the vector of the residuals

$$\boldsymbol{\varepsilon} = [\varepsilon(1), \ldots, \varepsilon(N)]^{\mathrm{T}} \tag{3.3.6}$$

where N is the number of measurements. The vector of the residuals can be expressed by

$$\boldsymbol{\varepsilon} = \boldsymbol{y} - \boldsymbol{\Phi}\hat{\boldsymbol{\theta}} \tag{3.3.7}$$

with the memory matrix

$$\mathbf{\Phi} = \begin{bmatrix} \boldsymbol{\phi}^{\mathrm{T}}(1) \\ \vdots \\ \boldsymbol{\phi}^{\mathrm{T}}(N) \end{bmatrix} \tag{3.3.8}$$

The aim of the parameter estimation is to minimize the sum of squares of the residuals

$$J = \tfrac{1}{2}\sum_{k=1}^{N} \varepsilon^2(k) = \tfrac{1}{2}\boldsymbol{\varepsilon}^{\mathrm{T}}\boldsymbol{\varepsilon} \rightarrow \min \tag{3.3.9}$$

3.3.1 Least squares method

Lemma 3.3.1

If the memory vector does not contain terms of the residual and the residuals are white noise independent from the input signal, then the parameters can be obtained by

$$\hat{\boldsymbol{\theta}}_{uy} = \left[\mathbf{\Phi}_{uy}^{\mathrm{T}}\mathbf{\Phi}_{uy}\right]^{-1}\mathbf{\Phi}_{uy}^{\mathrm{T}}\boldsymbol{y} \tag{3.3.10}$$

Proof. The minimization of (3.3.9)

$$\frac{\partial J}{\partial \hat{\boldsymbol{\theta}}_{uy}} = \frac{\partial}{\partial \hat{\boldsymbol{\theta}}_{uy}}\left[\tfrac{1}{2}\boldsymbol{\varepsilon}^{\mathrm{T}}\boldsymbol{\varepsilon}\right] = \frac{\partial \boldsymbol{\varepsilon}^{\mathrm{T}}}{\partial \hat{\boldsymbol{\theta}}_{uy}}\boldsymbol{\varepsilon} = -\mathbf{\Phi}_{uy}\left[\boldsymbol{y} - \mathbf{\Phi}_{uy}\hat{\boldsymbol{\theta}}_{uy}\right] = \mathbf{0}$$

leads to (3.3.10). ■

Lemma 3.3.2

The LS estimate is bias-free if the residual is independent of all model components.

Proof. The true parameter vector is denoted by $\boldsymbol{\theta}$ and the estimated by $\hat{\boldsymbol{\theta}}$. Then

$$\begin{aligned}
\operatorname*{plim}_{N\to\infty}\left\{\hat{\boldsymbol{\theta}}_{uy} - \boldsymbol{\theta}_{uy}\right\} &= \operatorname*{plim}_{N\to\infty}\left\{\left[\mathbf{\Phi}_{uy}^{\mathrm{T}}\mathbf{\Phi}_{uy}\right]^{-1}\mathbf{\Phi}_{uy}^{\mathrm{T}}\boldsymbol{y} - \boldsymbol{\theta}_{uy}\right\} \\
&= \operatorname*{plim}_{N\to\infty}\left\{\left[\mathbf{\Phi}_{uy}^{\mathrm{T}}\mathbf{\Phi}_{uy}\right]^{-1}\mathbf{\Phi}_{uy}^{\mathrm{T}}\left[\mathbf{\Phi}_{uy}\boldsymbol{\theta}_{uy} + \boldsymbol{\varepsilon}\right] - \boldsymbol{\theta}_{uy}\right\} \\
&= \operatorname*{plim}_{N\to\infty}\left\{\left[\mathbf{\Phi}_{uy}^{\mathrm{T}}\mathbf{\Phi}_{uy}\right]^{-1}\mathbf{\Phi}_{uy}^{\mathrm{T}}\boldsymbol{\varepsilon}\right\} = \mathbf{0}
\end{aligned}$$

This is fulfilled if

$$\operatorname*{plim}_{N\to\infty}\left\{\tfrac{1}{N}\mathbf{\Phi}_{uy}^{\mathrm{T}}\mathbf{\Phi}_{uy}\right\} < \infty \tag{3.3.11}$$

$$\operatorname*{plim}_{N\to\infty}\left\{\tfrac{1}{N}\mathbf{\Phi}_{uy}^{\mathrm{T}}\boldsymbol{\varepsilon}\right\} = \mathbf{0} \tag{3.3.12}$$

The existence of (3.3.11) is the condition for computing the estimate (inverting the matrix). (3.3.12) is valid because the memory vector does not contain terms of the residuals, and the residuals are white noise and independent of the process components. ■

Applications of the method for the following nonlinear models are known (among others):

- *generalized Hammerstein model* (Chang and Luus, 1971; Haber and Keviczky, 1974; Bányász *et al.*, 1974);
- *recursive polynomial difference equation model* (Billings and Voon, 1984);
- *bilinear model* (Beghelli and Guidorzi, 1976), etc.

3.3.2 Extended least squares method

The extended LS method is the application of the LS method in the case when residual terms exist in the memory vector as in (3.3.4). An iterative application of the LS method leads to a bias-free parameter estimation.

Algorithm 3.3.1 The extended least squares method contains the following iterative steps:

1. Assume the residuals are zero and omit all terms in the memory vector $\boldsymbol{\varphi}_{uy\varepsilon}$ that contains the residuals. The resulting memory vector is then $\boldsymbol{\varphi}_{uy}$;
2. Perform a LS estimation with the memory vector $\boldsymbol{\varphi}_{uy}$. Compute the residuals by (3.3.1) and (3.3.5);
3. Compute the memory vector $\boldsymbol{\varphi}_{uy\varepsilon}$ and matrix $\boldsymbol{\Phi}_{uy\varepsilon}$ by the measured input and output values and by the computed residuals;
4. Perform a LS estimation for the parameters $\hat{\boldsymbol{\theta}}_{uy\varepsilon}$;
5. Compute the residuals by (3.3.1) and (3.3.4) and the loss function by (3.3.9);
6. Check whether the last computed cost function is less than the previously computed one. If yes stop the algorithm otherwise go to Step 3.

The method was recommended first for linear systems by Talmon (1971).

Applications of the method for the following nonlinear models are known (among others):

- *generalized Hammerstein model* (Haber *et al.*, 1986);
- *recursive polynomial difference equation model* (Billings and Voon, 1984), etc.

3.3.3 Suboptimal least squares method

The noise-free output signal is given by

$$w(k) = \boldsymbol{\varphi}_{uw}^{\mathrm{T}}(k)\boldsymbol{\theta}_{uy} \tag{3.3.13}$$

If a white noise is superposed on the noise-free output signal then the residual model is given by

$$\varepsilon(k) = y(k) - w(k) = y(k) - \boldsymbol{\varphi}_{uw}^{\mathrm{T}}(k)\hat{\boldsymbol{\theta}}_{uy} \tag{3.3.14}$$

The memory vector $\boldsymbol{\varphi}_{uw}$ contains the noise-free output signal. This is not known but can be estimated as

$$\hat{w}(k) = \boldsymbol{\varphi}_{uw}^{\mathrm{T}}(k)\hat{\boldsymbol{\theta}}_{uy} \tag{3.3.15}$$

Algorithm 3.3.2 The suboptimal least squares method contains the following iterative steps:

1. Perform a LS estimation with the residual model (3.3.1) and (3.3.5);
2. Compute the noise-free output signal by (3.3.15);
3. Perform a LS estimation with the residual model (3.3.14);
4. Compute the residuals by (3.3.14) and the loss function by (3.3.9);
5. Check whether the last computed cost function is less than the previously computed one. If yes stop the algorithm, otherwise go to Step 2.

The method was recommended first for linear systems by Moore (1982). The algorithm was described and simulated for nonlinear systems in Billings and Voon (1984).

An application of the method for the following nonlinear model is known :

- *recursive polynomial difference equation model* (Billings and Voon, 1984).

3.3.4 Instrumental variable method

Suppose that the aim of the identification is now to estimate only the parameters of the deterministic process model. The LS estimation would lead to a biased estimation because in a general case the elements of the memory vector may be correlated with the residuals. The basic idea of the method is to introduce a new instrumental memory vector, which is not correlated with the residuals but is strongly correlated with the model components. Denote this so called instrumental memory vector by $\boldsymbol{\varphi}_{\imath uy}$ and the corresponding matrix by $\boldsymbol{\Phi}_{\imath uy}$

$$\boldsymbol{\Phi}_{\imath uy} = \begin{bmatrix} \boldsymbol{\varphi}_{\imath uy}^{\mathrm{T}}(1) \\ \vdots \\ \boldsymbol{\varphi}_{\imath uy}^{\mathrm{T}}(N) \end{bmatrix} \tag{3.3.16}$$

The dimension of $\boldsymbol{\varphi}_{\imath uy}$ cannot be less than that of $\boldsymbol{\varphi}_{uy}$.

Lemma 3.3.3

If the instrumental memory vector is correlated with the terms of the process model but not with the errors then the process parameters can be obtained by

$$\hat{\boldsymbol{\theta}}_{uy} = \left[\boldsymbol{\Phi}_{\imath uy}^{\mathrm{T}} \boldsymbol{\Phi}_{uy}\right]^{-1} \boldsymbol{\Phi}_{\imath uy}^{\mathrm{T}} \boldsymbol{y} \tag{3.3.17}$$

Proof. The vector of errors is defined as

$$\boldsymbol{\varepsilon} = \boldsymbol{y} - \boldsymbol{\Phi}_{uy}(k)\hat{\boldsymbol{\theta}}_{uy} \tag{3.3.18}$$

Multiply (3.3.17) by $\boldsymbol{\Phi}_{\imath uy}^{\mathrm{T}}$

$$\boldsymbol{\Phi}_{\imath uy}^{\mathrm{T}} \boldsymbol{\varepsilon} = \boldsymbol{\Phi}_{\imath uy}^{\mathrm{T}} \boldsymbol{y} - \boldsymbol{\Phi}_{\imath uy}^{\mathrm{T}} \boldsymbol{\Phi}_{uy}(k)\hat{\boldsymbol{\theta}}_{uy} \tag{3.3.19}$$

If

$$\operatorname*{plim}_{N\to\infty}\left\{\boldsymbol{\Phi}_{\imath uy}\boldsymbol{\varepsilon}\right\} = \mathbf{0}$$

then (3.3.16) is an unbiased estimate. ∎

Algorithm 3.3.3 The instrumental variable method contains the following iterative steps:

1. Perform a LS estimation with the residual model (3.3.1) and (3.3.5);
2. Compute the instrumental matrix noise-free output signal by (3.3.15);
3. Perform a LS estimation with the residual model (3.3.14);
4. Compute the residuals by (3.3.14) and the loss function by (3.3.9);
5. Check whether the last computed cost function is less than the previously computed one. If yes stop the algorithm, otherwise go to Step 2. ∎

Remarks:

1. With linear system identification the instrumental memory vector can have the same components as $\boldsymbol{\varphi}_{uy}$, with the only difference that the measured output signals are replaced by the computed noise-free ('shadow') output signals $\left(\boldsymbol{\varphi}_{uy} = \boldsymbol{\varphi}_{uw}\right)$.
2. With nonlinear system identification the same choice $\left(\boldsymbol{\varphi}_{uy} = \boldsymbol{\varphi}_{uw}\right)$ can lead to a biased estimate, as shown in Billings and Voon (1984). Assume, e.g., that the residual has the term $e^2(k-i)$ and the process model $u^2(k-j)$. Then their cross-correlation is $r_{uu}(0)r_{ee}(0)$ which is not zero even if $u(k)$ and $e(k)$ are independent and of zero mean.
3. If the noise terms are linear in the stochastic model, i.e., the residuals are linear function of the source noise then the choice of the computed output signal in the instrumental memory vector $\left(\boldsymbol{\varphi}_{uy} = \boldsymbol{\varphi}_{uw}\right)$ is correct (Billings and Voon, 1984).

Applications of the method for the following models are known (among others):

- *generalized Hammerstein model* (Stoica and Söderström, 1982);
- *recursive polynomial difference equation model* (Billings and Voon, 1984), etc.

3.3.5 Parametric correlation method

Assume that zero mean disturbances $n(k)$ are superposed on the noise-free output signal

$$y(k) = w(k) + n(k) \tag{3.3.20}$$

$n(k)$ has not to be a white noise but must be uncorrelated with the input signal. Suppose that the task is to estimate the parameters of the noise-free process model, as with the method of the instrumental variables. The basic idea of the method is to estimate the parameters by means of the correlation functions that filter the output noise.

Definition 3.3.3 Correlate the memory vector of the model (3.3.5) linear-in-parameters with the multiplier $x(k)$. The memory vector of the correlation functions is defined as

$$\boldsymbol{\varphi}_{cuy} = \mathrm{E}\left\{x(k-\kappa)\boldsymbol{\varphi}_{uy}(k)\right\}$$ ∎

Lemma 3.3.4
Assume that the noise-free output signal has the form (3.3.5) linear-in-parameters. Then the relation between the cross-correlation functions is

$$r_{xy}(\kappa)=\boldsymbol{\phi}_{cuy}^{\mathrm{T}}\hat{\boldsymbol{\theta}}_{uy} \tag{3.3.21}$$

Proof.

$$\begin{aligned} r_{xy}(\kappa)=r_{xw}(\kappa)&=\mathrm{E}\{x(k-\kappa)w(k)\}=\mathrm{E}\{x(k-\kappa)\boldsymbol{\phi}_{uw}^{\mathrm{T}}(k)\hat{\boldsymbol{\theta}}_{uy}\}\\ &=\mathrm{E}\{x(k-\kappa)\boldsymbol{\phi}_{uw}^{\mathrm{T}}(k)\}\hat{\boldsymbol{\theta}}_{uy}=\mathrm{E}\{x(k-\kappa)\boldsymbol{\phi}_{uy}^{\mathrm{T}}(k)\}\hat{\boldsymbol{\theta}}_{uy}=\boldsymbol{\phi}_{cuy}^{\mathrm{T}}(\kappa)\hat{\boldsymbol{\theta}}_{uy} \end{aligned}$$

■

Lemma 3.3.5
The estimation of the parameters can be obtained from the correlation functions by the LS method

$$\hat{\boldsymbol{\theta}}_{uy}=\left[\boldsymbol{R}_{xuy}^{\mathrm{T}}\boldsymbol{R}_{xuy}\right]^{-1}\boldsymbol{R}_{xuy}^{\mathrm{T}}\boldsymbol{r}_{xy} \tag{3.3.22}$$

where

$$\boldsymbol{r}_{xy}=\left[r_{xy}(\kappa_1),r_{xy}(\kappa_1+1),\ldots,r_{xy}(\kappa_2)\right] \tag{3.3.23}$$

$$\boldsymbol{R}_{xuy}=\begin{bmatrix}\boldsymbol{\phi}_{cuy}^{\mathrm{T}}(\kappa_1)\\ \vdots\\ \boldsymbol{\phi}_{cuy}^{\mathrm{T}}(\kappa_2)\end{bmatrix} \tag{3.3.24}$$

and $(\kappa_2-\kappa_1+1)\geq n_{\boldsymbol{\theta}}$. ■

Proof. The cross-correlation function between the multiplier and the noise output signal is the same as between the multiplier and the noise-free output signal

$$r_{xw}(\kappa)=r_{xy}(\kappa)=\operatorname*{plim}_{N\to\infty}\left\{\frac{1}{N}\sum_{k=1}^{N}x(k-\kappa)y(k)\right\}$$

if the multiplier is chosen thus so that it does not correlate with the noise

$$r_{xn}(\kappa)=\operatorname*{plim}_{N\to\infty}\left\{\frac{1}{N}\sum_{k=1}^{N}x(k-\kappa)n(k)\right\}=0$$

Then the equation

$$\boldsymbol{r}_{xy}=\boldsymbol{R}_{xuy}^{\mathrm{T}}\hat{\boldsymbol{\theta}}_{uy} \tag{3.3.25}$$

is at least theoretically a noise-free relation. ■

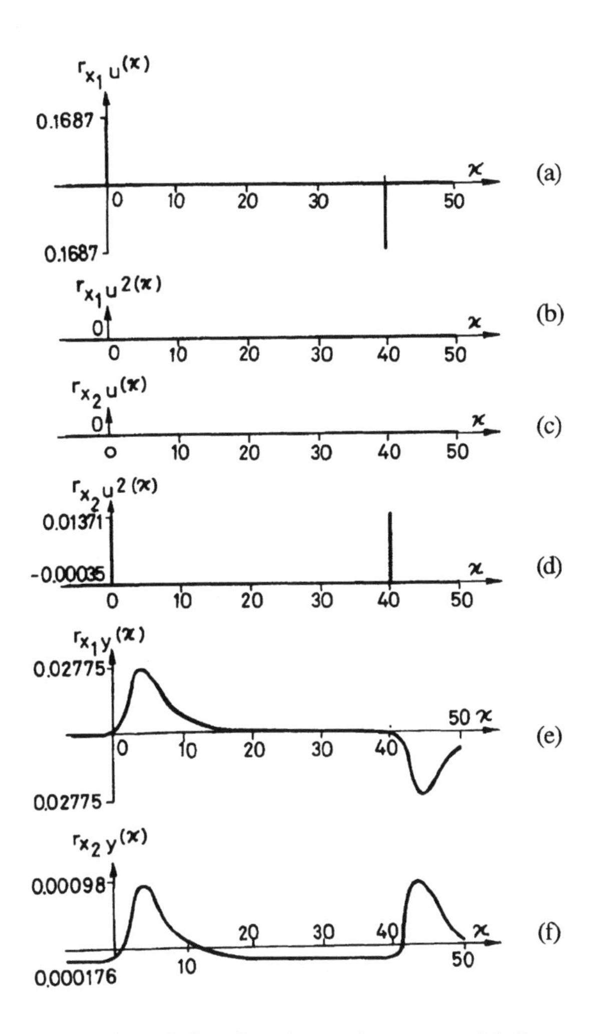

Fig. 3.3.1 Correlation functions using two multipliers:
(a) $r_{x_1u}(\kappa)$; (b) $r_{x_1u^2}(\kappa)$; (c) $r_{x_2u}(\kappa)$; (d) $r_{x_2u^2}(\kappa)$; (e) $r_{x_1y}(\kappa)$; (f) $r_{x_2y}(\kappa)$

Remarks:

1. If $(\kappa_2 - \kappa_1 + 1) = n_\theta$ then (3.3.25) can be solved exactly and (3.3.22) becomes a determined equation system.
2. Although (3.3.25) is deterministic after having eliminated the noises, it is practical to estimate the parameters in the domain $\kappa_1 \le \kappa \le \kappa_2$, where the difference $(\kappa_2 - \kappa_1 + 1)$ – that is the number of the data pairs of the correlation functions – should be greater than the number of the unknown parameters. An approximate 5 times factor is recommended. The correlation functions must not be periodic in the domain ($\kappa_1 \le \kappa \le \kappa_2$) because then they would contain less information than necessary.

3. Since the noise terms may contain nonlinear terms with nonlinear systems, the multiplier has to have a zero mean value.
4. The method described so far is equivalent to that applied to the linear systems (Isermann, 1988). In the nonlinear case it can quite easily happen that the chosen multiplier $x(k)$ does not correlate with every term in the memory vector. In this case one has to choose more multipliers (Haber, 1979a; 1979b; 1988).

Lemma 3.3.6
In case of more multipliers the estimation of the parameters can be obtained from the correlation functions by the LS method (3.3.22), where

$$\boldsymbol{r}_{xy} = \begin{bmatrix} \boldsymbol{r}_{x_1 y} \\ \vdots \\ \boldsymbol{r}_{x_j y} \end{bmatrix} \tag{3.3.26}$$

$$\boldsymbol{R}_{xuy} = \begin{bmatrix} \boldsymbol{R}_{x_1 uy} \\ \vdots \\ \boldsymbol{R}_{x_j uy} \end{bmatrix} \tag{3.3.27}$$

where each $\boldsymbol{r}_{x_j y}$ and $\boldsymbol{R}_{x_j uy}$ has the same structure as (3.3.23) and (3.3.24), respectively, for $x(k) = x_j(k)$.

Proof. Denote the different multipliers by $x_j(k)$; j=1,2,... then

$$\boldsymbol{r}_{x_j y}(\kappa) = \mathrm{E}\left\{x_j(k-\kappa)\boldsymbol{\phi}_{uy}^{\mathrm{T}}(k)\right\}\hat{\boldsymbol{\theta}}_{uy} \tag{3.3.28}$$

Writing the correlation functions $\boldsymbol{r}_{x_j y}(\kappa)$; $j = 1, 2, \ldots$ below each other from $\kappa = \kappa_1$ to κ_2 results in

$$\begin{bmatrix} r_{x_1 y}(\kappa_1) \\ r_{x_1 y}(\kappa_1+1) \\ \vdots \\ r_{x_1 y}(\kappa_2) \\ \vdots \\ r_{x_j y}(\kappa_1) \\ r_{x_j y}(\kappa_1+1) \\ \vdots \\ r_{x_j y}(\kappa_2) \end{bmatrix} = \begin{bmatrix} \mathrm{E}\left\{x_1(k-\kappa_1)\boldsymbol{\phi}_{uy}^{\mathrm{T}}(k)\right\} \\ \mathrm{E}\left\{x_1(k-\kappa_1-1)\boldsymbol{\phi}_{uy}^{\mathrm{T}}(k)\right\} \\ \vdots \\ \mathrm{E}\left\{x_1(k-\kappa_2)\boldsymbol{\phi}_{uy}^{\mathrm{T}}(k)\right\} \\ \vdots \\ \mathrm{E}\left\{x_j(k-\kappa_1)\boldsymbol{\phi}_{uy}^{\mathrm{T}}(k)\right\} \\ \mathrm{E}\left\{x_j(k-\kappa_1-1)\boldsymbol{\phi}_{uy}^{\mathrm{T}}(k)\right\} \\ \vdots \\ \mathrm{E}\left\{x_j(k-\kappa_2)\boldsymbol{\phi}_{uy}^{\mathrm{T}}(k)\right\} \end{bmatrix} \hat{\boldsymbol{\theta}}_{uy} \tag{3.3.29}$$

(3.3.29) corresponds to (3.3.25). The unknown parameters can be estimated by the LS method.

■

Example 3.3.1 *Identification of the noise-free quadratic generalized Hammerstein model by two multipliers (Haber, 1979b, 1988)*

The static nonlinear term is described by the equation

$$v(k) = 2 + u(k) + 0.5u^2(k)$$

The pulse transfer function of the linear dynamic term is:

$$\frac{B(q^{-1})}{A(q^{-1})} = \frac{0.065q^{-1} + 0.043q^{-2} - 0.008q^{-3}}{1 - 1.5q^{-1} + 0.705q^{-2} - 0.1q^{-3}} q^{-1}$$

The input signal is a PRTS with maximal length, period 80, mean value zero and peak-to-peak value ± 0.5. The multipliers are chosen to have zero mean value, so the estimation of the constant term is not possible. Let

$$x_1(k) = u(k) - \mathrm{E}\{u(k)\} \tag{3.3.30}$$

and

$$x_2(k) = u^2(k) - \mathrm{E}\{u^2(k)\} \tag{3.3.31}$$

The cross-correlation functions $r_{x_1u}(\kappa)$, $r_{x_1u^2}(\kappa)$, $r_{x_2u}(\kappa)$, $r_{x_2u^2}(\kappa)$, $r_{x_1y}(\kappa)$ and $r_{x_2y}(\kappa)$ are shown in Figure 3.3.1. It can be seen that $r_{x_1u^2}(\kappa) = r_{x_2u}(\kappa) = 0$, i.e., $x_1(k)$ does not estimate the quadratic and $x_2(k)$ does not estimate the linear terms. The structure of Equation (3.3.28) is the following:

$$\begin{bmatrix} \boldsymbol{r}_{x_1y} \\ \vdots \\ \boldsymbol{r}_{x_2y} \end{bmatrix} = \begin{bmatrix} \times & 0 & \times \\ 0 & \times & \times \\ \times & \times & \times \end{bmatrix} \begin{bmatrix} \boldsymbol{b}_1 \\ \boldsymbol{b}_2 \\ \boldsymbol{a} \end{bmatrix}$$

where $\boldsymbol{b}_1, \boldsymbol{b}_2$ and $\boldsymbol{a}$ contain the parameters of $B_1(q^{-1}), B_2(q^{-1})$ and $\tilde{A}(q^{-1})$. Having performed the LS estimation in the domain $(-8 \le \kappa \le 40)$, the true parameters were obtained.

Example 3.3.2 *Identification of the noise-free quadratic generalized Hammerstein model by one multiplier (Haber, 1979b, 1988)*

Since $x_1(k)$ in (3.3.30) and $x_2(k)$ in (3.3.31) are correlated only by the linear or the quadratic terms of the process model, respectively, it is practical to use the linear combination of $x_1(k)$ and $x_2(k)$ as a multiplier. Thus let

$$x_1(k) = u(k) - \mathrm{E}\{u(k)\} + u^2(k) - \mathrm{E}\{u^2(k)\}$$

The cross-correlation functions $r_{x_1u}(\kappa)$, $r_{x_1u^2}(\kappa)$ and $r_{x_1y}(\kappa)$ are drawn in Figure 3.3.2. $r_{x_1u}(\kappa)$ and $r_{x_1u^2}(\kappa)$ are linearly dependent in the domain $(-8 \le \kappa \le 39)$, but they are independent in $(-8 \le \kappa \le 79)$. Having performed the identification in the latter domain we obtain the true parameters.

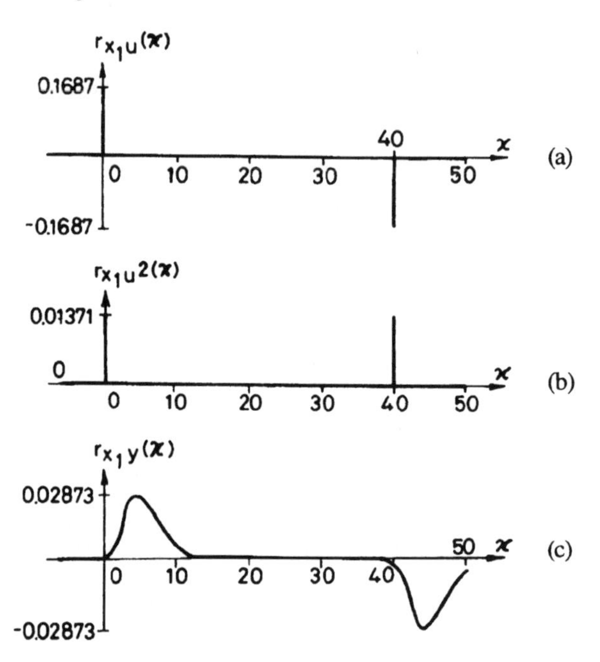

Fig. 3.3.2 Correlation functions using one multiplier:
(a) $r_{x_1u}(\kappa)$; (b) $r_{x_1u^2}(\kappa)$; (c) $r_{x_1y}(\kappa)$

Remarks: (Haber, 1979a, 1988)

1. The multipliers have to be selected so that the matrix $\boldsymbol{R}_{xuy}^{\mathrm{T}}\boldsymbol{R}_{xuy}$ does not become singular.
2. A necessary condition for Item 1 is that the cross-correlation function between each component of the memory vector and at least one of the multipliers differs from zero.
3. In the linear case the conditions of the one-step LS estimation is the linear independence of the signals. These have to be met now for the correlation functions $\mathrm{E}\{x_j(k-\kappa)\phi_{uy}(k)[i]\}$; $i = 1,..,n_\theta$, i.e., they
   - should not be zero,
   - must be linearly independent.
4. The more persistently exciting the correlation functions are, the better the parameter estimation is.
5. Different types of multipliers can be chosen for the identification of the same model structure. In Example 3.3.1 one can take, e.g., the absolute value instead of the

square function as both multipliers are even functions.

6. By the proper choice of the multipliers it is possible to identify only certain components of the whole model structure. In the examples presented earlier the constant term was not identified by the multiplier of zero mean.
7. If two correlation functions depend linearly in a given shifting time domain, a better choice of the domain may help to make same correlation functions linearly independent.
8. Instead of more multipliers, an equivalent one can be applied which is a linear combination of the individual ones.

Applications of the method for the following nonlinear models are known (among others):

- *generalized Hammerstein model* (Haber, 1979a, 1979b, 1988; Bamberger, 1978; Bamberger and Isermann, 1978);
- *bilinear model* (Baheti *et al.*, 1977);
- *quasi-linear model with signal dependent parameters* (Velev and Vuchkov, 1986), etc.

3.3.6 Prediction error method

The prediction error method estimates the parameters of the stochastic process model $\boldsymbol{\theta}_{uy\varepsilon}$ by minimizing the sum of squares of the residuals (3.3.9). Although the residual model is linear in the parameters, the solution needs iteration because the residual terms in the memory vector can be computed only in the knowledge of the (estimated) parameter vector. There are many ways how a quadratic functional can be minimized. A usual algorithm is the Newton–Raphson method.

Algorithm 3.3.4 The prediction error method based on the Newton–Raphson minimization of the loss function (3.3.9) becomes

$$\hat{\boldsymbol{\theta}}_{\ell+1} = \hat{\boldsymbol{\theta}}_{\ell} - \gamma_{\ell} \left[\left. \frac{\partial^2 J\left(\hat{\boldsymbol{\theta}}_{uy\varepsilon}\right)}{\partial \hat{\boldsymbol{\theta}}_{uy\varepsilon} \, \partial \hat{\boldsymbol{\theta}}_{uy\varepsilon}^{\mathrm{T}}} \right|_{\hat{\boldsymbol{\theta}}_{uy\varepsilon} = \hat{\boldsymbol{\theta}}_{uy\varepsilon,\ell}} \right]^{-1} \left. \frac{\partial J\left(\hat{\boldsymbol{\theta}}_{uy\varepsilon}\right)}{\partial \hat{\boldsymbol{\theta}}_{uy\varepsilon}} \right|_{\hat{\boldsymbol{\theta}}_{uy\varepsilon} = \hat{\boldsymbol{\theta}}_{uy\varepsilon,\ell}} \tag{3.3.32}$$

with $\gamma_{\ell} = 1$. The first derivative (gradient) of the cost function, the Jacobian is

$$\frac{\partial J\left(\hat{\boldsymbol{\theta}}_{uy\varepsilon}\right)}{\partial \hat{\boldsymbol{\theta}}_{uy\varepsilon}} = \frac{\partial}{\partial \hat{\boldsymbol{\theta}}_{uy\varepsilon}} \left[\tfrac{1}{2} \boldsymbol{\varepsilon}^{\mathrm{T}} \boldsymbol{\varepsilon} \right] = \frac{\partial \boldsymbol{\varepsilon}^{\mathrm{T}}}{\partial \hat{\boldsymbol{\theta}}_{uy\varepsilon}} \boldsymbol{\varepsilon} \tag{3.3.33}$$

and the second derivative, the Hessian matrix, is

$$\frac{\partial^2 J\left(\hat{\boldsymbol{\theta}}_{uy\varepsilon}\right)}{\partial \hat{\boldsymbol{\theta}}_{uy\varepsilon} \, \partial \hat{\boldsymbol{\theta}}_{uy\varepsilon}^{\mathrm{T}}} = \frac{\partial \boldsymbol{\varepsilon}^{\mathrm{T}}}{\partial \hat{\boldsymbol{\theta}}_{uy\varepsilon}} \frac{\partial \boldsymbol{\varepsilon}}{\partial \hat{\boldsymbol{\theta}}_{uy\varepsilon}^{\mathrm{T}}} + \frac{\partial}{\partial \hat{\boldsymbol{\theta}}_{uy\varepsilon}} \left[\frac{\partial \boldsymbol{\varepsilon}}{\partial \hat{\boldsymbol{\theta}}_{uy\varepsilon}^{\mathrm{T}}} \right]^{\mathrm{T}} \boldsymbol{\varepsilon} \tag{3.3.34}$$

■

Remarks:

1. There are different methods by which the exact computation can be simplified

- *Gauss–Newton method:*
  The Hessian is approximated by the first term on the LHS of (3.3.34). This approximation is good if the parameters are near to their true values, i.e., ε is approximately a white noise sequence. The advantage of this procedure is that only the first derivative of ε need to be calculated, the Hessian is certainly a positive definite matrix and the loss function decreases in every iteration step.
- *Gauss–Newton method with additional directional search:*
  The Newton–Raphson algorithm would find the minimum of the loss function (with the actually estimated residuals) if the loss function would be a quadratic function of the parameters. Since this is not the case the Gauss–Newton procedure can be done more effective if in every iteration step a search of the optimum parameter vector is used in the direction computed by (3.3.33). In other words $\gamma_\ell \neq 1$.

2. The maximum likelihood method minimizes the sum of squares of the residuals by the same algorithm. The maximum likelihood algorithm assumes a Gaussian distribution of the residuals and the prediction error method not. If the residuals have a normal distribution then the two methods are identical not only formally but from the points of view of estimation theory, as well.
3. The method was introduced by Åström and Bohlin (1965) for linear systems.
4. The method was introduced by Leontaritis and Billings (1988) for nonlinear systems described by difference equations. Application of the method for the following nonlinear models are known (among others):
   - *generalized Hammerstein model* (Bányász *et al.*, 1973);
   - *recursive polynomial difference equation model* (Billings and Fadzil, 1985; Leontaritis and Billings, 1988; Billings *et al.*, 1988);
   - *bilinear model* (Gabr and Rao, 1984);
   - *output affine model* (Chen and Billings, 1988);
   - *recursive rational difference equation model* (Billings and Chen, 1989), etc.

3.3.7 Orthogonal parameter estimation

All methods presented are based on the minimization of the sum of squares of certain residuals. Most of the methods use the LS estimation as a basis. With nonlinear systems the number of parameters is usually big and the matrix inversion may cause problems. This can be avoided by orthogonalizing the components of the memory vector.

Definition 3.3.4 Denote the components of the prediction model by $v_i(k)$

$$\boldsymbol{\phi}(k) = \left[v_1(k), \ldots, v_{n_\theta}(k)\right]^{\mathrm{T}}$$

The predicted output signal can be expressed by the form linear-in-parameters

$$\hat{y}(k) = \boldsymbol{\phi}^{\mathrm{T}}\boldsymbol{\theta}$$

■

Definition 3.3.5 Denote the components of the orthogonalized prediction model by $v_i^{\mathrm{o}}(k)$

$$\boldsymbol{\phi}^{o}(k) = \left[v_1^{o}(k), \ldots, v_{n_\theta}^{o}(k)\right]^{T}$$

where

$$\sum_{k=1}^{N} v_i^{o}(k) v_j^{o}(k) = \begin{cases} 0 & \text{if} \quad i \neq j \\ \neq 0 & \text{if} \quad i = j \end{cases}$$

The predicted output signal can be expressed by the orthogonalized form linear-in-parameters

$$\hat{y}(k) = \boldsymbol{\phi}^{o^{T}} \boldsymbol{\theta}^{o}$$

■

Lemma 3.3.7

The components $v_j^{o}(k)$, $j = 1, \ldots, n_\theta$ of the orthogonalized model can be calculated as

$$v_j^{o}(k) = v_j(k) - \sum_{i=1}^{j-1} \frac{\sum_{k=1}^{N} v_j(k) v_i^{o}(k)}{\sum_{k=1}^{N} \left[v_i^{o}(k)\right]^2} v_i^{o}(k), \qquad j = 1, \ldots, n_\theta, \qquad k = 1, \ldots, N \tag{3.3.35}$$

(3.3.35) can be rewritten in a vector form

$$\boldsymbol{v}_j^{o} = \boldsymbol{v}_j - \sum_{i=1}^{j-1} \frac{\boldsymbol{v}_j^{T} \boldsymbol{v}_i^{o}}{\boldsymbol{v}_i^{o^{T}} \boldsymbol{v}_i^{o}} \boldsymbol{v}_i^{o} \qquad j = 1, \ldots, n_\theta \tag{3.3.36}$$

by introducing the vectors

$$\boldsymbol{v}_j(k) = \left[v_j(1), \ldots, v_j(N)\right]^{T}$$

and

$$\boldsymbol{v}_j^{o}(k) = \left[v_j^{o}(1), \ldots, v_j^{o}(N)\right]^{T}$$

Proof. It has to be proved that the vector of an orthogonal component, e.g., $\boldsymbol{v}_j^{o}$ is orthogonal to any, previously computed vector of orthogonal components, e.g., $\boldsymbol{v}_\ell^{o}$. $\ell < j$:

$$\boldsymbol{v}_j^{o^{T}} \boldsymbol{v}_\ell^{o} = \left[\boldsymbol{v}_j^{T} - \sum_{i=1}^{j-1} \frac{\boldsymbol{v}_j^{T} \boldsymbol{v}_i^{o}}{\boldsymbol{v}_i^{o^{T}} \boldsymbol{v}_i^{o}} \boldsymbol{v}_i^{o^{T}}\right] \boldsymbol{v}_\ell^{o} = \boldsymbol{v}_j^{T} \boldsymbol{v}_\ell^{o} - \sum_{i=1}^{j-1} \frac{\boldsymbol{v}_j^{T} \boldsymbol{v}_i^{o}}{\boldsymbol{v}_i^{o^{T}} \boldsymbol{v}_i^{o}} \boldsymbol{v}_i^{o^{T}} \boldsymbol{v}_\ell^{o} \tag{3.3.37}$$

$\boldsymbol{v}_\ell^{o}$ is orthogonal to any already orthogonalized components

$$\boldsymbol{v}_i^{o^{T}} \boldsymbol{v}_\ell^{o} = \begin{cases} 0 & \text{if} \quad i \neq \ell \\ \boldsymbol{v}_\ell^{o^{T}} \boldsymbol{v}_\ell^{o} & \text{if} \quad i = \ell \end{cases}$$

Then (3.3.37) has the form

$$\boldsymbol{v}_j^{\mathrm{T}}\boldsymbol{v}_\ell^{\mathrm{o}} - \frac{\boldsymbol{v}_j^{\mathrm{T}}\boldsymbol{v}_\ell^{\mathrm{o}}}{\boldsymbol{v}_\ell^{\mathrm{o}^{\mathrm{T}}}\boldsymbol{v}_\ell^{\mathrm{o}}}\boldsymbol{v}_\ell^{\mathrm{o}^{\mathrm{T}}}\boldsymbol{v}_\ell^{\mathrm{o}} = \boldsymbol{v}_j^{\mathrm{T}}\boldsymbol{v}_\ell^{\mathrm{o}} - \boldsymbol{v}_j^{\mathrm{T}}\boldsymbol{v}_\ell^{\mathrm{o}} = 0$$

■

Lemma 3.3.7 is the Gram–Schmidt orthogonalization.

Lemma 3.3.8
The estimation of the parameters of the orthogonalized model is

$$\hat{\theta}_i^{\mathrm{o}} = \frac{\boldsymbol{v}_i^{\mathrm{o}^{\mathrm{T}}}\boldsymbol{y}}{\boldsymbol{v}_i^{\mathrm{o}^{\mathrm{T}}}\boldsymbol{v}_i^{\mathrm{o}}} \tag{3.3.38}$$

Proof. The LS estimate of the parameters is

$$\hat{\boldsymbol{\theta}}^{\mathrm{o}} = \left[\boldsymbol{\Phi}^{\mathrm{o}^{\mathrm{T}}}\boldsymbol{\Phi}^{\mathrm{o}}\right]^{-1}\boldsymbol{\Phi}^{\mathrm{o}^{\mathrm{T}}}\boldsymbol{y} \tag{3.3.39}$$

The matrix $\boldsymbol{\Phi}^{\mathrm{o}}$ has the form

$$\boldsymbol{\Phi}^{\mathrm{o}} = \begin{bmatrix} \boldsymbol{\varphi}^{\mathrm{o}^{\mathrm{T}}}(1) \\ \vdots \\ \boldsymbol{\varphi}^{\mathrm{o}^{\mathrm{T}}}(N) \end{bmatrix} = \begin{bmatrix} v_1^{\mathrm{o}}(1) & v_2^{\mathrm{o}}(1) & \cdots & v_{n_\theta}^{\mathrm{o}}(1) \\ \vdots & \vdots & \ddots & \vdots \\ v_1^{\mathrm{o}}(N) & v_2^{\mathrm{o}}(N) & \cdots & v_{n_\theta}^{\mathrm{o}}(N) \end{bmatrix} = \left[\boldsymbol{v}_1^{\mathrm{o}}, \ldots, \boldsymbol{v}_{n_\theta}^{\mathrm{o}}\right]$$

The i, jth element of the matrix $\boldsymbol{\Phi}^{\mathrm{o}^{\mathrm{T}}}\boldsymbol{\Phi}^{\mathrm{o}}$ is

$$\boldsymbol{\Phi}^{\mathrm{o}^{\mathrm{T}}}\boldsymbol{\Phi}^{\mathrm{o}}[i,j] = \boldsymbol{v}_i^{\mathrm{o}^{\mathrm{T}}}\boldsymbol{v}_j^{\mathrm{o}} \tag{3.3.40}$$

and the ith element of the vector $\boldsymbol{\Phi}^{\mathrm{o}^{\mathrm{T}}}\boldsymbol{y}$ is

$$\boldsymbol{\Phi}^{\mathrm{o}^{\mathrm{T}}}\boldsymbol{y}[i] = \boldsymbol{v}_i^{\mathrm{o}^{\mathrm{T}}}\boldsymbol{y} \tag{3.3.41}$$

Because of the orthogonality the covariance matrix is diagonal

$$\boldsymbol{\Phi}^{\mathrm{o}^{\mathrm{T}}}\boldsymbol{\Phi}^{\mathrm{o}} = \mathrm{diag}\langle \boldsymbol{v}_1^{\mathrm{o}^{\mathrm{T}}}\boldsymbol{v}_1^{\mathrm{o}}, \ldots, \boldsymbol{v}_{n_\theta}^{\mathrm{o}^{\mathrm{T}}}\boldsymbol{v}_{n_\theta}^{\mathrm{o}} \rangle \tag{3.3.42}$$

and (3.3.38) results from (3.3.39) – (3.3.42). ■

(3.3.38) means that the estimates of the individual parameters are independent of each other. In other words, the inclusion of a new orthogonal component into the model does not affect the already existing estimates.

Lemma 3.3.9
The sum of squares of the residuals is

$$\sum_{k=1}^{N} \varepsilon^2(k) = \sum_{k=1}^{N} y^2(k) - \sum_{i=1}^{n_\theta} \frac{\left[\boldsymbol{v}_i^{\text{o}^\text{T}} \boldsymbol{y}\right]^2}{\boldsymbol{v}_i^{\text{o}^\text{T}} \boldsymbol{v}_i^{\text{o}}} = \boldsymbol{y}^\text{T}\boldsymbol{y} - \hat{\boldsymbol{\theta}}^{\text{o}^\text{T}} \boldsymbol{\Phi}^{\text{o}^\text{T}} \boldsymbol{y} \tag{3.3.43}$$

Proof.

$$\sum_{k=1}^{N} \varepsilon^2(k) = \boldsymbol{\varepsilon}^\text{T}\boldsymbol{\varepsilon} = \left[y - \boldsymbol{\Phi}^\text{o}\hat{\boldsymbol{\theta}}^\text{o}\right]^\text{T}\left[y - \boldsymbol{\Phi}^\text{o}\hat{\boldsymbol{\theta}}^\text{o}\right] = y^\text{T}y - 2\hat{\boldsymbol{\theta}}^{\text{o}^\text{T}}\boldsymbol{\Phi}^{\text{o}^\text{T}}y$$

$$+\hat{\boldsymbol{\theta}}^{\text{o}^\text{T}}\boldsymbol{\Phi}^{\text{o}^\text{T}}\boldsymbol{\Phi}^\text{o}\hat{\boldsymbol{\theta}}^\text{o} = y^\text{T}y - 2\hat{\boldsymbol{\theta}}^{\text{o}^\text{T}}\boldsymbol{\Phi}^{\text{o}^\text{T}}y + \hat{\boldsymbol{\theta}}^{\text{o}^\text{T}}\boldsymbol{\Phi}^{\text{o}^\text{T}}y = y^\text{T}y - \hat{\boldsymbol{\theta}}^{\text{o}^\text{T}}\boldsymbol{\Phi}^{\text{o}^\text{T}}y$$

Taking (3.3.38) and (3.3.41) into account we get also the middle term in (3.3.43). ∎

Lemma 3.3.10
The parameters of the non-orthogonal model can be transformed from the parameters of the orthogonal model by the formula

$$\hat{\theta}_{n_\theta} = \hat{\theta}^\text{o}_{n_\theta} \tag{3.3.44a}$$

$$\hat{\theta}_i = \hat{\theta}_i^\text{o} - \sum_{j=i+1}^{n_\theta} \frac{\boldsymbol{v}_j^\text{T}\boldsymbol{v}_i^\text{o}}{\boldsymbol{v}_i^\text{T}\boldsymbol{v}_i^\text{o}} \hat{\theta}_j \qquad i = n_\theta - 1, \ldots, 1 \tag{3.3.44b}$$

Proof. The proof is given for $n_\theta = 3$ here and the same procedure can be followed for more parameters. Introduce

$$\gamma_{ji} = \frac{\boldsymbol{v}_j^\text{T}\boldsymbol{v}_i^\text{o}}{\boldsymbol{v}_i^\text{T}\boldsymbol{v}_i^\text{o}}$$

Then

$$\begin{aligned} y(k) &= \hat{\theta}_1^\text{o} v_1^\text{o} + \hat{\theta}_2^\text{o} v_2^\text{o} + \hat{\theta}_3^\text{o} v_3^\text{o} \\ &= \hat{\theta}_1^\text{o} v_1 + \hat{\theta}_2^\text{o}\left[v_2 - \gamma_{21}v_1^\text{o}\right] + \hat{\theta}_3^\text{o}\left[v_3 - \gamma_{31}v_1^\text{o} - \gamma_{21}v_2^\text{o}\right] \\ &= \hat{\theta}_1^\text{o} v_1 + \hat{\theta}_2^\text{o}\left[v_2 - \gamma_{21}v_1\right] + \hat{\theta}_3^\text{o}\left[v_3 - \gamma_{31}v_1 - \gamma_{32}\left(v_2 - \gamma_{21}v_1\right)\right] \\ &= v_3\hat{\theta}_3^\text{o} + v_2\left[\hat{\theta}_2^\text{o} - \gamma_{32}\hat{\theta}_3^\text{o}\right] + v_1\left[\hat{\theta}_1^\text{o} - \gamma_{21}\left(\hat{\theta}_2^\text{o} - \gamma_{32}\hat{\theta}_3^\text{o}\right) - \gamma_{31}\hat{\theta}_3^\text{o}\right] \end{aligned} \tag{3.3.45}$$

The model output can also be expressed by means of the original model components

$$y(k) = \sum_{j=1}^{n_\theta} \hat{\theta}_j v_j(k) = \hat{\theta}_1 v_1 + \hat{\theta}_2 v_2 + \hat{\theta}_3 v_3 \tag{3.3.46}$$

Equating the v_i-s; $i = 1, \ldots, n_\theta$ from (3.3.45) with (3.3.46) results in

$$\hat{\theta}_3 = \hat{\theta}_3^\text{o}$$

$$\hat{\theta}_2 = \hat{\theta}_2^{\text{o}} - \gamma_{32}\hat{\theta}_3^{\text{o}} = \hat{\theta}_2^{\text{o}} - \gamma_{32}\hat{\theta}_3 = \hat{\theta}_2^{\text{o}} - \sum_{j=2}^{3} \gamma_{j2}\hat{\theta}_j$$

$$\hat{\theta}_1 = \hat{\theta}_1^{\text{o}} - \gamma_{21}\left(\hat{\theta}_2^{\text{o}} - \gamma_{32}\hat{\theta}_3^{\text{o}}\right) - \gamma_{31}\hat{\theta}_3^{\text{o}} = \hat{\theta}_1^{\text{o}} - \gamma_{21}\hat{\theta}_2 - \gamma_{31}\hat{\theta}_3 = \hat{\theta}_1^{\text{o}} - \sum_{j=2}^{3} \gamma_{j1}\hat{\theta}_j$$

which is the form (3.3.44) for $n_{\boldsymbol{\theta}} = 3$. ■

Example 3.3.3 *Orthogonal estimation of a static parabola*
The equation of the static parabola is

$$y(k) = 2 + u(k) + 0.5u^2(k)$$

Let the measured input signal be the following sequence

$$\boldsymbol{u} = [-1, 1, -1, 3, 3, 1, 3, -1]^{\text{T}}$$

The noise-free output signal is then

$$\boldsymbol{y} = [1.5, 3.5, 1.5, 9.5, 9.5, 3.5, 9.5, 1.5]^{\text{T}}$$

The non-orthogonal model components are

$$\hat{y}(k) = 2 \cdot v_1(k) + 1 \cdot v_2(k) + 0.5 \cdot v_3(k)$$

i.e.,

$$v_1(k) = 1 \qquad v_2(k) = u(k) \qquad v_3(k) = u^2(k)$$

The vectors of the components are

$$\boldsymbol{v}_1 = [1, 1, 1, 1, 1, 1, 1, 1]^{\text{T}}$$

$$\boldsymbol{v}_2 = \boldsymbol{u} = [-1, 1, -1, 3, 3, 1, 3, -1]^{\text{T}}$$

$$\boldsymbol{v}_3 = [1, 1, 1, 9, 9, 1, 9, 1]^{\text{T}}$$

The vectors of the orthogonalized components can be calculated from (3.3.36)

$$\boldsymbol{v}_1^{\text{o}} = \boldsymbol{v}_1 = [1, 1, 1, 1, 1, 1, 1, 1]^{\text{T}}$$

$$\boldsymbol{v}_2^{\text{o}} = \boldsymbol{v}_2 - \frac{\boldsymbol{v}_2^{\text{T}}\boldsymbol{v}_1^{\text{o}}}{\boldsymbol{v}_1^{\text{o}\,\text{T}}\boldsymbol{v}_1^{\text{o}}}\boldsymbol{v}_1^{\text{o}} = \boldsymbol{v}_2 - \tfrac{8}{8}\boldsymbol{v}_1^{\text{o}} = \boldsymbol{v}_2 - \boldsymbol{v}_1^{\text{o}}$$

$$\boldsymbol{v}_2^{\text{o}} = [-2, 0, -2, 2, 2, 0, 2, -2]^{\text{T}}$$

$$\boldsymbol{v}_3^{\text{o}} = \boldsymbol{v}_3 - \frac{\boldsymbol{v}_3^{\text{T}}\boldsymbol{v}_1^{\text{o}}}{\boldsymbol{v}_1^{\text{o}\,\text{T}}\boldsymbol{v}_1^{\text{o}}}\boldsymbol{v}_1^{\text{o}} - \frac{\boldsymbol{v}_3^{\text{T}}\boldsymbol{v}_2^{\text{o}}}{\boldsymbol{v}_2^{\text{o}\,\text{T}}\boldsymbol{v}_2^{\text{o}}}\boldsymbol{v}_2^{\text{o}} = \boldsymbol{v}_3 - \tfrac{32}{8}\boldsymbol{v}_1^{\text{o}} - \tfrac{48}{24}\boldsymbol{v}_2^{\text{o}} = \boldsymbol{v}_3 - 4\boldsymbol{v}_1^{\text{o}} - 2\boldsymbol{v}_2^{\text{o}}$$

$$v_3^o = [1, -3, 1, 1, 1, -3, 1, 1]^T$$

The parameters of the orthogonalized model can be estimated by (3.3.24)

$$\hat{\theta}_1^o = \frac{v_1^{o^T} y}{v_1^{o^T} v_1^o} = \frac{40}{8} = 5$$

$$\hat{\theta}_2^o = \frac{v_2^{o^T} y}{v_2^{o^T} v_2^o} = \frac{48}{24} = 2$$

$$\hat{\theta}_3^o = \frac{v_3^{o^T} y}{v_3^{o^T} v_3^o} = \frac{12}{24} = 0.5$$

The parameters of the non-orthogonal model are calculated by the formulas (3.3.44)

$$\hat{\theta}_3 = \hat{\theta}_3^o = 0.5$$

$$\hat{\theta}_2 = \hat{\theta}_2^o - \frac{v_3^{o^T} v_2^o}{v_2^{o^T} v_2^o} \hat{\theta}_3 = 2 - \frac{48}{24} 0.5 = 1$$

$$\hat{\theta}_1 = \hat{\theta}_1^o - \frac{v_3^{o^T} v_1^o}{v_1^{o^T} v_1^o} \hat{\theta}_3 - \frac{v_2^{o^T} v_1^o}{v_1^{o^T} v_1^o} \hat{\theta}_2 = 5 - \frac{32}{8} 0.5 - \frac{8}{8} 1 = 2 \qquad \blacksquare$$

Remarks:

1. The orthogonalization method presented is known as the Gram–Schmidt orthogonalization procedure.
2. If the memory vector contains functions of the residuals then the orthogonalization of the model components has to be performed in each iteration step again. This is the case with the prediction error and with the extended matrix methods. (Korenberg *et al.*, 1988).

Applications of the method for the following nonlinear models are known (among others):

- *recursive polynomial difference equation model* (Korenberg, 1985, 1989; Janiszowski, 1986; Liu *et al.*, 1987; Billings *et al.*, 1989; Chen *et al.*, 1989; Kortmann and Unbehauen, 1987; 1988; Kortmann *et al.*, 1988);
- *output affine model* (Billings *et al.*, 1988), etc.

3.4 METHODS FOR RESIDUAL MODELS NONLINEAR-IN-PARAMETERS

3.4.1 Classification of prediction models of the output signal

The solution of the minimization of the loss function depends mainly on the way in which the prediction model of the output signal $\hat{y}(k, \hat{\theta})$ depends on the unknown parameters. The prediction model can have the following forms:

1. *linear-in-parameters and not depending on the residuals:*

A least squares method minimizes the cost function in one step.

2. *linear-in-parameters and depending on the residuals:*
   As the residuals depend on the estimated parameters the least squares method minimizes the cost function iteratively in several steps.
3. *nonlinear-in-parameters:*
   Any optimization method minimizes the cost function iteratively in several steps.

The parameter estimation algorithm can be simplified if the process model nonlinear-in-parameters can be transformed to a model linear-in-parameters.

Example 3.4.1 *Linear first-order system*
The first-order differential equation

$$T\dot{y}(t) + y(t) = Ku(t)$$

has the equivalent difference equation

$$y(k) = \exp(-\Delta T/T)y(k-1) + K[1-\exp(-\Delta T/T)]u(k-1) \tag{3.4.1}$$

if the input signal is constant between two samplings. Equation (3.4.1) is nonlinear in the parameters K and T. It can be transformed to an equation linear-in-parameters if the following new parameters are introduced

$$a_1 = \exp(-\Delta T/T) \qquad b_1 = K[1-\exp(-\Delta T/T)]$$

The difference equation with the transformed parameters is linear in the parameters

$$y(k) = a_1 y(k-1) + b_1 u(k-1)$$

The estimated parameters $\hat{K}$ and $\hat{T}$ can be calculated from the estimated parameters $\hat{a}_1$ and $\hat{b}_1$ as

$$\hat{T}_1 = \frac{-\Delta T}{\ln(\hat{a}_1)} \qquad \hat{K} = \frac{\hat{b}_1}{1+\hat{a}_1}$$

■

Definition 3.4.1 (Draper and Smith, 1966) A process model that cannot be transformed to a model linear-in-parameters is called an intrinsically nonlinear or a non-intrinsically linear model. ■

Definition 3.4.2 (Draper and Smith, 1966) A process model that can be transformed to a model linear-in-parameters is called an intrinsically linear model. ■

The linearizing transformation of a nonlinear process model can be done by the following methods:

- by introducing new parameters (see Example 3.4.1);

- by introducing new variables (see Example 3.4.2);
- by introducing redundant parameters (see Example 3.4.3).

Example 3.4.2 *Weighting function of first-order system*
The weighting function of a first-order linear system has the form

$$y(k) = \theta_1 \exp(-\theta_2 t)$$

The unknown parameters θ_1 and θ_2 can be estimated from the equation linear-in-parameters

$$\ln[y(k)] = \ln(\theta_1) - \theta_2 t$$

■

Example 3.4.3 *Third-order quadratic simple Hammerstein model*
The simple Hammerstein model has the equations

$$y(k) = -a_1 y(k-1) - a_2 y(k-2) - a_3 y(k-3) + b_1 v(k-1) + b_1 v(k-2) + b_3 v(k-3)$$

$$v(k) = c_0 + c_1 u(k) + c_2 u^2(k)$$

There are nine unknown parameters. The input–output equation

$$\begin{aligned} y(k) = &-a_1 y(k-1) - a_2 y(k-2) - a_3 y(k-3) \\ &+b_1\left[c_0 + c_1 u(k-1) + c_2 u^2(k-1)\right] \\ &+b_2\left[c_0 + c_1 u(k-2) + c_2 u^2(k-2)\right] \\ &+b_3\left[c_0 + c_1 u(k-3) + c_2 u^2(k-3)\right] \end{aligned} \tag{3.4.2}$$

is nonlinear in the parameters

$$\boldsymbol{\theta} = \left[a_1, a_2, a_3, b_1, b_2, b_3, c_0, c_1, c_2\right]^{\mathrm{T}}$$

It can be transformed to a form linear-in-parameters like

$$\begin{aligned} y(k) = &-a_1 y(k-1) - a_2 y(k-2) - a_3 y(k-3) + c_0(b_1 + b_2 + b_3) \\ &+c_1 b_1 u(k-1) + c_1 b_2 u(k-2) + c_1 b_3 u(k-3) + \\ &+c_2 b_1 u^2(k-1) + c_2 b_2 u^2(k-2) + c_2 b_3 u^2(k-3) \end{aligned} \tag{3.4.3}$$

Equation (3.4.3) is linear in the new parameters

$$\boldsymbol{\theta}^* = \left[a_1, a_2, a_3, c_0(b_1 + b_2 + b_3), c_1 b_1, c_1 b_2, c_1 b_3, c_2 b_1, c_2 b_2, c_2 b_3\right]^{\mathrm{T}}$$

but has one parameter more than (3.4.2). Such a description is called redundant. The original parameters $\boldsymbol{\theta}$ cannot be calculated unambiguously from the estimated parameters $\boldsymbol{\theta}^*$. ■

Definition 3.4.3 A model is redundant if it contains more parameters as needed. ■

3.4.2 Different minimization algorithms

Most parameter estimation methods use the gradient of the loss function (3.3.3)

$$\frac{\partial J(\hat{\boldsymbol{\theta}})}{\partial \hat{\boldsymbol{\theta}}} = \frac{\partial}{\partial \hat{\boldsymbol{\theta}}}\left[\tfrac{1}{2}\boldsymbol{\varepsilon}^{\mathrm{T}}\boldsymbol{\varepsilon}\right] = \frac{\partial \boldsymbol{\varepsilon}^{\mathrm{T}}(\hat{\boldsymbol{\theta}})}{\partial \hat{\boldsymbol{\theta}}}\boldsymbol{\varepsilon}$$

which needs the knowledge of the gradient of the predicted error signal according to the parameter. The process models may be classified according to the gradients of the predicted error signal according to the parameters:

- models linear-in-parameters;
- models nonlinear-in-parameters with calculable gradients;
- models nonlinear-in-parameters which cannot be differentiated.

Example 3.4.4 *Shifted sign function*

If the input-output relation is a shifted sign function (Figure 3.4.1)

$$y(t) = \theta_1 \operatorname{sgn}(u(t) - \theta_2)$$

then the predicted output signal cannot be differentiated with respect to θ_2. ■

Depending on whether the derivatives are calculable or not, one can distinguish:

- methods without derivatives;
- methods using only first-order gradients;
- methods using first- and second-order gradients.

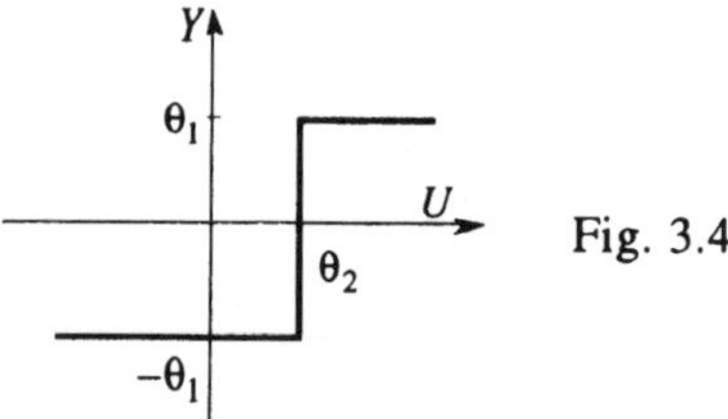

Fig. 3.4.1 Shifted sign function

1. Derivative free methods

(a) *Systematic search in a parameter domain (raster method):*

The algorithm contains the following steps:

1. The ranges between the parameter bounds have to be divided into some (e.g., eight) sections in every dimension.
2. The loss function at every raster point is to be computed.
3. The location of the smallest functional value (global minimum) is to be searched.
4. The range between the old boundary values of the parameters has to be divided (e.g., halved) for each parameter. The new boundary values are set symmetrically to the global optimum value obtained in the previous step.
5. The search is to be continued with step 1 until a given stopping criterion is satisfied.

Figure 3.4.2 shows the rasters of Steps 1, 2 and 3 for two parameters.

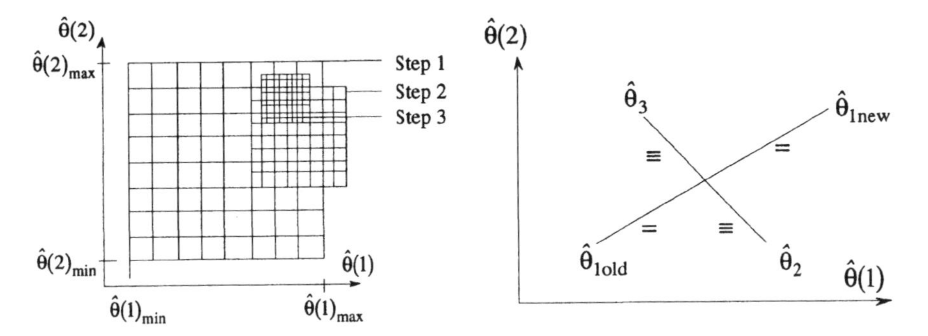

Fig. 3.4.2 Raster method of optimization

Fig. 3.4.3 Simplex optimization method: the first reflection

(b) *Simplex method (Nelder and Mead, 1965):*

This method is a systematic search with the following steps:

1. Select $(n_\theta + 1)$ initial parameters. They build a so called simplex.
2. Compute the loss functions for the selected parameters.
3. Reflect the parameter with the largest loss function through the *center of gravity* of all other parameters.
4. If the loss function of the reflected parameter is less than that of the original one stretch the simplex in the same direction until a minimum is found and finally go to 3.
5. If the loss function of the reflected parameter is greater than that of the original one try to reflect the next parameter with the largest loss function except those tried earlier. If none of the reflections lead to a smaller loss function go to 7.
6. Go to 3.
7. Contract the simplex to the parameter with the smallest loss function by halving the distances of the parameters from that with the smallest loss function.
8. Go to 3.

Figure 3.4.3 shows the initial simplex with the first reflection for two unknown parameters.

(c) *Using finite differences:*

The gradient methods can be used by approximating the gradient by finite differences as

$$\frac{\partial J(\hat{\boldsymbol{\theta}})}{\partial \hat{\theta}_i} \approx \frac{J(\hat{\boldsymbol{\theta}} + \Delta\hat{\theta}_i \boldsymbol{i}) - J(\hat{\boldsymbol{\theta}} + \Delta\hat{\theta}_i \boldsymbol{i})}{2\Delta\hat{\theta}_i}$$

$\Delta\hat{\theta}_i, i = 1, \ldots, n_\theta$ should be selected small and $\boldsymbol{i}$ is the ith unit vector.

(d) *Genetic algorithm:*

A genetic (or evolutionary) algorithm is a multi-dimensional, stochastic global search

- with randomly generated initial selection and
- with a systematical modification (recombination and mutation) of chosen components.

For details on different versions of genetic algorithms see e.g., Goldberg (1989), Baeck and Schwefel (1993).

2. Methods using only first-order error sensitivity

These methods use the algorithm

$$\hat{\boldsymbol{\theta}}_{\ell+1}=\hat{\boldsymbol{\theta}}_{\ell}-\boldsymbol{R}\left.\frac{\partial J(\hat{\boldsymbol{\theta}})}{\partial\hat{\boldsymbol{\theta}}}\right|_{\hat{\boldsymbol{\theta}}=\hat{\boldsymbol{\theta}}_{\ell}}$$

with a convergence matrix $\boldsymbol{R}$ not depending on the second-order derivatives of the loss function. Some methods are

(a) *Gauss–Seidel method:*

The loss function will be minimized in respect of only one parameter and this procedure will be continued for all parameters in turn.

(b) *minimization in the direction of the gradient:*

The optimum will be searched in the direction of the gradient with

$$\boldsymbol{R}=\frac{\boldsymbol{g}^{\mathrm{T}}\boldsymbol{g}}{\boldsymbol{g}^{\mathrm{T}}\boldsymbol{H}\boldsymbol{g}}\qquad\text{and}\qquad\boldsymbol{g}=\left.\frac{\partial J(\hat{\boldsymbol{\theta}})}{\partial\hat{\boldsymbol{\theta}}}\right|_{\hat{\boldsymbol{\theta}}=\hat{\boldsymbol{\theta}}_{\ell}}$$

Here $\boldsymbol{H}$ is the so called Hessian matrix

$$\frac{\partial^2 J(\hat{\boldsymbol{\theta}})}{\partial\hat{\boldsymbol{\theta}}\partial\hat{\boldsymbol{\theta}}^{\mathrm{T}}}=\frac{\partial}{\partial\hat{\boldsymbol{\theta}}}\left[\boldsymbol{\varepsilon}^{\mathrm{T}}\frac{\partial\boldsymbol{\varepsilon}}{\partial\hat{\boldsymbol{\theta}}^{\mathrm{T}}}\right]=\frac{\partial\boldsymbol{\varepsilon}^{\mathrm{T}}}{\partial\hat{\boldsymbol{\theta}}}\frac{\partial\boldsymbol{\varepsilon}}{\partial\hat{\boldsymbol{\theta}}^{\mathrm{T}}}+\frac{\partial}{\partial\hat{\boldsymbol{\theta}}}\frac{\partial\boldsymbol{\varepsilon}}{\partial\hat{\boldsymbol{\theta}}^{\mathrm{T}}}\boldsymbol{\varepsilon}\tag{3.4.5}$$

which can be well approximated in some cases by

$$\frac{\partial^2 J(\hat{\boldsymbol{\theta}})}{\partial\hat{\boldsymbol{\theta}}\partial\hat{\boldsymbol{\theta}}^{\mathrm{T}}}\cong\frac{\partial\boldsymbol{\varepsilon}^{\mathrm{T}}}{\partial\hat{\boldsymbol{\theta}}}\frac{\partial\boldsymbol{\varepsilon}}{\partial\hat{\boldsymbol{\theta}}^{\mathrm{T}}}\tag{3.4.6}$$

because the residuals are small near the optimum. In addition to this, (3.4.6) is always positive semi-definite, contrary to (3.4.5), which in some cases improves the convergence of the iterative algorithm.

(c) *Gauss–Newton method:*

In optimal case the convergence matrix is the inverse of the matrix of the second derivatives $\boldsymbol{R} = \boldsymbol{H}^{-1}$.

3. Methods using first- and second-order gradients

(a) *Newton–Raphson method:*

Lemma 3.4.1
The optimal (in the quadratic sense) minimization algorithm of the loss function (3.3.9) is

$$\hat{\boldsymbol{\theta}}_{\ell+1} = \hat{\boldsymbol{\theta}}_{\ell} - \left[\left. \frac{\partial^2 J(\hat{\boldsymbol{\theta}})}{\partial \hat{\boldsymbol{\theta}} \partial \hat{\boldsymbol{\theta}}^{\mathrm{T}}} \right|_{\hat{\boldsymbol{\theta}}=\hat{\boldsymbol{\theta}}_{\ell}} \right]^{-1} \left. \frac{\partial J(\hat{\boldsymbol{\theta}})}{\partial \hat{\boldsymbol{\theta}}} \right|_{\hat{\boldsymbol{\theta}}=\hat{\boldsymbol{\theta}}_{\ell}} \tag{3.4.7}$$

Proof. The loss function will be approximated by a quadratic function near the optimum:

$$J(\hat{\boldsymbol{\theta}}_{\ell+1}) \approx J(\hat{\boldsymbol{\theta}}_{\ell}) + \left. \frac{\partial J(\hat{\boldsymbol{\theta}})}{\partial \hat{\boldsymbol{\theta}}} \right|_{\hat{\boldsymbol{\theta}}=\hat{\boldsymbol{\theta}}_{\ell}} \left(\hat{\boldsymbol{\theta}}_{\ell+1} - \hat{\boldsymbol{\theta}}_{\ell}\right) + \tfrac{1}{2}\left(\hat{\boldsymbol{\theta}}_{\ell+1} - \hat{\boldsymbol{\theta}}_{\ell}\right)^{\mathrm{T}} \left. \frac{\partial^2 J(\hat{\boldsymbol{\theta}})}{\partial \hat{\boldsymbol{\theta}} \partial \hat{\boldsymbol{\theta}}^{\mathrm{T}}} \right|_{\hat{\boldsymbol{\theta}}=\hat{\boldsymbol{\theta}}_{\ell}} \left(\hat{\boldsymbol{\theta}}_{\ell+1} - \hat{\boldsymbol{\theta}}_{\ell}\right)$$

Derive $J(\hat{\boldsymbol{\theta}}_{\ell+1})$ according to $\hat{\boldsymbol{\theta}} = \hat{\boldsymbol{\theta}}_{\ell+1}$ and set the gradient to zero

$$\left. \frac{\partial J(\hat{\boldsymbol{\theta}})}{\partial \hat{\boldsymbol{\theta}}} \right|_{\hat{\boldsymbol{\theta}}=\hat{\boldsymbol{\theta}}_{\ell+1}} = \left. \frac{\partial J(\hat{\boldsymbol{\theta}})}{\partial \hat{\boldsymbol{\theta}}} \right|_{\hat{\boldsymbol{\theta}}=\hat{\boldsymbol{\theta}}_{\ell}} + \left. \frac{\partial^2 J(\hat{\boldsymbol{\theta}})}{\partial \hat{\boldsymbol{\theta}} \partial \hat{\boldsymbol{\theta}}^{\mathrm{T}}} \right|_{\hat{\boldsymbol{\theta}}=\hat{\boldsymbol{\theta}}_{\ell}} \left(\hat{\boldsymbol{\theta}}_{\ell+1} - \hat{\boldsymbol{\theta}}_{\ell}\right) = \mathbf{0}$$

which is equivalent to (3.4.7). ■

As is seen, this method uses a convergence matrix that is the inverse of (3.4.5). The method is fast but sometimes not convergent because the Hessian (3.4.5) may not be positive definite if the prediction of the output signal is nonlinear in the parameters.

(b) *Marquardt algorithm (Marquardt, 1963):*

One way to ensure a convergent algorithm is to restrict the step $\left(\hat{\boldsymbol{\theta}}_{\ell+1} - \hat{\boldsymbol{\theta}}_{\ell}\right)$, because then the quadratic approximation near $\hat{\boldsymbol{\theta}}_{\ell}$ is approximately kept valid. This is made by minimizing a modified loss function.

Lemma 3.4.2
The Marquardt algorithm minimizes

$$J^{*}(\hat{\boldsymbol{\theta}}) = \tfrac{1}{2}\boldsymbol{\varepsilon}^{\mathrm{T}}\boldsymbol{\varepsilon} + \lambda \Delta\hat{\boldsymbol{\theta}}^{\mathrm{T}} \Delta\hat{\boldsymbol{\theta}} \rightarrow \min \tag{3.4.8}$$

by

$$\hat{\boldsymbol{\theta}}_{\ell+1} = \hat{\boldsymbol{\theta}}_{\ell} - \left[\left. \frac{\partial^2 J(\hat{\boldsymbol{\theta}})}{\partial \hat{\boldsymbol{\theta}} \partial \hat{\boldsymbol{\theta}}^{\mathrm{T}}} \right|_{\hat{\boldsymbol{\theta}}=\hat{\boldsymbol{\theta}}_{\ell}} + \lambda \boldsymbol{I} \right]^{-1} \left. \frac{\partial J(\hat{\boldsymbol{\theta}})}{\partial \hat{\boldsymbol{\theta}}} \right|_{\hat{\boldsymbol{\theta}}=\hat{\boldsymbol{\theta}}_{\ell}} \tag{3.4.9}$$

Proof. Apply the Newton–Raphson method for the loss function (3.4.8). The first-order derivatives are

$$\left. \frac{\partial J^*(\hat{\boldsymbol{\theta}})}{\partial \hat{\boldsymbol{\theta}}} \right|_{\hat{\boldsymbol{\theta}}=\hat{\boldsymbol{\theta}}_{\ell}} = \left. \frac{\partial J(\hat{\boldsymbol{\theta}})}{\partial \hat{\boldsymbol{\theta}}} \right|_{\hat{\boldsymbol{\theta}}=\hat{\boldsymbol{\theta}}_{\ell}} + \lambda \left. \left(\hat{\boldsymbol{\theta}} - \hat{\boldsymbol{\theta}}_{\ell} \right) \right|_{\hat{\boldsymbol{\theta}}=\hat{\boldsymbol{\theta}}_{\ell}} = \left. \frac{\partial J(\hat{\boldsymbol{\theta}})}{\partial \hat{\boldsymbol{\theta}}} \right|_{\hat{\boldsymbol{\theta}}=\hat{\boldsymbol{\theta}}_{\ell}}$$

and the second-order derivatives are

$$\frac{\partial^2 J^*(\hat{\boldsymbol{\theta}})}{\partial \hat{\boldsymbol{\theta}} \partial \hat{\boldsymbol{\theta}}^{\mathrm{T}}} = \left. \frac{\partial^2 J(\hat{\boldsymbol{\theta}})}{\partial \hat{\boldsymbol{\theta}} \partial \hat{\boldsymbol{\theta}}^{\mathrm{T}}} \right|_{\hat{\boldsymbol{\theta}}=\hat{\boldsymbol{\theta}}_{\ell}} + \lambda \boldsymbol{I}$$

■

Marquardt (1963) recommended the following strategy for choosing λ:

1. Set $\lambda_0 = 0.01$ as an initial value.
2. Estimate $\hat{\boldsymbol{\theta}}_{\ell}$.
3. Set $\lambda_{\ell+1} = 0.1\lambda_{\ell}$ if $J(\hat{\boldsymbol{\theta}}_{\ell+1}) < J(\hat{\boldsymbol{\theta}}_{\ell})$ otherwise set $\lambda_{\ell+1} = 10\lambda_{\ell}$. Go to 2.

3.4.3 Handling restrictions to the parameters

Restrictions to the parameters can be

- of equality or
- of inequality types.

Example 3.4.5 *Mill process – a restriction of equality type*

Describe the relation between the inlet feed $u(k)$ and outlet feed $y(k)$ of a mill by a second-order process

$$y(k) = q^{-d} \frac{b_1 q^{-1} + b_2 q^{-2}}{1 + a_1 q^{-1} + a_2 q^{-2}} u(k)$$

In the steady state the outlet is equal to the inlet. Therefore the static gain is equal to unity:

$$K = \frac{b_1 + b_2}{1 + a_1 + a_2} = 1$$

or

$$b_1 + b_2 = 1 + a_1 + a_2$$

which is a restriction of equality type.

■

Example 3.4.6 *Thermal process – a restriction of inequality type*
The relation between the heating power $u(t)$ and the temperature $y(t)$ of an extruder can be approximated by a linear second-order aperiodic process

$$y(t)+\alpha_1\dot{y}(t)+\alpha_2\ddot{y}(t)=\beta_0 u(t)+\beta_1\dot{u}(t)$$

The transfer function

$$G(s)=\frac{\beta_0+\beta_1 s}{1+\alpha_1 s+\alpha_2 s^2}$$

cannot have complex poles, thus

$$\alpha_1^2-4\alpha_2\geq 0$$

should be fulfilled. ■

The following minimizing methods are known in these cases:

1. Equality constraints
   Assume the constraints are in the form

$$f_{\mathrm{ci}}(\hat{\boldsymbol{\theta}})=\varepsilon_{\mathrm{c}}(i)=0 \qquad i=1,\ldots,n_{\mathrm{c}} \tag{3.4.10}$$

(a) *Complementing the measurement equations with the equations of the constraints:*
Complement the series of computed residuals (3.3.1) with the equations of constraints (3.4.10) and build a vector of the modified residuals as

$$\boldsymbol{\varepsilon}^*=\left[\varepsilon(1),\ldots,\varepsilon(N),\varepsilon_{\mathrm{c}}(1),\ldots,\varepsilon_{\mathrm{c}}(n_{\mathrm{c}})\right]^{\mathrm{T}}$$

The parameter estimation is the minimization of the weighted sum

$$J^*=\tfrac{1}{2}\sum_{k=1}^{N+n_{\mathrm{c}}}\lambda_i\varepsilon^{*2}(k)$$

where λ_i should be chosen in such a way that the terms belonging to the constraints are much greater than those belonging to the noisy measurements

$$\lambda_i>>\lambda_j \qquad i=N+1,N+2,N+n_{\mathrm{c}} \qquad j=1,\ldots,N$$

(b) *Kuhn–Tucker method:*
The linear combinations of the equations of the constraints should be added to the original loss function (3.3.9)

$$J^*(\hat{\boldsymbol{\theta}})=J(\hat{\boldsymbol{\theta}})+\sum_{i=1}^{n_{\mathrm{c}}}\lambda_i f_{\mathrm{ci}}(\hat{\boldsymbol{\theta}})=\tfrac{1}{2}\sum_{k=1}^{N}\varepsilon^2(k)+\sum_{i=1}^{n_{\mathrm{c}}}\lambda_i f_{\mathrm{ci}}(\hat{\boldsymbol{\theta}}) \tag{3.4.11}$$

and (3.4.11) has to be minimized according to $\hat{\boldsymbol{\theta}}$ and λ_i, $\lambda_i > 0$, $i = 1, \ldots, n_c$.

(c) *Using penalty functions:*

If (3.4.10) is fulfilled then the square of $f_{ci}\left(\hat{\boldsymbol{\theta}}\right)$, $i = 1, \ldots, n_c$ are zero. Add $f_{ci}^2\left(\hat{\boldsymbol{\theta}}\right)$ to the original loss function

$$J^*\left(\hat{\boldsymbol{\theta}}\right) = J\left(\hat{\boldsymbol{\theta}}\right) + \sum_{i=1}^{n_c} \lambda_i f_{ci}^2\left(\hat{\boldsymbol{\theta}}\right) = \frac{1}{2}\sum_{k=1}^{N} \varepsilon^2(k) + \sum_{i=1}^{n_c} \lambda_i f_{ci}^2\left(\hat{\boldsymbol{\theta}}\right) \tag{3.4.12}$$

and minimize (3.4.12) according to the unknown parameters. The larger the factors $\lambda_i > 0$ the constraints are better taken into account.

2. Inequality constraints

Assume the constraints are in form

$$f_{ci}\left(\hat{\boldsymbol{\theta}}\right) = \varepsilon_c(i) \geq 0 \qquad i = 1, \ldots, n_c \tag{3.4.13}$$

(a) *Using penalty functions:*

Extend the original loss function (3.3.9) with terms inversely proportional to the constraints equations (3.4.13)

$$J^*\left(\hat{\boldsymbol{\theta}}\right) = J\left(\hat{\boldsymbol{\theta}}\right) + \sum_{i=1}^{n_c} \frac{\lambda_i}{f_{ci}\left(\hat{\boldsymbol{\theta}}\right)} = \frac{1}{2}\sum_{k=1}^{N} \varepsilon^2(k) + \sum_{i=1}^{n_c} \frac{\lambda_i}{f_{ci}\left(\hat{\boldsymbol{\theta}}\right)} \tag{3.4.14}$$

where $\lambda_i > 0$, $i = 1, \ldots, n_c$ (Carroll, 1961). If λ_i is equal to zero then the unconstrained minimum of (3.4.14) coincides with the minimum of the original loss function (3.3.9) under the inequality constraints (Fiacco and McCormick; 1964). The constrained minimum can be obtained by the following iterative steps (Fiacco and McCormick, 1964):

1. Select the initial parameter vector $\hat{\boldsymbol{\theta}}_i$ and the initial values λ_i, $\lambda_i > 0$, $i = 1, \ldots, n_c$.
2. Minimize the modified loss function (3.4.14).
3. Reduce λ_i, $\lambda_i > 0$, $i = 1, \ldots, n_c$ and minimize the modified loss function (3.4.14).
4. If the loss function is less than in the previous step then go to 3 otherwise stop the algorithm.

(b) *Contraction to the boundary of the region where the constraints are fulfilled:*

A strategy for minimizing the original loss function (3.3.9) under constraints is as follows (Rosen, 1960, 1961):

1. Choose an initial parameter vector in the feasible region.
2. Compute the next iteration towards the optimum of (3.3.9).

3. If the computed new parameter vector falls into the feasible region then accept it. Go to step 2.
4. If the old parameter vector falls in the feasible region and the new parameter vector falls into the unfeasible region contract the step in the direction of the optimum to the boundary of the feasible region. Go to 2.
5. If the old parameter vector is on the boundary and the new parameter vector is not acceptable step along the boundary. Go to 2.
6. Stop the algorithm if the loss function does not decrease any more.

This algorithm is not always convergent.

A good survey on the minimization method presented and others, with and without constraints is given in Bard (1974).

3.5 GATE FUNCTION METHOD

Gate functions divide the working domain of the process into different domains, which are described either by different model types or by different parameter sets with the same structure. Assume the gate function is a function of several signals summarized in a vector $\boldsymbol{x}$

$$\boldsymbol{x} = \left[x_1(t), \ldots, x_M(t)\right]^{\mathrm{T}}$$

The most common gate function models are summarized in Section 1.7:

- *Bose's static model:*

  $$M = 1,\ x_1(t) = u(t)$$

- *Volterra or Zadeh model with gate functions:*

  $$M = m+1,\ x_1(k) = u(k),\ x_2(k) = u(k-1),\ \ldots,\ x_M(k) = u(k-m+1)$$

- *Wiener's gate function model:*

  $$M = n,\ x_1(k) = v_0(k),\ x_2(k) = v_1(k),\ \ldots,\ x_M(k) = v_{m-1}(k)$$

  where $v_i(k),\ i = 0, 1, 2, \ldots$ are the output signals of the linear dynamic orthogonal (Laguerre) filters.
- *elementary gate function model or multi-model:*
  The signals $x_1, x_2, \ldots$ do not correspond to the input–output signals of the process.
- *spatial linear or nonlinear model:*

  $$M = 2n+1,\quad x_1(k) = u(k),\quad x_2(k) = u(k-1),\ \ldots,\ x_{n+1}(k) = u(k-n),$$
  $$x_{n+2}(k) = y(k-1),\quad x_{n+3}(k) = y(k-2),\ \ldots,\ x_M(k) = y(k-n)$$

As is seen, the Volterra model belongs to the spatial nonparametric nonlinear models.

The identification procedure is summarized in the following algorithm.

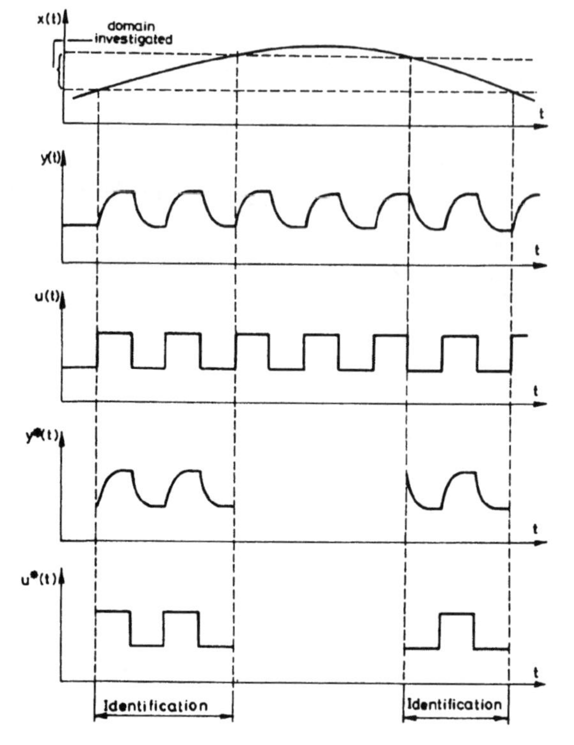

Fig. 3.5.1 Scheme of the identification by means of gate functions

Algorithm 3.5.1 (Schetzen, 1965b; Haber and Keviczky, 1985) The gate function identification method consists of the following steps:

1. Divide the whole working domain into parts. A signal (vector) $\boldsymbol{x}$ has to be selected which is able to appoint the domains. It can be assumed that the structure and the parameters of the models in each domain are constant.
2. Define the gate function as

$$Q_i[\boldsymbol{x}(t)] = \begin{cases} 1 & x_{1j_1} \le x_1(t) \le x_{1j_2} \qquad x_{2j_3} \le x_2(t) \le x_{2j_4} \; \ldots \\ & \ldots \; x_{Mj_{2M-1}} \le x_{2M}(t) \le x_{Mj_{2M}} \\ 0 & \text{otherwise} \end{cases}$$

which means that the gate function is equal to unity if the signals $x_1(t), x_2(t), \ldots$ fall between certain boundaries.

3. Cut those parts of the input and output signals from the measured ones that belong to the given domain, i.e., the corresponding gate function is equal to unity:

$$u_i^*(t) = Q_i[\boldsymbol{x}(t)]u(t)$$

$$y_i^*(t) = Q_i[\boldsymbol{x}(t)]u(t)$$

4. Estimate the parameters of the submodel valid in the corresponding domain from the cut input and output signals $u_i^*(t)$ and $y_i^*(t)$.
5. By taking all domains of interest into account, a series of submodels can be obtained.
6. The global valid model of the process is the sum of the models valid only in the individual domains. ■

Figure 3.5.1 shows the scheme of the method and the gated input and output signals.

Remarks:

1. The size of the domain is determined by the distance between the lower and the upper boundaries of the signals in the vector $\boldsymbol{x}$.
2. To have an accurate gate function model the domains have to be chosen small enough. This has the advantage that linear models with few parameters can be used.
3. The domains have to be chosen large enough to have a not too small signal to noise ratio.

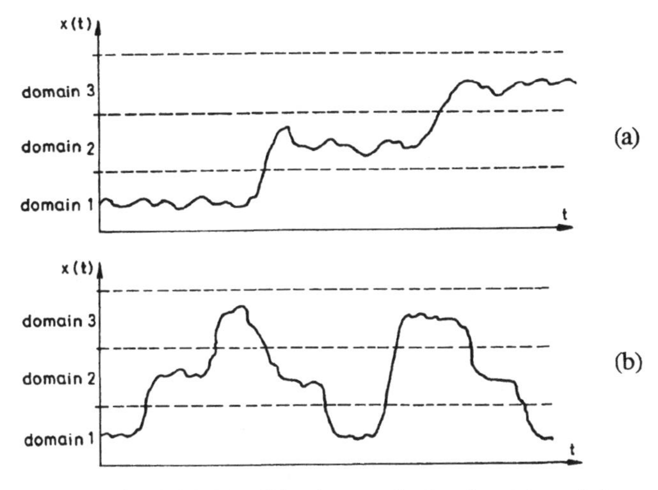

Fig. 3.5.2 Identification methods of multi-models: (a) sequential; (b) parallel

4. The domains have to be chosen large enough to have enough data pairs for a realistic parameter estimation.
5. In the case of estimating dynamic models attention should be paid to the following fact. When the gate function arrives into a new domain where estimation has to be performed, then earlier values of the input and output signal do not belong to the same domain. The regression may be performed only later when all data in the memory vector belong to the investigated domain.

Depending on the shape of the input signal we distinguish between two methods (Figure 3.5.2) (Diekmann and Unbehauen, 1985):

- *sequential method:*
  The input signal covers all domains in turn, i.e., the elementary gate models

are identified in turn. The gated input and output signals in each domain build uninterrupted sequences.

- *parallel method:*
  The input signal covers the whole input range with an arbitrary shape, which means that the gated input and output signals of the individual domains consist of several separate sequences. The sequences belonging to the same domain have to be joined. Care has to be taken that the regression is performed only between related data. This can be done by continuing the regression only for such further steps equal to the sum of the dead time plus the order of the process later than the domain is matched. The expression *parallel method* arises from the regression of the parameters of the different domains being able to be done in parallel. This possibility is important in real time identification.

Gate function methods for

- linear and nonlinear multi-models and
- linear and nonlinear spatial piecewice models

are treated in details later in this chapter .

3.6 TWO-STEP METHOD: FITTING THE BEST MODEL LINEAR-IN-PARAMETERS AND IDENTIFICATION OF THE ESTIMATED SIMULATED MODEL

In the practice active experiments can be very rarely performed. Therefore the following method can be recommended.

Algorithm 3.6.1 In order to spare with the real time experiments the following two-step method is recommended:

1. Test the process by a test signal which
   - cover the whole input signal range of interest and
   - excite the dynamic behavior of the process persistently enough.

2. Identify a model that fits the measured input and output signal very well. Any structure is allowed. A model linear-in-parameters has the advantage that the estimation of parameters is an easy task.
3. Choose appropriate test signal(s) and identify the previously estimated model with the desired structure. One has the possibility of performing a structure determination, as well. ■

The point of the above two-step method is that several identification experiments can be repeated without disturbing the real process for long. Namely, some special test signals, e.g.,

- excitation with a Gaussian white noise signal or
- steady state measurements

need a long experimental time.

To get good results it is important that the test signal given to the real process be informative enough and that the initially estimated model provides a very good fit.

Example 3.6.1 *Two-step identification of the simple Wiener model*

The process is a simple Wiener model with the static quadratic function

$$v(k) = 2 + u(t) + 0.5u^2(t)$$

and with the first-order linear transfer function

$$G(s) = \frac{1}{1 + 10s}$$

The sampling time is $\Delta T = 2$ [s], thus the pulse transfer function of the linear term is

$$H(z^{-1}) = \frac{0.1817z^{-1}}{1 - 0.81817z^{-1}}$$

Assume the process should be identified in the domain $-1 \le U \le 3$ of the input signal. A symmetrical PRTS signal with maximum length of 26, minimum switching time of 2 [s] and with lowest and uppermost levels -1 and 3, respectively is applied. $N = 26 \cdot 5 = 130$ data pairs are used for the identification.

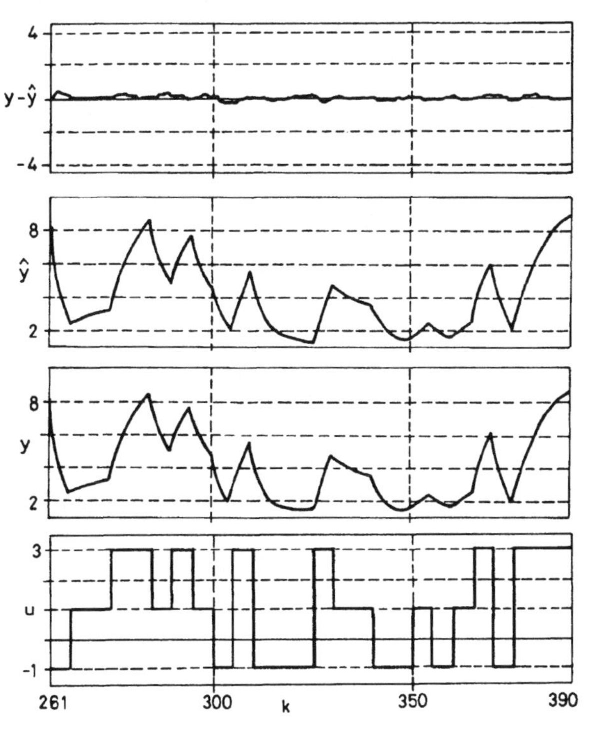

Fig. 3.6.1 Measured input (u), output (y) and computed output signal $(\hat{y})$ based on the estimated model at the identification of the simple Wiener model

Figure 3.6.1 shows the measured input and output signals. A second-order quadratic parametric Volterra model gives an excellent fit to the measured output signal

$$\hat{y}(k) = (1.458 \pm 0.02812)\hat{y}(k-1) - (0.5244 \pm 0.02325)\hat{y}(k-2)$$
$$+(0.128 \pm 0.01553) + (0.2228 \pm 0.01121)u(k-1)$$

$$-(0.1556 \pm 0.01308)u(k-2) + (0.0211 \pm 0.005024)u^2(k-1)$$
$$-(0.07345 \pm 0.005405)u^2(k-2) + (0.08429 \pm 0.003487)u(k-1)u(k-2) \quad (3.6.1)$$

The computed output signal $\hat{y}(k)$ based on the estimated model fits the measured one well.

Two further identification experiments were performed now on the simulated model described by the equation (3.6.1).

1. *Identification of the steady state behavior:*
   The model is excited by stepwise increasing the input signal from $U = -1$ till $U = 3$. Figure 3.6.2 shows the eight step responses. Table 3.6.1 summarizes the steady state values belonging to the input levels. The values lie on a parabola, as is seen in Figure 3.6.3. Its equation can be obtained by static regression with the LS method

   $$\hat{Y} = (1.928 \pm 0.0002) + (1.012 \pm 0.0002)U + (0.4809 \pm 0.0001)U^2 \quad (3.6.2)$$

   As is seen, this equation approximates the steady state relation of the simple Wiener model investigated.

TABLE 3.6.1 Steady state input and output values of the simple Wiener model

No	*U*	*Y*
1	-1	1.396
2	-0.5	1.542
3	0	1.928
4	0.5	2.554
5	1	3.420
6	1.5	4.528
7	2	5.875
8	2.5	7.464
9	3	9.292

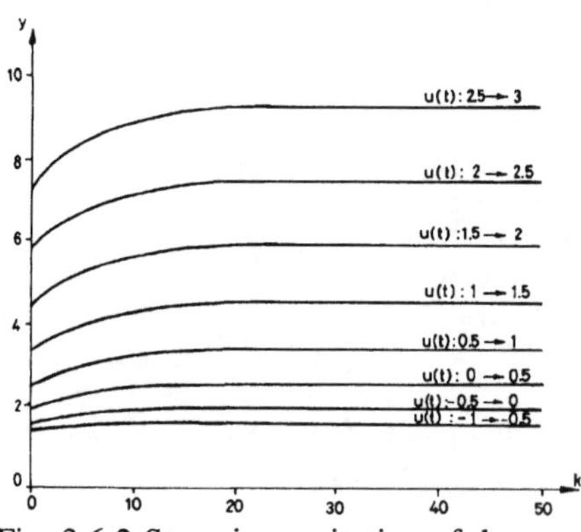

Fig. 3.6.2 Stepwise excitation of the simple Wiener model

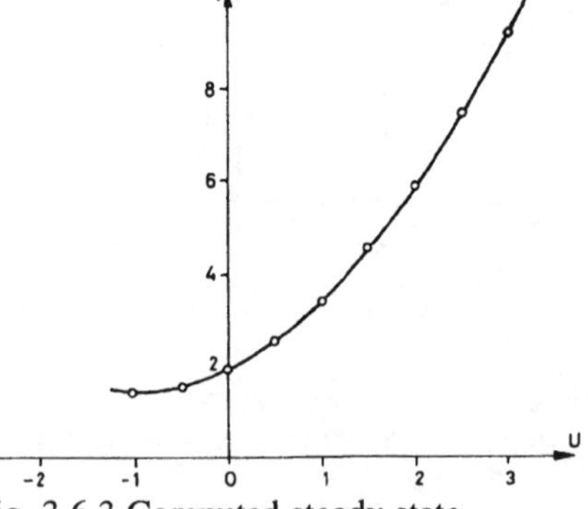

Fig. 3.6.3 Computed steady state relation of the estimated simple Wiener model

2. *Identification the linearized model at small excitation:*
   A similar test signal to the first one but with a small amplitude ±0.1 around the working point 1 was used on the simulated model. Higher-order models are overestimated and the following first-order model is obtained

$$\hat{y}(k) = (0.8233 \pm 0.00064)\hat{y}(k-1) + (0.2517 \pm 0.0019) + (0.3532 \pm 0.00076)u(k-1)$$

The corresponding pulse transfer function

$$H\left(z^{-1}\right) = \frac{0.3532z^{-1}}{1-0.8233z^{-1}}$$

has the equivalent transfer function

$$G(s) = \frac{1.9988}{1+10.29s}$$

The estimated time constant $\hat{T} = 10.29$ [s] is almost equal to the true time constant $T = 10$ [s]. An assumption that the process is of simple Wiener type could be successfully verified if the serial connection of the linear dynamic term (3.6.3) has a unit gain, i.e.

$$G(s) = \frac{1}{1+10.29s}$$

and has the static nonlinear function (3.6.2). ■

In the above example only parameter estimation of an assumed structure but no structure identification is performed on the simulated model. However, the second step of the two-step identification method can include a structure search as well. This makes the method especially advantageous because then, for example, different test signals can be applied on the simulated model instead of testing the real process. Examples for a two-step identification method with structure search are presented in Chapter 5 *Structure Identification*.

3.7 ESTIMATION OF THE VOLTERRA KERNELS

3.7.1 Computing the kernels from pulse and step responses

The response of a linear dynamic system to a pulse signal is the weighting function series. We would like to extend this fact to higher degree nonlinear systems. The equation of a degree-*n* homogenous system is

$$y_n(k) = y_n(u(k)) = \sum_{\kappa_1=0}^{k} \cdots \sum_{\kappa_n=0}^{k} h_n(\kappa_1, \ldots, \kappa_n) u(k-\kappa_1) \ldots u(k-\kappa_n) \tag{3.7.1}$$

The multi-dimensional kernel cannot be measured with either one or with more pulses. However, it would be possible if the input terms $u(k-\kappa_i)$, $\kappa_i = 1, \ldots, n$ were applied separately to the system. If the input terms can be separately set, then we get the so-called Volterra function.

Definition 3.7.1 (Schetzen, 1965a) The n-dimensional homogenous Volterra function is defined by

$$y_n(k) = y_n(u_1(k), \ldots, u_n(k)) = \sum_{\kappa_1=0}^{k} \cdots \sum_{\kappa_n=0}^{k} h_n(\kappa_1, \ldots, \kappa_n) u_1(k-\kappa_1) \ldots u_n(k-\kappa_n) \tag{3.7.2}$$

Schetzen (1965a) showed

- how the kernels can be measured if the Volterra functions are known (Lemma 3.7.1) and
- how the Volterra functions can be obtained from input–output measurements (Theorem 3.7.1).

Lemma 3.7.1
(Schetzen, 1965a) The response of a homogenous degree-n Volterra function to the pulse excitations

$$u_i(k) = U\left[1(k-k_{i-1}) - 1(k-k_{i-1}-1)\right] \qquad i = 1, \ldots, n \qquad k_0 = 0 \tag{3.7.3}$$

is

$$h_n(k, k-k_1, \ldots, k-k_{n-1})$$

Proof. Substituting (3.7.3) into (3.7.2) leads to

$$y_n(k) = U^2 h_n(k, k-k_1, \ldots, k-k_n) \tag{3.7.4}$$

■

Example 3.7.1 *Pulse response of a quadratic kernel*
Let $u_1(k)$ a pulse at $k = 0$ and $u_2(k)$ at $k = 2$, both with amplitudes of U. Calculate $y(3)$

$$\begin{aligned} y(3) = {} & h_2(0,0)u_1(3)u_2(3) + h_2(0,1)u_1(3)u_2(2) + h_2(0,2)u_1(3)u_2(1) \\ & + h_2(0,3)u_1(3)u_2(0) + h_2(1,0)u_1(2)u_2(3) + h_2(1,1)u_1(2)u_2(2) \\ & + h_2(1,2)u_1(2)u_2(1) + h_2(1,3)u_1(2)u_2(0) + h_2(2,0)u_1(1)u_2(3) \\ & + h_2(2,1)u_1(1)u_2(2) + h_2(2,2)u_1(1)u_2(1) + h_2(2,3)u_1(1)u_2(0) \\ & + h_2(3,0)u_1(0)u_2(3) + h_2(3,1)u_1(0)u_2(2) + h_2(3,2)u_1(0)u_2(1) \\ & + h_2(3,3)u_1(0)u_2(0) = U^2 h_2(3,1) \end{aligned}$$

because all products except $u_1(0)u_2(2)$ are zero. ■

The building of Volterra functions will be shown first for a quadratic system. (The complete, i.e., not homogenous, system is denoted by $V_{\text{nc}}[u(k)]$.)

Theorem 3.7.1 (Schetzen, 1965a) The response of the two-dimensional Volterra function can be obtained if the process reduced by the constant term is excited by $u_1(k)$

and $u_2(k)$ and the following operations are executed

$$
\begin{aligned}
y(u_1(k), u_2(k)) &= \sum_{\kappa_1=0}^{k} \sum_{\kappa_2=0}^{k} h_n(\kappa_1, \kappa_2) u_1(k-\kappa_1) u_2(k-\kappa_2) \\
&= \tfrac{1}{2}\left\{\left[V_{2c}[u_1(k)+u_2(k)] - h_0\right] - \left[V_{2c}[u_1(k)] - h_0\right] - \left[V_{2c}[u_2(k)] - h_0\right]\right\}
\end{aligned}
\tag{3.7.5}
$$

Proof. (3.7.5) has the form

$$
y(u_1(k), u_2(k)) = \tfrac{1}{2}\left\{\sum_{i=1}^{2} V_i[u_1(k)+u_2(k)] - \sum_{i=1}^{2} V_i[u_1(k)] - \sum_{i=1}^{2} V_i[u_2(k)]\right\}
$$

which can be separated into two equations, one with the linear terms and the second with the quadratic terms:

$$
\begin{aligned}
&\tfrac{1}{2}\left\{V_1[u_1(k)+u_2(k)] - V_1[u_1(k)] - V_1[u_2(k)]\right\} \\
&= \tfrac{1}{2}\left\{\sum_{\kappa_1=0}^{k} h_1(\kappa_1)[u_1(k-\kappa_1)+u_2(k-\kappa_1)] - \sum_{\kappa_1=0}^{k} h_1(\kappa_1)u_1(k-\kappa_1) - \sum_{\kappa_1=0}^{k} h_1(\kappa_1)u_2(k-\kappa_1)\right\} \\
&= \tfrac{1}{2}\left\{\sum_{\kappa_1=0}^{k} h_1(\kappa_1)[u_1(k-\kappa_1)+u_2(k-\kappa_1)] - u_1(k-\kappa_1) - u_2(k-\kappa_1)\right\} = 0
\end{aligned}
$$

$$
\begin{aligned}
&\tfrac{1}{2}\left\{V_2[u_1(k)+u_2(k)] - V_2[u_1(k)] - V_2[u_2(k)]\right\} \\
&= \tfrac{1}{2}\left\{\sum_{\kappa_1=0}^{k} \sum_{\kappa_2=0}^{k} h_2(\kappa_1, \kappa_2)[u_1(k-\kappa_1)+u_2(k-\kappa_1)][u_1(k-\kappa_2)+u_2(k-\kappa_2)]\right. \\
&\left. - \sum_{\kappa_1=0}^{k} \sum_{\kappa_2=0}^{k} h_2(\kappa_1, \kappa_2)u_1(k-\kappa_1)u_1(k-\kappa_2) - \sum_{\kappa_1=0}^{k} \sum_{\kappa_2=0}^{k} h_2(\kappa_1, \kappa_2)u_2(k-\kappa_1)u_2(k-\kappa_2)\right\} \\
&= \tfrac{1}{2}\left\{\sum_{\kappa_1=0}^{k} \sum_{\kappa_2=0}^{k} h_2(\kappa_1, \kappa_2)[u_1(k-\kappa_1)+u_2(k-\kappa_1)][u_1(k-\kappa_2)+u_2(k-\kappa_2)]\right. \\
&\left. - u_1(k-\kappa_1)u_1(k-\kappa_2) - u_2(k-\kappa_1)u_2(k-\kappa_2)\right\} \\
&= \sum_{\kappa_1=0}^{k} \sum_{\kappa_2=0}^{k} h_2(\kappa_1, \kappa_2)u_1(k-\kappa_1)u_2(k-\kappa_2)
\end{aligned}
$$

■

Remarks:

1. As mentioned, the process output has to be reduced by the constant term. This can be seen also from that (3.7.5) would not be zero for a static process if the output were not reduced by the constant term

$$
\tfrac{1}{2}\left\{V_0[u_1(k)+u_2(k)] - V_0[u_1(k)] - V_0[u_2(k)]\right\} = \tfrac{1}{2}[h_0 - h_0 - h_0] \neq 0
$$

2. The proof of Theorem 3.7.1 lies on the equalities

$$\frac{1}{2}\left[\left(x_1+x_2\right)^2-x_1^2-x_2^2\right]=x_1x_2$$

$$\frac{1}{2}\left[\left(x_1+x_2\right)^1-x_1^1-x_2^1\right]=0$$

We turn now to the cubic case.

Theorem 3.7.2 (Schetzen, 1965a) The three-dimensional Volterra function of a cubic system can be obtained as

$$\begin{aligned}
y(k) &= y\left[u_1(k),u_2(k),u_3(k)\right]\\
&= \sum_{\kappa_1=0}^{k}\sum_{\kappa_2=0}^{k}\sum_{\kappa_3=0}^{k} h_3(\kappa_1,\kappa_2,\kappa_3)u_1(k-\kappa_1)u_2(k-\kappa_2)u_3(k-\kappa_3)\\
&= \frac{1}{3!}\Big\{\left[V_{3c}\left[u_1(k)+u_2(k)+u_3(k)\right]-h_0\right]-\left[V_{3c}\left[u_1(k)+u_2(k)\right]-h_0\right]\\
&-\left[V_{3c}\left[u_2(k)+u_3(k)\right]-h_0\right]-\left[V_{3c}\left[u_1(k)+u_3(k)\right]-h_0\right]\\
&+\left[V_{3c}\left[u_1(k)\right]-h_0\right]+\left[V_{3c}\left[u_2(k)\right]-h_0\right]+\left[V_{3c}\left[u_3(k)\right]-h_0\right]\Big\}
\end{aligned} \tag{3.7.6}$$

Proof. The proof is based on the algebraic equalities

$$\frac{1}{3!}\left\{\left(x_1+x_2+x_3\right)^3-\left[\left(x_1+x_2\right)^3+\left(x_2+x_3\right)^3+\left(x_1+x_3\right)^3\right]+\left(x_1^3+x_2^3+x_3^3\right)\right\}=x_1x_2x_3$$

$$\frac{1}{3!}\left\{\left(x_1+x_2+x_3\right)^2-\left[\left(x_1+x_2\right)^2+\left(x_2+x_3\right)^2+\left(x_1+x_3\right)^2\right]+\left(x_1^2+x_2^2+x_3^2\right)\right\}=0$$

$$\frac{1}{3!}\left\{\left(x_1+x_2+x_3\right)^1-\left[\left(x_1+x_2\right)^1+\left(x_2+x_3\right)^1+\left(x_1+x_3\right)^1\right]+\left(x_1^1+x_2^1+x_3^1\right)\right\}=0$$

■

The method can be generalized to higher degree kernels.

Theorem 3.7.3 (Schetzen, 1965a) The n-dimensional Volterra function is

$$\begin{aligned}
y(k) = y\left(u_1(k),\ldots,u_n(k)\right) &= \sum_{\kappa_1=0}^{k}\cdots\sum_{\kappa_n=0}^{k} h_n(\kappa_1,\ldots,\kappa_n)u_1(k-\kappa_1)\ldots u_n(k-\kappa_n)\\
&= \frac{1}{n!}\Big\{\left[V_{nc}\left[u_1(k)+\ldots+u_n(k)\right]-h_0\right]\\
&-\left[\left[V_{nc}\left[u_1(k)+\ldots+u_{n-1}(k)\right]-h_0\right]+\left[V_{nc}\left[u_2(k)+\ldots+u_n(k)\right]-h_0\right]+\ldots\right]\\
&+\left[\left[V_{nc}\left[u_1(k)+\ldots+u_{n-2}(k)\right]-h_0\right]+\left[V_{nc}\left[u_2(k)+\ldots+u_{n-1}(k)\right]-h_0\right]+\ldots\right]\\
&+(-1)^{n-1}\left[\left[V_{nc}\left[u_1(k)\right]-h_0\right]+\left[V_{nc}\left[u_2(k)\right]-h_0\right]+\ldots+\left[V_{nc}\left[u_n(k)\right]-h_0\right]\right]\Big\}
\end{aligned} \tag{3.7.7}$$

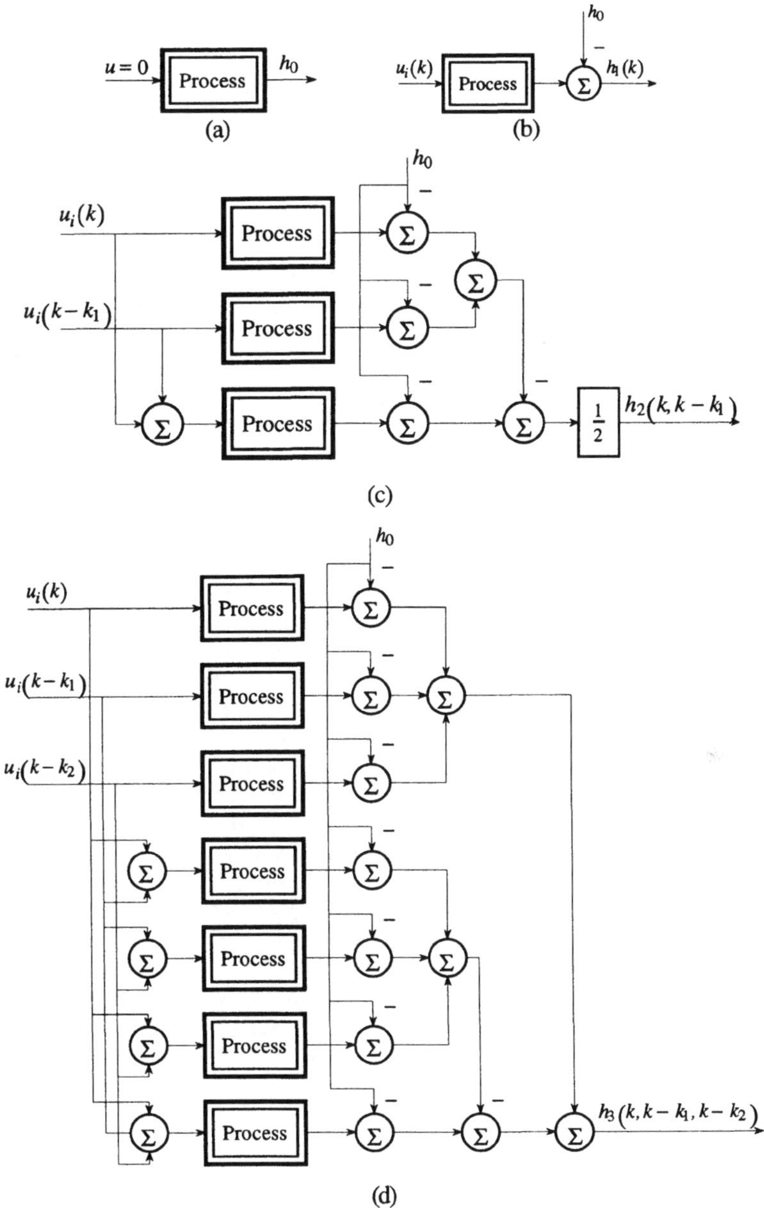

Fig. 3.7.1 Measuring the highest degree Volterra kernel from pulse excitations: (a) constant; (b) linear; (c) quadratic; (d) cubic system

Proof. The proof is based on the equalities

$$\tfrac{1}{n!}\left\{\left(x_1+\ldots+x_n\right)^n-\left[\left(x_1+\ldots+x_{n-1}\right)^n+\left(x_2+\ldots+x_n\right)^n+\ldots\right]+\left[\left(x_1+\ldots+x_{n-2}\right)^n\right.\right.$$

$$+\left(x_2+\ldots+x_{n-1}\right)^n\Big]+\ldots+(-1)^{n-1}\left(x_1^n+x_2^n+\ldots+x_n^n\right)\Big\}=x_1x_2\ldots x_n$$

$$\tfrac{1}{n!}\Big\{\left(x_1+\ldots+x_n\right)^i-\Big[\left(x_1+\ldots+x_{n-1}\right)^i+\left(x_2+\ldots+x_n\right)^i+\ldots\Big]$$

$$+\Big[\left(x_1+\ldots+x_{n-2}\right)^i+\left(x_2+\ldots+x_{n-1}\right)^i+\ldots\Big]+\ldots+(-1)^{i-1}\left(x_1^i+\ldots+x_n^i\right)\Big\}=0$$

$$i=1,\ldots,n-1$$

which are proven in Schetzen (1965a). ■

The scheme of measuring the Volterra kernels is presented in Figure 3.7.1 for constant, linear, quadratic and cubic systems. A nonlinear process with different degree channels can be estimated as follows.

Algorithm 3.7.1 A process that can be described by a degree-n Volterra model can be identified in the following steps:

1. Determine the degree of the highest power of the polynomial steady state characteristic (i.e., from steady state measurements). Denote the degree by n.
2. Let $i=n$.
3. Determine the ith degree Volterra kernel by exciting the process by i pulses using Theorem 3.7.3. The responses of the higher degree channels $i=i+1,\ldots,n$ have to be subtracted from the responses.
4. Reduce i by one.
5. Go to 3 if $i>1$.
6. The constant term is the output signal without excitation. ■

Remarks: (Schetzen, 1965a)

1. If calculating the response of an even degree channel then the effects of the odd degree channels can be avoided easily. (See with the methods of separating the responses of different degree channels in Chapter 5.)
2. If calculating the response of an odd degree channel then the effects of the even degree channels can be avoided easily. (See with the methods of separating the responses of different degree channels in Chapter 5.)

3.7.2 Correlation method using Gaussian white noise input

The isolated kernels of different degrees can be estimated by the cross-correlated method in a similar way as with the linear systems if the input signal is a white noise.

Assumption 3.7.1 The input signal is a Gaussian white noise with zero mean and standard deviation σ_u. ■

The following two lemmas show the theoretical background of the method.

Lemma 3.7.2

(Schetzen, 1980) The cross-correlation of the product of n shifted Gaussian white noise input signals and of the output signal of a degree-n channel of the Volterra model is

$$\mathrm{E}\left\{\left[\prod_{i=1}^{n} u\left(k-\kappa_i\right)\right]V_n\big(u(k)\big)\right\}=\frac{1}{n!\sigma_u^{2n}}h_n\left(\kappa_1,\ldots,\kappa_n\right) \qquad \text{if} \quad \kappa_i\neq\kappa_j \;\; i\neq j \qquad (3.7.8)$$

Proof. Put the output signal of the Volterra series (3.7.9)

$$V_n(u(k)) = \sum_{i_1=0}^{k} \cdots \sum_{i_n=0}^{k} h_n(i_1,\dots,i_n)\prod_{j=1}^{n} u(k-i_j) \tag{3.7.9}$$

into (3.7.8) then (3.7.8) becomes

$$\sum_{i_1=0}^{k} \cdots \sum_{i_n=0}^{k} h_n(i_1,\dots,i_n)\mathrm{E}\left\{\left[\prod_{j=1}^{n} u(k-i_j)\right]\left[\prod_{i=1}^{n} u(k-\kappa_i)\right]\right\} \tag{3.7.10}$$

The expected value in (3.7.10) is the sum of products of the expected values of the products of two input terms with different arguments, built over all distinct ways. Under the assumption that all κ_i-s are different only pairs with $i_i = \kappa_i$ build a correlation function that differs from zero. Then (3.7.10) becomes

$$\mathrm{E}\left\{\left[\prod_{i=1}^{n} u(k-\kappa_i)\right]\left[\prod_{i=1}^{n} u(k-i_j)\right]\right\} = n!\prod_{j=1}^{n} r_{uu}(\kappa_j - i_j) = n!\sigma_u^{2n}, \quad \kappa_j = i_j \quad \forall j \tag{3.7.11}$$

Substitution of (3.7.11) into (3.7.10) leads to (3.7.8). ∎

Lemma 3.7.3
(Schetzen, 1980).The cross-correlation of the product of n shifted Gaussian white noise input signals and of the output signal of a degree-m $(m<n)$ channel of the Volterra model is

$$\mathrm{E}\left\{\left[\prod_{i=1}^{n} u(k-\kappa_i)\right] V_m(u(k))\right\} = 0 \qquad \text{if} \qquad \kappa_i \neq \kappa_j, \qquad i \neq j \tag{3.7.12}$$

Proof. The proof is similar to that of Lemma 3.7.2. The main difference is that

$$\mathrm{E}\left\{\left[\prod_{i=1}^{n} u(k-\kappa_i)\right]\left[\prod_{j=1}^{m} u(k-i_j)\right]\right\} = m!\left[\prod_{j=1}^{m} r_{uu}(\kappa_j - i_j)\right]\left[\prod_{j=1}^{n-m} u(k-\kappa_j)\right]$$

$$= m!\sigma_u^{2m}\left[\prod_{j=1}^{n-m} u(k-\kappa_j)\right] = 0 \tag{3.7.13}$$

$$\kappa_j = i_j, \quad j = 1,\dots,m, \quad \kappa_j \neq \kappa_i \quad j \neq i$$

The identification equation is summed up in the next theorem.

Theorem 3.7.4 Let n the highest degree of the nonlinear polynomial system. With Assumption 3.7.1 the degree-n Volterra kernel can be estimated as

$$\mathrm{E}\left\{\left[\prod_{i=1}^{n} u(k-\kappa_i)\right] y(k)\right\} = \frac{1}{n!\sigma_u^{2n}} h_n(\kappa_1,\dots,\kappa_n) \qquad \text{if} \qquad \kappa_i \neq \kappa_j, \; i \neq j \tag{3.7.14}$$

Proof. A noise $n(k)$ additive to the output is filtered by the method

$$\begin{aligned} \mathrm{E}\left\{\left[\prod_{i=1}^{n} u(k-\kappa_i)\right] y(k)\right\} &= \mathrm{E}\left\{\left[\prod_{i=1}^{n} u(k-\kappa_i)\right] \left[w(k)+n(k)\right]\right\} \\ &= \mathrm{E}\left\{\left[\prod_{i=1}^{n} u(k-\kappa_i)\right] w(k)\right\} \end{aligned} \tag{3.7.15}$$

The noise-free output consists of the sum of homogenous degree-i systems $(i \leq n)$

$$w(k) = \sum_{i=0}^{n} V_n\left[u(k)\right] \tag{3.7.16}$$

Putting (3.7.16) into (3.7.15) and using the results of the Lemmas 3.7.2 and 3.7.3 leads to (3.7.14). ■

Figure 3.7.2 illustrates the method.

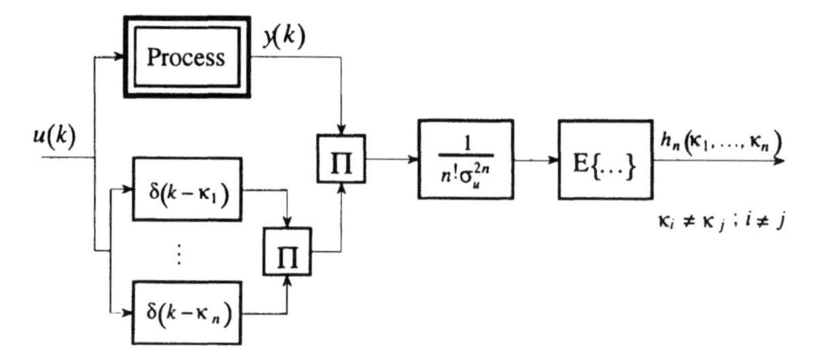

Fig. 3.7.2 Estimation of the highest degree Volterra kernel for different arguments to Gaussian white noise excitation by correlation method

The following estimation algorithm can be used if the upper degree of the polynomial nonlinearity is known.

Algorithm 3.7.2 A process that can be described by a degree-n Volterra model can be identified in the following steps:

1. Determine the degree of the highest power of the polynomial steady state characteristic (i.e., from steady state measurements). Denote the degree by i.
2. Excite the process by a Gaussian white noise.
3. Let the computed output signal $y_c(k)$ equal the measured output signal

   $y_c(k) = y(k)$

4. Estimate the degree-i kernels by correlating the computed output signal $y_c(k)$ by the product of i differently shifted input signals as given in (3.7.14)

$$\mathrm{E}\left\{\left[\prod_{j=1}^{i} u(k-\kappa_j)\right] y_c(k)\right\} = \frac{1}{i!\sigma_u^{2i}} h_i(\kappa_1,\ldots,\kappa_i) \qquad \text{if} \quad \kappa_\ell \neq \kappa_j, \;\; \ell \neq j$$

5. Reduce the computed output signal by the output signal of the degree-i channel
   $y_i(k) = V_i[u(k)]$

   $y_c(k) ::= y_c(k) - y_i(k)$ (here the assignment operator ::= means 'let equal').

6. Reduce i by one.
7. Go to 4 if $i > 0$.
8. The constant term is estimated as the average value of the output signal by zero excitation. ■

Remarks:

1. The output signals of the parallel channels of different degrees can be determined also by Gardiner's method by means of repeated excitations (Gardiner, 1973b).
2. In the case of non-white Gaussian signal the filtering technique presented in Section 3.8 with the Wiener method can be applied.

3.7.3 Correlation technique using exponentially filtered Gaussian white noise input

The correlation technique presented in the last section allows us to estimate the Volterra kernel of an isolated homogenous channel of the Volterra model. If the highest degree of the nonlinearity is known then the method can be used to identify the complete system. It would be better to have a method by which Volterra kernels of any degree can be estimated immediately from the output signal. Such a method was introduced by Korenberg (1973).

Assumption 3.7.2 The Volterra kernel decays as

$$h_n(\kappa_1,\ldots,\kappa_n) \leq C\exp[-\gamma_1(\kappa_1+\ldots+\kappa_n)]$$ ■

Assumption 3.7.3 The input signal is a bounded Gaussian white noise $x(k)$ filtered by an exponential filter

$$u(k) = \exp(-\gamma k)x(k), \quad x(k) \leq X, \quad \gamma < \gamma_1$$ ■

This means that the Volterra kernel tends to zero with the increasing argument faster than the filtered random signal decays with the increasing time.

Korenberg (1973) recommended to subtract the outputs of the lower degree channels from the process output before estimating the kernels.

Theorem 3.7.5 (Extension of Korenberg (1973).) The estimate of the degree-n Volterra kernel is

$$h_n(\kappa_1,\ldots,\kappa_n) = \frac{1}{n!\sigma_u^{2n}} \mathrm{E}\left\{\left[\prod_{i=1}^{n} x(k-\kappa_i)\exp[\gamma(k-\kappa_i)]\right]\left[y(k) - \sum_{i=0}^{n-1} V_i[u(k)]\right]\right\} \tag{3.7.17}$$

$$\kappa_i \neq \kappa_j \qquad i \neq j$$

Proof. The output signal of the process reduced by the outputs of the lower degree channels may have noise and higher degree terms.

1. *Filtering the additive output noise:*
   Any additive noise having zero mean value and being independent from the input signal is filtered by the cross-correlation method. Then $y(k)$ can be replaced by $w(k)$ in (3.7.17).

2. *The noise-free output is a homogenous system of degree-n :*

$$\mathrm{E}\left\{\left[\prod_{i=1}^{n} x(k-\kappa_i)\exp[\gamma(k-\kappa_i)]\right]\sum_{i_1=0}^{k}\cdots\sum_{i_n=0}^{k} h_n(i_1,\ldots,i_n)u(k-i_1)\ldots u(k-i_n)\right\}$$

$$=\sum_{i_1=0}^{k}\cdots\sum_{i_n=0}^{k} h_n(i_1,\ldots,i_n)\exp\left(\gamma\sum_{i=1}^{n}[i_i-\kappa_i]\right)\mathrm{E}\left\{\prod_{i=1}^{n} x(k-\kappa_i)x(k-i_i)\right\}$$

$$=n!\sigma_u^{2n}h_n(\kappa_1,\ldots,\kappa_n) \qquad \kappa_i\neq\kappa_j,\ \ i\neq j \tag{3.7.18}$$

because there are $n!$ pairs of products where $\kappa_i = i_i$ and $\kappa_i \neq \kappa_j$, $i \neq j$ (see (3.7.11) in Lemma 3.7.2).

3. *The output contains higher components than degree-n :*
   Investigate an isolated degree-m component $(m > n)$.

$$\mathrm{E}\left\{\left[\prod_{i=1}^{n} x(k-\kappa_i)\exp[\gamma(k-\kappa_i)]\right]\sum_{i_1=0}^{k}\cdots\sum_{i_m=0}^{k} h_m(i_1,\ldots,i_m)u(k-i_1)\ldots u(k-i_m)\right\}$$

$$=\mathrm{E}\left\{\left[\prod_{i=1}^{n} x(k-\kappa_i)\right]\exp\left(\gamma\sum_{i=1}^{n}[k-\kappa_i]\right)\right.$$

$$\left.\times\sum_{i_1=0}^{k}\cdots\sum_{i_m=0}^{k} h_m(i_1,\ldots,i_m)\exp\left(-\gamma\sum_{j=1}^{m}[k-i_j]\right)\left[\prod_{j=1}^{m} x(k-i_j)\right]\right\}$$

$$\leq\mathrm{E}\left\{\left[\prod_{i=1}^{n} x(k-\kappa_i)\right]\exp(-\gamma[m-n]k)\exp\left(-\gamma\sum_{i=1}^{n}\kappa_i\right)\right.$$

$$\left.\times\sum_{i_1=0}^{k}\cdots\sum_{i_m=0}^{k} C\exp\left(-\gamma_1\sum_{j=1}^{m} i_j\right)\exp\left(\gamma\sum_{j=1}^{m} i_j\right)\left[\prod_{j=1}^{m} x(k-i_j)\right]\right\}$$

$$\leq\mathrm{E}\left\{X^n\exp(-\gamma[m-n]k)\exp\left(-\gamma\sum_{i=1}^{n}\kappa_i\right)\sum_{i_1=0}^{k}\cdots\sum_{i_m=0}^{k} C\exp\left(-\sum_{j=1}^{m} i_j[\gamma_1-\gamma]\right)X^m\right\}$$

$$\leq\mathrm{E}\left\{X^n\exp(-\gamma[m-n]k)\exp\left(-\gamma\sum_{i=1}^{n}\kappa_i\right)C\left(\frac{1}{\gamma_1-\gamma}\right)^m X^m\right\}$$

$$= X^{n+m}\exp\left(-\gamma\sum_{i=1}^{n}\kappa_i\right)C\left(\frac{1}{\gamma_1-\gamma}\right)^m \mathrm{E}\{\exp(-\gamma[m-n]k)\}=0 \qquad (3.7.19)$$

■

The scheme of the method is drawn in Figure 3.7.3.

The identification procedure is summed up in the next algorithm.

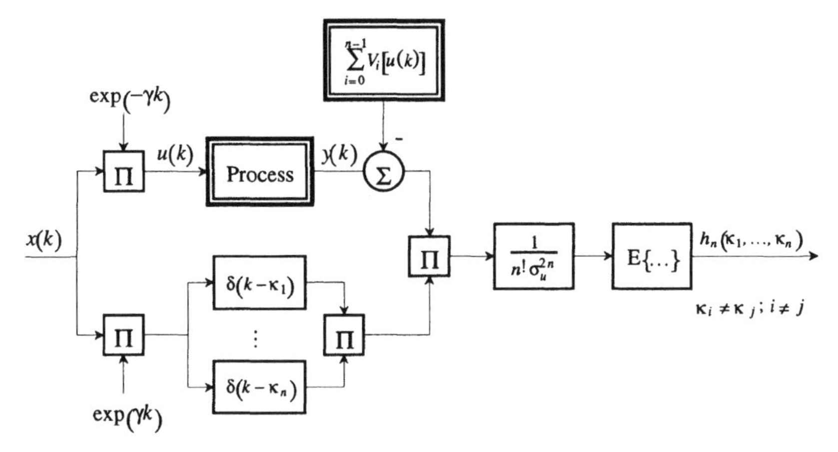

Fig. 3.7.3 Estimation of the degree-n Volterra kernel for different arguments of exponentially filtered Gaussian white noise excitation by correlation method

Algorithm 3.7.3 (Korenberg, (1973)) The estimation of the kernels can be performed successively as follows:

1. Estimate the constant term by zero input as being the mean value of the process output.
2. Subtract the constant term from the measured process output

$$y_c(k) = y(k) - c_0$$

3. Excite the process with an input signal fulfilling Assumption 3.7.3.
4. Let $i = 1$.
5. Estimate the degree-i Volterra kernels.
6. Compute the output signal of the degree-i channel $y_i(k) = V_i[u(k)]$.
7. Subtract the output signal of the degree-i channel from the reduced computed output signal $y_c(k)$

$$y_c(k) ::= y_c(k) - y_i(k)$$

8. If i is less than the degree of the process, i.e., the identified model does not fit the process well yet, increase i by one and go to Step 3. ■

3.7.4 Least squares based parameter estimation methods

The Volterra series is linear in the parameters. Considering for example a quadratic system the parameter vector is

$$\boldsymbol{\theta} = \left[h_0, h_1(0), \ldots, h_1(m), h_2(0,0), h_2(0,1), \ldots, h_2(m,m)\right]^{\mathrm{T}} \tag{3.7.20}$$

and the memory vector is

$$\boldsymbol{\phi} = \left[1, u(k), \ldots, u(k-m), u^2(k), u(k)u(k-1), \ldots, u^2(k-m)\right]^{\mathrm{T}} \tag{3.7.21}$$

The parameter vector contains the Volterra kernels of the upper triangular form. Using the symmetrical form would lead to a singular information matrix, because the same components would occur in the memory vector many-folds. (In the quadratic case, e.g., $u(k)u(k-1)$ and $u(k-1)u(k)$.)

If noise $n(k)$ is superposed on the output signal then the system equation is

$$y(k) = \boldsymbol{\phi}^{\mathrm{T}}\boldsymbol{\theta} + n(k)$$

and an LS parameter estimation leads to unbiased results even if $n(k)$ is not a white noise. The noise has to have a zero mean and be uncorrelated with the input signal. The problem of noise filtering will be illustrated in the following example.

Example 3.7.2 *Static system with noise acting at the input or at the output*
Consider the static polynomial

$$w(k) = c_0 + c_1 u(k) + c_2 u^2(k)$$

and assume that a white noise $e(k)$ effects the system before or behind the plant as drawn in Figure 3.7.4.

1. *Noise before the plant* (Figure 3.7.4a):
   The input–output equation is

$$\begin{aligned} y(k) &= c_0 + c_1\left[u(k) + e(k)\right] + c_2\left[u(k) + e(k)\right]^2 \\ &= c_0 + c_1 u(k) + c_1 e(k) + c_2 u^2(k) + 2c_2 u(k)e(k) + c_2 e^2(k) \\ &= \boldsymbol{\phi}^{\mathrm{T}}\boldsymbol{\theta} + \varepsilon(k) \end{aligned}$$

   with

$$\varepsilon(k) = c_1 e(k)$$

   and

$$\boldsymbol{\theta} = \left[c_0, c_1, c_2, 2c_2/c_1, c_2/c_1^2\right]^{\mathrm{T}}$$

$$\boldsymbol{\phi} = \left[1, u(k), u^2(k), u(k)\varepsilon(k), \varepsilon^2(k)\right]^{\mathrm{T}}$$

2. *Noise behind the plant* (Figure 3.7.4b):
   The input–output equation is

$$y(k) = c_0 + c_1 u(k) + c_2 u^2(k) + e(k) = \boldsymbol{\phi}^{\mathrm{T}}\boldsymbol{\theta} + \varepsilon(k)$$

with

$$\varepsilon(k) = e(k)$$

and

$$\boldsymbol{\theta} = \left[c_0, c_1, c_2\right]^{\mathrm{T}}$$

$$\boldsymbol{\phi} = \left[1, u(k), u^2(k)\right]^{\mathrm{T}}$$

■

$e(k)$ $u(k)$ Σ $v(k)$ $c_0 + c_1 v(k) + c_2 v^2(k)$ $y(k)$

(a)

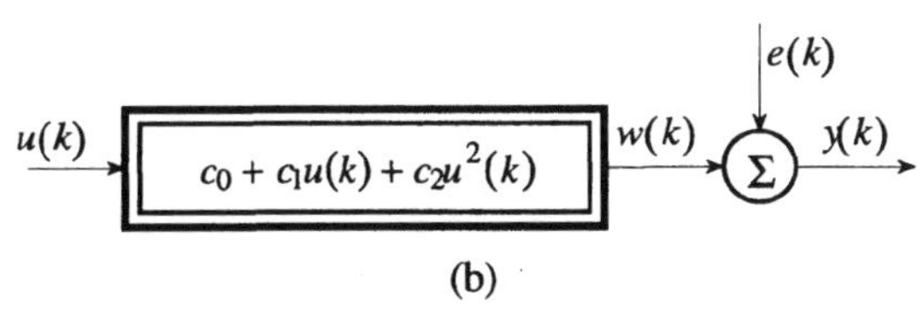

(b)

Fig. 3.7.4 Static parabolic with noise at the input (a) or at the output (b)

If no information about the noise exists, then the noisy process can be described by a system with two inputs, where the first input is the measured input signal of the process and the second is the noise. Then the memory vector becomes

$$\begin{aligned}\boldsymbol{\phi} = \Big[&1, u(k), u(k-1), \ldots, u(k-m), u^2(k), u(k)u(k-1), \ldots, u^2(k-m),\\ &e(k)e(k-1), \ldots, e(k-m), e^2(k), e(k)e(k-1), \ldots, e^2(k-m),\\ &u(k)e(k), u(k)e(k-1), u(k-1)e(k), \ldots, u(k-m)e(k-m),\\ &u^2(k)e(k), u^2(k)e(k-1), u^2(k-1)e(k), \ldots, u^2(k-m)e(k-m),\\ &u(k)e^2(k), u(k)e^2(k-1), u(k-1)e^2(k), \ldots, u(k-m)e^2(k-m),\\ &u^2(k)e^2(k), u^2(k)e^2(k-1), u^2(k-1)e^2(k), \ldots, u^2(k-m)e^2(k-m)\Big]^{\mathrm{T}}\end{aligned}$$

The parameter vector can be estimated by the extended matrix or the prediction error methods. Further on, only the case of an independent, zero mean, additive output noise will be dealt with. Then the LS parameter estimation can be applied (Eykhoff, 1963, 1974; Alper, 1965; Westenberg, 1969; Mosca, 1970, 1971, 1972).

The Volterra series model approximates a real process usually with many parameters. The number of terms in the different degree channels is:

- constant: 1;
- linear: $m+1$;
- quadratic: $(m+1)(m+2)/2$.

The number of the terms in the weighting function series is usually $m = 7$ to 20. Then a quadratic model including the linear and constant terms contains parameters from $1+(7+1)+(7+1)(7+2)/2=45$ to $1+(20+1)+(20+1)(20+2)/2=254$. The high number of parameters may cause numerical problems while performing the matrix inversion with the LS method. There are different ways to overcome the problem:

1. *Iterative matrix inversion* (Roy and Sherman, 1967a, 1967b; Westenberg, 1969; Diskin and Boneh, 1972).
2. *Orthogonalization of the model components* (Root, 1971; Figueiredo, 1984; Matthews, 1995).
3. *Combined parameter estimation and structure search*
   By this method the model components are considered as possible ones, and only some of them will be included in the model (Bard and Lapidus, 1970).

Applications of the methods are, e.g.:

- *human operator* (Taylor and Balakrishnan, 1967);
- *femoral artery of a dog* (Hubbell, 1969);
- *water tanks* (Bard and Lapidus, 1970);
- *streamflow process* (Diskin and Boneh, 1972).

3.7.5 Least squares method using Laguerre filters

The weighting function series needs too many parameters to describe a linear system. If the shifted input signals are replaced by the outputs of Laguerre filters then usually the process can be approximated by fewer parameters. Wiener (1958) recommended the same replacement in the Volterra series model. The noise-free Volterra model with Laguerre and Taylor series has the form

$$w(k)=c_0+\sum_{i=0}^{m} c_i^{\ell} v_i(k)+\sum_{i=0}^{m}\sum_{j=i}^{m} c_{ij}^{\ell} v_i(k) v_j(k)$$

$$+\sum_{i=0}^{m}\sum_{j=i}^{m}\sum_{p=j}^{m} c_{ijp}^{\ell} v_i(k) v_j(k) v_p(k)+\ldots$$

where the $v_i(k)$ is the output of the ith order Laguerre filter (Section 1.2.2).

The memory and parameter vectors for the quadratic case are

$$\boldsymbol{\theta}=\left[c_0, c_0^{\ell}, \ldots, c_m^{\ell}, c_{00}^{\ell}, c_{01}^{\ell}, \ldots, c_{mm}^{\ell}\right]^{\mathrm{T}} \tag{3.7.22}$$

$$\boldsymbol{\phi}=\left[1, v_0(k), \ldots, v_m(k), v_0^2(k), v_0(k) v_1(k), \ldots, v_m^2(k)\right]^{\mathrm{T}} \tag{3.7.23}$$

The memory vectors of the Volterra series model (3.7.21) and that of the Volterra model with Laguerre filters (3.7.23) are formally similar. Therefore those stated about the estimation of the parameters of the Volterra series model are valid now for the Wiener model. The advantage of using the Laguerre filters is that the process can usually be approximated with fewer parameters than with the Volterra series model.

Remarks:

1. Applications of the method are in the following papers:
   - *human cranio-rachidian system* (Gautier *et al.*, 1976);
   - *reversible exothermic, first-order chemical reaction in a continuously stirred tank reactor* (Zheng and Zafiriou, 1995);

- *rapid thermal processing in semiconductor manufacturing* (Zheng and Zafiriou, 1995);
- *flood process of a river* (Hatakeyama, 1997).

2. Kurth and Rake (1994) and Kurth (1996) recommended to use an oscillating basis function instead of the Laguerre functions.

3.7.6 Correlation method using Laguerre filters

The Volterra model and Wiener's model with Laguerre functions are formally similar. Consequently similar correlation method can be used for the estimation of the coefficients of Wiener's model as for the Volterra kernels shown in Theorem 3.7.4.

Theorem 3.7.6 (Wiener, 1958; Hung and Stark, 1977) Let n the highest degree of the nonlinear polynomial system. With Assumption 3.7.1 the coefficients of the degree-n terms of Wiener's model can be estimated as

$$\mathrm{E}\left\{\left[\prod_{j=1}^{n} v_{i_j}(k)\right] y(k)\right\} = \frac{1}{n!\sigma_u^{2n}} c_{i_1,\ldots,i_n}^{\ell} \qquad \text{if} \qquad i_p \neq i_j,\ p \neq j$$

Proof. The Volterra series model and Wiener's model with Laguerre functions and Taylor series have a similar structure. Both models can be separated into a dynamical part with more outputs as

- delayed inputs with the Volterra model and
- outputs of the Laguerre filters with Wiener's model.

The second part is a multivariable static model described with Taylor series. The inner variables between the dynamic and static part are independent Gaussian signals with the same standard deviation as the input signal. Consequently the correlation method used for Gaussian test signals with the Volterra series can be used with Wiener's model, as well. The only difference that instead of the shifted input signals now the outputs of the Laguerre filters have to be put into (3.7.14). ∎

The scheme of the parameter estimation is given in Figure 3.7.5.

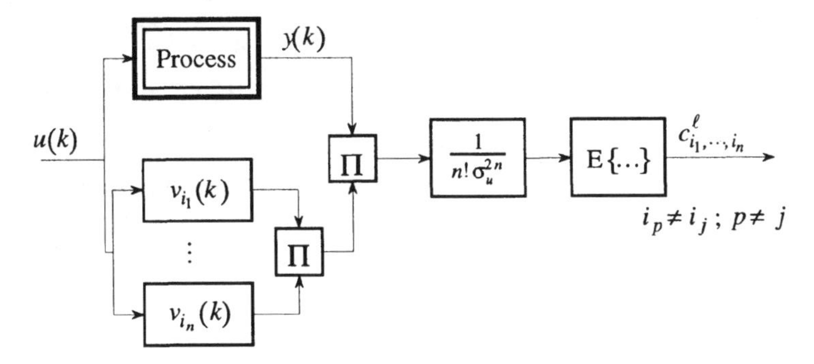

Fig. 3.7.5 Estimation of the highest degree coefficients of the orthogonal Wiener model with Laguerre functions and Taylor series for different arguments by correlation method to Gaussian white noise excitation

Remark:
Hung and Stark (1977) and Billings (1980) gave an other proof. They operated with the orthogonal features of the terms in the Wiener's model.

3.7.7 Two-step method: fitting a Wiener series model and computing the Volterra kernels

Assume the Wiener kernels of a process are known. These can be estimated, e.g., with the multi-dimensional correlation method if the input signal is white and Gaussian. The symmetrical Volterra kernels h^{sym} can be computed from the estimated Wiener kernels $h_i^{\mathrm{o}};\ i=0,1,2,\ldots$ as

$$h_n^{\mathrm{sym}}(\kappa_1,\ldots,\kappa_n)=\sum_{i=0}^{\infty}\frac{(-1)^i(n+2i)!}{i!n!}\left(\frac{\sigma_u^2}{2}\right)^i\sum_{\kappa_1^{\mathrm{o}}=0}^{\infty}\cdots\sum_{\kappa_i^{\mathrm{o}}=0}^{\infty}h_{n+2i}^{\mathrm{o}}(\kappa_1,\ldots,\kappa_n,\kappa_1^{\mathrm{o}},\ldots,\kappa_i^{\mathrm{o}},\kappa_i^{\mathrm{o}})$$

Remark:
Koukoulas and Kalouptsidis (1995) introduced for Gaussian color input signals two iterative algorithms for identifying the Volterra kernels from the cross-cumulants and their spectra in a similar way like calculating the Volterra kernels from the Wiener kernels estimated using the cross-correlation functions.

3.7.8 Two-step method: fitting a parametric model linear-in-parameters and computing the Volterra kernels

The method is based on the fact that any Volterra series model can be approximated by a recursive polynomial difference equation model.

Algorithm 3.7.4 (Diaz and Desrochers, 1988) The steps of the algorithm are as follows:

1. Estimate the parameters of the best fitting recursive polynomial difference equation model. This step needs usually structure search (e.g., stepwise regression).
2. Make the following substitution in the difference equation:

   $$u(k)\rightarrow\gamma u(k)$$
   $$y(k)\rightarrow y_0+\gamma y_1(k)+\gamma^2 y_2(k)+\ldots$$

   where $y_i(k)$ are the outputs of the different degree homogenous Volterra series models.
3. Equate the terms belonging to the same powers of γ.
4. Solve the equations in turn starting with $i=0$. The resulting expressions should contain only terms of the input signal.
5. The coefficients of $\prod_{j=1}^{i}u(k-\kappa_i)$ are the discrete time Volterra kernels.

Remarks:
1. Examples how the Volterra kernels can be calculated from a difference equation were given in Section 1.5.5.
2. As approximating parametric models often the parametric Volterra or the bilinear models give a good fit.

3. A result good enough can usually be achieved if the model includes the terms $u(k-i)$, $y(k-j)$, $u^2(k-i)$, $u(k-i)$, $y(k-j)$, $y^2(k-j)$, $i=j=1$ or eventually also $i=j=2$.

3.8 ESTIMATION OF THE WIENER KERNELS

The Wiener model consists of orthogonal channels with orthogonal terms for a special test signal. Consequently the coefficients in the Wiener model can be estimated independently.

3.8.1 Multi-dimensional correlation technique

This technique is a generalization of the correlation identification method known for linear systems with white noise excitation. To extend the method we have to deal with the problem of estimating an isolated degree-n channel and of estimating the degree-n channel in the presence of parallel channels of other degrees. Lemma 3.8.1 prepares Theorem 3.8.1.

Assumption 3.8.1 The input signal is a Gaussian white noise with zero mean and standard deviation σ_u. ■

Lemma 3.8.1

(Schetzen, 1980) The cross-correlation of the product of n shifted Gaussian white noise input signals and of the output signal of a degree-m $(m<n)$ channel of the Wiener model is, under Assumption 3.8.1,

$$\mathrm{E}\left\{\left[\prod_{i=1}^{n} u(k-\kappa_i)\right] V_m^{\mathrm{o}}[u(k)]\right\} = 0 \qquad \text{if} \quad \kappa_i \neq \kappa_j, \; i \neq j \tag{3.8.1}$$

Proof. A degree-n Wiener model consists of the sums of lower degree Volterra type models

$$V_n^{\mathrm{o}}[u(k)] = \sum_{\ell=0}^{n} V_\ell[u(k)] = \sum_{\ell=0}^{n} \sum_{i_1=0}^{k} \dots \sum_{i_\ell=0}^{k} h_\ell(i_1,\dots,i_n)\left[\prod_{j=1}^{\ell} u(k-i_j)\right] \tag{3.8.2}$$

Equation (3.8.1) becomes with (3.8.2)

$$\sum_{\ell=0}^{n} \sum_{i_1=0}^{k} \dots \sum_{i_\ell=0}^{k} h_\ell(i_1,\dots,i_\ell)\mathrm{E}\left\{\left[\prod_{j=1}^{\ell} u(k-i_j)\right]\left[\prod_{i=1}^{n} u(k-\kappa_i)\right]\right\} \tag{3.8.3}$$

The expected value in (3.8.3) is the sum of the products of the mean values of the products of two input terms with different arguments, built over all distinct routes. Under the assumption that all κ_i-s are different, only pairs with $i_i = \kappa_i$ build a correlation function differing from zero. Then (3.8.3) becomes

$$\mathrm{E}\left\{\left[\prod_{i=1}^{n} u(k-\kappa_i)\right]\left[\prod_{j=1}^{\ell} u(k-i_j)\right]\right\} = \ell!\sigma_u^{2\ell}\left[\prod_{j=1}^{n-\ell} u(k-\kappa_j)\right] = 0$$
$$\kappa_j = i_j, \quad j = 1,2,\ldots,\ell, \quad \ell < n, \quad \kappa_j \neq \kappa_i, \quad j \neq i \tag{3.8.4}$$

The substitution of (3.8.4) into (3.8.3) leads to (3.8.1). ■

Lemma 3.8.2
(Schetzen, 1980) The cross-correlation of the product of n shifted Gaussian white noise input signals and of the output signal of a degree-n channel of the Wiener model is, under Assumption 3.8.1,

$$\mathrm{E}\left\{\left[\prod_{i=1}^{n} u(k-\kappa_i)\right] V_n^{\mathrm{o}}[u(k)]\right\} = n!\sigma_u^{2n} h_n^{\mathrm{o}}(\kappa_1,\ldots,\kappa_n) \quad \text{if} \quad \kappa_i \neq \kappa_j, \quad j \neq i \tag{3.8.5}$$

Proof. According to Lemma 3.8.1 terms with degrees less than n has no contribution to the cross-correlation function. Therefore, it is enough to investigate (3.8.5) for the leading term of the Wiener model, which consists of the products of n shifted input signals

$$\sum_{i_1=0}^{k} \cdots \sum_{i_n=0}^{k} h_n^{\mathrm{o}}(i_1,\ldots,i_n)\,\mathrm{E}\left\{\left[\prod_{j=1}^{n} u(k-i_j)\right]\left[\prod_{i=1}^{n} u(k-\kappa_i)\right]\right\} \tag{3.8.6}$$

Under the assumption that all κ_i-s are different only pairs with $i_i = \kappa_i$ build a correlation function that differs from zero. Then the expected value in (3.8.6) becomes

$$\mathrm{E}\left\{\left[\prod_{i=1}^{n} u(k-\kappa_i)\right]\left[\prod_{j=1}^{n} u(k-i_j)\right]\right\} = n!\prod_{j=1}^{n} r_{uu}(\kappa_j - i_j) = n!\sigma_u^{2n}, \quad \kappa_j = i_j, \quad \forall j \tag{3.8.7}$$

The substitution of (3.8.7) into (3.8.6) leads to (3.8.5). ■

In the next lemma the restriction $\kappa_i \neq \kappa_j$, $i \neq j$ is removed.

Lemma 3.8.3
(Schetzen, 1980) The cross-correlation of the product of n shifted Gaussian white noise input signals and of the output signal of a degree-n channel of the Wiener model is under Assumption 3.8.1

$$\mathrm{E}\left\{\left[\prod_{i=1}^{n} u(k-\kappa_i)\right] V_n^{\mathrm{o}}[u(k)]\right\} = n!\sigma_u^{2n} h_n^{\mathrm{o}}(\kappa_1,\ldots,\kappa_n) \tag{3.8.8}$$

Proof. The proof is presented only for $n = 0,1,2$. For higher degree cases the literature is cited (Schetzen, 1980).

1. *case* $n = 0$:

$$\mathrm{E}\left\{ V_0^{\mathrm{o}}[u(k)]\right\} = h_0^{\mathrm{o}} = h_0 \tag{3.8.9}$$

2. *case* $n=1$:

$$\begin{aligned}\mathrm{E}\left\{u(k-\kappa_1)V_1^{\mathrm{o}}[u(k)]\right\} &= \sum_{i_1=0}^{k} h_1^{\mathrm{o}}(i_1)\mathrm{E}\left\{u(k-i_1)u(k-\kappa_1)\right\} \\ &= \sum_{i_1=0}^{k} h_1^{\mathrm{o}}(i_1)\mathrm{E}\left\{u(k-i_1)u(k-\kappa_1)\right\} = h_1^{\mathrm{o}}(\kappa_1)\end{aligned} \tag{3.8.10}$$

3. *case* $n=2$:

$$\begin{aligned}&\mathrm{E}\left\{u(k-\kappa_1)u(k-\kappa_2)V_2^{\mathrm{o}}[u(k)]\right\} \\ &= \mathrm{E}\left\{\left[\sum_{i_1=0}^{k}\sum_{i_2=0}^{k} h_2^{\mathrm{o}}(i_1,i_2)u(k-i_1)u(k-i_2) - \sigma_u^2\sum_{i_1=0}^{\infty} h_2^{\mathrm{o}}(i_1,i_1)\right]u(k-\kappa_1)u(k-\kappa_2)\right\} \\ &= \sum_{i_1=0}^{k}\sum_{i_2=0}^{k} h_2^{\mathrm{o}}(i_1,i_2)\mathrm{E}\left\{u(k-i_1)u(k-i_2)u(k-\kappa_1)u(k-\kappa_2)\right\} \\ &-\sigma_u^2\left[\sum_{i_1=0}^{\infty} h_2^{\mathrm{o}}(i_1,i_1)\right]\mathrm{E}\left\{u(k-\kappa_1)u(k-\kappa_2)\right\} = \sum_{i_1=0}^{k}\sum_{i_2=0}^{k} h_2^{\mathrm{o}}(i_1,i_2)r_{uu}(\kappa_1-i_1)r_{uu}(\kappa_2-i_2) \\ &+\sum_{i_1=0}^{k}\sum_{i_2=0}^{k} h_2^{\mathrm{o}}(\kappa_2,\kappa_1)r_{uu}(\kappa_2-i_1)r_{uu}(\kappa_1-i_2) \\ &+\left[\sum_{i_1=0}^{k}\sum_{i_2=0}^{k} h_2^{\mathrm{o}}(i_1,i_2)r_{uu}(i_2-i_1)\right]r_{uu}(\kappa_2-\kappa_1) \\ &-\sigma_u^2\left[\sum_{i_1=0}^{\infty} h_2^{\mathrm{o}}(i_1,i_1)\right]r_{uu}(\kappa_2-\kappa_1) = \sigma_u^4 h_2^{\mathrm{o}}(\kappa_1,\kappa_2)+\sigma_u^4 h_2^{\mathrm{o}}(\kappa_2,\kappa_1) \\ &= 2\sigma_u^4 h_2^{\mathrm{o}}(\kappa_1,\kappa_2)\end{aligned} \tag{3.8.11}$$

because the degree-n kernels can be considered as symmetrical ones. ■

Theorem 3.8.1 (Lee and Schetzen, 1965) With Assumption 3.8.1 the estimate of the discrete time Wiener kernels is

$$h_n^{\mathrm{o}}(\kappa_1,\ldots,\kappa_n) = \frac{1}{n!\sigma_u^{2n}}\mathrm{E}\left\{\left[\prod_{i=1}^{n} u(k-\kappa_i)\right]y(k)\right\} \quad \text{if} \quad \kappa_i \neq \kappa_j, \quad i \neq j \tag{3.8.12}$$

Proof. The cross-correlation function in (3.8.12) filters the additive output noise $n(k)$ having zero mean and being independent from the input signal:

$$\mathrm{E}\left\{\left[\prod_{i=1}^{n} u(k-\kappa_i)\right]y(k)\right\} = \mathrm{E}\left\{\left[\prod_{i=1}^{n} u(k-\kappa_i)\right][w(k)+n(k)]\right\} = \mathrm{E}\left\{\left[\prod_{i=1}^{n} u(k-\kappa_i)\right]w(k)\right\} \tag{3.8.13}$$

The noise-free process output can be written by its Wiener series

$$w(k)=\sum_{i=0}^{n-1}V_i^{o}[u(k)]+V_n^{o}[u(k)]+\sum_{i=n+1}^{\infty}V_i^{o}[u(k)] \tag{3.8.14}$$

The product of the input signals can be expressed as the output of a degree-n Volterra kernel:

$$\prod_{i=1}^{n}u(k-\kappa_i)=\sum_{\kappa_1=0}^{k}\cdots\sum_{\kappa_n=0}^{k}\prod_{i=1}^{n}\delta(\kappa_i)\prod_{i=1}^{n}u(k-\kappa_i)=V_n^{t}[u(k)] \tag{3.8.15}$$

The superscript t shows that (3.8.15) is a function of the test signal. Using (3.8.14) and (3.18.15) (3.8.13) becomes

$$\mathrm{E}\left\{V_n^{t}[u(k)]\left[\sum_{i=0}^{n-1}V_i^{o}[u(k)]+V_n^{o}[u(k)]+\sum_{i=n+1}^{\infty}V_i^{o}[u(k)]\right]\right\}$$
$$=\mathrm{E}\left\{V_n^{t}[u(k)]\left[\sum_{i=0}^{n-1}V_i^{o}[u(k)]+V_n^{o}[u(k)]\right]\right\} \tag{3.8.16}$$

because any Wiener kernel is uncorrelated to any Volterra kernels of lower degree. (3.8.16) leads to (3.8.12) with considering Lemmas 3.8.1 and 3.8.2. ■

Figure 3.8.1 shows the scheme of the correlation identification method for a degree-n system.

In the next theorem the restriction $\kappa_i \neq \kappa_j$, $i \neq j$ is removed.

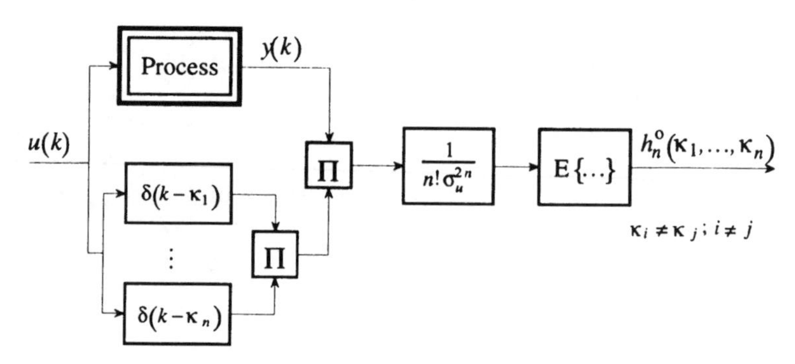

Fig. 3.8.1 Estimation of the highest degree Wiener kernel for different arguments by correlation method of Gaussian white noise excitation

Theorem 3.8.2 (Lee and Schetzen, 1965). With Assumption 3.8.1 the estimate of the discrete time Wiener kernels is

$$h_n^{o}(\kappa_1,\ldots,\kappa_n)=\frac{1}{n!\sigma_u^{2n}}\mathrm{E}\left\{\left[\prod_{i=1}^{n}u(k-\kappa_i)\right]\left[y(k)-\sum_{i=0}^{n-1}V_i^{o}[u(k)]\right]\right\}\quad\forall\kappa_i \tag{3.8.17}$$

In (3.8.17) $V_i^{\mathrm{o}}[u(k)]$ is the degree-i component of the Wiener series.

Proof. $y(k)$ can be replaced by the noise-free $w(k)$ in (3.8.17) as shown in Theorem 3.8.1. Using (3.8.15) the expected value in (3.8.17) can be written

$$\mathrm{E}\left\{V_n^t[u(k)]\left[V_n^{\mathrm{o}}[u(k)]+\sum_{i=n+1}^{\infty}V_i^{\mathrm{o}}[u(k)]\right]\right\}=\mathrm{E}\left\{V_n^t[u(k)]V_n^{\mathrm{o}}[u(k)]\right\} \qquad (3.8.18)$$

because any Wiener kernel is uncorrelated to any Volterra kernel of lower degree. (3.8.18) leads to (3.8.17) if Lemma 3.8.3 is considered. ■

Figure 3.8.2 shows the scheme of the correlation identification method for a degree-n system.

The correlation method can also be applied if the input signal is a non-white Gaussian signal. The essence of the procedure is that a Gaussian white noise is computed which can be considered as the source of the measured input signal (Figure 3.8.3). The correlation method is applied between the white source noise and the process output.

Assumption 3.8.2 The input signal is a Gaussian nonwhite noise with zero mean and standard deviation σ_u. ■

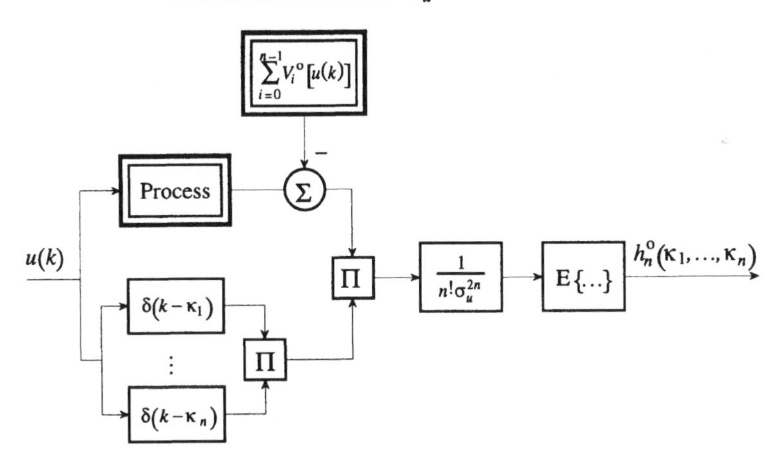

Fig. 3.8.2 Estimation of the highest degree Wiener kernel of Gaussian white noise excitation by correlation method

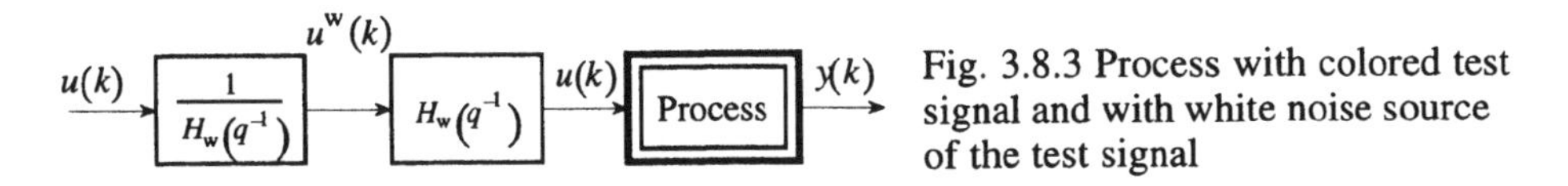

Fig. 3.8.3 Process with colored test signal and with white noise source of the test signal

Algorithm 3.8.1 (Lee and Schetzen, 1965; Schetzen, 1974). The Wiener kernels of a process can be estimated with Assumption 3.8.2 in the following way:

1. Estimate the weighting function series $h_w(k)$ of the filter that produces from the white noise the measured input signal with the measured power density spectrum.
2. Determine the inverse weighting function series $h_w^{-1}(k)$.
3. Simulate from the measured input signal by means of the inverse weighting function series $h_w^{-1}(k)$ the white noise sequence $u^w(k)$

$$u^w(k) = \sum_{\kappa=0}^{k} h_w^{-1}(k) u(k-\kappa)$$

4. Estimate from the Gaussian white noise $u^w(k)$ and from the output signal of the process the Wiener kernels $h_n^o(\kappa_1, \ldots, \kappa_n)$.
5. The equations of the process are

$$V_0^o = h_0^o$$

$$V_1^o = \sum_{\kappa_1=0}^{\infty} h_1^o(\kappa_1) u^w(k-\kappa_1) = \sum_{\kappa_1=0}^{\infty} \sum_{i_1=0}^{\infty} h_1^o(\kappa_1) h_w^{-1}(i_1-\kappa_1) u(k-i_1)$$

$$V_2^o = \sum_{\kappa_1=0}^{\infty} \sum_{\kappa_2=0}^{\infty} h_2^o(\kappa_1,\kappa_2) u^w(k-\kappa_1) u^w(k-\kappa_2) - \sigma_u^2 \sum_{\kappa=0}^{\infty} h_2^o(\kappa,\kappa)$$

$$= \sum_{\kappa_1=0}^{\infty} \sum_{\kappa_2=0}^{\infty} \sum_{i_1=0}^{\infty} \sum_{i_2=0}^{\infty} h_2^o(\kappa_1,\kappa_2) h_w^{-1}(i_1-\kappa_1) h_w^{-1}(i_2-\kappa_2)$$

$$\times u(k-\kappa_1) u(k-\kappa_2) - \sigma_u^2 \sum_{\kappa=0}^{\infty} h_2^o(\kappa,\kappa)$$

$$V_3^o = \sum_{\kappa_1=0}^{\infty} \sum_{\kappa_2=0}^{\infty} \sum_{\kappa_3=0}^{\infty} h_3^o(\kappa_1,\kappa_2,\kappa_3) u^w(k-\kappa_1) u^w(k-\kappa_2) u^w(k-\kappa_3)$$

$$-3\sigma_u^2 \sum_{\kappa=0}^{\infty} \sum_{\kappa_1=0}^{\infty} h_3^o(\kappa_1,\kappa_1,\kappa) u^w(k-\kappa)$$

$$= \sum_{\kappa_1=0}^{\infty} \sum_{\kappa_2=0}^{\infty} \sum_{\kappa_3=0}^{\infty} \sum_{i_1=0}^{\infty} \sum_{i_2=0}^{\infty} \sum_{i_3=0}^{\infty} h_3^o(\kappa_1,\kappa_2,\kappa_3) h_w^{-1}(i_1-\kappa_1)$$

$$\times h_w^{-1}(i_2-\kappa_2) h_w^{-1}(i_3-\kappa_3) u(k-\kappa_1) u(k-\kappa_2) u(k-\kappa_3)$$

$$-3\sigma_u^2 \sum_{\kappa=0}^{\infty} \sum_{\kappa_1=0}^{\infty} \sum_{i=0}^{\infty} h_3^o(\kappa_1,\kappa_1,\kappa) h_w^{-1}(i-\kappa) u(k-\kappa)$$

etc. ■

Remarks:

1. The continuous time Wiener kernels can be determined analogously to the discrete time kernels:

$$g_n^o(\tau_1, \ldots, \tau_n) = \frac{1}{n! \sigma_u^{2n}} \mathrm{E}\left\{ \left[\prod_{i=1}^{n} u(t-\tau_i) \right] y(t) \right\} \quad \text{if} \quad \tau_i \neq \tau_j, \quad i \neq j \tag{3.8.19}$$

$$g_n^{\mathrm{o}}(\tau_1,\ldots,\tau_n)=\frac{1}{n!\sigma_u^{2n}}\mathrm{E}\left\{\left[\prod_{i=1}^{n}u(t-\tau_i)\right]\left[y(t)-\sum_{i=0}^{n-1}V_i^{\mathrm{o}}[u(t)]\right]\right\}\quad\forall\tau_i$$

2. Several authors investigated the problem of to what extent the Gaussian white noise can be replaced by a pseudo-random signal:
   - *PRBS signal* (Aracil, 1970; Kadri, 1971, 1972; Kadri and Lamb, 1973; Barker and Obidegwu, 1973a; Barker *et al.*, 1972);
   - *PRTS signal* (Gyftopoulos and Hooper, 1965; Hooper and Gyftopoulos, 1967; Barker *et al.*, 1972; Barker and Obidegwu, 1973a, 1973b; Barker and Davy, 1978);
   - *PRQS signal* (Barker *et al.*, 1972);
   - *PRMS signal* (Ream, 1970; Sutter, 1987);
   - *constant-switching-pace symmetric random signal* (Marmarelis, 1978, 1979), etc.

   Applications of the method to real systems are in the following publications:
   - *pupillary control system* (Stark, 1969; Sandberg and Stark, 1968);
   - *neural chain in catfish retina* (Marmerelis and Naka, 1972, 1973);
   - *neural encoding by an insect mechano-receptor* (Korenberg *et al.*, 1988b), etc.

   Some further applications are cited in Korenberg and Hunter (1990).

3.8.2 Multi-dimensional spectral density method

French and Butz (1973) have shown that the estimation of the frequency transform of the Wiener kernels saves time, contrary to the correlation method, if the fast Fourier transform is used.

Theorem 3.8.3 (French and Butz, 1973) With Assumption 3.8.1 the estimate of the multi-dimensional frequency transform of the continuous time Wiener kernels is

$$G_n^{\mathrm{o}}(j\omega_1,\ldots,j\omega_n)=\int_{\tau_1=0}^{\infty}\cdots\int_{\tau_n=0}^{\infty}g_n^{\mathrm{o}}(\tau_1,\ldots,\tau_n)d\tau_1\ldots d\tau_n$$

$$=\frac{1}{n!\sigma_u^{2n}}\left[\prod_{i=1}^{n}\overline{U}(j\omega_i)\right]Y(j\omega_1+\ldots+j\omega_n)\quad\text{if}\quad\omega_i\neq\omega_j,\quad i\neq j\tag{3.8.21}$$

In (3.8.21) $Y(j\omega)$ is the Fourier transform of the output signal

$$Y(j\omega)=\int_{t=-\infty}^{\infty}u(t)\exp(-j\omega t)dt$$

and $\overline{U}(j\omega)$ is the complex conjugate of the Fourier transform of the input signal

$$\overline{U}(j\omega)=\int_{t=-\infty}^{\infty}u(t)\exp(j\omega t)dt$$

Proof. The estimate of the continuous time Wiener kernel is given by (3.8.19) for $\tau_i \neq \tau_j,\ i \neq j$. The input signal is the inverse Fourier transform of $U(j\omega)$

$$u(t) = \frac{1}{2\pi j} \int_{\omega=-\infty}^{\infty} U(j\omega)\exp(j\omega t)d\omega \tag{3.8.22}$$

and the output signal is the inverse Fourier transform of $Y(j\omega)$

$$y(t) = \frac{1}{2\pi j} \int_{\omega=-\infty}^{\infty} Y(j\omega)\exp(j\omega t)d\omega \tag{3.8.23}$$

Put (3.8.22) and (3.8.23) into the multi-dimensional cross-correlation function (3.8.19)

$$\mathrm{E}\left\{\left[\prod_{i=1}^{n} u(t-\tau_i)\right]y(t)\right\}$$

$$= \mathrm{E}\left\{\left[\prod_{i=1}^{n} \frac{1}{2\pi j} \int_{\omega_i=-\infty}^{\infty} U(j\omega_i)\exp[j\omega_i(t-\tau_i)]d\omega_i\right]\frac{1}{2\pi j}\int_{\omega_i=-\infty}^{\infty} Y(j\omega)\exp(j\omega t)d\omega\right\}$$

$$= \left(\frac{1}{2\pi j}\right)^{n+1} \int_{\omega_1=-\infty}^{\infty} \cdots \int_{\omega_n=-\infty}^{\infty} \int_{\omega=-\infty}^{\infty} \left[\prod_{i=1}^{n} U(j\omega_i)\right] Y(j\omega)\exp\left[-j\sum_{i\cdot r}^{n} \omega_i\tau_i\right]$$

$$\times \lim_{T\to\infty} \frac{1}{2T} \int_{t=-T}^{T} \exp\left[jt\left(\omega + \sum_{i=1}^{n}\omega_i\right)\right] dt\, d\omega_1 \ldots d\omega_n\, d\omega$$

$$= \left(\frac{1}{2\pi j}\right)^{n+1} \int_{\omega_1=-\infty}^{\infty} \cdots \int_{\omega_n=-\infty}^{\infty} \int_{\omega=-\infty}^{\infty} \left[\prod_{i=1}^{n} U(j\omega_i)\right] Y(j\omega)\exp\left[-j\sum_{i=1}^{n} \omega_i\tau_i\right]$$

$$\times \delta\left(\omega + \sum_{i=1}^{n}\omega_i\right) d\omega_1 \ldots d\omega_n\, d\omega$$

$$= \left(\frac{1}{2\pi j}\right)^{n} \int_{\omega_1=-\infty}^{\infty} \cdots \int_{\omega_n=-\infty}^{\infty} \left[\prod_{i=1}^{n} U(j\omega_i)\right] Y\left(-j\sum_{i=1}^{n}\omega_i\right)\exp\left[-j\sum_{i=1}^{n} \omega_i\tau_i\right] d\omega_1 \ldots d\omega_n$$

$$= \left(\frac{1}{2\pi j}\right)^{n} \int_{\omega_1=-\infty}^{\infty} \cdots \int_{\omega_n=-\infty}^{\infty} \left[\prod_{i=1}^{n} U(-j\omega_i)\right] Y\left(j\sum_{i=1}^{n}\omega_i\right)\exp\left[j\sum_{i=1}^{n} \omega_i\tau_i\right] d\omega_1 \ldots d\omega_n \tag{3.8.24}$$

The continuous time Wiener kernel can be expressed by its multi-dimensional Fourier transform (e.g., George, 1959)

$$g_n^{\circ}(\tau_1, \ldots, \tau_n) = \left(\frac{1}{2\pi j}\right)^{n} \int_{\omega_1=\infty}^{\infty} \cdots \int_{\omega_n=\infty}^{\infty} G_n^{\circ}(j\omega_1, \ldots, j\omega_n)\exp\left[j\sum_{i=1}^{n} \omega_i\tau_i\right] d\omega_1 \ldots d\omega_n \tag{3.8.25}$$

Substituting (3.8.25) and (3.8.24) into the LHS and RHS of (3.8.19), respectively, leads to (3.8.21). ■

The scheme of the method is shown in Figure 3.8.4 for a degree-n kernel.

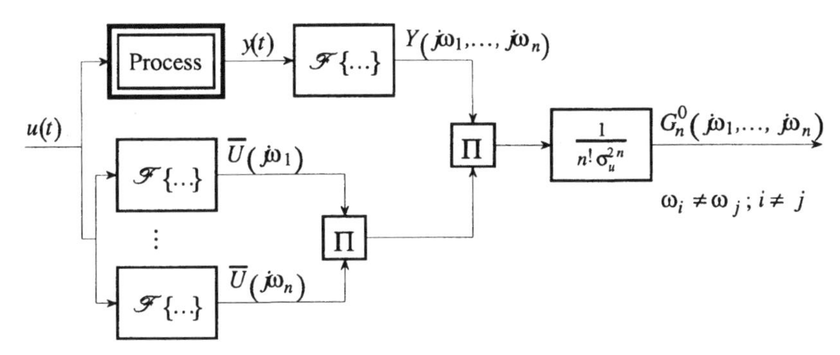

Fig. 3.8.4 Frequency domain estimation of the highest degree Wiener kernel for different arguments by correlation method of Gaussian white noise excitation

Remarks:

1. The estimation of the constant term is the expected value of the output signal without excitation

$$g_0^{\mathrm{o}} = \mathrm{E}\{y(t)\} = Y(j0)$$

2. The estimation of the linear term is, from (3.8.21) with $n = 1$,

$$G_1^{\mathrm{o}}(j\omega_1) = \frac{1}{\sigma_u^2}\overline{U}(j\omega_1)Y(j\omega_1)$$

3. The estimation of the quadratic term is, from (3.8.21) with $n = 2$,

$$G_2^{\mathrm{o}}(j\omega_1, j\omega_2) = \frac{1}{2\sigma_u^4}\overline{U}(j\omega_1)\overline{U}(j\omega_2)Y(j\omega_1 + j\omega_2), \qquad \omega_1 \neq \omega_2 \tag{3.8.26}$$

4. French and Butz (1973) derived that the restriction in (3.8.26) can be removed and the formula valid for all frequencies is

$$G_2^{\mathrm{o}}(j\omega_1, j\omega_2) = \frac{1}{2\sigma_u^4}\overline{U}(j\omega_1)\overline{U}(j\omega_2)Y(j\omega_1 + j\omega_2) - \frac{1}{2\sigma_u^2}\delta(\omega_1 + \omega_2) \tag{3.8.27}$$

5. French and Butz (1973) showed that a $(Nn!/2)$-fold saving in the computations can be achieved, contrary to the correlation method, by estimating the kernels in the frequency domain using a fast Fourier transform.
6. In case of a colored Gaussian noise the filtering technique presented in Algorithm 3.8.1 can be used.

3.8.3 Orthogonal estimation of the coefficients of the Wiener model built from Laguerre and Hermite series

The Wiener model can be expressed with the Laguerre and Hermite series as given in Section 1.2.3:

$$\begin{aligned} y(k) = c_0 &+ \sum_{i=0}^{m} c_i^{\mathrm{H}} P_1^{\mathrm{H}}(v_i(k)) + \sum_{i=0}^{m} c_{ii}^{\mathrm{H}} P_2^{\mathrm{H}}(v_i(k)) \\ &+ \sum_{i=0}^{m} \sum_{j=i+1}^{m} c_{ij}^{\mathrm{H}} P_1^{\mathrm{H}}(v_i(k)) P_1^{\mathrm{H}}(v_j(k)) + \sum_{i=0}^{m} c_{iii}^{\mathrm{H}} P_3^{\mathrm{H}}(v_i(k)) \\ &+ \sum_{i=0}^{m} \sum_{j=0}^{m} c_{iij}^{\mathrm{H}} P_2^{\mathrm{H}}(v_i(k)) P_1^{\mathrm{H}}(v_j(k)) \\ &+ \sum_{i=0}^{m} \sum_{j=i+1}^{m} \sum_{\ell=j+1}^{m} c_{ij\ell}^{\mathrm{H}} P_1^{\mathrm{H}}(v_i(k)) P_1^{\mathrm{H}}(v_j(k)) P_1^{\mathrm{H}}(v_\ell(k)) + \ldots \end{aligned} \tag{3.8.28}$$

As mentioned in Section 1.2.3 the components are orthogonal if a Gaussian white noise test signal is used. This will be shown now for a cubic system. The known fact that the average value of the product of two Hermite functionals with the same Gaussian white noise argument is zero will be used.

Lemma 3.8.4

(Schetzen, (1980)) The average value of two different components of the Wiener model with Laguerre and Hermite series is zero under Assumption 3.8.1.

Proof. The arguments of the Hermite functions are the outputs of the Laguerre filters. Therefore, they are independent Gaussian white noises. The proof is given for linear and quadratic systems, for higher degree systems the extension is similar.

1. *Linear system:*

$$\overline{P_1^{\mathrm{H}}(v_i(k)) \cdot 1} = \overline{P_1^{\mathrm{H}}(v_i(k))} = 0$$

$$\overline{P_1^{\mathrm{H}}(v_i(k)) P_1^{\mathrm{H}}(v_j(k))} = \overline{P_1^{\mathrm{H}}(v_i(k))}\ \overline{P_1^{\mathrm{H}}(v_j(k))} = 0, \quad i \neq j$$

2. *Quadratic system:*

$$\overline{P_1^{\mathrm{H}}(v_i(k)) P_1^{\mathrm{H}}(v_j(k)) \cdot 1} = \overline{P_1^{\mathrm{H}}(v_i(k))}\ \overline{P_1^{\mathrm{H}}(v_j(k))} = 0, \quad i \neq j$$

$$\overline{P_1^{\mathrm{H}}(v_i(k)) P_1^{\mathrm{H}}(v_j(k)) P_1^{\mathrm{H}}(v_i(k))} = \overline{P_1^{\mathrm{H}}(v_i(k)) P_1^{\mathrm{H}}(v_i(k))}\ \overline{P_1^{\mathrm{H}}(v_j(k))} = 0, \quad i \neq j$$

$$\overline{P_1^{\mathrm{H}}(v_i(k)) P_1^{\mathrm{H}}(v_j(k)) P_1^{\mathrm{H}}(v_j(k))} = \overline{P_1^{\mathrm{H}}(v_i(k))}\ \overline{P_1^{\mathrm{H}}(v_j(k)) P_1^{\mathrm{H}}(v_j(k))} = 0, \quad i \neq j$$

$$\overline{P_1^{\mathrm{H}}(v_i(k)) P_1^{\mathrm{H}}(v_j(k)) P_1^{\mathrm{H}}(v_\ell(k))} = \overline{P_1^{\mathrm{H}}(v_i(k))}\ \overline{P_1^{\mathrm{H}}(v_j(k))}\ \overline{P_1^{\mathrm{H}}(v_\ell(k))} = 0$$

$$i \neq j \neq \ell \neq i$$

$$\overline{P_2^{\mathrm{H}}(v_i(k)) \cdot 1} = \overline{P_2^{\mathrm{H}}(v_i(k))} = 0$$

$$\overline{P_2^{\mathrm{H}}(v_i(k))P_1^{\mathrm{H}}(v_j(k))} = \overline{P_2^{\mathrm{H}}(v_i(k))}\;\overline{P_1^{\mathrm{H}}(v_j(k))} = 0, \quad i \neq j$$

$$\overline{P_2^{\mathrm{H}}(v_i(k))P_1^{\mathrm{H}}(v_i(k))} = \overline{P_2^{\mathrm{H}}(v_i(k))}\;\overline{P_1^{\mathrm{H}}(v_i(k))} = 0$$

$$\overline{P_2^{\mathrm{H}}(v_i(k))P_2^{\mathrm{H}}(v_j(k))} = \overline{P_2^{\mathrm{H}}(v_i(k))}\;\overline{P_2^{\mathrm{H}}(v_j(k))} = 0, \quad i \neq j$$

$$\overline{P_1^{\mathrm{H}}(v_i(k))P_1^{\mathrm{H}}(v_j(k))P_2^{\mathrm{H}}(v_i(k))} = \overline{P_1^{\mathrm{H}}(v_i(k))P_2^{\mathrm{H}}(v_i(k))}\;\overline{P_1^{\mathrm{H}}(v_j(k))} = 0$$
$$i \neq j$$

$$\overline{P_1^{\mathrm{H}}(v_i(k))P_1^{\mathrm{H}}(v_j(k))P_2^{\mathrm{H}}(v_j(k))} = \overline{P_1^{\mathrm{H}}(v_i(k))}\;\overline{P_1^{\mathrm{H}}(v_j(k))P_2^{\mathrm{H}}(v_j(k))} = 0$$
$$i \neq j$$

$$\overline{P_1^{\mathrm{H}}(v_i(k))P_1^{\mathrm{H}}(v_j(k))P_2^{\mathrm{H}}(v_\ell(k))} = \overline{P_1^{\mathrm{H}}(v_i(k))}\;\overline{P_1^{\mathrm{H}}(v_j(k))}\;\overline{P_2^{\mathrm{H}}(v_\ell(k))} = 0$$
$$i \neq j \neq \ell \neq i$$

■

Theorem 3.8.4 (Wiener, 1958; Schetzen, 1980) The coefficients of the Wiener model with Laguerre and Hermite series can be estimated orthogonally if the test signal fulfills Assumption 3.8.1. The parameter estimation equations for the lower degree channels are:

1. *constant term:*

$$c_0 = \mathrm{E}\{y(k)\}$$

2. *linear channel:*

$$c_i^{\mathrm{H}} = \frac{1}{\sigma_u^2}\mathrm{E}\left\{P_1^{\mathrm{H}}(v_i(k))y(k)\right\} \tag{3.8.29}$$

3. *quadratic channel:*

$$c_{ii}^{\mathrm{H}} = \frac{1}{2\sigma_u^4}\mathrm{E}\left\{P_2^{\mathrm{H}}(v_i(k))y(k)\right\} \tag{3.8.30}$$

$$c_{ij}^{\mathrm{H}} = \frac{1}{\sigma_u^4}\mathrm{E}\left\{P_1^{\mathrm{H}}(v_i(k))P_1^{\mathrm{H}}(v_j(k))y(k)\right\}, \quad i \neq j \tag{3.8.31}$$

4. *cubic channel:*

$$c_{iii}^{\mathrm{H}} = \frac{1}{3!\sigma_u^6}\mathrm{E}\left\{P_3^{\mathrm{H}}(v_i(k))y(k)\right\} \tag{3.8.32}$$

$$c_{iij}^{\mathrm{H}} = \frac{1}{2!\sigma_u^6}\mathrm{E}\left\{P_2^{\mathrm{H}}(v_i(k))P_1^{\mathrm{H}}(v_j(k))y(k)\right\}, \quad i \neq j \tag{3.8.33}$$

$$c_{ij\ell}^{\mathrm{H}} = \frac{1}{\sigma_u^6}\mathrm{E}\left\{P_1^{\mathrm{H}}(v_i(k))P_1^{\mathrm{H}}(v_j(k))P_1^{\mathrm{H}}(v_\ell(k))y(k)\right\}, \quad i \neq j \neq \ell \neq i \tag{3.8.34}$$
,

Proof.

1. *Orthogonality of the parameter estimation:*
   All model components are orthogonal to each other. Multiply the model equation (3.8.28) with any of the components and compute the average values. Then all other terms tend to zero, and only the coefficients belonging to the multiplier component remain in the equation. This parameter can also be estimated independently of the others.

2. *Estimation of the constant term:*
   Assume an additive noise $n(k)$. Since the average value both of the not degree-0 components and of the noise are zero, the average value of the output signal gives the constant term

$$\mathrm{E}\{y(k)\} = \mathrm{E}\{w(k) + n(k)\} = \mathrm{E}\{w(k)\} = c_0$$

3. *Filtering the additive output noise:*
   Multiply the output signal by any non zero-degree component and compute the average value. This procedure filters any additive noise with zero mean.

4. *Estimation of linear terms:*

$$\mathrm{E}\left\{P_1^{\mathrm{H}}(v_i(k))y(k)\right\} = \mathrm{E}\left\{P_1^{\mathrm{H}}(v_i(k))w(k)\right\} = c_i^{\mathrm{H}}\mathrm{E}\left\{\left[P_1^{\mathrm{H}}(v_i(k))\right]^2\right\}$$
$$= c_i^{\mathrm{H}}\mathrm{E}\left\{u^2(k)\right\} = c_i^{\mathrm{H}}\sigma_u^2$$

   which is equivalent to (3.8.29).

5. *Estimation of quadratic terms:*

$$\mathrm{E}\left\{P_2^{\mathrm{H}}(v_i(k))y(k)\right\} = \mathrm{E}\left\{P_2^{\mathrm{H}}(v_i(k))w(k)\right\} = c_{ii}^{\mathrm{H}}\mathrm{E}\left\{\left[P_2^{\mathrm{H}}(v_i(k))\right]^2\right\}$$
$$= c_{ii}^{\mathrm{H}}\mathrm{E}\left\{\left[u^2(k) - \sigma_u^2\right]^2\right\} = c_{ii}^{\mathrm{H}}\left[\mathrm{E}\left\{u^4(k)\right\} - 2\mathrm{E}\left\{\sigma_u^2 u^2(k)\right\} + \sigma_u^4\right]$$
$$= c_{ii}^{\mathrm{H}}\left[3\sigma_u^4 - 2\sigma_u^4 + \sigma_u^4\right] = 2\sigma_u^4 c_{ii}^{\mathrm{H}}$$

   which is equivalent to (3.8.30).

$$\mathrm{E}\left\{P_1^{\mathrm{H}}(v_i(k))P_1^{\mathrm{H}}(v_j(k))y(k)\right\} = \mathrm{E}\left\{P_1^{\mathrm{H}}(v_i(k))P_1^{\mathrm{H}}(v_j(k))w(k)\right\}$$
$$= c_{ij}^{\mathrm{H}}\mathrm{E}\left\{\left[P_1^{\mathrm{H}}(v_i(k))P_1^{\mathrm{H}}(v_j(k))\right]^2\right\} = c_{ij}^{\mathrm{H}}\mathrm{E}\left\{\left[P_1^{\mathrm{H}}(v_i(k))\right]^2\right\}\mathrm{E}\left\{\left[P_1^{\mathrm{H}}(v_j(k))\right]^2\right\}$$
$$= c_{ij}^{\mathrm{H}}\mathrm{E}\left\{u^2(k)\right\}\mathrm{E}\left\{u^2(k)\right\} = \sigma_u^4 c_{ij}^{\mathrm{H}} \qquad i \neq j$$

   which is equivalent to (3.8.31).

6. *Estimation of cubic terms:*

$$\mathrm{E}\left\{P_3^{\mathrm{H}}(v_i(k))y(k)\right\}=\mathrm{E}\left\{P_3^{\mathrm{H}}(v_i(k))w(k)\right\}=c_{iii}^{\mathrm{H}}\mathrm{E}\left\{\left[P_3^{\mathrm{H}}(v_i(k))\right]^2\right\}$$
$$=c_{iii}^{\mathrm{H}}\mathrm{E}\left\{\left[u^3(k)-3\sigma_u^2u(k)\right]^2\right\}=c_{iii}^{\mathrm{H}}\left[\mathrm{E}\left\{u^6(k)\right\}-6\sigma_u^2\mathrm{E}\left\{u^4(k)\right\}+9\sigma_u^4\mathrm{E}\left\{u^2(k)\right\}\right]$$
$$=c_{iii}^{\mathrm{H}}\left[15\sigma_u^6-18\sigma_u^6+9\sigma_u^6\right]=6\sigma_u^6c_{iii}^{\mathrm{H}}$$

which is equivalent to (3.8.32).

$$\mathrm{E}\left\{P_2^{\mathrm{H}}(v_i(k))P_1^{\mathrm{H}}(v_j(k))y(k)\right\}=\mathrm{E}\left\{P_2^{\mathrm{H}}(v_i(k))P_1^{\mathrm{H}}(v_j(k))w(k)\right\}$$
$$=c_{iij}^{\mathrm{H}}\mathrm{E}\left\{\left[P_2^{\mathrm{H}}(v_i(k))P_1^{\mathrm{H}}(v_j(k))\right]^2\right\}=c_{iij}^{\mathrm{H}}\mathrm{E}\left\{\left[P_2^{\mathrm{H}}(v_i(k))\right]^2\right\}\mathrm{E}\left\{\left[P_1^{\mathrm{H}}(v_j(k))\right]^2\right\}$$
$$=c_{iij}^{\mathrm{H}}\mathrm{E}\left\{\left[u^2(k)-\sigma_u^2\right]^2\right\}\mathrm{E}\left\{u^2(k)\right\}=c_{iij}^{\mathrm{H}}2\sigma_u^4\sigma_u^2=2\sigma_u^6c_{iij}^{\mathrm{H}}\qquad i\neq j$$

which is equivalent to (3.8.33).

$$\mathrm{E}\left\{P_1^{\mathrm{H}}(v_i(k))P_2^{\mathrm{H}}(v_j(k))P_1^{\mathrm{H}}(v_\ell(k))y(k)\right\}=\mathrm{E}\left\{P_1^{\mathrm{H}}(v_i(k))P_1^{\mathrm{H}}(v_j(k))P_1^{\mathrm{H}}(v_\ell(k))w(k)\right\}$$
$$=c_{ij\ell}^{\mathrm{H}}\mathrm{E}\left\{\left[P_1^{\mathrm{H}}(v_i(k))P_1^{\mathrm{H}}(v_j(k))P_1^{\mathrm{H}}(v_\ell(k))\right]^2\right\}$$
$$=c_{ij\ell}^{\mathrm{H}}\mathrm{E}\left\{\left[P_1^{\mathrm{H}}(v_i(k))\right]^2\left[P_1^{\mathrm{H}}(v_j(k))\right]^2\left[P_1^{\mathrm{H}}(v_\ell(k))\right]^2\right\}$$
$$=c_{ij\ell}^{\mathrm{H}}\mathrm{E}\left\{u^2(k)\right\}\mathrm{E}\left\{u^2(k)\right\}\mathrm{E}\left\{u^2(k)\right\}=\sigma_u^6c_{ij\ell}^{\mathrm{H}}\qquad i\neq j\neq \ell\neq i$$

which is equivalent to (3.8.34). ■

The independent estimation of the kernels of the constant, linear, quadratic and cubic terms is shown in Figure 3.8.5 to 3.8.8, respectively.

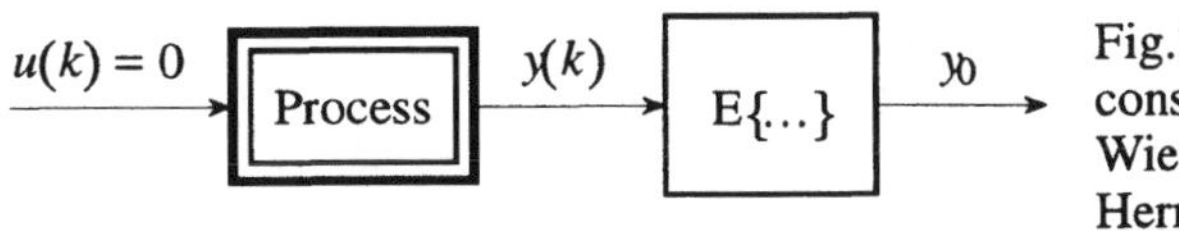

Fig.3.8.5 Estimation of the constant term in the orthogonal Wiener model with Laguerre and Hermite series of Gaussian white noise excitation

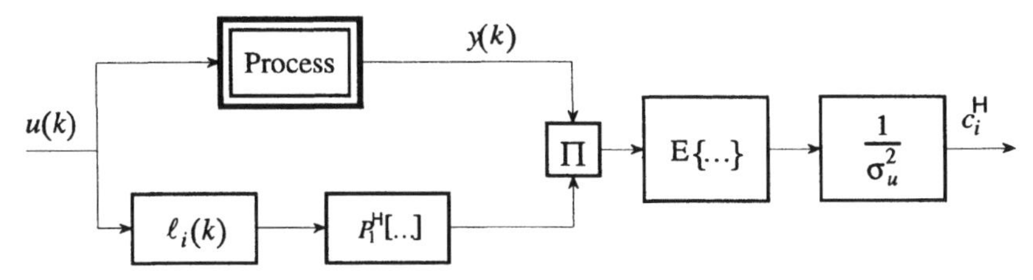

Fig. 3.8.6 Estimation of the coefficients of the linear terms in the orthogonal Wiener model with Laguerre and Hermite series of Gaussian white noise excitation

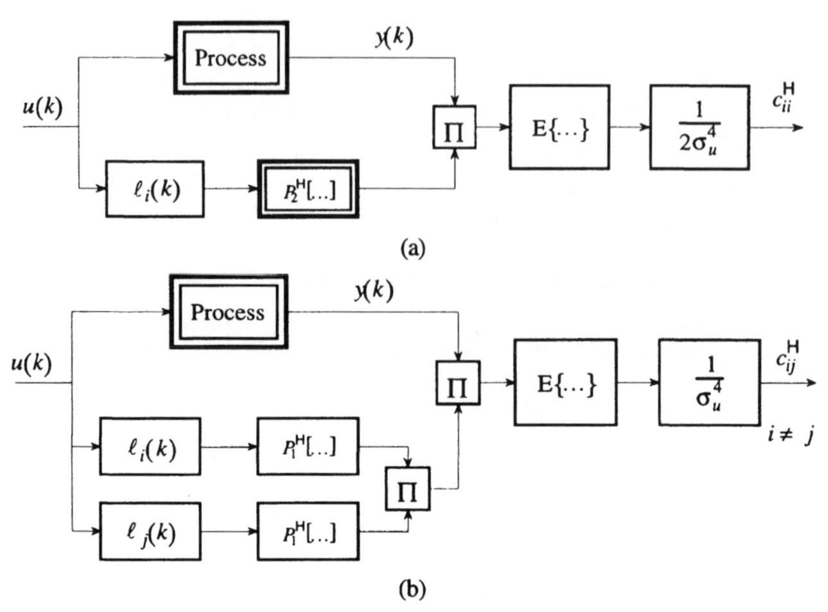

Fig. 3.8.7 Estimation of the coefficients of the quadratic terms in the orthogonal Wiener model with Laguerre and Hermite series of Gaussian white noise excitation: (a) auto-terms; (b) cross product terms

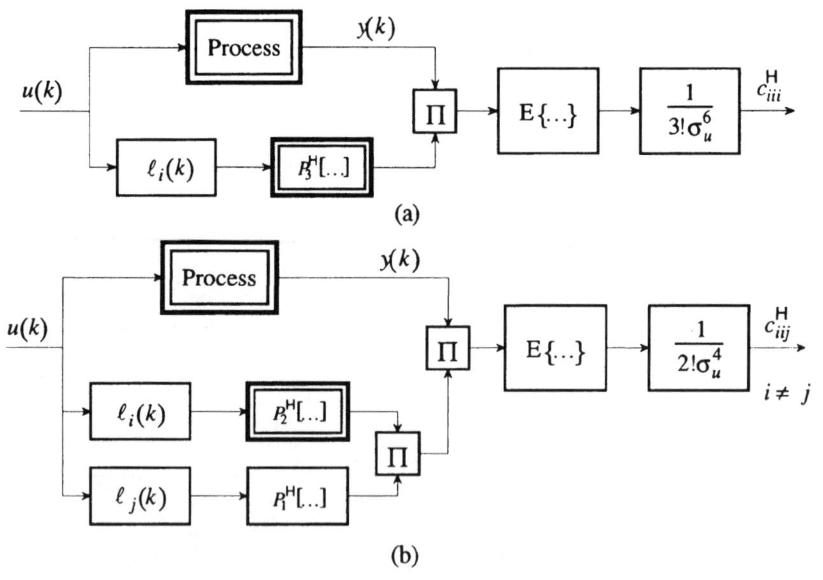

Fig. 3.8.8 Estimation of the coefficients of the cubic terms in the orthogonal Wiener model with Laguerre and Hermite series of Gaussian white noise excitation: (a) auto-terms; (b) cross product terms of two filter outputs;

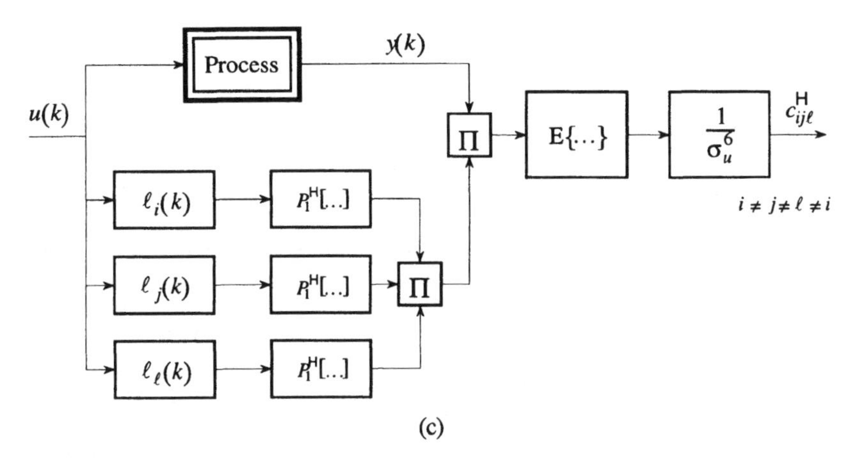

Fig. 3.8.8 Estimation of the coefficients of the cubic terms in the orthogonal Wiener model with Laguerre and Hermite series of Gaussian white noise excitation: (c) cross product terms of three filter outputs

3.9 METHODS APPLIED TO THE SIMPLE HAMMERSTEIN MODEL

Figure 3.9.1a shows the simple Hammerstein model consisting of a nonlinear static and a linear dynamic term. For simplicity a quadratic characteristic will be assumed. The continuous time model is seen in Figure 3.9.1b and the discrete time model in Figure 3.9.1c. As the delay time of the linear dynamic term plays an important role in the algorithms, its value relative to the sampling time is given exactly by d. Thus the numerator of the pulse transfer function is

$$q^{-d}B\left(q^{-1}\right) = q^{-d}\left(b_0 + b_1 q^{-1} + b_2 q^{-2} + \ldots + b_{nb} q^{-nb}\right)$$

The linear channel of the model is

$$y_1(k) = \frac{B\left(q^{-1}\right)}{A\left(q^{-1}\right)} c_1 u(k), \quad A\left(q^{-1}\right) = 1 + \tilde{A}\left(q^{-1}\right), \quad B\left(q^{-1}\right) = b_0 + \tilde{B}\left(q^{-1}\right)$$

This shows that one parameter of the linear dynamic part can be chosen freely. The usual choices are

- either $b_0 = 1$ or
- unit static gain

$$K = \frac{\sum_{i=0}^{nb} b_i}{1 + \sum_{j=1}^{na} a_j} = 1$$

3.9.1 Non-iterative method starting with the estimation of the steady state characteristic

Assume we know the nonlinear static term. Then the inner variable $v(k)$ can be calculated from the input signal and the linear dynamic term can be estimated from the sequences of the inner variable and the output signal.

The equation of the nonlinear static term is equal to the steady state characteristic of the process if the static gain of the linear dynamic part is chosen equal to one. The task is to estimate the steady state characteristic of the process, which can be done by different methods, e.g.:

- from repeated steady state measurements in different working points;
- from the estimation of the parameters of a generalized Hammerstein model by putting $q = 1$ in the difference equation.

The first method is only suitable if the linear dynamic term is a proportional one. If, e.g., it is of differentiating type then all steady state values are zero and if it is of integrating type then the responses tend to infinity. However, by choosing a proper test signal the output signal in the steady state can be forced to be a non-zero constant value. This can be achieved by a ramp signal $t \cdot 1(t)$ signal for the differentiating systems and by a pulse signal for the integrating systems.

The estimation of the steady state characteristic of a dynamic process is studied later in Section 3.18.

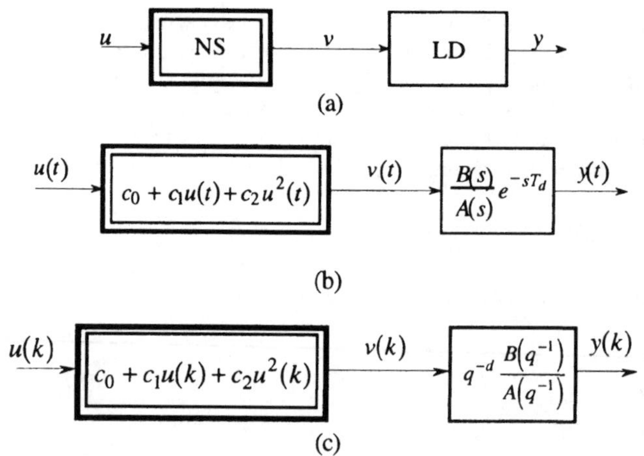

Fig. 3.9.1 Scheme of the simple Hammerstein model
(a) general; (b) continuous time; (c) discrete time

Algorithm 3.9.1 The non-iterative identification method of the simple Hammerstein model starting with the estimation of the steady state characteristic consists of the following steps:

1. Estimate the steady state characteristic from input–output measurements.
2. Equate the estimated parameters to the nonlinear static term.
3. Compute the sequence of the inner variable from the input signal by means of the estimated nonlinear static term.
4. Estimate the linear dynamic part from the sequences of the inner variable and the measured output signal. ■

3.9.2 Non-iterative method starting with the estimation of the parameters of the linear dynamic term

Assume now that we know the linear dynamic term. Then the inner variable $v(k)$ can be calculated from the output signal by means of the inverse of the (pulse) transfer function and the nonlinear static term can be estimated from the sequences of the input signal and the inner variable by a static regression.

There are different ways in which the linear dynamic term can be estimated:

1. identification by a two-valued signal;
2. identification with a small excitation;
3. correlation analysis by means of a Gaussian white noise.

1. *Identification by a two-valued signal:*

If the nonlinear static term is excited by a two-valued signal then the inner variable is two-valued, as well, and the nonlinear characteristic is linearized by a secant (Figure 3.9.2). A parameter estimation from the input and output signals leads to the correct parameters of the linear dynamic terms except a multiplicative factor. Then choosing, e.g., the static gain of the linear term equal to one, the linear dynamic term is estimated unambiguously. The following test signals and parameter estimation methods can be used:

- step function: graphoanalitical or LS parameter estimation;
- pulse function: graphoanalitical, deconvolution or LS parameter estimation;
- two-valued signal: LS parameter estimation;
- PRBS: LS parameter estimation or correlation method.

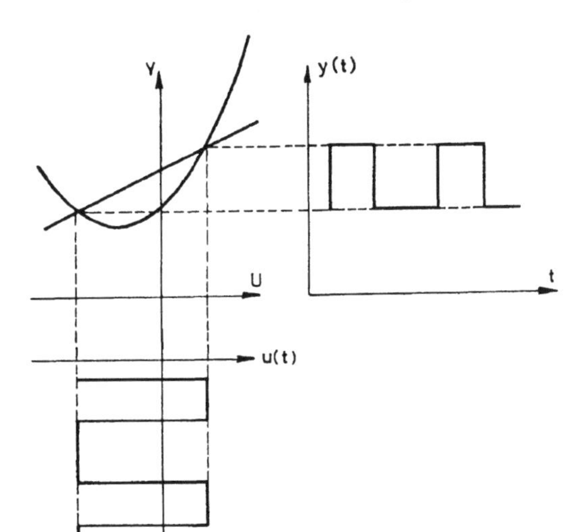

Fig. 3.9.2 Substitution of the nonlinear characteristic by a linear one for a two-valued excitation

2. *Identification with a small excitation:*

If the amplitude range of the test signal is small, then the static nonlinear term is excited in a small vicinity of the working point and the nonlinear characteristic can be considered as almost linear. Then the pulse transfer function of an estimated linear model is proportional to that of the linear term in the simple Hammerstein model. The model to be estimated consists of a constant and linear dynamic term (Figure 3.9.3)

$$y(k) = y_0 + \Delta y(k)$$

$$y_0 = \frac{B(1)}{A(1)}\left[c_0 + c_1 u_0 + c_2\left(u_0^2 + \sigma_u^2\right)\right]$$

$$\Delta y(k) = \left(c_1 + 2c_2 u_0\right) q^{-d} \frac{B\left(q^{-1}\right)}{A\left(q^{-1}\right)} \Delta u(k)$$

because

$$v_0 = \mathrm{E}\{v(k)\} = \mathrm{E}\left\{c_0 + c_1\left[u_0 + \Delta u(k)\right] + c_2\left[u_0 + \Delta u(k)\right]^2\right\}$$

$$= c_0 + c_1 u_0 + c_2\left[u_0^2 + \sigma_u^2\right]$$

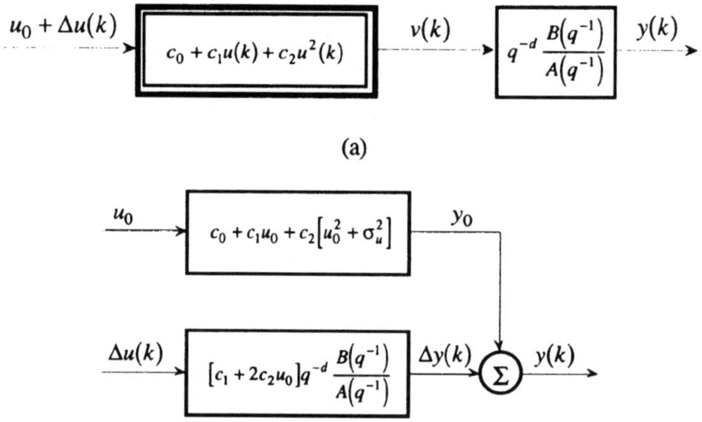

(a)

(b)

Fig. 3.9.3 Linearized model of the simple Hammerstein model for a small excitation

3. *Correlation analysis by means of a Gaussian white noise:*
   Several authors have shown that the continuous (and the discrete time) cross-correlation function (series) between the input and output signals of the simple Hammerstein model are proportional to the weighting function (series) of the linear dynamic part

$$r_{xy'}(\tau) = A_1 h(\tau), \quad r_{xy'}(\kappa) = A_1 h(\kappa)$$

where

$$x(k) = u(k) - \mathrm{E}\{u(k)\}, \quad y'(k) = y(k) - \mathrm{E}\{y(k)\}$$

(Korenberg 1973b; Billings and Fakhouri, 1978a, 1979c, 1979d, 1980, 1982; Schweizer, 1987).

The pulse transfer function can be estimated in the knowledge that the weighting function series is the response to a unit pulse:

- by graphoanalytical method, or
- by LS parameter estimation.

The resulting pulse transfer function has to be normalized for a static gain equal to one.

Algorithm 3.9.2 The non-iterative identification method of the simple Hammerstein model starting with the estimation of the linear dynamic term consists of the following steps:

1. Estimate the parameters of the linear dynamic model from a two-valued, small range or Gaussian excitation of the process.
2. Compute or estimate the parameters of the inverse pulse transfer function.
3. Compute the sequence of the inner variable from the output signal by means of the inverse pulse transfer function of the linear dynamic term.
4. Estimate the nonlinear static part from the sequences of the measured input signal and the computed inner variable. ∎

The problem of computing the inverse model of a linear dynamic system is dealt with later in this Section.

Example 3.9.1 *Identification of the linear dynamic term by means of a small range excitation*

The simple Hammerstein model has the system equations:

$$v(k) = 2 + u(k) + 0.5u^2(k), \qquad 10\dot{y}(t) + y(t) = v(t)$$

The process was excited by a PRTS with maximum length of 26 and minimum switching time of 10 [s]. The working point was $u_0 = 1$ and the amplitude of the PRTS was $\Delta u = \pm 0.1$. The input–output noise-free measurements were sampled by 2 [s]. The difference equation of the dynamic term obtained by SRE transformation is

$$y(k) = 0.8187y(k-1) + 0.1813u(k-1)$$

$N = 3 \cdot 26 \cdot 5 = 390$ data were recorded. Linear identification was performed between the data records from $2 \cdot 26 \cdot 5 + 1 = 262$ to $N = 390$. The estimated parameters are

$$\hat{y}(k) = (0.8175 \pm 0.0024)y(k-1) + (0.3632 \pm 0.0030)u(k-1)$$
$$+(0.2741 \pm 0.0075)$$

or

$$\hat{y}(k) = \frac{0.3632q^{-1}}{1 - 0.8175q^{-1}} \Delta u(k) + 1.684$$

Both the computed static gain

$$\hat{c}_1 = 0.3632(1 - 0.8175) = 1.99$$

and the estimated time constant,

$$\hat{T}_1 = -\Delta T / \ln(-a_1) = -2/\ln(0.8175) = 9.93 \text{ [s]}$$

approximate their true values $T = 10$ [s] and

$$c_1 = 1 + 2 \cdot 0.5 \cdot U\big|_{U=1} = 2$$

well. ■

3.9.3 Correlation method

In the last section it was shown how the linear dynamic term can be estimated by means of correlation analysis and how the static part can be estimated in the knowledge of the inner variable and the output signal. Now a method that uses only correlation functions is presented.

Lemma 3.9.1

(Krempl, 1973a, 1973b) Excite the simple Hammerstein model by a PRTS signal with zero mean and with a half period $N_p/2$ greater than the settling time of the process. Then the followings are true:

$$r_{uy}(\kappa) = \left[C_1 \sum_{i=1}^{M'} c_{2i-1} U^{2i} \right] h(\kappa) \qquad 0 \le \kappa \le N_p/2 \qquad M' \le (M+1)/2$$

$$r_{xy}(\kappa) = \left[C_2 \sum_{i=1}^{M'} c_{2i} U^{2(i+1)} \right] h(\kappa) \qquad 0 \le \kappa \le N_p/2 \qquad M' \le M/2$$

with

$$x(k) = u^2(k) - \mathrm{E}\{u^2(k)\}$$

C_1, C_2 are constants and depend on the test signal used.

Proof. See in Krempl (1973a, 1973b).

Algorithm 3.9.3 (Krempl, (1973a; 1973b)) The simple Hammerstein model can be identified by means of correlation analysis in the following way:

1. Excite the process by a PRTS signal with zero mean and with a half period greater than the settling time of the process.
2. Estimate the linear dynamic term from the cross-correlation function $r_{uy}(k)$ because they are proportional to each other as shown in Lemma 3.9.1. Choose the static gain of the linear dynamic term unit.
3. Estimate the odd-degree coefficients c_{2i-1}, $i = 1, \ldots, M'$ of the static nonlinear polynomial from the relation between them and $r_{uy}(k)$ given in Lemma 3.9.1. Observe that

$$\sum_{\kappa=0}^{N_p/2} r_{uy}(\kappa) = \left[C_1 \sum_{i=1}^{M'} c_{2i-1} U^{2i} \right] \sum_{\kappa=0}^{N_p/2} h(\kappa) = C_1 \sum_{i=1}^{M'} c_{2i-1} U^{2i}$$

if the static gain of the dynamic term was chosen unit. To estimate M' coefficients the test with M' different amplitudes has to be repeated.

4. Estimate the even-degree coefficients c_{2i}, $i=1,\dots,M'$ of the static nonlinear polynom from the relation between them and $r_{xy}(k)$ given in Lemma 3.9.1. Observe that

$$\sum_{\kappa=0}^{N_p/2} r_{xy}(\kappa)=\left[C_2\sum_{i=1}^{M'} c_{2i}U^{2(i+1)}\right]\sum_{\kappa=0}^{N_p/2} h(\kappa)=C_2\sum_{i=1}^{M'} c_{2i}U^{2(i+1)}$$

if the static gain of the dynamic term was chosen unit. To estimate M' coefficients the test with M' different amplitudes has to be repeated. ■

Remark: (Krempl, 1973b)
The static coefficients can be estimated from one equation system instead of two, as well.

3.9.4 Using a process model linear-in-parameters
The equation of the discrete time, noise-free, simple Hammerstein model is

$$A(q^{-1})w(k)=q^{-d}B(q^{-1})\left[c_0+c_1u(k)+c_2u^2(k)\right]$$

which can be rewritten to a form linear-in parameters

$$A(q^{-1})w(k)=c_0^*+B_1(q^{-1})u(k-d)+B_2(q^{-1})u^2(k-d) \tag{3.9.1}$$

with

$$c_0^*=c_0B(1),\quad B_1(q^{-1})=c_1B(q^{-1}),\quad B_2(q^{-1})=c_2B(q^{-1})$$

Assume, e.g., an additive output noise $n(k)$, then

$$y(k)=w(k)+n(k)$$

and

$$y(k)=\boldsymbol{\varphi}^{\mathrm{T}}(k)\boldsymbol{\theta}+A(q^{-1})n(k) \tag{3.9.2}$$

where

$$\boldsymbol{\varphi}(k)=\big[1,u(k-d),\dots,u(k-d-nb),u^2(k-d),\dots,$$
$$u^2(k-d-nb),-y(k-1),\dots,-y(k-na)\big]^{\mathrm{T}}$$

$$\boldsymbol{\theta}=\left[c_0^*,b_{10},\dots,b_{1nb},b_{20},\dots,b_{2nb},a_1,\dots,a_{na}\right]^{\mathrm{T}} \tag{3.9.3}$$

Both (3.9.1) and (3.9.2) show that

- the estimation problem is traced back to the identification of the a MISO

system with common denominators in the channels;
- that the linearity in the parameters of the residual model depends only on the noise model assumed.

The simple LS parameter estimation can be applied if either the noise can be neglected or

$$n(k) = \frac{1}{A(q^{-1})} e(k)$$

where $e(k)$ is the white source noise.

Figure 3.9.4 shows the redrawing of the simple Hammerstein model to the form linear-in-parameters which is the generalized Hammerstein model. The generalized Hammerstein model is a redundant form of the simple Hammerstein model.

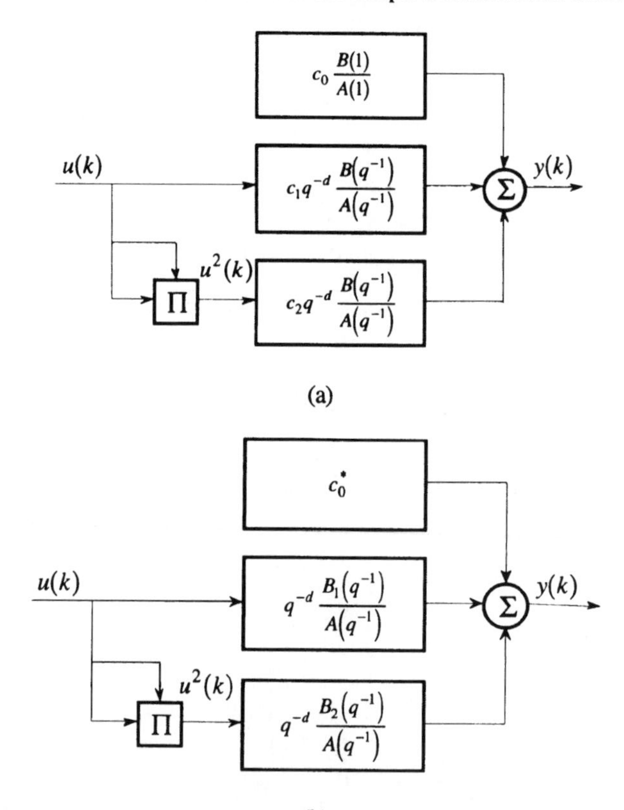

Fig. 3.9.4 Transformation of the simple Hammerstein model to the generalized model

Algorithm 3.9.4 The parameters of the simple Hammerstein model can be obtained by the further procedure from the estimated parameter vector $\hat{\boldsymbol{\theta}}$ (3.9.3) of a fitted generalized Hammerstein model:

1. The parameters of the denominator a_j, $j = 1, \ldots, n$ are explicitly in the estimated parameter vector $\hat{\boldsymbol{\theta}}$.
2. The coefficients of the static curve are $\hat{c}_1 = \hat{b}_{10}$, $\hat{c}_2 = \hat{b}_{20}$.
3. The first parameter of the numerator is set to one: $\hat{b}_0 = 1$ and the further elements can be obtained in different ways:

$$\hat{b}_1 = \hat{b}_{ji} / \hat{b}_{j0}, \quad i = 0, \ldots, nb, \quad j = 1, 2 \tag{3.9.4}$$

Gallman (1976) recommended to simulate the system with all possible parameter sets $(j = 1, 2)$ and those parameters are selected which give the least value of a bias function, e.g., of the mean square error.

4. The constant term is calculated as $c_0 = c_0^* / B(1)$. ■

Remarks:

1. By this algorithm the static gain of the pulse transfer function is not equal to one. This can be achieved by a further normalization.
2. The difference between the same differently calculated parameters of the numerator in (3.9.4) is a measure, whether the model is a simple Hammerstein model or not.

Example 3.9.2 *Estimation of the parameters by LS and prediction error methods (Bányász et al., 1973)*

The discrete time simple Hammerstein model had the parameters:

$$c_0 = 0, \quad c_1 = 2, \quad c_2 = 1$$

$$\frac{B\left(q^{-1}\right)}{A\left(q^{-1}\right)} = \frac{1 + 0.5q^{-1}}{1 - 1.5q^{-1} + 0.7q^{-2}}$$

The input signal was a random sequence with normal distribution, zero mean and variance one. Stochastic noise was added to the output. The source noise was a Gaussian white noise. The linear noise filter had the form $C\left(q^{-1}\right)/A\left(q^{-1}\right)$ and the following cases were investigated:

- *autoregressive filter:* $C\left(q^{-1}\right) = 1/\left(1 - 0.6q^{-1}\right)$
- *autoregressive moving average filter:* $C\left(q^{-1}\right) = 1 - q^{-1} + 0.2q^{-2}$
- *white noise:* $C\left(q^{-1}\right) = A\left(q^{-1}\right)$

The ratio of the standard deviations of the noise and of the noise-free output signal was 30 percent, which is a large enough value. $N = 500$ data were used in the estimation algorithms. The following methods were applied:

- least squares (LS);
- prediction error (PE).

The equivalent generalized Hammerstein process model has the equation

$$y(k) = c_0^* + b_{10}u(k) + b_{11}u(k-1) + b_{20}u^2(k) + b_{21}u^2(k-1)$$
$$-a_1 y(k-1) - a_2 y(k-2) + d_1 e(k-1) + d_2 e(k-2)$$

Table 3.9.1 shows the results of the parameter estimation. The results are quite good in view of the large noise to signal ratio. ■

Applications of the method are in the following publications (among others):

- *cement rotary kiln* (Li and Chen, 1988);
- *cement mill* (Haber *et al.*, 1986);
- *stream flow process* (Hu and Yuan, 1988), etc.

TABLE 3.9.1 Estimation of the parameters of a Hammerstein model by LS and prediction error method

Method	$C(q^{-1})$	σ_e	$\hat{c}_0^*$ (3)	$\hat{b}_{10}$ (2)	$\hat{b}_{11}$ (1)	$\hat{b}_{20}$ (1)	$\hat{b}_{21}$ (0.5)	$\hat{a}_1$ (-1.5)	$\hat{a}_2$ (0.7)			$\hat{c}_0$ (2)	$\hat{b}_1 = \frac{\hat{b}_{11}}{\hat{b}_{10}}$ (0.5)	$\hat{b}_1 = \frac{\hat{b}_{21}}{\hat{b}_{20}}$ (0.5)	$\hat{\sigma}_\varepsilon$	$\hat{\sigma}_{\hat{n}}$	Iter.
LS	$\frac{1}{1-0.6q^{-1}}$	0.5	2.961	1.952	0.9707	1.007	0.4967	-1.517	0.7136	-	-	1.977	0.4973	0.4931	0.519	0.4737	1
prediction error			2.973	1.961	1.008	1.000	0.5065	-1.509	0.7061	0.5023	-	1.964	0.5140	0.5065	0.5343	0.3634	6
prediction error			2.985	1.973	1.025	1.003	0.5087	-1.503	0.7012	0.5683	0.2359	1.965	0.5195	0.5072	0.5185	0.2555	7
LS	$1-q^{-1}+0.2q^{-2}$	3.0	3.469	1.830	2.110	0.9586	0.9134	-1.142	0.3787	-	-	1.611	1.152	0.9529	3.903	3.915	1
prediction error			3.085	1.847	1.242	1.014	0.5264	-1.483	0.6875	-1.004	0.1918	1.845	0.6724	0.5191	3.068	0.3156	7
LS	$1-1.5q^{-1}+0.7q^{-2}$	3.0	3.410	1.856	2.639	0.9728	1.163	-0.9234	0.1686	-	-	1.408	1.422	1.195	4.537	5.034	1
prediction error			3.150	1.869	1.198	0.9606	0.5531	-1.491	0.6969	-1.524	0.7533	1.920	0.6410	0.5757	3.063	0.3158	10

3.9.5 Using a process model linear-in-parameters with limitations

The form (3.9.2) linear-in-parameters can be extended by the known condition between the estimated parameters:

$$\frac{\hat{\theta}_{3+nb+i}}{\hat{\theta}_{2+i}} = \frac{\hat{\theta}_{3+nb}}{\hat{\theta}_2}, \quad i = 0, \ldots, nb \tag{3.9.5}$$

which is based on the relation

$$\frac{b_{2i}}{b_{1i}} = \frac{b_{20}}{b_{10}}, \quad i = 0, \ldots, nb$$

The parameter estimation can be performed with considering the constrain (3.9.5).

3.9.6 Using a process model nonlinear-in-parameters

The noise-free output signal is described by the equations

$$v(k) = c_0 + c_1 u(k) + c_2 u^2(k)$$

$$w(k) = \frac{B(q^{-1})}{A(q^{-1})} v(k-d)$$

An additive colored noise

$$n(k) = \frac{E(q^{-1})}{D(q^{-1})} e(k)$$

is assumed at the output

$$y(k) = w(k) + n(k)$$

The residual can be expressed by the form

$$\varepsilon(k) = \frac{D(q^{-1})}{E(q^{-1})} [y(k) - w(k)]$$

The prediction error method can be applied, which needs the first-order derivative of the residual according to the unknown parameters. The sensitivity functions are given now.

$$\frac{\partial \varepsilon(k)}{\partial c_0} = -\frac{D(1)}{E(1)} \frac{B(1)}{A(1)} \qquad \frac{\partial \varepsilon(k)}{\partial c_1} = -\frac{D(q^{-1})}{E(q^{-1})} \frac{B(q^{-1})}{A(q^{-1})} u(k-d)$$

$$\frac{\partial \varepsilon(k)}{\partial c_2} = -\frac{D(q^{-1})}{E(q^{-1})} \frac{B(q^{-1})}{A(q^{-1})} u^2(k-d) \qquad \frac{\partial \varepsilon(k)}{\partial b_i} = -\frac{D(q^{-1})}{E(q^{-1})} \frac{1}{A(q^{-1})} v(k-d-i)$$

$$\frac{\partial \varepsilon(k)}{\partial a_i} = \frac{D(q^{-1})}{E(q^{-1})} \frac{1}{A(q^{-1})} w(k-i) \qquad \frac{\partial \varepsilon(k)}{\partial d_i} = \frac{1}{E(q^{-1})} [y(k-i) - w(k-i)]$$

$$\frac{\partial \varepsilon(k)}{\partial e_i} = -\frac{1}{E(q^{-1})} \varepsilon(k-i)$$

As is known, the prediction error method is an iterative procedure.

3.9.7 Iterative parameter estimation technique using the inverse of the linear model

The iterative techniques rely on the relaxation technique, which means that parameter groups are estimated in turn iteratively.

There are several measures that can detect when the iteration has to be stopped:

- the deviation between the parameters of two subsequent steps;
- the deviation between the loss functions of two subsequent steps, etc.

Algorithm 3.9.5 Haber and Keviczky (1974) used the following iteration procedure:

1. Assume the inner variable is equal to the measured output signal.
2. Estimate the parameters of the static term by regression from the input signal and from the inner variable.
3. Compute the inner variable from the input signal by the estimated nonlinear static model.
4. Estimate the pulse transfer function of the linear dynamic model from the inner variable and from the measured output signal.
5. Compute the inner variable from the measured output signal by means of the inverse

model of the estimated linear dynamic term.
6. Check whether the iteration has to be stopped. If not go to Step 2. ■

Algorithm 3.9.6 Hunter and Korenberg (1986) suggested the following iteration procedure:
1. Assume the inner variable is equal to the measured input signal.
2. Estimate the linear inverse model from the inner variable and from the measured output signal.
3. Compute the inner variable from the measured output signal by means of the linear inverse model.
4. Estimate the parameters of the static term by regression from the input signal and from the inner variable.
5. Compute the inner variable from the input signal by the estimated nonlinear static model.
6. Check whether the iteration has to be stopped. If not go to Step 2.
7. Estimate the parameters of the linear dynamic model from the computed inner variable and from the measured output signal. ■

Algorithms 3.9.5 and 3.9.6 differ in two points:
1. The iteration is started with the estimation of the parameters
   - of the nonlinear static term (Haber and Keviczky, 1974);
   - of the linear dynamic term (Hunter and Korenberg, 1986).
2. Haber and Keviczky (1974) estimated in every step the linear dynamic model and computed the inverse model based on this. Hunter and Korenberg (1986) estimated the inverse model in each step.

The submodels whose parameters have to be estimated in the iteration steps are linear in the parameters, at least if proper noise models are assumed.

1. *Static part* (noise-free):

$$
\begin{aligned}
v(k) &= c_0 + c_1 u(k) + c_2 u^2(k) \\
v(k) &= \boldsymbol{\varphi}_s^T(k)\boldsymbol{\theta}_s \\
\boldsymbol{\varphi}_s(k) &= \left[1, u(k), u^2(k)\right]^T \\
\boldsymbol{\theta}_s &= \left[c_0, c_1, c_2\right]^T
\end{aligned}
\tag{3.9.6}
$$

2. *Linear dynamic part* (noise-free):

$$
\begin{aligned}
w(k) &= B\left(q^{-1}\right)u(k-d) - q\left[A\left(q^{-1}\right) - 1\right]w(k-1) \\
w(k) &= \boldsymbol{\varphi}_\ell^T(k)\boldsymbol{\theta}_\ell \\
\boldsymbol{\varphi}_\ell(k) &= \left[v(k-d), \ldots, v(k-d-nb), -w(k-1), \ldots, -w(k-na)\right]^T \\
\boldsymbol{\theta}_\ell &= \left[b_0, \ldots, b_{nb}, a_1, \ldots, a_{na}\right]^T
\end{aligned}
$$

3. *Inverse linear dynamic part* (noise-free):

$$u(k) = (1/b_0)A(q^{-1})w(k+d) - q(1/b_0)[B(q^{-1}) - b_0]u(k-1) = (1/b_0)w(k+d) + \ldots$$
$$+(a_{na}/b_0)w(k+d-na) - (b_1/b_0)u(k-1) - \ldots - (b_{nb}/b_0)u(k-nb)$$
$$u(k) = \boldsymbol{\phi}_{i\ell}^{\mathrm{T}}(k)\boldsymbol{\theta}_{i\ell}$$
$$\boldsymbol{\phi}_{i\ell}(k) = [w(k+d), w(k+d-1), \ldots, w(k+d-na), -u(k-1), \ldots, -u(k-nb)]^{\mathrm{T}}$$
$$\boldsymbol{\theta}_{i\ell} = [1/b_0, a_1/b_0, \ldots, a_{na}/b_0, -b_1/b_0, \ldots, -b_{nb}/b_0]^{\mathrm{T}}$$

Remarks:
1. d-step ahead terms occuring in the memory vector cause no problem with the off-line identification, where the whole record is known.
2. To obtain unambiguous results the parameters of the linear dynamic term have to be normalized, as shown at the beginning of this chapter.
3. An iterative method (Vörös, 1985) which can but not need use the inverse of the linear dynamic term is presented in the next section.
4. Stoica (1981) showed that the relaxation technique failed to converge in some cases. He investigated the case when the linear dynamic term was described by weighting function series.

Example 3.9.3 *Iterative estimation using the inverse model (Haber and Keviczky, 1974)*

The discrete time simple Hammerstein model had the parameters:

$$c_0 = 2, \quad c_1 = 2, \quad c_2 = 1$$
$$\frac{B(q^{-1})}{A(q^{-1})} = \frac{1 + 0.5q^{-1}}{1 - 1.5q^{-1} + 0.7q^{-2}}$$

The input signal was a random sequence with normal distribution, zero mean and variance one. Stochastic noise was added to the output. The source noise was a Gaussian white noise. The linear noise filter had the form $1/A(q^{-1})$. $N = 300$ data were used in the estimation algorithms. Both noise-free and noisy cases were simulated. In the latter case the ratio of the standard deviations of the noise and of the noise-free output signal was $\psi = 30\%$.

In both cases the iteration was started with the static regression (SR). Figure 3.9.5 shows the mean square error (MSE)

$$\sigma_{\hat{n}} = \mathrm{MSE} = \tfrac{1}{N}\sum_{k=1}^{N}[\hat{y}(k) - y(k)]^2$$

for the estimated model output and the true noise-free output signals as a function of the iteration steps. For comparison a generalized Hammerstein model was also fitted by the LS method. This *direct estimation* (DE) gave the same MSE to which value the iterative method converged. Further improvement was not achieved.

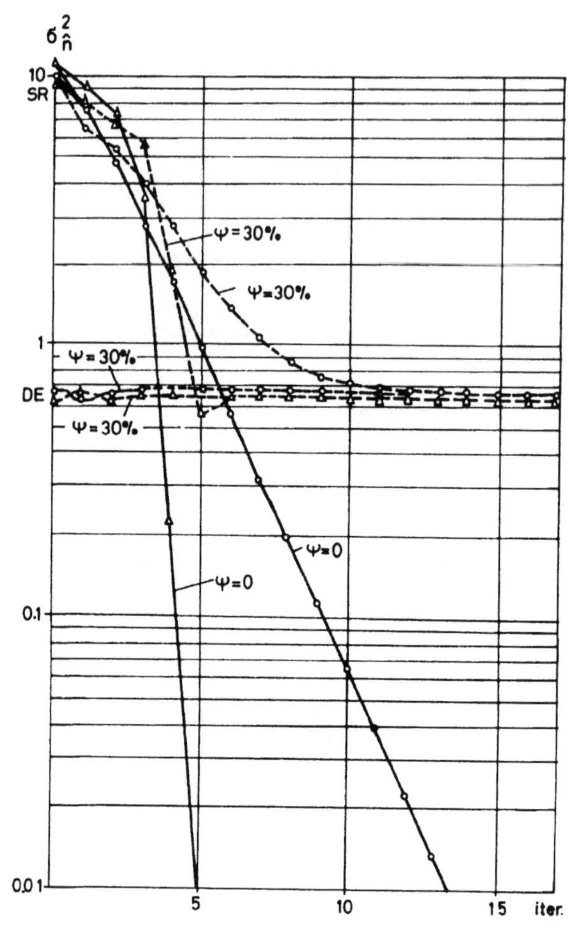

Fig. 3.9.5 Mean square error as a function of the iteration steps. DE is the non-iterative LS method based on the generalized Hammerstein model: Algorithm 3.9.5 (o); Algorithm 3.9.7 (Δ)

3.9.8 Iterative parameter estimation technique without using the inverse of the linear model

Gallman (1976) and Narendra and Gallman (1966) suggested an iterative technique without using the inverse model of the linear dynamic term. The method is presented here according to (Haber and Keviczky, 1974).

The simple Hammerstein model can be redrawn on the Figure 3.9.6 This corresponds to a noise-free model linear-in-parameters

$$w(k) = \boldsymbol{\phi}_*^{\mathrm{T}}(k)\boldsymbol{\theta}_* \tag{3.9.7}$$

with

$$\boldsymbol{\phi}_*(k) = \left[1^*, u^*(k), u^{2*}(k), -y(k-1), \ldots, -y(k-na)\right]^{\mathrm{T}} \tag{3.9.8}$$

$$\boldsymbol{\theta}_* = \left[c_0, c_1, c_2, a_1, \ldots, a_{na}\right]^{\mathrm{T}} \tag{3.9.9}$$

with

$$1^* = B(1) \tag{3.9.10}$$

$$u^*(k) = B\left(q^{-1}\right)u(k-d) \tag{3.9.11}$$

$$u^{2*}(k) = B\left(q^{-1}\right)u^2(k-d) \tag{3.9.12}$$

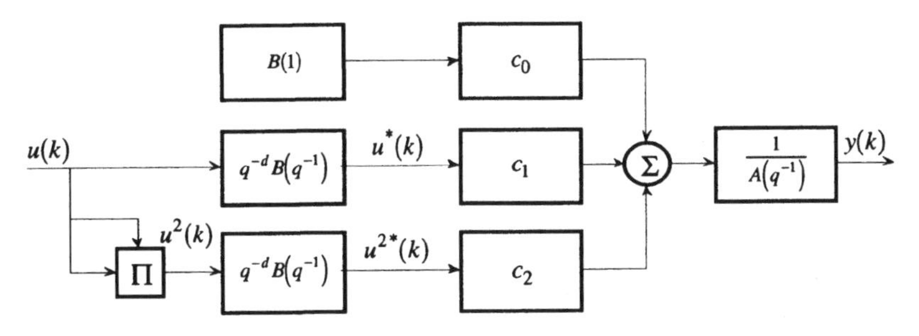

Fig. 3.9.6 Equivalent scheme of the simple Hammerstein model

Algorithm 3.9.7 Haber and Keviczky (1974) used the following algorithm, which is based on the works (Gallman, 1976; Narendra and Gallman, 1966):

1. Assume the inner variable is equal to the measured output signal.
2. Estimate the parameters of the nonlinear static term from the measured input signal and from the inner variable.
3. Compute the inner variable from the measured input signal by means of the estimated nonlinear static term.
4. Estimate the parameters of the linear dynamic term from the inner variable and from the output signal.
5. Filter the input signal and its square by the estimated $\hat{B}\left(q^{-1}\right)$ according to (3.9.11) and (3.9.12).
6. Estimate the parameter vector (3.9.9) from the filtered values and from the measured output signals. The estimated parameter vector (3.9.9) contains the coefficients of the static nonlinear term.
7. Check whether the iteration has to be stopped. If not go to Step 3. ■

Vörös (1985, 1995) proposed a similar method. He constructed an equivalent model linear-in-parameters of the simple Hammerstein model as given in Figure 3.9.7. The input–output model is linear in the parameters and includes all parameters of the process:

$$w(k) = B\left(q^{-1}\right)v(k-d) - \tilde{A}\left(q^{-1}\right)w(k)$$

$$= b_0\left[c_0 + c_1 u(k-d) + c_2 u^2(k-d) + \tilde{B}\left(q^{-1}\right)v(k-d) - \tilde{A}\left(q^{-1}\right)w(k)\right]$$

The noise-free process can be written into a form linear-in-parameters:

$$w(k) = \boldsymbol{\phi}^{\mathrm{T}}(k)\boldsymbol{\theta} \tag{3.9.13}$$

$$\boldsymbol{\varphi}(k)=\left[1, u(k-d), u^2(k-d), v(k-d-1),\ldots,\right.$$
$$\left. v(k-d-nb), -y(k-1),\ldots,-y(k-na)\right]^{\mathrm{T}} \tag{3.9.14}$$

$$\boldsymbol{\theta}=\left[b_0c_0, b_0c_1, b_0c_2, b_1,\ldots, b_{nb}, a_1,\ldots, a_{na}\right]^{\mathrm{T}} \tag{3.9.15}$$

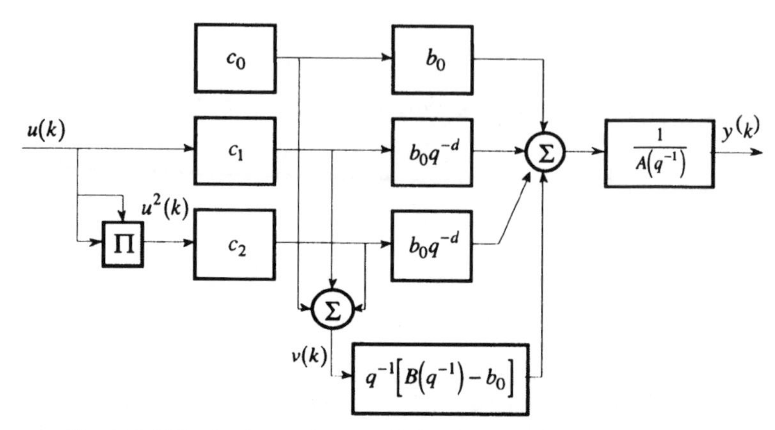

Fig. 3.9.7 Equivalent scheme of the simple Hammerstein model

Algorithm 3.9.8 Vörös (1985) proposed the following iterative algorithm:

1. Assume the inner variable is equal to the measured output signal.
2. Estimate the parameters of the nonlinear static term from the measured input signal and from the inner variable.
3. Compute the inner variable from the measured input signal by means of the estimated nonlinear static model.
4. Estimate the parameters (3.9.15) of the nonlinear static and the linear dynamic terms according to the model of (3.9.13) to (3.9.15) from the input, output and the computed inner variable. Assume that $b_0=1$.
5. Check whether the iteration has to be stopped. If not go to Step 3. ■

Remarks:

1. The inner variable can also be computed from the output signal by means of the inverse model of the linear dynamic term (Vörös, 1985).
2. Vörös (1995) extended the algorithm for the case when the output signal is disturbed by color noise. In Vörös (1995) simulation results are also given.

Example 3.9.4 *Iterative estimation without using the inverse model (Haber and Keviczky, 1974)*

The same model as in Example 3.9.3 was identified under the same circumstances by Algorithm 3.9.7.

Figure 3.9.5 shows the mean square error (MSE) for the estimated model output and the true noise-free output signals as a function of the iteration steps. The convergence is faster than in Example 3.9.3 achieved by means of Algorithm 3.9.6. For comparison also a generalized Hammerstein model was fitted by the LS method. This *direct estimation* (DE) gave the same MSE to which value the iterative method converged. Further improvement was not achieved. ■

3.9.9 Gate function method

The working point of the nonlinear term is determined by the input signal of the process. Therefore the process can be described by the sum of linear multi-models belonging to different domains of the input signal.

Algorithm 3.9.9 The identification algorithm using gate functions consists of the following steps:

1. Excite the system by a test signal that covers the whole range of the input signal.
2. Divide the range of the input signal into small intervals.
3. Perform the followings for all intervals:
   - Cut those input and output signals from the whole records where the input signal is in the actual interval.
   - Estimate a linear pulse transfer function.
   - Calculate the static gain.
4. In the case of a simple Hammerstein model the pulse transfer functions divided by the static gains are equal to each other. This is the pulse transfer function of the linear part of the simple Hammerstein model.
5. The static nonlinear term is given by the static gains as a function of the mean values of the domains of the input signal. An analytic function can be computed by a static regression. ■

An application of the method is the identification of a cement mill in Keviczky (1976).

3.9.10 Identification by graphical plotting

The method is based on the fact that the plot of the output signal as a function of the input signal $y(u(t))$ is single valued if the process is static and has a single valued characteristic. To get an unambiguous result assume the static gain of the linear dynamic part is unit.

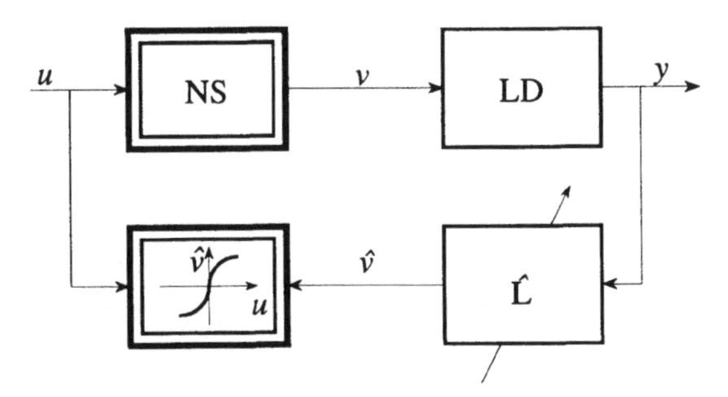

Fig 3.9.8 Identification of the simple Hammerstein model by graphical plotting

Algorithm 3.9.10 Lammers *et al.*, (1979) recommended the following procedure for the identification of the simple Hammerstein model (Figure 3.9.8):

1. Connect the measured input signal through a linear dynamic filter $\hat{L}$ to the x-position of an x-y plotter or oscilloscope.
2. Connect the measured output signal to the y-position of an x-y plotter or oscilloscope.
3. Tune the filter $\hat{L}$ until the plot is single valued one. Then the plot is the

characteristic curve of the static nonlinearity (NS) and $\hat{\mathrm{L}}$ is the estimate of the linear dynamic term (LD) ■

3.9.11 Frequency method

The identification procedure uses the results of two lemmas.

Lemma 3.9.2

Excite a static system with a characteristic of power function

$$y(t) = u^M(t)$$

by a sinusoid test signal

$$u(t) = U\cos\omega t$$

The highest-order harmonic is of order M and has the time function

$$y^{(M)}(t) = \frac{U^M}{2^{M-1}}\cos(M\omega t) \tag{3.9.16}$$

Proof. Assume (3.9.16) is valid for an $(M-1)$-degree system

$$y^{(M-1)}(t) = \frac{U^{M-1}}{2^{M-2}}\cos((M-1)\omega t)$$

Then for a degree-M system

$$y^{(M)}(t) = y^{(M-1)}(t)U\cos(\omega t) = \frac{U^{M-1}}{2^{M-2}}\cos((M-1)\omega t)U\cos(\omega t)$$

$$= \frac{U^M}{2^{M-1}}\left[\cos(M\omega t) + \cos((M-2)\omega t)\right]$$

Consequently, the highest (degree-M) harmonics is given by (3.9.16). ■

Lemma 3.9.3

Excite the continuous time simple Hammerstein model by a sinusoidal test signal

$$u(t) = U\cos\omega t$$

The highest order harmonics is

$$y^{(M)}(t) = \frac{U^M}{2^{M-1}}c_M|G(M\omega)|\cos(M\omega t + \varphi(M\omega)) \tag{3.9.17}$$

where M is the highest degree of the nonlinear power and the frequency function of the linear dynamic term is

$$G(j\omega) = |G(\omega)| \exp(j\varphi(\omega))$$

Proof. The highest order harmonics will be generated by the higher degree nonlinear term $c_M u^M(t)$. According to 3.9.16 this is

$$y^{(M)}(t) = \frac{U^M}{2^{M-1}} c_M \cos(M\omega t)$$

This signal will be filtered by the linear dynamic term at frequency $M\omega$, thus the highest harmonic of the output signal is as given by (3.9.17). ■

Denote the degree-M frequency characteristic of the time function (3.9.17) multiplied by $2^{M-1}/U^M$ by $G_M(j\omega_M)$

$$G_M(j\omega_M) = |G_M(\omega_M)| \exp(j\varphi(\omega_M))$$

where

$$G_M(\omega_M) = \frac{2^{M-1}}{U^M} |y^M(\omega_M)| = c_M |G_M(\omega_M)|$$

$$\varphi(\omega_M) = \varphi(y^M(\omega_M)) = \varphi(M\omega)$$

Algorithm 3.9.11 The procedure of the frequency domain identification is the following:

1. Excite the process by $U\cos\omega t$ at several frequencies.
2. Determine the highest order harmonics $y^{(M)}$ from the responses by filtering out the lower order harmonics.
3. Compute the frequency characteristics of $(2^{M-1}/U^M) y^{(M)}(t)$.
4. Rescale the angular frequencies by their M-times values.
5. Identify a linear dynamical system from the frequency characteristics.
6. Assume the linear dynamic term has a unit static gain. Then the estimated static gain is c_M.
7. Compute the inner variable from the measured output signal by means of the linear inverse model.
8. Estimate the parameters of the static term by regression from the input signal and from the inner variable. ■

If the amplitude characteristic holds enough information to identify the transfer function, i.e., the process is of minimum phase and has no delay time, then a simplified procedure can be used (Gardiner, 1971, 1973a).

Algorithm 3.9.12 (Gardiner, 1971, 1973a) In the case of a minimum phase and delay time free system a simplified procedure of the frequency domain identification can be applied:

1-3. As in Algorithm 3.9.10.

4. Determine the breakpoints of the angular frequencies from the Bode amplitude plot.

5. Transform all breakpoints to their M-times values.
6. Calculate the poles and zeros of the transfer function from breakpoints transformed as with the linear system identification.
7-8. As Items 6-8 in Algorithm 3.9.11. ■

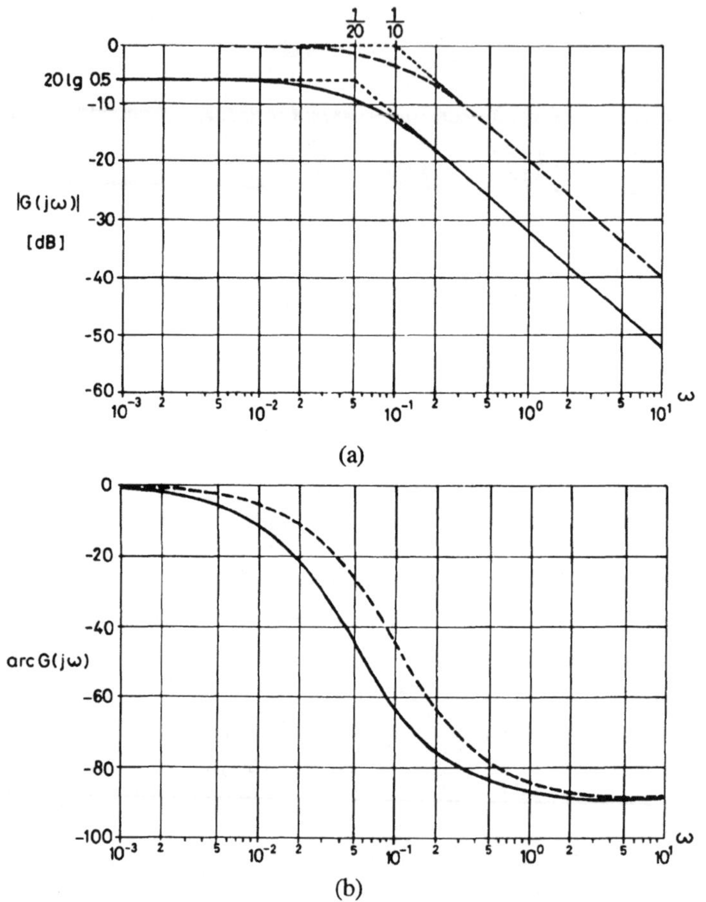

Fig. 3.9.9 Frequency characteristics of the second harmonic (–) and of the linear dynamic (- - -) term of the simple Hammerstein model: (a) amplitude plot; (b) phase plot

Example 3.9.5 *Identification of a first-order, quadratic, simple Hammerstein model by the frequency method*

The process is as in Example 3.9.1. The asymptotically approximating Bode amplitude and phase plots of the second harmonics $\left(2/U^2\right)y^{(2)}(t)$ are given in Figures 3.9.9a and 3.9.9b by continuous lines, respectively. The angular frequencies on the plot are the frequencies of the exciting sinusoidal signal. The amplitude and phase plots of the linear dynamic term with the constant coefficient c_2 are calculated by replacing the angular frequencies by their double values. They are also drawn in Figures 3.9.9a and 3.9.9b by dashed lines, respectively. As is seen, the Bode plot has the form of a linear proportional first-order system; it starts with $20\lg\hat{c}_2 = 20\lg 0.5 = -6.02$ and the breakpoint is at $1/\hat{T} = 0.1$. The estimated transfer function of the linear dynamic term is

$$\hat{G}(s) = \frac{1}{1+10s}$$

The nonlinear static part can be estimated from stationary measurements as the steady state input–output relation. ■

A further simulation example is presented with the structure identification of the simple Hammerstein model by the frequency method.

The frequency method was applied also by Gertner and Zagurskii (1996) for the parameter estimation of the simple Hammerstein model.

3.10 METHODS APPLIED TO THE SIMPLE WIENER MODEL

Figure 3.10.1a shows the simple Wiener model consisting of a nonlinear static and a linear dynamic term. For simplicity a quadratic characteristic will be assumed. The continuous time model is seen in Figure 3.10.1b and the discrete time model in Figure 3.10.1c. The discrete delay time relative to the sampling time of the linear dynamic term is d. The linear channel of the model is

$$y_1(k) = q^{-d}\frac{B(q^{-1})}{A(q^{-1})}c_1 u(k)$$

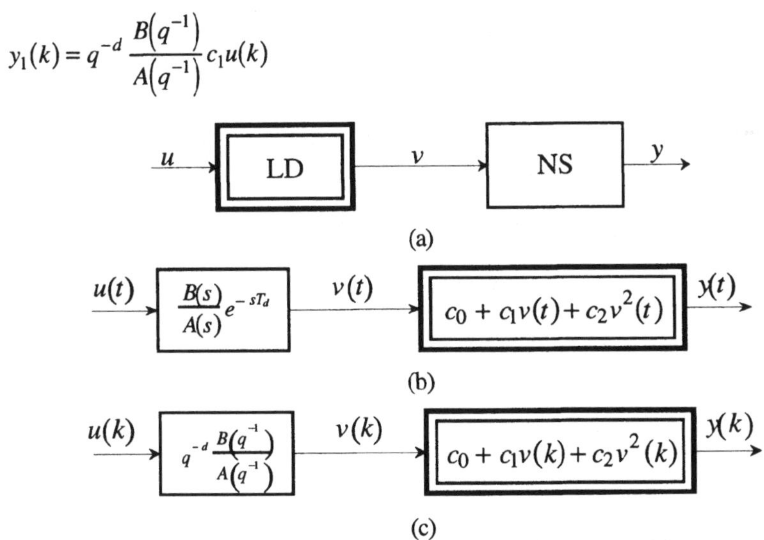

Fig. 3.10.1 Scheme of the simple Wiener model
(a) general; (b) continuous time; (c) discrete time

This shows that one parameter of the linear dynamic part can be chosen free as with the simple Hammerstein model. The usual choices are

- $b_0 = 1$, or
- unit static gain

$$K = \frac{\sum_{i=0}^{nb} b_i}{1 + \sum_{j=1}^{na} a_j} = 1$$

3.10.1 Non-iterative method starting with the estimation of the steady state characteristic

Assume we know the nonlinear static term. Then the inner variable $v(k)$ can be calculated from the output signal by the inverse characteristic and the linear dynamic term can be estimated from the sequences of the input signal and the inner variable.

The equation of the nonlinear static term is equal to the steady state characteristic of the process if the static gain of the linear dynamic part is chosen one. The task is to estimate the steady state characteristic of the process, which can be done by different methods, e.g.:

- from repeated steady state measurements in different working points;
- form the estimation of the parameters of an approximating model, whose parameters are easy to obtain (e.g., model linear-in-parameters), and by putting $q = 1$ in the difference equation.

Care has to be taken to the choice of the test signal if the linear dynamic term is not a proportional one, as mentioned with the identification of the simple Hammerstein model.

The estimation of the steady state characteristic of a dynamic process is studied later in Section 3.18.

Algorithm 3.10.1 The non-iterative method starting with the estimation of the steady state characteristic is as follows:

1. Estimate the steady state characteristic from input–output measurements.
2. Equal the estimated parameters to the nonlinear static term.
3. Compute the sequence of the inner variable from the output signal by means of the inverse of the estimated nonlinear static term.
4. Estimate the linear dynamic part from the sequences of the measured input signal and the inner variable. ∎

Remark:

The method assumes that the nonlinear characteristic has an unambiguously inverse function in the working domain investigated.

3.10.2 Non-iterative method starting with the estimation of the parameters of the linear dynamic term

Assume now that we know the linear dynamic term. Then the inner variable $v(k)$ can be calculated from the input signal by means of the (pulse) transfer function and the nonlinear static term can be estimated from the sequences of the inner variable and the output signal by a static regression.

There are different ways for the estimation of the linear dynamic term:

1. identification with a small excitation;
2. correlation analysis by means of a Gaussian white noise.

1. *Identification with a small excitation:*
   If the amplitude range of the test signal is small, then the static nonlinear term is excited in a small vicinity of the working point and the nonlinear characteristic can be considered as almost linear. Then the pulse transfer function of an estimated linear model is proportional to that of the linear term in the simple Wiener model. The model to be estimated consists of a constant and linear dynamic term (Figure 3.10.2)

$$y(k) = y_0 + \Delta y(k)$$

$$y_0 = c_0 + c_1 \frac{B(1)}{A(1)} u_0 + c_2 \left[\left(\frac{B(1)}{A(1)} u_0 \right)^2 + \sigma_v^2 \right]$$

$$\Delta y(k) = \left[c_1 + 2c_2 \frac{B(1)}{A(1)} u_0 \right] q^{-d} \frac{B(q^{-1})}{A(q^{-1})} \Delta u(k)$$

where σ_v is the standard deviation of the inner variable $v(k)$. In the derivation we used that

$$v_0 = \mathrm{E}\{v(k)\} = \frac{B(1)}{A(1)} u_0$$

$$y_0 = \mathrm{E}\{c_0 + c_1 v(k) + c_2 v^2(k)\} = c_0 + c_1 v_0 + c_2 [v_0 + \sigma_v^2]$$

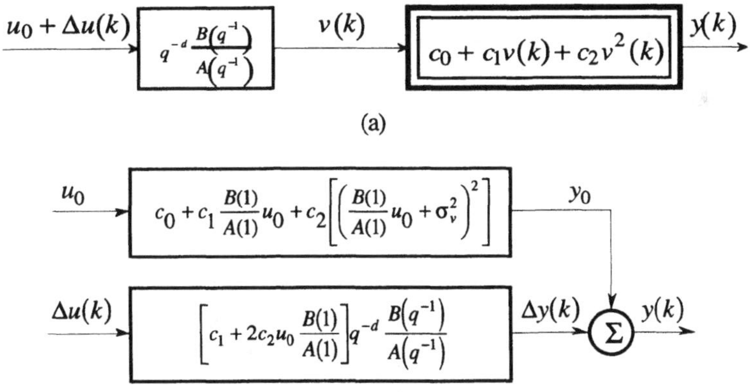

$u_0 + \Delta u(k)$ → $q^{-d} \frac{B(q^{-1})}{A(q^{-1})}$ → $v(k)$ → $c_0 + c_1 v(k) + c_2 v^2(k)$ → $y(k)$

(a)

u_0 → $c_0 + c_1 \frac{B(1)}{A(1)} u_0 + c_2 \left[\left(\frac{B(1)}{A(1)} u_0 + \sigma_v^2 \right)^2 \right]$ → y_0

$\Delta u(k)$ → $\left[c_1 + 2c_2 u_0 \frac{B(1)}{A(1)} \right] q^{-d} \frac{B(q^{-1})}{A(q^{-1})}$ → $\Delta y(k)$ → Σ → $y(k)$

(b)

Fig. 3.10.2 Linearized model of the simple Wiener model for a small excitation

2. *Correlation analysis by means of a Gaussian white noise:*
   Several authors have shown that the continuous (and the discrete time) cross-correlation function (series) between the input and output signals of the simple Wiener model are proportional to the weighting function (series) of the linear dynamic part

$$r_{xy'}(\tau) = A_1 h(\tau), \quad r_{xy'}(\kappa) = A_1 h(\kappa)$$

where

$$x(k) = u(k) - \mathrm{E}\{u(k)\}, \quad y'(k) = y(k) - \mathrm{E}\{y(k)\}$$

(Korenberg, 1973b; Billings and Fakhouri, 1977, 1978a, 1979d, 1980, 1982; Schweizer, 1987).

The pulse transfer function can be estimated in the knowledge that the weighting function series is the response to a unit pulse:

- by graphoanalytical method, or
- by LS parameter estimation.

The resulting pulse transfer function has to be normalized for a static gain equal to one.

Algorithm 3.10.2 The non-iterative method starting with the estimation of the linear dynamic term is as follows:

1. Estimate the parameters of the linear dynamic model using correlation analysis (exciting the process by a Gaussian white noise or using LS method with small range excitation).
2. Compute the sequence of the inner variable from the input signal by means of the pulse transfer function of the linear dynamic term.
3. Estimate the nonlinear static part from the sequences of the computed inner variable and of the measured output signal by static regression.

Remark:
This method does not assume that the static nonlinearity is invertible.

Example 3.10.1 *Identification of the linear dynamic term by means of a small excitation*

The simple Wiener model has the system equations:

$$\begin{aligned} v(k) &= 2 + u(k) + 0.5u^2(k) \\ 10\dot{y}(t) + y(t) &= u(t) \end{aligned} \tag{3.10.1}$$

The process was excited by a PRTS with maximum length of 26 and minimum switching time 10 [s]. The working point was $u_0 = 1$ and the amplitude of the PRTS $\Delta u = \pm 0.1$. The input–output noise-free measurements were sampled by 2 [s]. The difference equation of the dynamic term obtained by SRE transformation is

$$v(k) = 0.8187v(k-1) + 0.1813u(k-1)$$

$N = 3 \cdot 26 \cdot 5 = 390$ data were recorded. Linear identification was performed between the data records from $2 \cdot 26 \cdot 5 + 1 = 262$ to $N = 390$. The estimated parameters are

$$\begin{aligned} \hat{y}(k) = {} & (0.8184 \pm 0.00049)y(k-1) + (0.3627 \pm 0.0015)\Delta u(k-1) \\ & +(0.2728 \pm 0.00059) \end{aligned}$$

or

$$\hat{y}(k) = \frac{0.3627q^{-1}}{1 - 0.8184q^{-1}}\Delta u(k) + 1.683$$

Both the computed static gain

$$\hat{c}_1 = 0.3627/(1-0.8184) = 2.179$$

and the estimated time constant

$$\hat{T}_1 = -\Delta T/\ln(-\hat{a}_1) = -2/\ln(0.8184) = 9.97 \text{ [s]}$$

approximate their true values $T = 10$ [s] and

$$c_1 = 1 + 2 \cdot 0.5 \cdot U|_{U=1} = 2$$

well. ■

3.10.3 Correlation method

In the last section it was shown how the linear dynamic term can be estimated by means of correlation analysis and how the static part can be estimated in the knowledge of the inner variable and the output signal. Now a method that uses only correlation functions is presented.

Lemma 3.10.1

(Krempl, 1973, 1974) Excite the simple Hammerstein model by a PRTS signal with zero mean and with a half period $N_p/2$ greater than the settling time of the process. Then the following are true:

$$r_{uv}(\kappa) = C_1 h(\kappa) \qquad 0 \le \kappa \le N_p/2$$

$$r_{uy}(\kappa) = \sum_{i=1}^{M'} c_{2i-1} r_{uv^{2i-1}}(\kappa) \qquad M' \le (M+1)/2 \tag{3.10.2}$$

$$r_{xy}(\kappa) = \sum_{i=1}^{M'} c_{2i} r_{xv^{2i}}(\kappa) \qquad M' \le M/2 \tag{3.10.3}$$

with

$$x(k) = u^2(k) - \mathrm{E}\{u^2(k)\}$$

C_1 is a constant and depends on the test signal used.

Proof. The correlation equations are obtained by correlating the equation of the static polynomial

$$y(k) = c_0 + c_1 v(k) + c_2 v^2(k) + c_3 v^3(k) + \ldots + c_M v^M(k)$$

by $u(k)$ and $x(k)$, respectively. The correlation with the term c_0 is zero, because both multipliers have zero mean. ■

Algorithm 3.10.3 (Krempl, 1973, 1974) The simple Wiener model can be identified by means of correlation analysis in the following way:

1. Excite the process by $M' = \text{entier}(M/2)$ similar PRTS signals with zero mean and with different amplitudes. The half period of the sequences has to be greater than the settling time of the process.
2. Determine $r_{uv}(\kappa)$ from the measured cross-correlation functions by Gardiner's method (Gardiner, 1973b).
3. Estimate the linear dynamic term from the cross-correlation function $r_{uv}(\kappa)$ because they are proportional to each other as shown in Lemma 3.10.1. Choose the static gain of the linear dynamic term unit.
4. Estimate the odd-degree coefficients c_{2i-1}, $i = 1, \ldots, M'$ of the static nonlinear polynomial from the relation between them and $r_{uy}(\kappa)$ given in Lemma 3.10.1:

$$r_{uy}(\kappa) = \sum_{i=1}^{M'} c_{2i-1} r_{uv^{2i-1}}(\kappa) \qquad M' \leq (M+1)/2$$

   To estimate M' coefficients the test with M' different amplitudes has to be repeated. The cross-correlation functions $r_{uv^{2i-1}}(\kappa)$ can be calculated in the knowledge of the pulse transfer function of the linear dynamic part. It is enough to calculate them for a single shifting time.
5. Estimate the even degree coefficients c_{2i}, $i = 1, \ldots, M'$ of the static nonlinear polynomial from the relation between them and $r_{uy}(\kappa)$ given in Lemma 3.10.1:

$$r_{xy}(\kappa) = \sum_{i=1}^{M'} c_{2i} r_{xv^{2i}}(\kappa) \qquad M' \leq M/2$$

   To estimate M' coefficients the test with M' different amplitudes has to be repeated. The cross-correlation functions $r_{xv^{2i}}(\kappa)$ can be calculated in the knowledge of the pulse transfer function of the linear dynamic part. It is enough to calculate them for a single shifting time. ■

Hu and Wang (1991) applied PRTS for the parameter estimation of the weighting function of the linear dynamic term, of the coefficients of the odd and even degree terms of the nonlinear static polynomial.

3.10.4 Estimation by the Volterra series and by the parametric Volterra model

Since the simple Wiener model is not linear in the parameters, its parameters cannot be estimated by a parametric model linear-in-parameters. If we would like to use the advantages of a model linear-in-parameters then a nonparametric one has to be fitted to the measurements. This is the Volterra series. The disadvantage is that too many parameters have to be estimated to obtain a good fit. The parametric Volterra model is a parametric approximation of the Volterra series model. Its advantage over the nonparametric model is that a good input–output fit can be achieved by few parameters.

Example 3.10.2 *Identification of a first-order quadratic simple Wiener model by parametric Volterra models of different order*

The process and the test signal are the same as in Example 3.10.1. First- and second-order quadratic parametric Volterra models were fitted to the input–output data:

- *first-order quadratic parametric Volterra model* (V1):

$$\hat{y}(k) = (0.8208 \pm 0.009805)\hat{y}(k-1) + (0.2647 \pm 0.04216)$$
$$+(0.2549 \pm 0.02246)u(k-1) + (0.05268 \pm 0.009614)u^2(k-1)$$

- *second-order quadratic parametric Volterra model* (V2):

$$\hat{y}(k) = (1.458 \pm 0.02812)\hat{y}(k-1) - (0.5242 \pm 0.02325)\hat{y}(k-2)$$
$$+(0.1281 \pm 0.01553) + (0.2228 \pm 0.01121)u(k-1)$$
$$-(0.1556 \pm 0.01308)u(k-2) + (0.211 \pm 0.005024)u^2(k-1)$$
$$-(0.07344 \pm 0.005405)u^2(k-2) + (0.08429 \pm 0.03487)u(k-1)u(k-2)$$

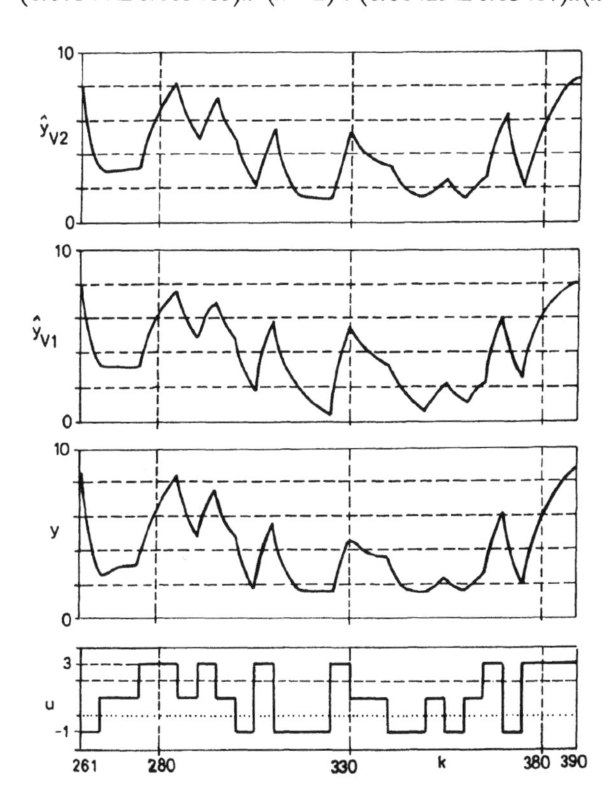

Fig. 3.10.3 Measured input (u), output (y) signals of the simple Wiener model and computed output signals $(\hat{y}_{V1}, \hat{y}_{V2})$ of the estimated quadratic first- and second-order parametric Volterra model

The input and output signals of the Wiener model and the computed output signal of the estimated model are drawn in Figure 3.10.3. The fit is much better in the second-order case. This fact can also be seen from the standard deviations of the residuals, which were $\sigma_\varepsilon = 0.2018$ in the first-order and $\sigma_\varepsilon = 0.06224$ in the second-order case. ■

3.10.5 Estimation by means of a process model linear-in-parameters

Even though the simple Wiener model is not linear in the parameters, a model linear-in-parameters can also be derived. This will be shown now for a noise-free, quadratic, first-order process.

1. Output–affine difference equation model:

The equation of the linear part of the model is

$$v(k) = b_0 u(k-1) - a_1 v(k-1)$$

and that of the nonlinear static part is

$$y(k) = c_0 + c_1 v(k) + c_2 v^2(k)$$

Substitute $v(k)$ from the dynamic equation into the nonlinear one

$$\begin{aligned} y(k) &= c_0 + c_1\left[b_0 u(k-1) - a_1 v(k-1)\right] + c_2\left[b_0 u(k-1) - a_1 v(k-1)\right]^2 \\ &= c_0 + c_1 b_0 u(k-1) - c_1 a_1 v(k-1) + c_2 b_0^2 u^2(k-1) \\ &\quad -2c_2 b_0 a_1 u(k-1) v(k-1) + c_2 a_1^2 v^2(k-1) \end{aligned}$$

Substitute $c_2 v^2(k)$ from the static nonlinear equation

$$\begin{aligned} y(k) &= c_0 + c_1 b_0 u(k-1) - c_1 a_1 v(k-1) + c_2 b_0^2 u^2(k-1) \\ &\quad -2c_2 b_0 a_1 u(k-1) v(k-1) + a_1^2 y(k-1) - a_1^2 c_0 - a_1^2 c_1 v(k-1) \\ &= c_0' + c_1 b_0 u(k-1) + c_2 b_0^2 u^2(k-1) + a_1^2 y(k-1) - a_1 v(k-1)\left[c_1' + 2c_2 b_0 u(k-1)\right] \end{aligned}$$

with

$$c_0' = c_0\left(1 - a_1^2\right), \qquad c_1' = c_1\left(1 + a_1\right)$$

Substitute $v(k-1)$ from the difference equation of the linear term

$$\begin{aligned} y(k) &= c_0' + c_1 b_0 u(k-1) + c_2 b_0^2 u^2(k-1) + a_1^2 y(k-1) \\ &\quad -a_1\left[b_0 u(k-2) - a_1 v(k-2)\right]\left[c_1' + 2c_2 b_0 u(k-1)\right] \end{aligned}$$

Substitute $a_1 v(k-2)$

$$a_1 v(k-2) = \frac{c_0' + c_1 b_0 u(k-2) + c_2 b_0^2 u^2(k-2) + a_1^2 y(k-2) - y(k-1)}{\left[c_1' + 2c_2 b_0 u(k-2)\right]}$$

into the nonlinear difference equation

$$y(k) = c_0' + c_1 b_0 u(k-1) + c_2 b_0^2 u^2(k-1) + a_1^2 y(k-1)$$
$$-a_1 \left[b_0 u(k-2) - \frac{c_0' + c_1 b_0 u(k-2) + c_2 b_0^2 u^2(k-2) + a_1^2 y(k-2) - y(k-1)}{\left[c_1' + 2c_2 b_0 u(k-2)\right]} \right]$$
$$\times \left[c_1' + 2c_2 b_0 u(k-1)\right]$$

A rearrangement leads to

$$c_1' y(k) + 2c_2 b_0 u(k-2) y(k)$$
$$= \left[c_0' + c_1 b_0 u(k-1) + c_2 b_0^2 u^2(k-1) + a_1^2 y(k-1)\right]\left[c_1' + 2c_2 b_0 u(k-2)\right]$$
$$-a_1 b_0 u(k-2)\left[c_1' + 2c_2 b_0 u(k-2)\right]\left[c_1' + 2c_2 b_0 u(k-1)\right]$$
$$+a_1\left[c_1' + 2c_2 b_0 u(k-1)\right]\left[c_0' + c_1 b_0 u(k-2) + c_2 b_0^2 u^2(k-2) + a_1^2 y(k-2) - y(k-1)\right]$$

or, further,

$$\left[c_1' + 2c_2 b_0 u(k-2)\right] y(k) - a_1\left[a_1 c_1' + 2a_1 c_2 b_0 u(k-2) + c_1' + 2c_2 b_0 u(k-1)\right] y(k-1)$$
$$-a_1^3\left[c_1' + 2c_2 b_0 u(k-1)\right] y(k-2) = \left[c_0' + c_1 b_0 u(k-1) + c_2 b_0^2 u^2(k-1)\right]$$
$$\times\left[c_1' + 2c_2 b_0 u(k-2)\right] - a_1 b_0 u(k-2)\left[c_1' + 2c_2 b_0 u(k-2)\right]\left[c_1' + 2c_2 b_0 u(k-1)\right]$$
$$+a_1\left[c_1' + 2c_2 b_0 u(k-1)\right]\left[c_0' + c_1 b_0 u(k-2) + c_2 b_0^2 u^2(k-2)\right]$$

which is an output–affine form linear-in-parameters.

2. Linear-in-parameters, recursive, polynomial model:

Express the inner variable $v(k)$ from the output signal

$$v(k) = \frac{-c_1}{2c_2} \pm \sqrt{\frac{c_1^2}{4c_2} - \frac{c_0 - y(k)}{c_2}} = -c_1' \pm \sqrt{c_1'^2 - y'(k)}$$

with

$$c_1' = \frac{c_1}{2c_2} \qquad y'(k) = \frac{c_0 - y(k)}{c_2}$$

Substitute $v(k)$ into the difference equation of the linear term:

$$-c_1' \pm \sqrt{c_1'^2 - y'(k)} = b_0 u(k-1) - a_1 c_1' \mp \sqrt{c_1'^2 - y'(k-1)}$$

where

$$y'(k-1) = \frac{c_0 - y(k-1)}{c_2}$$

Rearrange the equation

$$\pm\sqrt{c_1'^2 - y'(k)} \pm \sqrt{c_1'^2 - y'(k-1)} = b_0 u(k-1) + (1-a_1)c_1'$$

Raise both sides to the second power

$$\left[c_1'^2 - y'(k)\right] \pm \sqrt{\left[c_1'^2 - y'(k)\right]\left[c_1'^2 - y'(k-1)\right]} + \left[c_1'^2 - y'(k-1)\right]$$
$$= b_0^2 u^2(k-1) + 2b_0(1-a_1)c_1' u(k-1) + (1-a_1)^2 c_1'^2$$

Rearrange the equation again

$$\sqrt{\left[c_1'^2 - y'(k)\right]\left[c_1'^2 - y'(k-1)\right]} = b_0^2 u^2(k-1) + 2b_0(1-a_1)c_1' u(k-1)$$
$$+\left[(1-a_1)^2 - 2\right]c_1'^2 + y'(k) + y'(k-1)$$

Raise both sides to the second power again,

$$\left[c_1'^2 - y'(k)\right]\left[c_1'^2 - y'(k-1)\right] = \left[b_0^2 u^2(k-1) + 2b_0(1-a_1)c_1' u(k-1)\right.$$
$$\left. +\left[(1-a_1)^2 - 2\right]c_1'^2 + y'(k) + y'(k-1)\right]^2$$

The resulting difference equation is linear in the parameters.

Remark:
The forms linear-in-parameters have many more parameters than the original model and are redundant in the parameters.

3.10.6 Estimation of the inverse process model by a model linear-in-parameters
The inverse of the simple Wiener model is the simple Hammerstein model. All those techniques available for the identification of the simple Hammerstein model can be applied also for the simple Wiener model if the nonlinear static term can be approximated by an inverse characteristic.

Approximate, e.g., the inverse of the static characteristic by a quadratic polynomial:

$$v(k) = \left[c_{i0} + c_{i1} w(k) + c_{i2} w^2(k)\right]$$

where $w(k)$ is the noise-free output signal. The inverse model of the linear part is

$$u(k) = q^d \frac{A(q^{-1})}{B(q^{-1})} v(k)$$

and need not to be a realizable one if an off-line identification is performed. Assume $b_0 = 1$.

The inverse process model can be described by a generalized Hammerstein model

$$B(q^{-1})u(k) = q^d A(q^{-1})\left[c_{i0} + c_{i1}w(k) + c_{i2}w^2(k)\right]$$

which can be rewritten to a form linear-in-parameters

$$B(q^{-1})u(k) = c_{i0}^* + A_{i1}(q^{-1})w(k+d) + A_{i2}(q^{-1})w^2(k+d) \tag{3.10.4}$$

with

$$c_{i0}^* = c_{i0}A(1)$$
$$A_{i1}(q^{-1}) = c_{i1}A(q^{-1})$$
$$A_{i2}(q^{-1}) = c_{i2}A(q^{-1})$$

The input signal is linear in the parameters

$$u(k) = \boldsymbol{\phi}^{\mathrm{T}}(k)\boldsymbol{\theta} \tag{3.10.5}$$

where

$$\boldsymbol{\phi}(k) = \left[-u(k-1), \ldots, -u(k-nb), 1, w(k+d), \ldots, w(k+d-na), \\ w^2(k+d), \ldots, w^2(k+d-na)\right]^{\mathrm{T}} \tag{3.10.6}$$

$$\boldsymbol{\theta} = \left[b_1, \ldots, b_{nb}, c_{i0}^*, a_{10}, \ldots a_{1na}, a_{20}, \ldots, a_{2na}\right]^{\mathrm{T}} \tag{3.10.7}$$

The process can be estimated by the LS method if:

- there is no noise at the output $(y(k) = w(k))$; and
- the measurement noise in the input signal is a special autoregressive sequence

$$n(k) = \frac{e(k)}{B(q^{-1})}$$

Then

$$\boldsymbol{\phi}(k) = \left[-u(k-1), \ldots, -u(k-nb), 1, y(k+d), \ldots, y(k+d-na), \\ y^2(k+d), \ldots, y^2(k+d-na)\right]^{\mathrm{T}}$$

The application of the LS method to (3.10.5) means the assumption of the residual model

$$\varepsilon(k) = u(k) - \boldsymbol{\phi}^{\mathrm{T}}(k)\boldsymbol{\theta}$$

The case of the output noise leads to an iterative estimation. Consider, e.g., the case of an additive white noise at the output

$$y(k) = w(k) + e(k)$$

Substitution of $w(k)$ into (3.10.4) leads to

$$B(q^{-1})u(k) = c_{i0}^{*} + A_{i1}(q^{-1})y(k+d) - A_{i1}(q^{-1})e(k+d)$$
$$+A_{i2}(q^{-1})y^2(k+d) - 2A_{i2}(q^{-1})y(k+d)e(k+d) + A_{i2}(q^{-1})e^2(k+d)$$

which is a stochastic model linear-in-parameters and the residuals can be computed after prior parameter estimation, which means in an iterative way.

The method presented is a one-shot method if the noise is not filtered iteratively. The disadvantage of the method is that the inverse model is assumed to be a generalized Hammerstein one and not a simple Hammerstein one.

Example 3.10.3 *Identification of the simple Wiener model (Haber and Zierfuss, 1988)*

The parameters of the process and that of the excitation are the same as in Example 3.10.1 except that the amplitude of the excitation was $\Delta U = \pm 2$. The simulated noise-free input and output signals and the inner variable are seen in Figure 3.10.4.

The nonlinear inverse characteristic is approximated by a quadratic polynomial

$$v(k) = f^{-1}(y(k)) = c_{i0} + c_{i1}y(k) + c_{i2}y^2(k) \tag{3.10.8}$$

The first-order differential equation of the linear dynamic term

$$T_1\dot{v}(t) + v(t) = Ku(t)$$

can be expressed by means of the inverse characteristic of the nonlinear static term

$$T_1 \frac{\partial f^{-1}(y(t))}{\partial t} + f^{-1}(y(t)) = Ku(t) \tag{3.10.9}$$

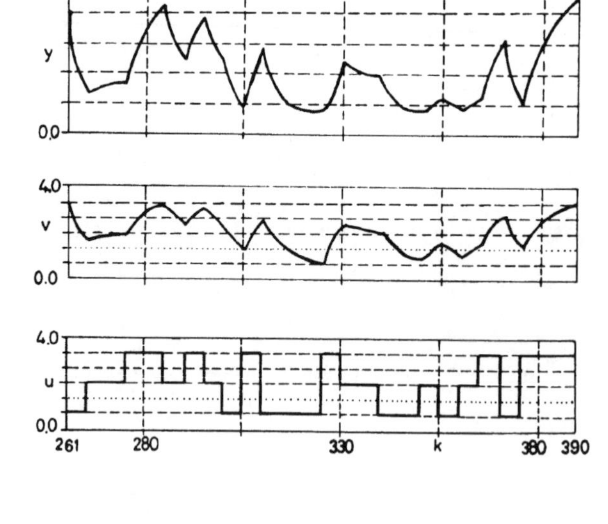

Fig. 3.10.4 Measured input (u) and output (y) signals and the inner variable (v) at the identification of the simple Wiener model by means of the inverse nonlinear characteristic

Using (3.10.8) we get

$$T_1\left[c_{i1} + 2c_{i2}y(k)\right]\dot{y}(t) + \left[c_{i0} + c_{i1}y(t) + c_{i2}y^2(t)\right] = Ku(t) \tag{3.10.10}$$

Discretize (3.10.10) by means of the Euler transformation

$$T_1\left[c_{i1} + 2c_{i2}y(k)\right]\frac{y(k+1) - y(k)}{\Delta T} + \left[c_{i0} + c_{i1}y(k) + c_{i2}y^2(k)\right] = Ku(k)$$

and rearrange the difference equation to $y(k)$

$$y(k) = -a_1^* y(k-1) + c_0^* + b_1^* u(k-1) + b_2^* y^2(k-1) + b_3^* y(k)y(k-1)$$

with

$$c_0^* = -c_0\Delta T/(T_1c_{i1}) \qquad b_1^* = K\Delta T/(T_1c_{i1}) \qquad a_1^* = \Delta T/T_1 - 1$$

$$b_2^* = c_{i2}(2 - \Delta T/T_1)/c_{i1} \qquad b_3^* = -2c_{i2}/c_{i1}$$

LS parameter estimation was used. The estimated parameters of the difference equation were

$$c_0^* = 0.3696 \pm 0.0345 \qquad b_1^* = 0.1977 \pm 0.0074 \qquad a_1^* = -0.8065 \pm 0.0173$$

$$b_2^* = -0.0848 \pm 0.0036 \qquad b_3^* = 0.0959 \pm 0.0033$$

which resulted in the following parameters of the differential equation

$$T_1 = 10.366, \quad c_{i0} = -1.87, \quad c_{i1} = 0.979, \quad c_{i2} = -0.0469$$

Four parameters had to be computed from five estimated ones, because there were two equations for calculating c_{i2}. Both resulted in almost the same value. Figure 3.10.5 shows the characteristic of the nonlinear term of the simulated simple Wiener model (3.10.1) and the inverse of the estimated inverse characteristic:

$$V = -1.87 + 0.979Y - 0.0469Y^2, \quad Y = 10.437 - \sqrt{69.061 - 21.322V}$$

The coincidence is acceptable in the excited region $-1 \le V \le 3$. ■

An application of the method is the identification of a pH process in Pajunen (1985a, 1985b).

3.10.7 Iterative estimation of the inverse process model

In the last section a generalized Hammerstein model was fitted to the inverse process model. Now it will be shown how a simple Hammerstein inverse model can be fitted to the data by an iterative algorithm. The steps of the procedure are similar to those applied to the identification of the simple Hammerstein model.

Algorithm 3.10.4 Haber and Zierfuss (1988) used the following iterative procedure for the identification of the inverse process model:

1. Fit a linear dynamic model with constant term between the measured output signal (as input) and the measured input signal (as output).
2. Choose the pulse transfer function of the linear dynamic term of the simple Wiener model the estimated one normalized so that the static gain becomes equal to one.
3. Compute the inner variable from the input signal by the inverse pulse transfer function of the estimated linear dynamic model.
4. Estimate the inverse model of the nonlinear static part from the measured output signal (as input) and from the computed inner variable (as output) by static regression.
5. Compute the inner variable from the measured output signal by means of the estimated parameters of the inverse nonlinear static model.
6. Estimate the pulse transfer function of the linear term from the measured input signal and from the computed inner variable.
7. Check whether the iteration has to be stopped. If not go to Step 2.
8. Estimate the parameters of the static term from the inner variable (as input) and from the measured output signal (as output) by regression. ■

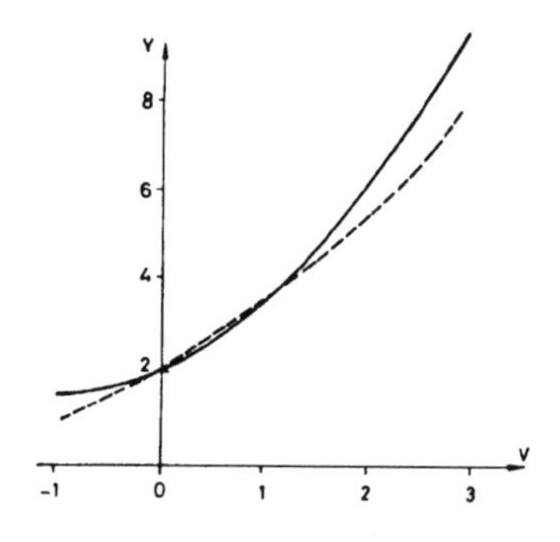

Fig. 3.10.5 True (–) and estimated (- - -) nonlinear characteristics of the simple Wiener model

Example 3.10.4 *Identification of the simple Wiener model (Haber and Zierfuss, 1988)*

The parameters of the process and that of the excitation are the same as in Example 3.10.1 except that the amplitude of the excitation was ±2. The simulated noise-free input and output signals are seen in Figure 3.10.4.

The inverse characteristic of the nonlinear term is approximated by a cubic polynomial. After two iteration steps the estimated parameters are in the close vicinity of the true ones:

$$b_1 = 0.1875 \pm 0.001 \qquad a_1 = -0.8125 \pm 0.002881$$

$$c_0 = 2.014 \pm 0.005 \qquad c_1 = 0.995 \pm 0.007 \qquad c_2 = 0.489 \pm 0.003$$

The estimated cubic polynomial of the inverse static characteristic is (Figure 3.10.5)

$$\hat{v}(k) = (-2.883 \pm 0.0821) + (1.757 \pm 0.0637)y(k) + (-0.2278 \pm 0.0144)y^2(k) + (0.0117 \pm 0.001)y^3(k)$$

Figure 3.10.6 shows the computed inner variable, and the computed input and output signals that were computed from the inner variable by means of the estimated models of the linear dynamic and static nonlinear parts, respectively.

3.10.8 Using a process model nonlinear-in-parameters

The noise-free output signal is described by the equations

$$v(k) = \frac{B(q^{-1})}{A(q^{-1})} u(k-d)$$

$$w(k) = c_0 + c_1 v(k) + c_2 v^2(k)$$

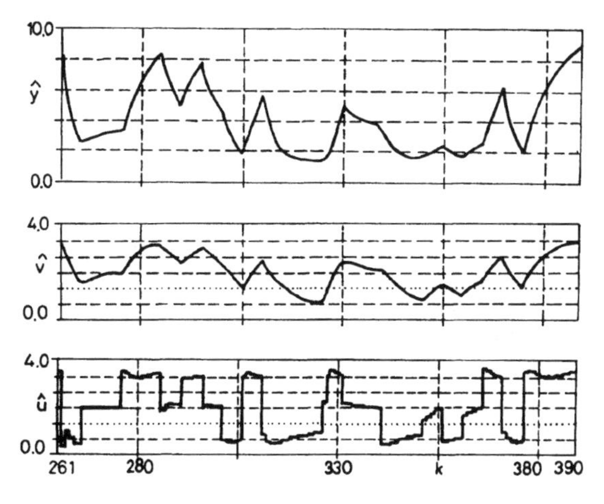

Fig. 3.10.6 Computed inner variable $(\hat{v})$, input $(\hat{u})$ and output $(\hat{y})$ of the simple Wiener model at the iterative estimation

An additive colored noise

$$n(k) = \frac{E(q^{-1})}{D(q^{-1})} e(k)$$

is assumed at the output

$$y(k) = w(k) + n(k)$$

The residual can be expressed by the form

$$\varepsilon(k) = \frac{D(q^{-1})}{E(q^{-1})} [y(k) - w(k)]$$

The prediction error method can be apr
residual by the unknown parameters. Tl

$$\frac{\partial \varepsilon(k)}{\partial c_0} = -\frac{D(1)}{E(1)}$$

$$\frac{\partial \varepsilon(k)}{\partial c_1} = -\frac{D(q^{-1})}{E(q^{-1})} v(k)$$

$$\frac{\partial \varepsilon(k)}{\partial c_2} = -\frac{D(q^{-1})}{E(q^{-1})} v^2(k)$$

$$\frac{\partial \varepsilon(k)}{\partial b_i} = -\frac{D(q^{-1})}{E(q^{-1})}\left[c_1 + 2c_2 v(k)\right]\frac{1}{A(q^{-1})}u(k-d-i)$$

$$\frac{\partial \varepsilon(k)}{\partial a_i} = \frac{D(q^{-1})}{E(q^{-1})}\left[c_1 + 2c_2 v(k)\right]\frac{1}{A(q^{-1})}v(k-i)$$

$$\frac{\partial \varepsilon(k)}{\partial d_i} = \frac{1}{E(q^{-1})}\left[y(k-i) - w(k-i)\right]$$

$$\frac{\partial \varepsilon(k)}{\partial e_i} = -\frac{1}{E(q^{-1})}\varepsilon(k-i)$$

Remarks:
1. As is known, the prediction-error method is an iterative procedure.
2. The above equations were used with the on-line identification by Kortmann and Unbehauen (1986).

3.10.9 Iterative parameter estimation technique using the inverse of the nonlinear static term

The iterative techniques (e.g., Algorithm 3.10.4) rely on the relaxation technique which means that the parameter groups are estimated in turn iteratively.

Algorithm 3.10.5 Hunter and Korenberg (1986) suggested the following iterative procedure:
1. Assume the inner variable is equal to the measured output signal.
2. Estimate the linear model from the measured input signal and from the inner variable.
3. Compute the inner variable from the measured input signal by means of the estimated pulse transfer function of the linear model.
4. Estimate the parameters of the static term by regression from the inner variable and from the measured output signal.
5. Compute the inner variable from the input signal by the inverse of the estimated nonlinear static model.
6. Check whether the iteration has to be stopped. If not go to Step 2.

Algorithm 3.10.6 Haber and Zierfuss (1988) suggested the following iterative procedure:
1. Assume the inner variable is equal to the measured input signal.
2. Estimate the parameters of the inverse of the nonlinear static term by regression from the output signal and from the inner variable.
3. Compute the inner variable from the output signal by the estimated inverse nonlinear static model.
4. Estimate the pulse transfer function of the linear dynamic model from the measured input signal and from the inner variable.

5. Compute the inner variable from the measured input signal by means of the estimated pulse transfer function of the linear dynamic term.
6. Check whether the iteration has to be stopped. If not go to Step 2.
7. Estimate the parameters of the static term by regression from the inner variable and from the measured output signal.

Remarks:
The above two methods differ in two points:
1. The iteration is started with the estimation of the parameters of the:
   - inverse of the nonlinear static term (Haber and Zierfuss, 1988);
   - linear dynamic term (Hunter and Korenberg, 1986).
2. Hunter and Korenberg (1986) estimated the parameters of the nonlinear term in every step and computed the inverse model based on this. Haber and Zierfuss (1988) estimated the inverse model in each step directly.

The submodels whose parameters have to be estimated are linear in the parameters, at least if proper noise models are assumed. This was already shown with the identification of the simple Hammerstein model in Section 3.9. The inverse of the nonlinear static term can also be assumed to be described by a polynomial, as shown now for a quadratic approximation of the inverse characteristic:

$$\hat{v}(k) = c_{i0} + c_{i1}y(k) + c_{i2}y^2(k)$$

$$\hat{v}(k) = \boldsymbol{\varphi}_{is}^{T}(k)\boldsymbol{\theta}_{is}$$

$$\boldsymbol{\varphi}_{is}(k) = \left[1, y(k), y^2(k)\right]^{T}$$

$$\boldsymbol{\theta}_{is} = \left[c_{i0}, c_{i1}, c_{i2}\right]^{T}$$

To obtain unambiguous results, the parameters of the linear dynamic term have to be normalized, as shown in the introduction of this chapter.

An iterative method (Vörös, 1985) which can but does not need to use the inverse of the linear dynamic term is presented in the next section.

Example 3.10.5 *Identification of the simple Wiener model (Haber and Zierfuss, 1988)*

The parameters of the process and that of the excitation are the same as in Example 3.10.1 except that the amplitude of the excitation was ± 2. The simulated noise-free input and output signals were seen in Figure 3.10.3.

The estimated parameters during the iteration are seen in Table 3.10.1. The inner variable was computed from the output signal by means of the inverse of the estimated nonlinear term

$$\hat{v}(k) = \frac{c_1}{2c_2} + \sqrt{\frac{c_1^2}{4c_2^2} - \frac{c_0 - y(k)}{c_2}}$$

If the term under the square was negative the constant term was increased so far that it became positive. A correction was needed only in the first two iteration steps. The corrected constant terms are also given in Table 3.10.1. The parameters converged to the true values in 5 steps .

TABLE 3.10.1 Iterative estimation of the simple Wiener model using the inverse nonlinear term

Iter.	$\hat{c}_0$	$\hat{c}_0$ corrected	$\hat{c}_1$	$\hat{c}_2$	$\hat{b}_1$	$\hat{a}_1$
1	3.315	2.515	1.207	0.318	0.1994	-0.8005
2	2.040	2.025	1.084	0.567	0.1712	-0.8160
3	2.006	2.006	1.000	0.498	0.1820	-0.8180
4	2.002	2.002	1.000	0.500	0.1815	-0.8185
5	2.001	2.001	1.000	0.500	0.1814	-0.8186

3.10.10 Iterative parameter estimation technique without using the inverse of the nonlinear static term

Vörös (1985, 1995) proposed a procedure by which all process parameters can be estimated simultaneously but iteratively, and there is no need to compute the inner variable from the output signal by means of the inverse nonlinear characteristic. The input–output model is linear in the parameters and includes all parameters of the process. Consider first the nonlinear static model

$$w(k) = c_0 + c_1 v(k) + c_2 v^2(k)$$

and substitute $v(k)$ by the output of the linear dynamic term

$$v(k) = B\left(q^{-1}\right)u(k-d) - \tilde{A}\left(q^{-1}\right)v(k)$$

only in the linear channel

$$\begin{aligned} w(k) &= c_0 + c_1 B\left(q^{-1}\right)u(k-d) - c_1\tilde{A}\left(q^{-1}\right)v(k) + c_2 v^2(k) \\ &= c_0 + B^*\left(q^{-1}\right)u(k-d) + c_2 v^2(k) - \sum_{j=1}^{na} a_1^* v(k-j) \end{aligned} \quad (3.10.11)$$

with

$$B^*\left(q^{-1}\right) = c_1 B\left(q^{-1}\right), \quad A^*\left(q^{-1}\right) = c_1 A\left(q^{-1}\right)$$

The noise-free process can be written into a form linear-in-parameters:

$$w(k) = \boldsymbol{\phi}_*^{\mathrm{T}}(k)\boldsymbol{\theta}_* \quad (3.10.12)$$

$$\boldsymbol{\phi}_*(k) = \left[1, u(k-d), \ldots, u(k-d-nb), v^2(k), -v(k-1), \ldots, -v(k-na)\right]^{\mathrm{T}} \quad (3.10.13)$$

$$\boldsymbol{\theta}_* = \left[c_0, b_1^*, \ldots, b_{nb}^*, c_2, a_1^*, \ldots, a_{na}^*\right]^{\mathrm{T}} \quad (3.10.14)$$

Algorithm 3.10.7 (Vörös, 1985) The steps of the iterative algorithm that do not compute the inner variable are as follows:

1. Assume the process is linear and $v(k-j)=w(k-j),\ \ j=1,\dots,na$.
2. Eliminate $v^2(k)$ in (3.10.13) and estimate all parameters except c_2.
3. Compute the inner variable $v(k)$ from the measured input signal by means of the estimated pulse transfer function of the linear term

$$\frac{q^{-d}B^*\left(q^{-1}\right)}{A^*\left(q^{-1}\right)}=\frac{q^{-d}B\left(q^{-1}\right)}{A\left(q^{-1}\right)}$$

   All parameters are in (3.10.14).
4. Estimate the parameters (3.10.14) of the nonlinear static and the linear dynamic terms according to the model of (3.10.12) by regression from the input, output and the computed inner variable.
5. Check whether the iteration has to be stopped. If not go to Step 3. ■

Remarks:

1. Vörös (1995) extended the algorithm for the case when the output signal is disturbed by color noise.
2. In Vörös (1985, 1995) simulation results are also given.

3.10.11 Gate function method

The working point of the nonlinear term is the inner variable. As the nonlinear term is a static one the working point can be transformed to the output signal. Therefore the process can be described by the sum of multi-models belonging to different domains of the output signal.

Algorithm 3.10.8 The identification algorithm using the gate functions consists of the following steps:

1. Excite the system by a test signal that covers the whole range of the output signal.
2. Divide the range of the output signal into small intervals.
3. Perform the following for all intervals:
   - Cut those input and output signals from the whole records where the output signal is in the actual interval;
   - Estimate a linear pulse transfer function;
   - Calculate the static gain.
4. In the case of a simple Wiener model the pulse transfer functions normalized to unit static gains are equal to each other. This is the pulse transfer function of the simple Wiener model.
5. The static nonlinear term is given by the static gains as a function of the mean values of the domains of the output signal. An analytic function can be computed by static regression. ■

Remarks:

1. If the chosen domains of the output signal are not small enough then it is practical to fit simple Hammerstein models to the measured values as inverse models. That

means the measured output signal is the input and the measured input signal is the output of the fitted model. The parameter estimation is easy if generalized Hammerstein models are fitted because then the models are linear in the parameters. This method was used by Pajunen (1985a, 1985b).

2. Application of the method is the identification of a pH process in Pajunen (1985a, 1985b).

3.10.12 Identification by graphical plotting

The method is based on the fact that the plot of the output function of a system as a function of the input signal $y(u(t))$ is single valued if the process is static with a single valued characteristic. To get an unambiguous result assume that the static gain of the linear dynamic part is unit.

Algorithm 3.10.9 Lammers *et al.* (1979) recommended the following procedure for the identification of the simple Wiener model (Figure 3.10.7):

1. Connect the measured input signal to the x-position of an x–y plotter or oscilloscope.
2. Connect the measured output signal through a linear dynamic filter $\hat{L}$ to the y-position of an x–y plotter or oscilloscope.
3. Tune the filter $\hat{L}$ till the plot is single valued one. Then the plot is the characteristic curve of the static nonlinearity (NS) and $\hat{L}$ is the estimate of the inverse of the linear dynamic term (LD).

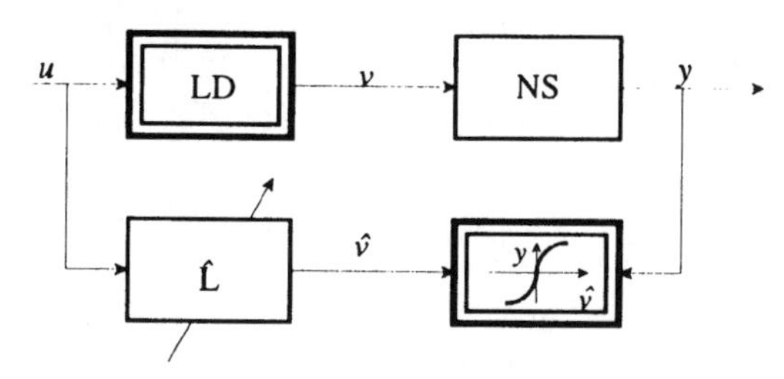

Fig. 3.10.7 Identification of the simple Wiener model by graphical plotting

3.10.13 Frequency method

The identification procedure uses the result of the following lemma.

Lemma 3.10.2

Excite the continuous time simple Wiener model by a sinusoidal test signal

$$u(t) = U\cos\omega t$$

The highest order harmonic is

$$y^{(M)}(t) = \frac{U^M}{2^{M-1}} c_M |G(\omega)|^M \cos(M\omega t + M\varphi(\omega)) \tag{3.10.15}$$

where M is the highest degree of the nonlinear power and the frequency function of the linear dynamic term is

$$G(j\omega) = |G(\omega)|\exp(j\varphi(\omega))$$

Proof. The inner variable has the same frequency as the input signal. Its time function is

$$v(t) = U|G(\omega)|\cos(\omega t + \varphi(\omega)) \tag{3.10.16}$$

(3.10.16) is the input to the static nonlinearity. The highest order harmonics will be generated by the nonlinear term $c_M v^M(t)$. Using the result of Lemma 3.9.2, the highest order harmonic has the form of (3.10.15). ■

Algorithm 3.10.10 The identification of the simple Wiener model by the frequency method consists of the following procedure:

1. Excite the process by $U\cos\omega t$ at several frequencies.
2. Determine the highest order harmonics $y^{(M)}$ from the responses by filtering out the lower order harmonics.
3. Compute the frequency characteristic of $\left(2^{M-1}/U^M\right)y^{(M)}(t)$.
4. Transform the amplitude values by extracting an Mth order root from them.
5. Transform the phase values by dividing them by M.
6. Identify a linear dynamic system from the frequency characteristics transformed.
7. Assume the linear dynamic term has a unit static gain. Then the estimated static gain is c_M.
8. In the knowledge of the linear dynamic term the static nonlinear term can be identified from the inner variable and from the output signal by static regression. ■

If the amplitude characteristic holds enough information to identify the transfer function, i.e., the process is of minimum phase and has no delay time, then a simplified procedure can be performed (Gardiner, 1973a).

Algorithm 3.10.11 (Gardiner, 1973a) The identification of the minimum phase and delay-time free simple Wiener model by the simplified frequency method consists of the following procedure:

1-3. As in Algorithm 3.10.10.

4. Determine the poles and zeros with their multiplicities from the breakpoints of the angular frequencies from the Bode amplitude plot.

5. Divide the observed multiplicities by M.

6-7. As Items 7–8 in Algorithm 3.10.10. ■

Example 3.10.6 *Identification of a first-order, quadratic simple Wiener model by frequency method*

The process is as in Example 3.10.1. The Bode amplitude and phase plots of the second harmonics $\left(2/U^2\right)y^{(2)}(t)$ are given in Figures 3.10.8a and 3.10.8b by continuous lines, respectively. The angular frequencies on the plot are the frequencies of the exciting sinusoidal signal. The value of the coefficient $\hat{c}_2$ at zero frequency is $\hat{c}_2 = 0.5$ from $20 \lg \hat{c}_2 = -6.02$. Divide the values of the amplitude plot by $\hat{c}_2$ and extract a square root from the amplitude values. The resulting amplitude plot of the linear dynamic term is also drawn in Figure 3.10.8a by dashed line. The phase characteristic of the linear dynamic term is obtained by dividing the measured phase values by 2 (Figure 3.10.8b dashed line). As seen, the Bode plots have the form of a linear first-order system; the breakpoints of both characteristics are at $1/\hat{T} = 0.1$. The estimated transfer function of the linear dynamic term is

$$\hat{G}(s) = \frac{1}{1+10s}$$

The nonlinear static part can be estimated from the stationary measurements as the steady state input–output characteristics. ■

A further simulation example is presented with the structure identification of the simple Wiener model by the frequency method in Section 5.5.3.

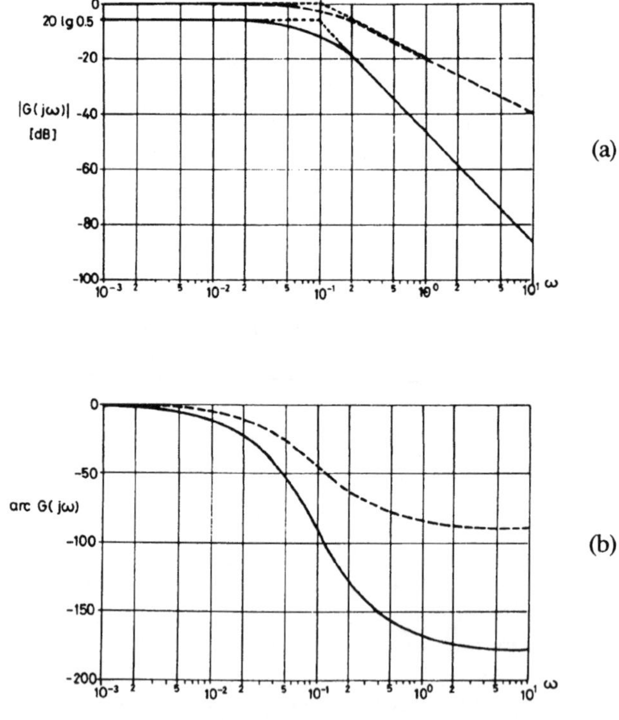

Fig. 3.10.8 Frequency characteristics of the second harmonic (–) and of the linear dynamic (- - -) term of the simple Wiener model: (a) amplitude plot; (b) phase plot

3.11 METHODS APPLIED TO THE SIMPLE WIENER–HAMMERSTEIN MODEL

Figure 3.11.1a shows the simple Wiener–Hammerstein model consisting of a nonlinear static part preceded and followed by linear dynamic terms. For simplicity a quadratic characteristic will be assumed. The continuous time model is seen in Figure 3.11.1b and the discrete time model in Figure 3.11.1c. The discrete delay times of the linear dynamic terms relative to the sampling time are d_1 and d_2, respectively. The linear channel of the model is

$$y_1(k) = q^{-d_1} \frac{B_1(q^{-1})}{A_1(q^{-1})} c_1 q^{-d_2} \frac{B_2(q^{-1})}{A_2(q^{-1})} u(k)$$

This shows that one parameter of both linear dynamic parts can be chosen freely. It is usual to choose the static gains equal to one

$$K = \frac{\sum_{i=0}^{nb_1} b_{1i}}{1+\sum_{j=1}^{na_1} a_{1j}} = \frac{\sum_{i=0}^{nb_2} b_{2i}}{1+\sum_{j=1}^{na_2} a_{2j}} = 1$$

The next lemma shows how the coefficients of the static term have to be transformed if the static gains of the linear dynamic terms is fixed.

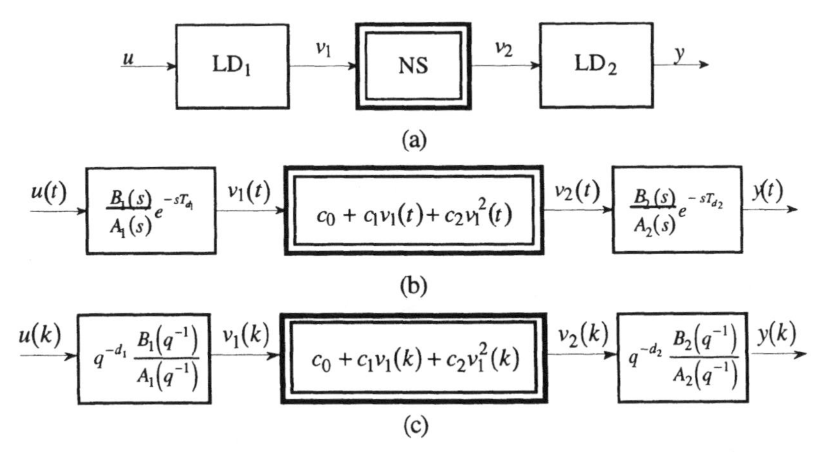

Fig. 3.11.1 Scheme of the simple Wiener–Hammerstein model: (a) general; (b) continuous time; (c) discrete time

Lemma 3.11.1

Assume that the simple Wiener–Hammerstein model with the static gains of the linear terms K_1 and K_2 before and behind the nonlinear term, respectively, and with the nonlinear term

$$V_2 = c_0 + c_1 V_1 + c_2 V_1^2$$

is given. If the static gains of the linear dynamic terms are changed to K_1^* and K_2^* then the coefficients of the nonlinear static term should be changed to

$$c_0^* = \frac{K_2 c_0}{K_2^*} \qquad c_1^* = \frac{K_2 c_1 K_1}{K_2^* K_1^*} \qquad c_2^* = \frac{K_2 c_2 K_1^2}{K_2^* K_1^{*2}}$$

to keep the steady state characteristic unaltered.
Proof. The steady state characteristic is in the first case

$$Y = K_2\left[c_0 + c_1 K_1 + c_2 (K_1 U)^2\right]$$

and in the second case

$$Y = K_2^*\left[c_0^* + c_1^* K_1^* + c_2^* (K_1^* U)^2\right]$$

The relation between the coefficients is obtained from equating the two equations. ■

3.11.1 Estimation with a process model linear-in-parameters

The simple Wiener–Hammerstein model can be divided into a linear dynamic part and a simple Hammerstein model, both being linear in the parameters. If the signal $v_1(k)$ is expressed by a function of the input signal in a form linear-in-parameters, then the whole model can be given by a form linear-in-parameters.

Use the nonparametric description of LD_1:

$$v_1(k) = b_{10} u(k - d_1) + \ldots + b_{1m_1} u(k - d_1 - m_1) = \sum_{i=0}^{m_1} b_{1i} u(k - d_1 - i) \tag{3.11.1}$$

and substitute it into the equation of the simple Hammerstein model given by

$$v_2(k) = c_0 + c_1 v_1(k) + c_2 v_1^2(k)$$

$$w(k) = b_{20} v_2(k - d_2) + \ldots + b_{2nb_2} v_2(k - d_2 - nb_2) - a_{21} w(k-1) - \ldots - a_{2n} w(k - na_2)$$

To have a form linear-in-parameters use the form of the generalized Hammerstein model

$$\begin{aligned} w(k) = c_0^* + b_{10}^* v_1(k - d_2) + \ldots + b_{1nb_2}^* v_1(k - d_2 - nb_2) + b_{22}^* v_1^2(k - d_2) + \ldots \\ + b_{2nb_2} v_1^2(k - d_2 - nb_2) - a_{21} w(k-1) - \ldots - a_{2na_2} w(k - na_2) \end{aligned} \tag{3.11.2}$$

with

$$c_0^* = c_0 \sum_{i=0}^{nb_2} b_{2i}$$

$$b_{1i}^* = c_1 b_{2i}, \quad i = 0, \ldots, nb_2$$

$$b_{2i}^* = c_1 b_{2i}, \quad i = 0, \ldots, nb_2$$

Put (3.11.1) into (3.11.2)

$$w(k) = c_0^* + b_{10}^* \sum_{i=0}^{m_1} b_{1i} u(k - d_1 - d_2 - i) + \ldots + b_{1nb_2}^* \sum_{i=0}^{m_1} b_{1i} u(k - d_1 - d_2 - i)$$
$$+ b_{20}^* \left[\sum_{i=0}^{m_1} b_{1i} u(k - d_1 - d_2 - i) \right]^2 + \ldots + b_{2nb_2}^* \left[\sum_{i=0}^{m_1} b_{1i} u(k - d_1 - d_2 - i) \right]^2 \quad (3.11.3)$$
$$- a_{21} w(k-1) - \ldots - a_{2na_2} w(k - na_2)$$

which is a form linear-in-parameters.

Remarks:
This form has many disadvantages, e.g.:
- there are too many parameters to be estimated;
- the part $NS - LD_2$ is modeled by a general Hammerstein model that may be redundant in the parameters of the corresponding simple Hammerstein model.

3.11.2 Iterative parameter estimation
Observe that the simple Wiener–Hammerstein model can be divided into partial models in two ways. Consequently two iterative parameter estimation algorithms can be imagined.

Algorithm 3.11.1 The simple Wiener–Hammerstein model can be separated into a linear dynamic model (LD_1) and a simple Hammerstein model ($NS - LD_2$). The iterative parameter estimation follows this partitioning:
1. Make an initial estimate for LD_1.
2. Compute $v_1(k)$ from the measured input signal by the estimated LD_1.
3. Estimate the parameters of NS and LD_2 from the inner variable $v_1(k)$ and from the measured output signal $y(k)$ by a method elaborated for the simple Hammerstein model.
4. Compute $v_1(k)$ from $y(k)$ by means the inverse of the simple Hammerstein model.
5. Estimate the parameters of LD_1 from the measured input signal and from the computed variable $v_1(k)$.
6. Check whether the iteration has to be stopped. If not go to Step 2. ■

Algorithm 3.11.2 The simple Wiener–Hammerstein model can be separated into a simple Wiener model ($LD_1 - NS$) and a linear dynamic model LD_2. The iterative parameter estimation follows this partitioning:
1. Make an initial estimate for LD_2.
2. Compute $v_2(k)$ from the measured output signal by the inverse of the estimated LD_2.
3. Estimate the parameters of LD_1 and NS from the measured input signal $u(k)$ and from the computed inner variable $v_2(k)$ by a method elaborated for the simple Wiener model.

4. Compute $v_2(k)$ from the measured input signal $u(k)$ with the estimated simple Wiener model.
5. Estimate the parameters of LD_2 from the computed inner variable and from the measured output signal $y(k)$.
6. Check whether the iteration has to be stopped. If not go to Step 2. ■

Remarks:
1. As is seen, both iterative algorithms have a secondary step that is also iterative itself. There are either no theoretical considerations or simulations available that would show the convergence of the methods.
2. Korenberg and Hunter (1986) present an algorithm similar to the first one. The main difference is that they estimate the parameters of LD_1 not by means of the computed inner variable but by minimizing directly the output error.
3. A simpler algorithm that does not use a two-loop-iteration can be obtained if the simple Hammerstein model part is estimated by the generalized Hammerstein model that is linear in the parameters. This is presented in the next algorithm.

Algorithm 3.11.3 The simple Wiener–Hammerstein model can be separated into a linear dynamic model (LD_1) and a simple Hammerstein model ($NS - LD_2$). The first part of the simple Wiener–Hammerstein can be described by the generalized Hammerstein model. The following algorithm uses this fact:
1. Make an initial estimate for LD_1.
2. Compute $v_1(k)$ from the measured input signal by the estimated LD_1.
3. Estimate the parameters of NS and LD_2 from the inner variable $v_1(k)$ and from the measured output signal $y(k)$ by a method elaborated for the generalized Hammerstein model.
4. Compute the parameters of the simple Hammerstein model from the (perhaps redundant) parameters of the generalized Hammerstein model.
5. Compute $v_1(k)$ from $y(k)$ by means of the inverse of the simple Hammerstein model.
6. Estimate the parameters of LD_1 from the measured input signal and from the computed variable $v_1(k)$.
7. Check whether the iteration has to be stopped. If not go to Step 2. ■

3.11.3 Using a process model nonlinear-in-parameters

The noise-free output signal is described by the equations

$$v_1(k) = \frac{B_1(q^{-1})}{A_1(q^{-1})} u(k - d_1)$$

$$v_2(k) = c_0 + c_1 v_1(k) + c_2 v_1^2(k)$$

$$w(k) = \frac{B_2(q^{-1})}{A_2(q^{-1})} u(k - d_2)$$

An additive colored noise

$$n(k) = \frac{E(q^{-1})}{D(q^{-1})} e(k)$$

is assumed at the output

$$y(k) = w(k) + n(k)$$

The residual can be expressed by the form

$$\varepsilon(k) = \frac{D(q^{-1})}{E(q^{-1})} [y(k) - w(k)]$$

The prediction error method can be applied, which needs the first-order derivative of the residual according to the unknown parameters. These sensitivity functions are:

$$\frac{\partial\varepsilon(k)}{\partial c_0} = -\frac{D(1)}{E(1)} \frac{B_2(1)}{A_2(1)} \qquad \frac{\partial\varepsilon(k)}{\partial c_1} = -\frac{D(q^{-1})}{E(q^{-1})} \frac{B_2(q^{-1})}{A_2(q^{-1})} v_1(k - d_2)$$

$$\frac{\partial\varepsilon(k)}{\partial c_2} = -\frac{D(q^{-1})}{E(q^{-1})} \frac{B_2(q^{-1})}{A_2(q^{-1})} v_1^2(k - d_2)$$

$$\frac{\partial\varepsilon(k)}{\partial b_{2i}} = -\frac{D(q^{-1})}{E(q^{-1})} \frac{1}{A_2(q^{-1})} v_2(k - d_2 - i) \qquad \frac{\partial\varepsilon(k)}{\partial a_{2i}} = \frac{D(q^{-1})}{E(q^{-1})} \frac{1}{A_2(q^{-1})} w(k - i)$$

$$\frac{\partial\varepsilon(k)}{\partial b_{1i}} = -\frac{D(q^{-1})}{E(q^{-1})} \frac{B_2(q^{-1})}{A_2(q^{-1})} [c_1 + 2c_2 v_1(k - d_2)] \frac{1}{A_1(q^{-1})} u(k - d_2 - d_1 - i)$$

$$\frac{\partial\varepsilon(k)}{\partial a_{1i}} = \frac{D(q^{-1})}{E(q^{-1})} \frac{B_2(q^{-1})}{A_2(q^{-1})} [c_1 + 2c_2 v_1(k - d_2)] \frac{1}{A_1(q^{-1})} v_1(k - d_2 - i)$$

$$\frac{\partial\varepsilon(k)}{\partial d_i} = \frac{1}{E(q^{-1})} [y(k - i) - w(k - i)] \qquad \frac{\partial\varepsilon(k)}{\partial e_i} = -\frac{1}{E(q^{-1})} \varepsilon(k - i)$$

As is known, the prediction error method is an iterative procedure.

3.11.4 Correlation method

This method is based on the following two lemmas.

Lemma 3.11.2

(Billings and Fakhouri, 1978a, 1980) Excite the simple Wiener–Hammerstein model with a Gaussian white noise. Then the cross-correlation functions between the normalized variables

$$y'(k) = y(k) - \mathrm{E}\{y(k)\}$$
$$x_1(k) = u(k) - \mathrm{E}\{u(k)\}$$
$$x_2(k) = \left[u(k) - \mathrm{E}\{u(k)\}\right]^2$$

are

$$r_{x_1 y'}(\kappa) = C_1 \sum_{\kappa_1=0}^{\kappa} h_1(\kappa_1) h_{12}(\kappa - \kappa_1) = C_1 h_1(\kappa) * h_2(\kappa)$$

$$r_{x_2 y'}(\kappa) = C_2 \sum_{\kappa_1=0}^{\kappa} h_{11}^2(\kappa_1) h_{12}(\kappa - \kappa_1) = C_2 h_1^2(\kappa) * h_2(\kappa)$$

where $h_1(\kappa)$ and $h_2(\kappa)$ are the weighting function series inclusive the zero values belonging to the dead times of the linear dynamic parts (* means the convolution). The constants C_1 and C_2 are functions of the parameters of the test signal and the static nonlinear element. This result is independent of the shape of the static nonlinearity, i.e., a polynomial form is not an assumption.

Lemma 3.11.3

(Billings and Fakhouri, 1980) Excite the simple Wiener–Hammerstein model by the sum of two uncorrelated pseudo-random signals $u_1(k)$ and $u_2(k)$. Each of them should have zero mean value

$$\mathrm{E}\{u_1(k)\} = 0 \qquad \mathrm{E}\{u_2(k)\} = 0$$

and a delta function-like auto-correlation function. Then the cross-correlation functions between the variables

$$y_i'(k) = y_i(k) - \mathrm{E}\{y_i(k)\}, \qquad i = 1, 2$$
$$x_1(k) = u_1(k)$$
$$x_2(k) = u_1(k) u_2(k)$$

are

$$r_{x_1 y_1'}(\kappa) = C_1 \sum_{\kappa_1=0}^{\kappa} h_1(\kappa_1) h_2(\kappa - \kappa_1) = C_1 h_1(\kappa) * h_2(\kappa)$$

$$r_{x_2 y_2'}(\kappa) = C_2 \sum_{\kappa_1=0}^{\kappa} h_1^2(\kappa_1) h_2(\kappa - \kappa_1) = C_2 h_1^2(\kappa) * h_2(\kappa)$$

where $h_1(\kappa)$ and $h_2(\kappa)$ are the weighting function series inclusive the zero values belonging to the dead times of the linear dynamic parts (* means the convolution). The constants C_1 and C_2 are functions of the parameters of the test signal and the static nonlinear element. The output signals of the linear and quadratic subsystems can be obtained by repeated excitations using Gardiner's method. ■

The estimation of the linear dynamic terms is given in Theorem 3.11.1.

Theorem 3.11.1 (Billings and Fakhouri, 1978a) Define the following pulse transfer functions:

$$q^{-(d_1+d_2)}\frac{B_1^c(q^{-1})}{A_1^c(q^{-1})}=\mathcal{Z}\{C_1 h_{11}(\kappa)*h_{12}(\kappa)\} \tag{3.11.4}$$

$$q^{-(d_1+d_2)}\frac{B_2^c(q^{-1})}{A_2^c(q^{-1})}=\mathcal{Z}\{C_2 h_{11}^2(\kappa)*h_{12}(\kappa)\} \tag{3.11.5}$$

$$q^{-d_1}\frac{B_3^c(q^{-1})}{A_3^c(q^{-1})}=\mathcal{Z}\{C_3 h_{11}^2(\kappa)\}$$

where $\mathcal{Z}$ denotes the discrete time Laplace (z-) transform. Let us see that

$$\frac{B_1^c(q^{-1})}{A_1^c(q^{-1})}=C_1\frac{B_1(q^{-1})}{A_1(q^{-1})}\frac{B_2(q^{-1})}{A_2(q^{-1})} \tag{3.11.6}$$

and

$$\frac{B_2^c(q^{-1})}{A_2^c(q^{-1})}=C_2\frac{B_3(q^{-1})}{A_3(q^{-1})}\frac{B_2(q^{-1})}{A_2(q^{-1})} \tag{3.11.7}$$

Divide (3.11.6) by (3.11.7) and rearrange it

$$\frac{B_1^c(q^{-1})}{A_1^c(q^{-1})}\frac{B_3(q^{-1})}{A_3(q^{-1})}=\frac{C_1}{C_2}\frac{B_2^c(q^{-1})}{A_2^c(q^{-1})}\frac{B_1(q^{-1})}{A_1(q^{-1})} \tag{3.11.8}$$

which is equivalent to

$$B_1^c(q^{-1})B_3(q^{-1})=\frac{C_1}{C_2}B_2^c(q^{-1})B_1(q^{-1}) \tag{3.11.9}$$

$$A_1^c(q^{-1})A_3(q^{-1})=A_2^c(q^{-1})A_1(q^{-1}) \tag{3.11.10}$$

$B_1(q^{-1})$ and $B_3(q^{-1})$ can be calculated by equating the coefficients of the same degrees of q^{-1} in (3.11.9) and using the LS method if the equations are over determined. The same is valid for $A_1(q^{-1})$ and $A_3(q^{-1})$ in (3.11.10).

From (3.11.6) we obtain

$$B_1^c\left(q^{-1}\right) = C_1 B_1\left(q^{-1}\right) B_2\left(q^{-1}\right) \tag{3.11.11}$$

$$A_1^c\left(q^{-1}\right) = A_1\left(q^{-1}\right) A_2\left(q^{-1}\right) \tag{3.11.12}$$

and $B_2\left(q^{-1}\right)$ and $A_2\left(q^{-1}\right)$ can be estimated from the knowledge of all other polynomials in (3.11.11) and (3.11.12). ■

The identification procedure is summarized in Algorithm 3.11.4.

Algorithm 3.11.4 Excite the simple Wiener–Hammerstein model by a zero mean Gaussian white noise or by the sum of two zero mean pseudo-random signals with delta function like auto-correlation functions. The steps of the parameter estimation are as follows:

1. Estimate the cross-correlation functions

$$r_{x_1 y'}(\kappa) = C_1 h_{11}(\kappa) * h_{12}(\kappa) \qquad \text{and} \qquad r_{x_2 y'}(\kappa) = C_2 h_{11}^2(\kappa) * h_{12}(\kappa)$$

2. Estimate the pulse transfer functions $B_1^c\left(q^{-1}\right)/A_1^c\left(q^{-1}\right)$ and $B_2^c\left(q^{-1}\right)/A_2^c\left(q^{-1}\right)$ from (3.11.4) and (3.11.5) by the LS method, assuming a pulse excitation as an input signal and the measured cross-correlation functions as the output signals.
3. Estimate the coefficients $B_1\left(q^{-1}\right)$ [and $B_3\left(q^{-1}\right)$] from (3.11.9) by the LS method.
4. Estimate the coefficients $A_1\left(q^{-1}\right)$ [and $A_3\left(q^{-1}\right)$] from (3.11.10) by the LS method.
5. Estimate the coefficients $B_2\left(q^{-1}\right)$ from (3.11.11) by the LS method.
6. Estimate the coefficients $A_2\left(q^{-1}\right)$ from (3.11.12) by the LS method.
7. Normalize $B_1\left(q^{-1}\right)/A_1\left(q^{-1}\right)$ and $B_2\left(q^{-1}\right)/A_2\left(q^{-1}\right)$ so that the static gains become unit.
8. Calculate the inner variable $v_1(k)$ from the input signal by the normalized transfer function of LD_1

$$v_1(k) = \frac{A_1(1)}{B_1(1)} \frac{B_1\left(q^{-1}\right)}{A_1\left(q^{-1}\right)} u(k - d_1)$$

9. Calculate the inner variable $v_2(k)$ from the output signal by the inverse of the normalized transfer function of LD_2

$$v_2(k) = \frac{B_2(1)}{A_2(1)} \frac{A_2\left(q^{-1}\right)}{B_2\left(q^{-1}\right)} y(k + d_2)$$

10. Fit a static nonlinear function between the sequences $\{v_1(k)\}$ as input and $\{v_2(k)\}$ as output signals by static regression. ■

Remark:
The coefficients of the static nonlinear part can be estimated from the signals $v_2(k)$ and $y(k)$, which are the input and output signals of the simple Hammerstein model part

$$y(k) = q^{-d_2} \frac{B_2(q^{-1})}{A_2(q^{-1})} \left[c_0 + c_1 v_1(k) + c_2 v_1^2(k)\right]$$

A rearrangement leads to

$$y(k) = c_0 \frac{B_2(1)}{A_2(1)} + c_1 \frac{B_2(q^{-1})}{A_2(q^{-1})} v_1(k-d_2) + c_2 \frac{B_2(q^{-1})}{A_2(q^{-1})} v_1^2(k-d_2) \tag{3.11.13}$$

(3.11.13) is an equation linear in the unknown coefficients c_i, $i = 0, 1, 2$.

$$y(k) = \boldsymbol{\varphi}_s^T \boldsymbol{\theta}_s \tag{3.11.14}$$

$$\boldsymbol{\varphi}_s = \left[\frac{B_2(1)}{A_2(1)}, \frac{B_2(q^{-1})}{A_2(q^{-1})} v_1(k-d_2), \frac{B_2(q^{-1})}{A_2(q^{-1})} v_1^2(k-d_2)\right]^T$$

$$\boldsymbol{\theta}_s = \left[c_0, c_1, c_2\right]^T$$

This procedure is applied in the following algorithm.

Algorithm 3.11.5 (Billings and Fakhouri, 1978a) Excite the simple Wiener–Hammerstein model by a zero mean Gaussian white noise or by the sum of two zero mean pseudo-random signals with delta function like auto-correlation functions. The steps of the parameter estimation are as follows:
1–8. As in Algorithm 3.11.4.
9. Estimate the coefficients of the static nonlinear function from the model (3.11.13) or (3.11.14) linear-in-parameters. ■

Simulation results are given in Billings and Fakhouri (1980).

Remark:
Tuis (1975, 1976) also identified the simple Wiener–Hammerstein model from correlation functions. He, however, could measure one of the inner variables. Thus the problem can be reduced to the correlation methods applied for the identification of the simple Hammerstein or the simple Wiener model.

3.11.5 Identification by repeated pulse excitations
With linear systems the relation between a pulse input and the weighting function series is the same as the relation between the auto-correlation and cross-correlation functions in the case of a white noise input signal. This idea leads to a procedure that is based on pulse inputs. First two lemmas show the theoretical backgrounds.

Lemma 3.11.4
The simple Wiener–Hammerstein model consists of two linear dynamic parts with the weighting function series $h_1(\kappa)$ and $h_2(\kappa)$, respectively and of a static polynomial between them with

$$v_2(k) = \sum_{i=0}^{M} c_i v_1^i(k)$$

The response of the linear channel to a pulse input $u(k) = U[1(k) - 1(k-1)]$ is

$$y_1(k) = c_1 U h_1(\kappa) * h_2(\kappa) \tag{3.11.15}$$

Proof. The inner variable $v_1(k)$ has the form

$$v_1(k) = U h_1(k), \quad k = 0, 1, 2, \ldots$$

because of the pulse-like excitation. (3.11.15) follows from the convolution sum of the signal

$$v_2(k) = c_1 v_1(k)$$

and the weighting function series of LD_2. ■

Lemma 3.11.5
The simple Wiener–Hammerstein model consists of two linear dynamic parts with the weighting function series $h_1(\kappa)$ and $h_2(\kappa)$, respectively, and of a static polynomial between them, with

$$v_2(k) = \sum_{i=0}^{M} c_i v_1^i(k)$$

The response of the quadratic channel to a pulse input $u(k) = U[1(k) - 1(k-1)]$ is

$$y_1(k) = c_1 U^2 h_1^2(\kappa) * h_2(\kappa) \tag{3.11.16}$$

Proof. The inner variable $v_2(k)$ of the quadratic channel is

$$v_2(k) = c_2 U^2 h_1^2(k), \quad k = 0, 1, 2, \ldots$$

(3.11.16) follows from the convolution sum of the signal $v_2(k)$ and the weighting function series of LD_2. ■

The identification procedure is summarized in Algorithm 3.11.6.

Algorithm 3.11.6 Excite the simple Wiener–Hammerstein model by pulses with $(M+1)$ different amplitudes, where M is the highest degree of the steady state polynomial relation. The steps of the parameter estimation are as follows:

1. Separate the responses of the channels of different degree by means of Gardiner's

method (Gardiner, 1973b), see Chapter 5.

2. Consider the output signal of the linear channel as

$$y_1(k) = C_1 h_1(\kappa) * h_2(\kappa)$$

and the output of the quadratic channel as

$$y_2(k) = C_2 h_1^2(\kappa) * h_2(\kappa)$$

3–11. As Steps 2 to 10 in Algorithm 3.11.4. ∎

Remark:
The modification of Algorithm 3.11.4 presented as Algorithm 3.11.5 can be done also in Algorithm 3.11.6.

3.11.6 Identification by graphical plotting
The method is based on the fact that the plot of the output signal as a function of the input signal $y(u(k))$ is single valued if the process is static with a single valued characteristic. As the inner variables are not measurable, the experimenter has to try to reconstruct them from the measured input and output variables. To get an unambiguous result assume that the static gains of the linear dynamic parts are unit.

Algorithm 3.11.7 Lammers *et al.* (1979) recommended the following procedure for the identification of the simple Wiener–Hammerstein model (Figure 3.11.2a):

1. Connect the measured input signal through a linear dynamic filter $\hat{L}_1$ to the x-position of an x–y plotter or oscilloscope.
2. Connect the measured output signal through a linear dynamic filter $\hat{L}_2$ to the y-position of an x–y plotter or oscilloscope.
3. Tune both filters until the plot is a single valued plot. Then the plot is the characteristic curve of the nonlinear element, and $\hat{L}_1$ is the estimate of LD_1 and $\hat{L}_2$ is the estimate of the inverse of LD_2. ∎

As is seen, the parameters of two linear dynamic filters have to be tuned simultaneously. The method can be modified that the parameters of only one filter have to be tuned. To do it we have to know the transfer function of the product of the two linear dynamic parts LD_1 and LD_2 of the process. This can be estimated either

- by correlation method presented in Section 3.11.4, or
- by pulse excitation presented in Section 3.11.5, or
- by parameter estimation of a linearized model from input–output measurements for a small excitation.

The latter method is illustrated in the next example.

Example 3.11.1 *Identification of the linear dynamic term by means of a small excitation*

The simple Wiener–Hammerstein model has the system equations:

$$10\dot{v}_1(t) + v_1(t) = u(t)$$
$$v_2(k) = 2 + v_1(k) + 0.5v_1^2(k)$$
$$5\dot{y}(t) + y(t) = v_2(t)$$

A sampling and holding device is between the two dynamic terms. The process was excited by a PRTS with maximum length of 26 and minimum switching time 10 [s]. The working point was $u_0 = 1$ and the amplitude of the PRTS $\Delta u = \pm 0.1$. The noise-free measurements were sampled by 2 [s]. The pulse transfer function of the two connected linear dynamic terms with the sampling and holding device between them is

$$\frac{0.1813q^{-1}}{1-0.8187q^{-1}} \frac{0.3297q^{-1}}{1-0.6703q^{-1}} = \frac{0.0598q^{-2}}{1-1.489q^{-1}+0.5488q^{-2}}$$

$N = 3 \cdot 26 \cdot 5 = 390$ data were recorded. Linear identification was performed between the data records from $2 \cdot 26 \cdot 5 + 1 = 262$ to $N = 390$. The estimated parameters are

$$\hat{y}(k) = (1.45 \pm 0.04928)\hat{y}(k-1) - (0.5086 \pm 0.04781)\hat{y}(k-2)$$
$$+(0.09011 \pm 0.02486) + (0.1250 \pm 0.01061)u(k-2)$$

The poles of the pulse transfer function are $p_1 = -0.8555$ and $p_2 = -0.5945$. Both the computed static gain

$$\hat{c}_1 = 0.1250/(1-1.45+0.5086) = 2.133$$

and the time constants

$$\hat{T}_1 = -\Delta T / \ln(-p_1) = -2/\ln(0.8555) = 12.81 \text{ [s]}$$
$$\hat{T}_2 = -\Delta T / \ln(-p_2) = -2/\ln(0.5945) = 3.85 \text{ [s]}$$

approximate their true values. ■

Knowing the product of the transfer functions of the linear dynamic terms there are two ways in which we can proceed.

Algorithm 3.11.8 Lammers *et al.* (1979) recommended a procedure for the identification of the simple Wiener–Hammerstein model by which only the linear dynamic term at the output has to be tuned (Figure 3.11.2b):

1. Identify the transfer function of the serially connected two linear dynamic terms.
2. Filter the measured input signal by the estimated overall linear term $\hat{\mathrm{L}}_{12}$.
3. Connect the above filtered signal through an unknown linear dynamic filter $\hat{\mathrm{L}}$ to the x-position of the oscilloscope.
4. Connect the measured output signal through the same unknown linear dynamic filter $\hat{\mathrm{L}}$ to the y-position of the oscilloscope.
5. Tune the filter $\hat{\mathrm{L}}$ till the plot is a single valued one. Then the plot is the characteristic curve of the nonlinear element, and $\hat{\mathrm{L}}$ is the estimate of the inverse of LD_2. The transfer function of LD_1 is the product of the overall transfer function with the transfer function of $\hat{\mathrm{L}}$. ■

Algorithm 3.11.9 Lammers *et al.* (1979) recommended a procedure for the identification of the simple Wiener–Hammerstein model by which only a linear dynamic term at the input has to be tuned (Figure 3.11.2c):

1. Identify the transfer function of the serially connected two linear dynamic terms.
2. Filter the measured output signal by the inverse of the estimated overall transfer function $\hat{L}_{12}^{-1}$.
3. Connect the above filtered signal through an unknown linear dynamic filter $\hat{L}$ to the y-position of the oscilloscope.
4. Connect the measured input signal through the same unknown linear dynamic filter $\hat{L}$ to the x-position of the oscilloscope.
5. Tune the filter $\hat{L}$ till the plot is a single valued one. Then the plot is the characteristic curve of the nonlinear element, and $\hat{L}$ is the estimate of LD_1. The transfer function of LD_2 is the inverse of the product of the inverse overall transfer function with the transfer function of $\hat{L}$. ■

3.11.7 Frequency method

The frequency method is based on the following lemma.

Lemma 3.11.6

Excite the continuous time simple Wiener–Hammerstein model by a sinusoidal test signal

$$u(t) = U\cos\omega t$$

The highest order harmonic is

$$y^{(M)}(t) = \frac{U^M}{2^{M-1}} c_M \left|G_1(\omega)\right|^M \left|G_2(M\omega)\right| \cos\left(M\omega t + M\varphi_1(\omega) + \varphi_2(M\omega)\right) \qquad (3.11.17)$$

where M is the highest degree of the nonlinear power and the frequency functions of the linear dynamic terms are

$$G_1(j\omega) = \left|G_1(\omega)\right|\exp\left(j\varphi_1(\omega)\right)$$
$$G_2(j\omega) = \left|G_2(\omega)\right|\exp\left(j\varphi_2(\omega)\right)$$

Proof. The partial model between the input signal and the inner variable $v_2(k)$ is a simple Wiener model. Thus the highest order harmonics of $v_2(k)$ can be computed by Lemma 3.10.1

$$y^{(M)}(t) = \frac{U^M}{2^{M-1}} c_M \left|G_1(\omega)\right|^M \cos\left(M\omega t + M\varphi_1(\omega)\right)$$

This signal will be filtered by the second linear dynamic term LD_2 at frequency $M\omega$, thus the highest harmonic of the output is as given in (3.11.17). ■

The transfer functions of the two linear dynamic parts of the simple Wiener–

Hammerstein model cannot unambiguously be obtained in certain cases. Such a case can occur if the transfer functions have multiple poles or zeros, or if the poles or zeros of the transfer functions of the linear dynamic terms are equal to each other or multiple of each other. Two examples illustrate the problem.

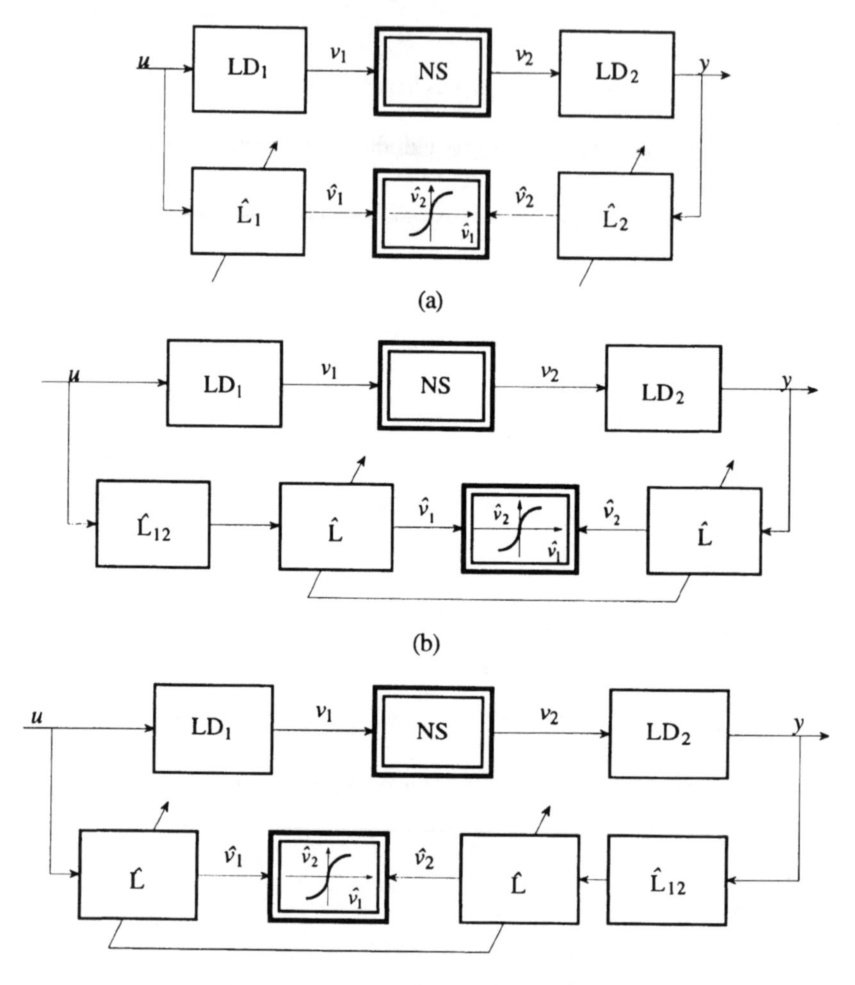

Fig. 3.11.2 Identification of the simple Wiener–Hammerstein model by graphical plotting: (a) by tuning two filters; (b) by tuning one filter at the input; (c) by tuning one filter at the output

Example 3.11.2 *Not unambiguously identifiable system*
The transfer functions of the linear dynamic terms are

$$G_1(s) = \frac{1}{1+sT_1}, \qquad G_2(s) = \frac{1+sT_2}{(1+sT_3)(1+sT_4)}$$

The frequency characteristic at the second harmonic is

$$y^{(2)}(j2\omega) = \frac{U^2}{2} c_2 \left[\frac{1}{1 + j\omega T_1} \right]^2 \frac{1 + j2\omega T_2}{(1 + j2\omega T_3)(1 + j2\omega T_4)} \tag{3.11.18}$$

In the special case of $T_1 = 2T_2$ (3.11.18) reduces to

$$y^{(2)}(j2\omega) = \frac{U^2}{2} c_2 \frac{1}{(1 + j2\omega T_1)(1 + j2\omega T_3)(1 + j2\omega T_4)}$$

which could lead also to the identification result:

$$G_1(s) = 1$$

$$G_2(s) = \frac{1}{(1 + sT_1/2)(1 + sT_3)(1 + sT_4)}$$

Example 3.11.3 *Not unambiguously identifiable system*
Assume the measured second-order frequency characteristic is

$$y^{(2)}(j2\omega) = \frac{U^2}{2} c_2 \left[\frac{1}{1 + j\omega T} \right]^3 \tag{3.11.19}$$

Two possible combinations of LD_1 and LD_2 can result in (3.11.19):

- *multiple poles are allowed:*

$$G_1(s) = 1, \qquad G_2(s) = \frac{1}{(1 + sT)^3}$$

- *multiple poles are not allowed:*

$$G_1(s) = \frac{1}{1 + sT}, \qquad G_2(s) = \frac{1}{1 + sT/2}$$

■

The above examples show why the case of multiple poles or zeros has to be excluded.

The frequency method, based on the pole and zero distribution, can be applied under certain conditions.

Assumption 3.11.1 The linear dynamic parts of the simple Wiener–Hammerstein model have the following features (Rugh, 1981):

- there are no multiple poles or zeros of the transfer functions ;
- the linear dynamic terms have minimum phase characteristics;
- they have no delay time;
- the second linear dynamic term LD_2 does not have poles (or zeros) with multiplicity M equal to the zeros (or poles) of the first linear dynamic term LD_1.

■

Algorithm 3.11.10 (Rugh, 1981; Gardiner, 1973a) The simple Wiener–Hammerstein model can be identified under Assumption 3.11.1 by the following procedure:

1. Excite the process by $U\cos\omega t$ at several frequencies.
2. Determine the highest order harmonics $y^{(M)}$ from the responses by filtering out the lower order harmonics.
3. Compute the frequency characteristic of $\left(2^{M-1}/U^M\right)y^{(M)}(t)$.
4. Determine the breakpoints of the angular frequencies from the Bode amplitude plot.
5. Breakpoints with integer times of the multiplicity M give the poles and zeros of LD_1 by a multiplicity divided by M.
6. The remaining breakpoints multiplied by M give the poles and zeros of LD_2 with their multiplicity of occurrence.
7. Assume the linear dynamic terms have unit static gains. Then the estimated static gain is c_M.
8. In the knowledge of the linear dynamic terms the static nonlinear term can be identified from the computed inner variables by static regression. ■

Example 3.11.4 *Identification of a first-order, quadratic simple Wiener–Hammerstein model*

The process has two linear dynamic terms with the transfer functions

$$G_1(s) = \frac{1}{1+5s} \qquad G_2(s) = \frac{1}{1+10s}$$

respectively. The nonlinear static term is given by the parabolic equation

$$v_2(t) = 2 + v_1(t) + 0.5v_1^2(t)$$

The Bode amplitude and phase plots of the second harmonic $\left(2/U^2\right)y^{(2)}(t)$ are given in Figures 3.11.3a and 3.11.3b, respectively by continuous lines. The amplitude plot starts with $\hat{c}_2 = 0.5$ and the breakpoints are:

- $1/\hat{T}_1 = 0.2$, double;
- $(1/2)\left(1/\hat{T}_2\right) = (1/2)\cdot 0.1 = 0.05$, single

Therefore both linear dynamic terms are of proportional, first-order type with unit static gains and time constants $\hat{T}_1 = 10$ and $\hat{T}_2 = 5$, respectively. The nonlinear static part can be estimated from stationary measurements as the steady state input–output relation. The amplitude and the phase plots of LD_1 and LD_2 are also given in Figures 3.11.3a and 3.11.3b by dashed and dot-and-dashed lines, respectively. ■

Application of the method is with Jelonek and Economakos (1972) the identification of a heating process. They applied also a technique with two-frequency excitation.

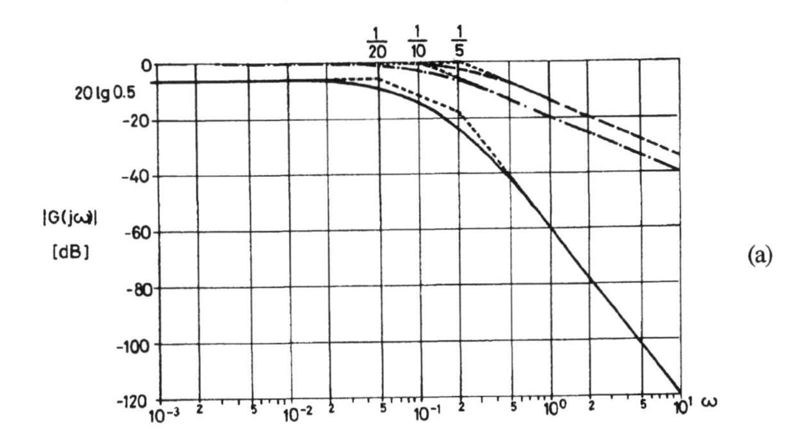

(a)

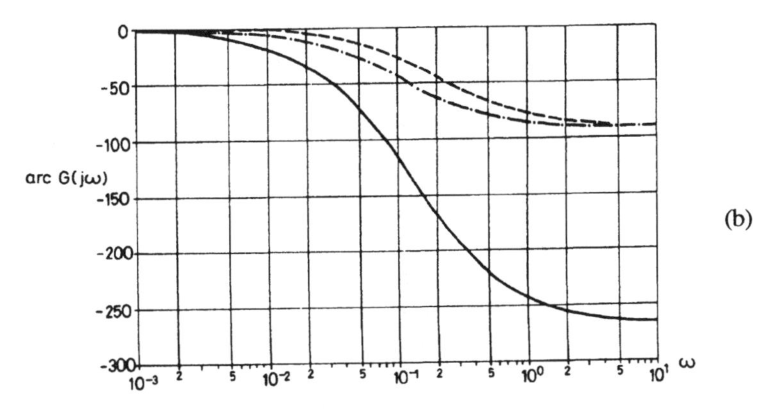

(b)

Fig. 3.11.3 Frequency characteristics of the second harmonic (—) and of the first (- - -) and second linear dynamic (-·-·-) terms of the simple Wiener–Hammerstein model: (a) amplitude plot; (b) phase plot

3.12 METHODS APPLIED TO THE S_M MODELS

Figure 3.12.1 shows the discrete time S_M model consisting of a constant, a linear dynamic and more simple Wiener–Hammerstein models all connected parallel. The steady state nonlinearity is a polynomial and is assumed to have the degree of M. The discrete delay times of the linear dynamic terms are d_{i1} and d_{i2}, respectively. The output signals of the parallel channels are denoted by $y_i(k)$, $i = 0, \ldots, M$. In each nonlinear channel two linear dynamic terms and a homogeneous polynomial term are connected serially, thus one parameter of two elements can be chosen freely. For example, the static gains of the dynamic terms may be selected to be unity. It is usual to choose the static gains of the dynamical terms equal to unity.

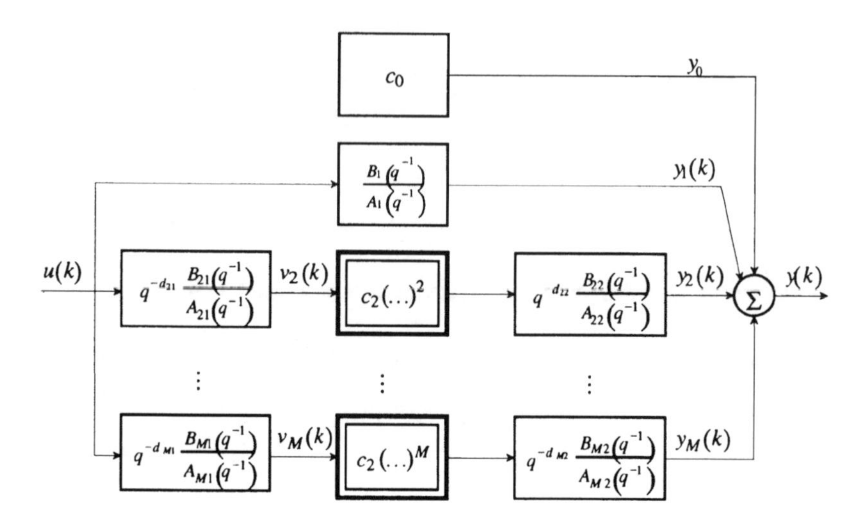

Fig. 3.12.1 Scheme of the discrete time S_M model

3.12.1 Iterative estimation

Theoretically a sequential estimation of the channels can be imagined according to the following algorithm.

Algorithm 3.12.1 The S_M model can be identified by a sequential iterative estimation of the channels as follows:

1. Assume an initial parameter vector for each subsystem except the ith $(0 \le i \le M)$.
2. Compute the output signals of all channels except i.
3. Compute the output signal of channel i as the difference of the measured output signal and the sum of the output signals of all submodels except the ith one.
4. Estimate the parameters of channel i from the measured input signal and from the computed output signal of this channel. (If $i = 0$, then a constant; if $i = 1$ then a linear dynamic term; if $i \ge 2$ then a simple Wiener–Hammerstein model has to be identified.)
5. Compute the output signal of channel i.
6. Check whether the iteration has to be stopped. If not then first increase i by 1 and if $i > M$ let $i ::= 0$ and then go to Step 2. ■

Remark: In Step 4 the estimation of the parameters of a nonlinear submodel requires also iterations.

3.12.2 Using a process model nonlinear-in-parameters

The noise-free output signal is described by the equations

$$w_0 = c_0$$

$$w_1(k) = \frac{B_1(q^{-1})}{A_1(q^{-1})} u(k - d_1)$$

$$v_i(k)=\frac{B_{i1}\left(q^{-1}\right)}{A_{i1}\left(q^{-1}\right)}u(k-d_{i1}) \qquad 2\le i\le M$$

$$w_i(k)=\frac{B_{i2}\left(q^{-1}\right)}{A_{i2}\left(q^{-1}\right)}v_i^i(k-d_{i2}) \qquad 2\le i\le M$$

$$w(k)=\sum_{i=0}^{M}w_i(k)$$

An additive colored noise

$$n(k)=\frac{E\left(q^{-1}\right)}{D\left(q^{-1}\right)}e(k)$$

is assumed at the output

$$y(k)=w(k)+n(k)$$

The residual can be expressed by the form

$$\varepsilon(k)=\frac{D\left(q^{-1}\right)}{E\left(q^{-1}\right)}\left[y(k)-w(k)\right]$$

The prediction error method can be applied, which needs the first derivative of the residual according to the unknown parameters. The sensitivity functions are now given.

$$\frac{\partial\varepsilon(k)}{\partial c_0}=-\frac{D(1)}{E(1)}$$

$$\frac{\partial\varepsilon(k)}{\partial b_{1j}}=-\frac{D\left(q^{-1}\right)}{E\left(q^{-1}\right)}\frac{1}{A_1\left(q^{-1}\right)}u(k-d_1-j), \qquad \frac{\partial\varepsilon(k)}{\partial a_{1j}}=\frac{D\left(q^{-1}\right)}{E\left(q^{-1}\right)}\frac{1}{A_1\left(q^{-1}\right)}w_1(k-j)$$

$$\frac{\partial\varepsilon(k)}{\partial b_{i2j}}=-\frac{D\left(q^{-1}\right)}{E\left(q^{-1}\right)}\frac{1}{A_{i2}\left(q^{-1}\right)}v_i^2(k-d_{i2}-j) \quad 2\le i\le M$$

$$\frac{\partial\varepsilon(k)}{\partial a_{i2j}}=\frac{D\left(q^{-1}\right)}{E\left(q^{-1}\right)}\frac{1}{A_{i2}\left(q^{-1}\right)}w_i(k-j) \qquad 2\le i\le M$$

$$\frac{\partial\varepsilon(k)}{\partial b_{i1j}}=-\frac{D\left(q^{-1}\right)}{E\left(q^{-1}\right)}\frac{B_{i2}\left(q^{-1}\right)}{A_{i2}\left(q^{-1}\right)}2v_i(k-d_{i2})\frac{1}{A_{i1}\left(q^{-1}\right)}u(k-d_{i2}-d_{i1}-j) \qquad 2\le i\le M$$

$$\frac{\partial\varepsilon(k)}{\partial a_{i1j}}=\frac{D\left(q^{-1}\right)}{E\left(q^{-1}\right)}\frac{B_{i2}\left(q^{-1}\right)}{A_{i2}\left(q^{-1}\right)}2v_i(k-d_{i2})\frac{1}{A_{i1}\left(q^{-1}\right)}v_i(k-d_{i2}-j) \qquad 2\le i\le M$$

$$\frac{\partial \varepsilon(k)}{\partial d_i} = \frac{1}{E(q^{-1})}[y(k-i) - w(k-i)] \quad ; \quad \frac{\partial \varepsilon(k)}{\partial e_i} = -\frac{1}{E(q^{-1})}\varepsilon(k-i)$$

As is known, the prediction-error method is an iterative procedure.

3.12.3 Separate identification of the parallel channels

The identification is based on tracing the problem back to the uncorrelated identification of individual simple Wiener–Hammerstein models.

Algorithm 3.12.2 Excite the S_M model by a test signal and repeat the excitation M-times with different amplitudes, where M is the highest degree of the steady-state polynomial relation. The steps of the parameter estimation are as follows:

1. Separate the responses of the channels of different degrees by means of Gardiner's method (Gardiner, 1973b), see Chapter 5.
2. The constant term is directly given from the separation of the channels' outputs.
3. The transfer function in the linear channel can be estimated from the input signal and from the output of the linear channel.
4. All further channels can be identified from the input signal and from their output signals as simple Wiener–Hammerstein models by the methods learned in Section 3.11.

■

3.12.4 Correlation method

Gardiner's method can be applied for separating the outputs of the channels. If correlation functions are computed then they filter the noise, and it is better to separate the correlation functions of the different channels than the noisy measured time functions by Gardiner's method. This will be applied now. The algorithm is prepared by two lemmas.

Lemma 3.12.1

(Billings and Fakhouri, 1979a) Excite the S_M model with a Gaussian white noise $u(k)$ and repeat the experiment by an input signal $\gamma u(k)$. An additive noise $n(k)$ of zero mean is assumed at the output. Denote the output signal measured at the excitation by the input signal $\gamma u(k)$ by $y_\gamma(k)$. Then the cross-correlation functions between the normalized variables

$$y'(k) = y(k) - \mathrm{E}\{y(k)\} \tag{3.12.1}$$

$$x_1(k) = u(k) - \mathrm{E}\{u(k)\} \tag{3.12.2}$$

$$x_2(k) = [u(k) - \mathrm{E}\{u(k)\}]^2 \tag{3.12.3}$$

are

$$r_{x_1 y'_\gamma}(\kappa) = \sum_{i=0}^{M} \gamma^i r_{x_1 w'_i}(\kappa) \tag{3.12.4}$$

$$r_{x_2 y'_\gamma}(\kappa) = \sum_{i=0}^{M} \gamma^i r_{x_2 w'_i}(\kappa) \tag{3.1.2.5}$$

where $w_i(k)$ is the noise-free output signal of the degree-i channel.

Proof. The noisy output signal is a sum of the following terms

$$y(k) = w(k) + n(k) = \sum_{i=0}^{M} w_i(k) + n(k)$$

Apply first the input signal $u(k)$. The linear cross-correlation function is

$$r_{x_1 y'}(\kappa) = r_{x_1 w'}(\kappa) + r_{x_1 n'}(\kappa) = r_{x_1 w'}(\kappa) = \sum_{i=0}^{M} r_{x_1 w_i'}(\kappa)$$

Apply now the input signal $\gamma u(k)$. The linear cross-correlation function is now

$$r_{x_1 y_\gamma'}(\kappa) = r_{x_1 w_\gamma'}(\kappa) = \sum_{i=0}^{M} r_{x_1 w_{i\gamma}'}(\kappa) = \sum_{i=0}^{M} \gamma^i r_{x_1 w_i'}(\kappa)$$

because the outputs of the parallel channels $w_i(k)$ are degree-i functions of the input signal.

Since nothing was stated about $x_1(k)$, the proof is valid also for $x_2(k)$. ■

Lemma 3.12.2
(Billings and Fakhouri, 1980) Excite the S_M model with the sum of two uncorrelated pseudo-random signals $u_1(k)$ and $u_2(k)$. Each of them should have zero mean value

$$\mathrm{E}\{u_1(k)\} = 0, \qquad \mathrm{E}\{u_2(k)\} = 0$$

and a Dirac delta function-like auto-correlation function. Then the cross-correlation functions between the variables (3.12.1) and

$$x_1(k) = u_1(k) \tag{3.12.6}$$
$$x_2(k) = u_1(k)u_2(k) \tag{3.12.7}$$

fulfill (3.12.4) and (3.12.5).

Proof. As shown in the proof of Lemma 3.12.1, (3.12.4) and (3.12.5) are generally valid to any excitations, thus also for those in this lemma. ■

Algorithm 3.12.3 Billings and Fakhouri (1979a) applied the following algorithm to the S_M model:

1. Excite the process either
   - by a Gaussian white noise signal $u(k)$ or
   - by the sum of two uncorrelated pseudo-random signals $u_1(k)$ and $u_2(k)$ with zero mean and having Dirac delta function-like auto-correlation functions
2. Repeat the experiment M-times by the input signals $\gamma_i u(k)$, $(\gamma_i \neq \gamma_j \neq 1,\ i \neq j)$
3. Compute the cross-correlation functions (3.12.4) and (3.12.5) defined
   - for the Gaussian excitation by (3.12.1) to (3.12.3) and

- for the pseudo-random excitation by (3.12.1), (3.12.6) and (3.12.7)

4. Calculate the cross-correlation functions $r_{x_1 w_i'}(\kappa)$, $i = 1, \ldots, M$ from the $(M+1)$ records $r_{x_1 y_{\gamma_i}'}(\kappa)$, $i = 1, \ldots, M+1$ and $r_{x_2 w_i'}(\kappa)$ $i = 1, \ldots, M$ from the $(M+1)$ records $r_{x_2 y_{\gamma_i}'}(\kappa)$, $\gamma = 1, \ldots, M+1$ by means of Gardiner's method (Gardiner, 1973b), elaborated for separation of the responses of a system to their components corresponding to different nonlinear degrees. (See in Chapter 5.)
5. Estimate the parameters of the individual channels from the linear and nonlinear cross-correlation functions. The models to be fitted depend on the type of the channel:
   - *constant term* $(i = 0)$*:*
     The constant term is the output at zero excitation. It results from the Gardiner's method for separation of the parallel channels if $(M+1)$ excitations were performed.
   - *linear dynamic term* $(i = 1)$*:*
     The linear pulse transfer function $q^{-d_1} B_1(q^{-1}) / A(q^{-1})$ has to be fitted to $r_{x_1 x_1}(\kappa)$ and $r_{x_1 w_1'}(\kappa)$ as input and output signals, respectively.
   - *simple Wiener–Hammerstein model* $(i \geq 2)$*:*
     The two linear pulse transfer functions in each channel can be calculated as learned with the identification of the simple Wiener–Hammerstein model in Section 3.11.4. ■

3.12.5 Frequency method

The output signals of the parallel channels can be separated by Gardiner's method (Gardiner, 1973b). The individual channels can be identified by the method learned with the simple Wiener–Hammerstein model.

Algorithm 3.12.4 The identification procedure of the S_M model consists of the following steps:

1. Excite the process by a sinusoidal signal.
2. Determine the order M of the highest order harmonic.
3. Choose $(M+1)$ different amplitudes for the test signal. Denote the first one by U.
4. Excite the process by a sinusoidal signal $U \cos \omega t$ at different frequencies. Repeat the experiments with M further different amplitudes.
5. Compute the output signals of the channels to $U \cos \omega t$ at all frequencies using Gardiner's method.
6. The constant term is the output of the degree-0 channel that can be also measured by zero excitation.
7. Identify the transfer function of the degree-1 channel by one of the known methods for linear systems.
8. Identify the transfer functions of the linear dynamic terms and the static coefficient in the nonlinear channels using the methods elaborated for the simple Wiener–Hammerstein model. ■

Remarks:

1. The restriction, introduced in Assumption 3.11.1 with the identification of the simple Wiener–Hammerstein model, has to be valid furthermore for the linear dynamic parts of the nonlinear channels.
2. A more detailed analysis of the problem and some other procedures are in (Baumgarten and Rugh, 1975; Wysocki and Rugh, 1976).
3. Klippel (1996) identified a horn loudspeaker by a two-tone sinusoidal excitation.

3.13 METHODS APPLIED TO THE FACTORABLE VOLTERRA MODEL

Figure 3.13.1 shows the discrete time factorable Volterra model. It consists of a constant, linear, quadratic and higher degree channels connected in parallel. The output of each nonlinear channel is the product of linear dynamic terms, therefore the static gain in each except one pulse transfer function can be chosen freely.

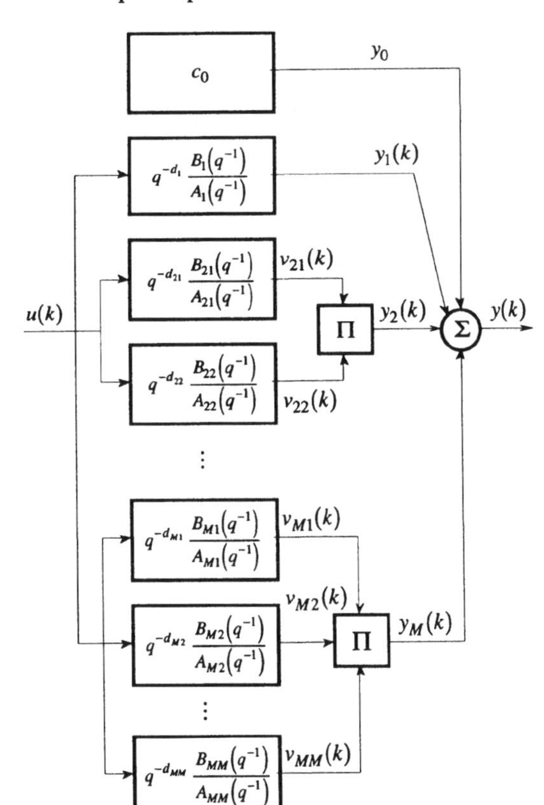

Fig. 3.13.1 Scheme of the discrete time factorable Volterra model

3.13.1 Iterative estimation

Algorithm 3.13.1 The following iterative procedure can be formed:

1. Excite the process by $(M+1)$ similar test signals which differ only in the amplitude scaling.
2. Separate the output signals of the parallel channels from the measured output signal

by Gardiner's method (Gardiner, 1973b).
3. The output of the constant channel is the constant term itself.
4. A linear dynamic system can be estimated between the input signal and the output signal of the linear channel.
5. The M transfer functions in the nonlinear channel of degree-M can be identified iteratively:
   5.1. Assume an initial parameter vector for each subsystem except the ith $(1 \le i \le M)$.
   5.2. Compute the output signals of all components except i.
   5.3. Compute the output signal of component i as the quotient of the measured output signal and the product of the output signals of all components except the ith one.
   5.4. Estimate the parameters of the linear dynamic term in the channel i from the measured input signal and from the computed output signal of this term.
   5.5. Compute the output signal of the component i.
   5.6. Check whether the iteration has to be stopped. If not then first increase i by 1 and if $i > M$ let $i ::= 1$ and then go to Step 5.2. ■

3.13.2 Using a process model nonlinear-in-parameters

The noise-free output signal is described by the equations

$$w(k) = \sum_{i=0}^{M} w_i(k), \qquad w_i(k) = \prod_{j=1}^{i} v_{ij}(k), \qquad w_0(k) = v_0(k) = c_0$$

$$v_1(k) = \frac{B_1\left(q^{-1}\right)}{A_1\left(q^{-1}\right)} u\left(k - d_1\right)$$

$$v_{ij}(k) = \frac{B_{ij}\left(q^{-1}\right)}{A_{ij}\left(q^{-1}\right)} u\left(k - d_{ij}\right), \qquad 2 \le i \le M, \qquad j = 1, \ldots, i$$

An additive colored noise

$$n(k) = \frac{E\left(q^{-1}\right)}{D\left(q^{-1}\right)} e(k)$$

is assumed at the output

$$y(k) = w(k) + n(k)$$

The residual can be expressed by the form

$$\varepsilon(k) = \frac{D\left(q^{-1}\right)}{E\left(q^{-1}\right)} \left[y(k) - w(k)\right]$$

The prediction error method can be applied, which needs the first derivative of the residual according to the unknown parameters. These sensitivity functions are:

$$\frac{\partial \varepsilon(k)}{\partial c_0} = -\frac{D(1)}{E(1)}$$

$$\frac{\partial \varepsilon(k)}{\partial b_{1\ell}} = -\frac{D\left(q^{-1}\right)}{E\left(q^{-1}\right)} \frac{1}{A_1\left(q^{-1}\right)} u\left(k-d_1-\ell\right), \quad \frac{\partial \varepsilon(k)}{\partial a_{1\ell}} = \frac{D\left(q^{-1}\right)}{E\left(q^{-1}\right)} \frac{1}{A_1\left(q^{-1}\right)} w_1(k-\ell)$$

$$\frac{\partial \varepsilon(k)}{\partial b_{ij\ell}} = -\frac{D\left(q^{-1}\right)}{E\left(q^{-1}\right)} \left[\prod_{j_1=1, j_1 \neq i}^{i} v_{ij_1}(k) \right] \frac{1}{A_{ij}\left(q^{-1}\right)} u\left(k-d_{ij}-\ell\right)$$

$$\frac{\partial \varepsilon(k)}{\partial a_{ij\ell}} = \frac{D\left(q^{-1}\right)}{E\left(q^{-1}\right)} \left[\prod_{j_1=1, j_1 \neq i}^{i} v_{ij_1}(k) \right] \frac{1}{A_{ij}\left(q^{-1}\right)} v_{ij}(k-\ell)$$

$$\frac{\partial \varepsilon(k)}{\partial d_i} = \frac{1}{E\left(q^{-1}\right)} \left[y(k-i) - w(k-i) \right], \quad \frac{\partial \varepsilon(k)}{\partial e_i} = -\frac{1}{E\left(q^{-1}\right)} \varepsilon(k-i)$$

As is known, the prediction error method is an iterative procedure.

3.13.3 Correlation method

The problem of the parameter estimation can be separated into four parts:

- computing a special cross-correlation function between the input signal and the output signal of a given homogenous channel,
- computing a function of the weighting functions of the linear subsystems from the above correlation function,
- estimation of the parameters of the weighting functions of the linear dynamic subsystems.
- estimating the pulse transfer functions from the weighting functions.

Assume that the outputs of the parallel channels of different nonlinear degrees are known. Assume, furthermore, that one has succeeded in identifying a special function of the weighting function series of the linear dynamic terms. The question is to reconstruct the pulse transfer function. The following lemmas treat the problem in detail.

Lemma 3.13.1

(Billings and Fakhouri, 1979b) Assume the function

$$f(\kappa_1, \kappa) = \sum_{i=1}^{M} h_{Mi}(\kappa_1) \prod_{j=1, j \neq i}^{M} h_{Mj}(\kappa) \tag{3.13.1}$$

is known for different shifting times $\kappa_1 = 0, \ldots, m$, $\kappa = 0, \ldots, m$. The elements of the weighting functions denoted by

$$\boldsymbol{h}_{Mi} = \left[h_{Mi}(0), \ldots, h_{Mi}(m) \right]^{\mathrm{T}}$$

can be determined from $(m+1)^2$ equations by minimizing the cost function

$$J\left(\boldsymbol{h}_{M1}^{\mathrm{T}}, \ldots, \boldsymbol{h}_{MM}^{\mathrm{T}}\right) = \left[f(\kappa_1, \kappa) - \sum_{i=1}^{M} h_{Mi}(\kappa_1) \prod_{j=1, j \neq i}^{M} h_{Mj}(\kappa) \right]^2$$

Remark:

Indeed, the number of independent equations is less than $(m+1)^2$ if the degree is two. This and the cubic case is shown in the following two examples.

Example 3.13.1 *Determining the quadratic kernels*
(3.13.1) has now the form:

$$
\begin{aligned}
f(0,0) &= h_{21}(0)h_{22}(0) + h_{21}(0)h_{22}(0) \\
f(0,1) &= h_{21}(0)h_{22}(1) + h_{21}(1)h_{22}(0) \\
&\vdots \\
f(0,m) &= h_{21}(0)h_{22}(m) + h_{21}(m)h_{22}(0) \\
&\vdots \\
f(1,0) &= h_{21}(1)h_{22}(0) + h_{21}(0)h_{22}(1) \\
f(2,0) &= h_{21}(2)h_{22}(0) + h_{21}(0)h_{22}(2) \\
&\vdots \\
f(m,m-1) &= h_{21}(m)h_{22}(m-1) + h_{21}(m-1)h_{22}(m) \\
f(m,m) &= h_{21}(m)h_{22}(m) + h_{21}(m)h_{22}(m)
\end{aligned}
$$

The number of independent equations is $(m+1)^{m+2}/2$. For $m \geq 3$ this value is more than the number of the unknowns $2(m+1)$. ■

Example 3.13.2 *Determining the cubic kernels*
(3.13.1) has now the form:

$$
\begin{aligned}
f(0,0) &= h_{31}(0)h_{32}(0)h_{33}(0) + h_{31}(0)h_{32}(0)h_{33}(0) + h_{31}(0)h_{32}(0)h_{33}(0) \\
f(0,1) &= h_{31}(0)h_{32}(1)h_{33}(1) + h_{31}(1)h_{32}(0)h_{33}(1) + h_{31}(1)h_{32}(1)h_{33}(0) \\
f(0,m) &= h_{31}(0)h_{32}(m)h_{33}(m) + h_{31}(m)h_{32}(0)h_{33}(m) + h_{31}(m)h_{32}(m)h_{33}(0) \\
&\vdots \\
f(1,0) &= h_{31}(1)h_{32}(0)h_{33}(0) + h_{31}(0)h_{32}(1)h_{33}(0) + h_{31}(0)h_{32}(0)h_{33}(1) \\
f(2,0) &= h_{31}(2)h_{32}(0)h_{33}(0) + h_{31}(0)h_{32}(2)h_{33}(0) + h_{31}(0)h_{32}(0)h_{33}(2) \\
&\vdots \\
f(m,m-1) &= h_{31}(m)h_{32}(m-1)h_{33}(m-1) + h_{31}(m-1)h_{32}(m)h_{33}(m-1) + \\
&\quad + h_{31}(m-1)h_{32}(m-1)h_{33}(m) \\
f(m,m) &= h_{31}(m)h_{32}(m)h_{33}(m) + h_{31}(m)h_{32}(m)h_{33}(m) + h_{31}(m)h_{32}(m)h_{33}(m)
\end{aligned}
$$

Now, the number of independent equations is $(m+1)^2$. ■

The next task is to show how the function (3.13.1) can be estimated. There are two ways:

- by a Gaussian test signal; or
- by a compound signal consisting of the sum of pseudo-random signals.

Theorem 3.13.1 (Billings and Fakhouri, 1979b) Excite the process with a

Gaussian white noise $u(k)$ with mean value of zero and standard deviation of σ_u. An additive noise $n(k)$ of zero mean is assumed at the output. Subtract the mean value from the output signal of the degree-M channel

$$y'_M(k) = y_M(k) - \mathrm{E}\{y_M(k)\}$$

The function of the weighting function series (3.13.1) can be determined by cross-correlation

- *for odd M:*

$$\sum_{i=1}^{M} h_{Mi}(\kappa_1) \prod_{j=1, j\neq i}^{M} h_{Mj}(\kappa) = \frac{1}{(M-1)!\sigma_u^{2M}} \Bigg[\mathrm{E}\{u(k-\kappa_1)u^{M-1}(k-\kappa)y'_M(k)\}$$
$$- \sum_{i=0}^{(M-3)/2} \frac{(M-1)!}{(M-2i-1)!(2i)!} \mathrm{E}\{\mathrm{E}\{u(k-\kappa_1)u^{2i}(k-\kappa)y'_M(k)\}u^{M-2i-1}(k-\kappa)\}$$
$$- \sum_{i=0}^{(M-3)/2} \frac{(M-1)!}{(M-2i-1)!(2i+1)!} \mathrm{E}\{\mathrm{E}\{u^{M-2i}(k-\kappa)y'_M(k)\}u^{M-2i-2}(k-\kappa)u(k-\kappa_1)\}\Bigg] \tag{3.13.3}$$

- *for even M:*

$$\sum_{i=1}^{M} h_{Mi}(\kappa_1) \prod_{j=1, j\neq i}^{M} h_{Mj}(\kappa) = \frac{1}{(M-1)!\sigma_u^{2M}} \Bigg[\mathrm{E}\{u(k-\kappa_1)u^{M-1}(k-\kappa)y'_M(k)\}$$
$$- \sum_{i=1}^{(M-2)/2} \frac{(M-1)!}{(M-2i+1)!(2i-1)!} \mathrm{E}\{\mathrm{E}\{u(k-\kappa_1)u^{2i-1}(k-\kappa)y'_M(k)\}u^{M-2i}(k-\kappa)\}$$
$$- \sum_{i=1}^{(M-2)/2} \frac{(M-1)!}{(M-2i+1)!(2i)!} \mathrm{E}\{\mathrm{E}\{u^{2i}(k-\kappa)y'_M(k)\}u^{M-2i}(k-\kappa)u(k-\kappa_1)\}\Bigg] \tag{3.13.4}$$

Proof. Because the noise $n(k)$ is independent, then its correlation with any function of the input signal is zero. Therefore the normalized output signal $y'_M(k)$ of the degree-M channel can be replaced by the normalized noise-free output signal $w'_M(k)$ in the expressions of the cross-correlations.

1. *Case* $M = 1$*:*
   The relation is known from the theory of linear systems

$$\mathrm{E}\{u(k-\kappa_1)y'_M(k)\} = \mathrm{E}\{u(k-\kappa_1)w'_M(k)\}$$
$$= \sum_{i=0}^{\infty} h_1(i)\mathrm{E}\{u(k-\kappa_1)[u(k-i) - \mathrm{E}\{u(k-i)\}]\}$$
$$= \sum_{i=0}^{\infty} h_1(i)\mathrm{E}\{u(k-\kappa_1)u(k-i)\} = \sum_{i=0}^{\infty} h_1(i)r_{uu}(\kappa_1 - i) = \sigma_u^2 h_1(\kappa_1)$$

It was taken into account that the sum differs from zero if $\kappa_1 = i$.

2. *Case* $M = 2$:

$$\mathrm{E}\{u(k-\kappa_1)u(k-\kappa)y_2'(k)\} = \mathrm{E}\{u(k-\kappa_1)u(k-\kappa)w_2'(k)\}$$

$$= \sum_{i_1=0}^{\infty}\sum_{i_2=0}^{\infty} h_{21}(i_1)h_{22}(i_2)\mathrm{E}\{[u(k-i_1)u(k-i_2) - \mathrm{E}\{u(k-i_1)u(k-i_2)\}]u(k-\kappa_1)u(k-\kappa)\}$$

$$= \sigma_u^4[h_{21}(\kappa_1)h_{22}(\kappa) + h_{21}(\kappa)h_{22}(\kappa_1)] = \sigma_u^4\sum_{i=1}^{2} h_{2i}(\kappa_1)\prod_{j=1,\, j\neq i}^{2} h_{2j}(\kappa)$$

because

$$\mathrm{E}\{u(k-i_1)u(k-i_2)u(k-\kappa_1)u(k-\kappa)\} - \mathrm{E}\{u(k-i_1)u(k-i_2)\}\mathrm{E}\{u(k-\kappa_1)u(k-\kappa)\}$$

$$= [r_{uu}(i_2-i_1)r_{uu}(\kappa-\kappa_1) + r_{uu}(\kappa_1-i_1)r_{uu}(\kappa-i_2) + r_{uu}(\kappa-i_1)r_{uu}(\kappa_1-i_2)]$$

$$-r_{uu}(i_2-i_1)r_{uu}(\kappa-\kappa_1) = r_{uu}(\kappa_1-i_1)r_{uu}(\kappa-i_2) + r_{uu}(\kappa-i_1)r_{uu}(\kappa_1-i_2) = \sigma_u^4$$

It was taken into account that the sum differs from zero if ($\kappa_1 = i_1$ and $\kappa = i_2$) or ($\kappa = i_1$ and $\kappa_1 = i_2$).

3. *Case* $M = 3$:

$$\mathrm{E}\{u(k-\kappa_1)u^2(k-\kappa)y_3'(k)\} = \mathrm{E}\{u(k-\kappa_1)u^2(k-\kappa)w_3'(k)\}$$

$$= \sum_{i_1=0}^{\infty}\sum_{i_2=0}^{\infty}\sum_{i_3=0}^{\infty} h_{31}(i_1)h_{32}(i_2)h_{33}(i_3)$$

$$\times\mathrm{E}\{u(k-i_1)u^2(k-\kappa)[u(k-i_1)u(k-i_2)u(k-i_3) - \mathrm{E}\{u(k-i_1)u(k-i_2)u(k-i_3)\}]\}$$

$$= \sum_{i_1=0}^{\infty}\sum_{i_2=0}^{\infty}\sum_{i_3=0}^{\infty} h_{31}(i_1)h_{32}(i_2)h_{33}(i_3)\mathrm{E}\{u(k-\kappa_1)u^2(k-\kappa)u(k-i_1)u(k-i_2)u(k-i_3)\}$$

$$= \sum_{i_1=0}^{\infty}\sum_{i_2=0}^{\infty}\sum_{i_3=0}^{\infty} 2h_{31}(i_1)h_{32}(i_2)h_{33}(i_3)$$

$$\times[r_{uu}(\kappa_1-i_1)r_{uu}(\kappa-i_2)r_{uu}(\kappa-i_3) + r_{uu}(\kappa_1-i_1)r_{uu}(\kappa-i_3)r_{uu}(\kappa-i_2)$$

$$+r_{uu}(\kappa_1-i_2)r_{uu}(\kappa-i_1)r_{uu}(\kappa-i_3) + r_{uu}(\kappa_1-i_2)r_{uu}(\kappa-i_3)r_{uu}(\kappa-i_1)$$

$$+r_{uu}(\kappa_1-i_3)r_{uu}(\kappa-i_1)r_{uu}(\kappa-i_2) + r_{uu}(\kappa_1-i_3)r_{uu}(\kappa-i_2)r_{uu}(\kappa-i_1)]$$

$$= 2\sigma_u^6[h_{31}(\kappa_1)h_{32}(\kappa)h_{33}(\kappa) + h_{31}(\kappa)h_{32}(\kappa_1)h_{33}(\kappa)] + h_{31}(\kappa)h_{32}(\kappa)h_{33}(\kappa_1)$$

$$= \sum_{i=1}^{3} h_{3i}(\kappa_1)\prod_{j=1,\, j\neq i}^{3} h_{3j}(\kappa)$$

It was taken into account that the sum differs from zero if ($\kappa_1 = i_1$ and $\kappa = i_2 = i_3$)

or ($\kappa_1 = i_2$ and $\kappa = i_1 = i_3$) or ($\kappa_1 = i_3$ and $\kappa = i_1 = i_2$).

The proof is similar to the higher degree cases. ■

Lemma 3.13.2

(Billings and Fakhouri, 1979b) Excite the process with the sum of independent pseudo-random (e.g., Gaussian) white noises $x_i(k)$ with mean value of zero and standard deviation of σ_u

$$u(k) = \sum_{i=1}^{M} x_i(k) \tag{3.13.5}$$

An additive noise $n(k)$ of zero mean is assumed at the output. Subtract the mean value from the output signal of the degree-M channel (3.13.2). A function of the weighting function series can be determined by cross-correlation

$$\sum_{i=1}^{M} h_{M1}(\kappa_1) \prod_{j=1, j\neq i}^{M} h_{M2}(\kappa) = \frac{1}{(M-1)!\sigma_u^{2M}} \mathrm{E}\left\{\left[x_1(k-\kappa_1)\prod_{i=2}^{M} x_i(k-\kappa)\right] y'_M(k)\right\} \tag{3.13.6}$$

Proof. If the noise $n(k)$ has a zero mean, then its correlation with any function of the input signal is zero. Therefore the normalized output signal $y'_M(k)$ of the degree-M channel can be replaced by the normalized noise-free output signal $w'_M(k)$ in the expressions of the cross-correlations.

$$\begin{aligned} &\frac{1}{(M-1)!\sigma_u^{2M}} \mathrm{E}\left\{x_1(k-\kappa_1)\prod_{i=2}^{M} x_i(k-\kappa) y'_M(k)\right\} \\ &= \frac{1}{(M-1)!\sigma_u^{2M}} \mathrm{E}\left\{x_1(k-\kappa_1)\prod_{i=2}^{M} x_i(k-\kappa) w'_M(k)\right\} \end{aligned} \tag{3.13.7}$$

The proof will be shown for $M = 2$, for higher degree systems a similar proof can be derived.

$$\frac{1}{\sigma_u^4} \sum_{i_1=0}^{\infty} \sum_{i_2=0}^{\infty} h_{21}(i_1) h_{22}(i_2) \mathrm{E}\{\left[\left[x_1(k-i_1) + x_2(k-i_1)\right]\left[x_1(k-i_2) + x_2(k-i_2)\right] - \cdots\right.$$

$$\left. - \mathrm{E}\{\left[x_1(k-i_1) + x_2(k-i_1)\right]\left[x_1(k-i_2) + x_2(k-i_2)\right]\}\, x_1(k-\kappa_1) x_2(k-\kappa)\right\}$$

The expected value of the inner product is

$$\mathrm{E}\{\left[x_1(k-i_1) + x_2(k-i_1)\right]\left[x_1(k-i_2) + x_2(k-i_2)\right]\} = 2r_{uu}(i_2 - i_1)$$

The expected value of its product with $x_1(k-\kappa_1)x_2(k-\kappa)$ is

$$\mathrm{E}\{2r_{uu}(i_2 - i_1) x_1(k-\kappa_1) x_2(k-\kappa)\} = 0$$

The remaining terms are

$$\mathrm{E}\{[x_1(k-i_1)+x_2(k-i_1)][x_1(k-i_2)+x_2(k-i_2)]x_1(k-\kappa_1)x_2(k-\kappa)\}$$
$$=\mathrm{E}\{x_1(k-i_1)x_1(k-i_2)x_1(k-\kappa_1)x_2(k-\kappa)\}$$
$$+\mathrm{E}\{x_1(k-i_1)x_2(k-i_2)x_1(k-\kappa_1)x_2(k-\kappa)\}$$
$$+\mathrm{E}\{x_2(k-i_1)x_1(k-i_2)x_1(k-\kappa_1)x_2(k-\kappa)\}$$
$$+\mathrm{E}\{x_2(k-i_1)x_2(k-i_2)x_1(k-\kappa_1)x_2(k-\kappa)\}$$
$$=r_{uu}(\kappa_1-i_1)r_{uu}(\kappa-i_2)+r_{uu}(\kappa_1-i_2)r_{uu}(\kappa-i_1)=\sigma_u^4$$

Since it was taken into account that the sum differs from zero if ($\kappa_1=i_1$ and $\kappa=i_2$) or ($\kappa_1=i_2$ and $\kappa=i_1$).

Therefore, (3.13.7) becomes

$$\frac{1}{\sigma_u^4}\sigma_u^4[h_{21}(\kappa_1)h_{22}(\kappa)+h_{21}(\kappa)h_{22}(\kappa_1)]=\sum_{i=1}^{2}h_{2i}(\kappa_1)\prod_{j=1,j\neq i}^{2}h_{2j}(\kappa)$$ ■

Lemma 3.13.3
(Billings and Fakhouri, 1979b) Excite the process with the sum of independent PRBS signals $x_i(k)$ with very long periods N_i. Subtract the mean values from $x_i(k)$

$$x_i'(k)=x_i(k)-\mathrm{E}\{x_i(k)\}\quad i=1,2$$

and from the output signals of the linear and quadratic channels according to (3.13.2). Then it is approximately true that

$$h_1(\kappa_1)\approx\gamma_1\mathrm{E}\{x_i'(k-\kappa_1)y_1'(k)\}$$

$$\sum_{i=1}^{2}h_{2i}(\kappa_1)\prod_{j=1,j\neq i}^{2}h_{2j}(\kappa)\approx\gamma_2\mathrm{E}\{x_1'(k-\kappa_1)x_2'(k-\kappa)y_2'(k)\}$$

The values of the constants γ_i depend on the PRBS signal used.

Proof. See in Billings and Fakhouri (1980). ■

The next question is how to calculate the cross-correlation function with the output signal of the parallel channels. There are two ways:

- *Gardiner's method* (Gardiner, 1973b):
  The output signals of the individual parallel channels of different degrees can be calculated from repeated excitations where the input signals differ only in its scales (amplitudes).
- *Sequential correlation analysis* (Billings and Fakhouri, 1979b):
  The special cross-correlation functions can be computed from only one excitation as shown in Theorem 3.13.2.

Theorem 3.13.2 (Billings and Fakhouri, 1979b) Excite the process by the sum of M independent white noises $x_i(k)$ with zero mean. Assume the degree M of the highest power term in the nonlinearity is known. An additive noise $n(k)$ of zero mean is assumed at the output. Subtract the mean value from the output signal

$$y'(k) = y(k) - \mathrm{E}\{y(k)\}$$

The cross-correlation functions of the outputs of the individual channels can be calculated by the following procedure:

1. The special cross-correlation function of the output of the degree-M $(M \geq 2)$ channel $y_M(k)$ is the same as of the output of the process

$$\mathrm{E}\left\{x_1(k-\kappa_1)\prod_{i=2}^{M} x_i(k-\kappa)y'_M(k)\right\} = \mathrm{E}\left\{x_1(k-\kappa_1)\prod_{i=2}^{M} x_i(k-\kappa)y'(k)\right\} \tag{3.13.8}$$

2. Identify the parameters of the channel with degree-M by the methods given in Lemmas 3.13.2 and 3.13.1.
3. Compute the noise-free output signal $w_M(k)$ of the degree-M channel

$$w_M(k) = \prod_{j=1}^{M} h_{Mj}(i_j)u(k-i_j) = \prod_{j=1}^{M} \frac{B_{ij}(q^{-1})}{A_{ij}(q^{-1})}u(k-d_{ij}) \quad \text{if} \quad M \geq 2$$

4. Compute the output signal of the whole system except the degree-M channel as

$$\sum_{i=0}^{M-1} w_i(k) = y(k) - w_M(k)$$

5. Substitute M by $(M-1)$. If $M \geq 2$ go to Step 1, otherwise there is a linear system with a constant term

$$y(k) - \sum_{i=2}^{M} w_i(k) = c_0 + \frac{B_1(q^{-1})}{A_1(q^{-1})}u(k-d_1)$$

whose parameters are to be estimated.

Proof. It is enough to prove (3.13.8) in Step 1. The output signal reduced by its mean value is the sum of the outputs of the parallel channels

$$y'(k) = \sum_{\ell=0}^{M-1} y'_\ell(k) + y'_M(k)$$

(3.13.8) is valid if

$$\mathrm{E}\left\{x_1(k-\kappa_1)\left[\prod_{i=2}^{M} x_i(k-\kappa)\right]y'_\ell(k)\right\} = 0 \quad 0 \leq \ell \leq M \tag{3.13.9}$$

The noisy outputs $y'_\ell(k)$ contain the noise terms $n_\ell(k)$ with zero mean

$$y'_\ell(k) = w_\ell(k) - \mathrm{E}\{w_\ell(k)\} + n_\ell(k) \tag{3.13.10}$$

(3.13.10) can be partitioned into three parts. The correlation with the noise term is zero because the noise is independent from the multipliers. The correlation with the constant term is also zero, because the multiplier has a zero mean value. The only term that remains is

$$\mathrm{E}\left\{x_1(k-\kappa_1)\left[\prod_{i=2}^{M} x_i(k-\kappa)\right]w_\ell(k)\right\} = 0 \quad 0 \le \ell \le M \tag{3.13.11}$$

(3.13.11) is, in detail,

$$\sum_{i_1=0}^{\infty} \cdots \sum_{i_\ell=0}^{\infty} \prod_{i=1}^{\ell} h_{\ell j}(i_j)\mathrm{E}\left\{\prod_{j=1}^{\ell}\left[\sum_{i=1}^{\ell} x_1(k-i_j)\right]x_1(k-\kappa_1)\prod_{i=2}^{M} x_i(k-\kappa_i)\right\}$$

which is zero, because the mean value of the product of independent components can be computed one by one, and the term $x_M(k-\kappa)$ always stays single. ■

The presented facts are summarized in the following two algorithms.

Algorithm 3.13.2 Billings and Fakhouri (1979b) recommended the following procedure for the identification of the factorable Volterra model:

1. Excite the degree-M process by the sum of M independent white noises $x_i(k)$ with zero mean value.
2. Compute the cross-correlation functions (3.13.8) of the output signals of the parallel channels of different degrees according to Theorem 3.13.2.
3. Estimate the weighting function series from (3.13.1) computed for different shifting times according to Lemma 3.13.1.
4. Estimate the equivalent pulse transfer functions of all weighting functions in the model. ■

Algorithm 3.13.3 Billings and Fakhouri (1979b) recommended the following procedure for the identification of the factorable Volterra model:

1. Excite the degree-M process by $(M+1)$ similar Gaussian test signals, thus only the amplitude scalings differ.
2. Separate the output signals of the parallel channels of different degrees by Gardiner's method.
3. Compute the cross-correlation functions (3.13.6) of the output signals of the parallel channels of different degrees according to Lemma 3.13.2.
4. Estimate the weighting function series from (3.13.1) computed for different shifting times according to Lemma 3.13.1.
5. Estimate the equivalent pulse transfer functions of all weighting functions in the model. ■

Remarks:

1. A combination of the two methods is if Gardiner's separation method is applied with

a compound signal in Algorithm 3.13.2.
2. The sequential correlation method requires only one test contrary to the Gardiner's method which requires $(M+1)$ experiments for a degree-M system.

3.13.4 Frequency method

The output signals of the parallel channels can be separated by Gardiner's method (Gardiner, 1973b). The individual channels can be identified then by the frequency method.

Algorithm 3.13.4 The identification procedure of the factorable Volterra model consists of the following steps:
1. Excite the process by a sinusoidal signal .
2. Determine the order M of the highest-order harmonics.
3. Choose $(M+1)$ different amplitudes for the test signal. Denote the first one by U.
3. Excite the process by sinusoidal signal $U\cos\omega t$ at different frequencies. Repeat the experiments with the other amplitudes selected.
4. Compute the output signals of the channels to $U\cos\omega t$ at all frequencies measured using the Gardiner's method.
5. The constant term is the output of the degree-0 channel.
6. Identify the transfer function of the degree-1 channel by the known methods for linear systems.
7. Identify the transfer functions of the linear dynamic terms in the nonlinear channels using the special multitone method elaborated for the homogenous factorable Volterra systems (Rugh, 1981; Harper and Rugh, 1976). ■

The identification of homogenous factorable Volterra systems is dealt with in (Harper and Rugh, 1976) in details. There it was shown that a single sinusoidal excitation is not sufficient for the estimation of the parameters — in contrast to the case of a channel of the S_M model.

Further on we restrict our investigations to the quadratic channel.

Example 3.13.3 *Identification of the product of two linear dynamic terms*

The output signals of the linear dynamic terms to sinusoidal test signals are in steady state

$$v_i(t) = U|G_i(\omega)|\cos(\omega t + \varphi_i(\omega)), \quad i = 1, 2$$

The product of the outputs of the two linear dynamic terms is

$$v(t) = v_1(t)v_2(t) =$$
$$= \frac{U^2}{2}|G_1(\omega)||G_2(\omega)|\left[\cos(2\omega t + \varphi_1(\omega) + \varphi_2(\omega)) + \cos(\varphi_1(\omega) - \varphi_2(\omega))\right]$$

The degree-2 harmonic is

$$y^{(2)}(t) = \frac{U^2}{2}|G_1(\omega)||G_2(\omega)|\cos(2\omega t + \varphi_1(\omega) + \varphi_2(\omega))$$

which does not show how the breakpoints in the angular frequency lie, which means that one can not separate the poles and zeros of the different linear dynamic terms. For example, a system with the transfer functions of the linear dynamic terms in the quadratic channel

$$G_1(s) = \frac{1}{1+sT_1} \qquad G_2(s) = \frac{1}{1+sT_2}$$

can also be identified as

$$G_1(s) = 1 \qquad G_2(s) = \frac{1}{(1+sT_1)(1+sT_2)}$$

■

Lemma 3.13.4
(Harper and Rugh, 1976) A degree-2 homogenous channel of a factorable Volterra model can be identified with a two-frequency signal

$$u(t) = U_1\cos\omega_1 t + U_2\cos\omega_2 t \quad ; \quad \text{if } \omega_1 \neq \omega_2,\ \omega_1 \neq 2\omega_2 \text{ and } \omega_1 \neq 0.5\omega_2$$

Proof. The output signals of the sub channels are as follows

$$v_1(t) = U_1|G_1(\omega_1)|\cos(\omega_1 t + \varphi_1(\omega_1)) + U_2|G_1(\omega_2)|\cos(\omega_2 t + \varphi_1(\omega_2))$$
$$v_2(t) = U_1|G_2(\omega_1)|\cos(\omega_1 t + \varphi_2(\omega_1)) + U_2|G_2(\omega_2)|\cos(\omega_2 t + \varphi_2(\omega_2))$$

The output signal of the whole channel is the product

$$\begin{aligned} v(t) = v_1(t)v_2(t) &= U_1^2|G_1(\omega_1)||G_2(\omega_1)|\cos(\omega_1 t + \varphi_1(\omega_1))\cos(\omega_1 t + \varphi_2(\omega_1)) \\ &+ U_2^2|G_1(\omega_2)||G_2(\omega_2)|\cos(\omega_2 t + \varphi_1(\omega_2))\cos(\omega_2 t + \varphi_2(\omega_2)) \\ &+ [U_1U_2|G_1(\omega_1)||G_2(\omega_2)|\cos(\omega_1 t + \varphi_1(\omega_1))\cos(\omega_2 t + \varphi_2(\omega_2)) \\ &+ U_1U_2|G_1(\omega_2)||G_2(\omega_1)|\cos(\omega_2 t + \varphi_1(\omega_2))\cos(\omega_1 t + \varphi_2(\omega_1))] \end{aligned}$$

The first two terms cause harmonics with $2\omega_1$ and $2\omega_2$:

$$2\omega_1: \qquad y_2^{(2\omega_1)} = \frac{U_1^2}{2}|G_1(\omega_1)||G_2(\omega_1)|\cos(2\omega_1 t + \varphi_1(\omega_1) + \varphi_2(\omega_1))$$
$$2\omega_2: \qquad y_2^{(2\omega_2)} = \frac{U_2^2}{2}|G_1(\omega_2)||G_2(\omega_2)|\cos(2\omega_2 t + \varphi_1(\omega_2) + \varphi_2(\omega_2))$$

The terms in the bracket cause harmonics with $\omega_1 \pm \omega_2$

$$\begin{aligned} \omega_1+\omega_2: \qquad y_2^{(\omega_1+\omega_2)} &= \frac{U_1U_2}{2}|G_1(\omega_1)||G_2(\omega_2)|\cos((\omega_1+\omega_2)t + \varphi_1(\omega_1) + \varphi_2(\omega_2)) \\ &+ \frac{U_1U_2}{2}|G_1(\omega_2)||G_2(\omega_1)|\cos((\omega_1+\omega_2)t + \varphi_1(\omega_2) + \varphi_2(\omega_1)) \end{aligned}$$

$$\omega_1-\omega_2: \qquad y_2^{(\omega_1-\omega_2)}=\frac{U_1U_2}{2}|G_1(\omega_1)||G_2(\omega_2)|\cos((\omega_1-\omega_2)t+\varphi_1(\omega_1)-\varphi_2(\omega_2))$$
$$+\frac{U_1U_2}{2}|G_1(\omega_2)||G_2(\omega_1)|\cos((\omega_2-\omega_1)t+\varphi_1(\omega_2)-\varphi_2(\omega_1))$$

We have eight measurements to any ω_1, ω_2 combinations: the amplitudes and phases at $2\omega_1, 2\omega_2, \omega_1 \pm \omega_2$. The number of unknowns is also eight: the absolute values and the phases of $G_1(j\omega)$ and $G_2(j\omega)$ at the frequencies ω_1 and ω_2, i.e., the system is identifiable. ■

3.14 METHODS APPLIED TO OTHER BLOCK ORIENTED MODELS

From the manifold of the possible structures three types of models will be considered:

- quadratic extended Wiener–Hammerstein model;
- nonlinear model related to linear multi-input single-output model;
- closed-loop models.

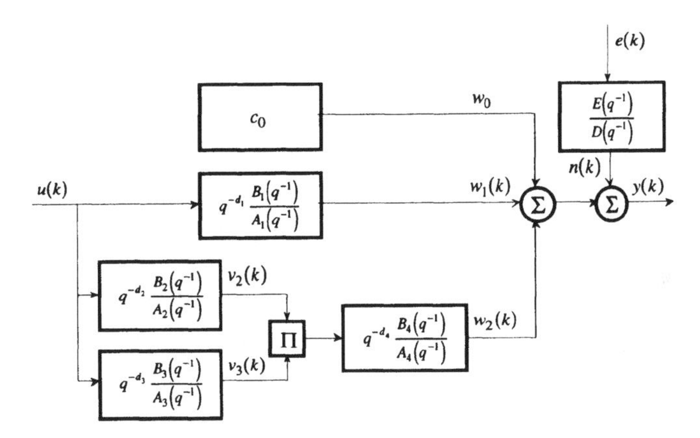

Fig. 3.14.1 The scheme of the discrete time quadratic extended Wiener–Hammerstein model

1. Quadratic extended Wiener–Hammerstein cascade model:

The quadratic extended Wiener–Hammerstein model includes the simple Wiener–Hammerstein model but also the generalized Hammerstein and the generalized and extended Wiener models as was shown in Section 1.3. If such a model is used as process model then one does not have to fix the structure among the listed models in advance. The scheme of the model is drawn in Figure 3.14.1. The equations of the model are as follows:

$$w_0=c_0$$
$$w_1(k)=\frac{B_1(q^{-1})}{A_1(q^{-1})}u(k-d_1), \qquad v_2(k)=\frac{B_2(q^{-1})}{A_2(q^{-1})}u(k-d_2)$$

$$v_3(k) = \frac{B_3(q^{-1})}{A_3(q^{-1})} u(k - d_3), \qquad w_2(k) = \frac{B_4(q^{-1})}{A_4(q^{-1})} [v_2(k - d_4) v_3(k - d_4)]$$

$$w(k) = w_0(k) + w_1(k) + w_2(k)$$

An additive colored noise

$$n(k) = \frac{E(q^{-1})}{D(q^{-1})} e(k)$$

is assumed at the output

$$y(k) = w(k) + n(k)$$

The residual can be expressed by the form

$$\varepsilon(k) = \frac{D(q^{-1})}{E(q^{-1})} [y(k) - w(k)]$$

The prediction error method can be applied, which needs the first derivative of the residual according to the unknown parameters. These are given now.

$$\frac{\partial \varepsilon(k)}{\partial c_0} = -\frac{D(1)}{E(1)}$$

$$\frac{\partial \varepsilon(k)}{\partial b_{1j}} = -\frac{D(q^{-1})}{E(q^{-1})} \frac{1}{A_1(q^{-1})} u(k - d_1 - j), \qquad \frac{\partial \varepsilon(k)}{\partial a_{1j}} = \frac{D(q^{-1})}{E(q^{-1})} \frac{1}{A_1(q^{-1})} w_1(k - j)$$

$$\frac{\partial \varepsilon(k)}{\partial b_{2j}} = -\frac{D(q^{-1})}{E(q^{-1})} \frac{B_4(q^{-1})}{A_4(q^{-1})} \left[v_3(k) \left(\frac{1}{A_2(q^{-1})} u(k - d_2 - j) \right) \right]$$

$$\frac{\partial \varepsilon(k)}{\partial a_{2j}} = -\frac{D(q^{-1})}{E(q^{-1})} \frac{B_4(q^{-1})}{A_4(q^{-1})} \left[v_3(k) \left(\frac{1}{A_2(q^{-1})} v_2(k - j) \right) \right]$$

$$\frac{\partial \varepsilon(k)}{\partial b_{3j}} = -\frac{D(q^{-1})}{E(q^{-1})} \frac{B_4(q^{-1})}{A_4(q^{-1})} \left[v_2(k) \left(\frac{1}{A_3(q^{-1})} u(k - d_3 - j) \right) \right]$$

$$\frac{\partial \varepsilon(k)}{\partial a_{3j}} = -\frac{D(q^{-1})}{E(q^{-1})} \frac{B_4(q^{-1})}{A_4(q^{-1})} \left[v_2(k) \left(\frac{1}{A_3(q^{-1})} v_3(k - j) \right) \right]$$

$$\frac{\partial \varepsilon(k)}{\partial b_{4j}} = -\frac{D(q^{-1})}{E(q^{-1})} \frac{1}{A_4(q^{-1})} [v_2(k - d_4 - j) v_3(k - d_4 - j)]$$

$$\frac{\partial \varepsilon(k)}{\partial a_{4j}} = \frac{D(q^{-1})}{E(q^{-1})} \frac{1}{A_4(q^{-1})} w_2(k-j) , \qquad \frac{\partial \varepsilon(k)}{\partial d_i} = \frac{1}{E(q^{-1})} [y(k-i) - w(k-i)]$$

$$\frac{\partial \varepsilon(k)}{\partial e_i} = -\frac{1}{E(q^{-1})} \varepsilon(k-i)$$

As is known, the prediction error method is an iterative procedure.

Example 3.14.1 *Quadratic, first-order generalized Wiener model (Haber et al., 1986)*

The parameters of the quadratic, first-order generalized Wiener model were

$$c_0 = 2, \qquad \frac{B_1(q^{-1})}{A_1(q^{-1})} = \frac{2}{1-0.5q^{-1}} , \qquad \frac{B_2(q^{-1})}{A_2(q^{-1})} = \frac{B_3(q^{-1})}{A_3(q^{-1})} = \frac{1}{1-0.5q^{-1}}$$

The noise-free process was excited by a Gaussian white noise signal with zero mean value and unit standard deviation. The number of data pairs was $N = 100$. Instead of a Gauss–Newton optimization the Fletcher–Powell algorithm was used. The parameters are summarized in Table 3.14.1 as a function of the iteration steps. The correct values were achieved in 11 steps. ■

TABLE 3.14.1 Estimated parameters of the first-order generalized Wiener model during the iteration

Iter.	$\hat{c}_0(1+\hat{a}_1)$ (1)	$\hat{b}_{10}$ (2)	$\hat{b}_{20}$ (1)	$\hat{a}_1$ (-0.5)	σ_ε
0	1.5	2.5	1.5	-0.6	617.8
1	1.445	2.537	0.713	0.431	713.7
2	1.317	2.509	0.42	-0.975	1681.2
3	1.215	2.433	0.812	-0.729	75.52
4	0.851	2.031	1.02	-0.536	2.78
5	1.027	2.023	0.978	-0.509	0.38
6	0.991	1.968	1.013	-0.483	0.35
7	1.003	2.014	1.001	-0.498	0.01
8	1.001	2	1	-0.499	10^{-4}
9	0.999	2	1	-0.5	10^{-7}
10	0.999	2	1	-0.5	10^{-10}
11	1	2	1	-0.5	10^{-17}

Example 3.14.2 *Quadratic, second-order generalized Wiener model (Haber et al., 1986)*

The parameters of the quadratic, second-order generalized Wiener model were

$$c_0 = 2$$

$$\frac{B_1(q^{-1})}{A_1(q^{-1})} = \frac{2+q^{-1}}{1+1.5q^{-1}-0.7q^{-2}}$$

$$\frac{B_2(q^{-1})}{A_2(q^{-1})}=\frac{B_3(q^{-1})}{A_3(q^{-1})}=\frac{1+0.5q^{-1}}{1+1.5q^{-1}-0.7q^{-2}}$$

The noise-free process was excited by a Gaussian white noise signal with zero mean value and unit standard deviation. The number of data pairs was $N=100$. Instead of a Gauss–Newton optimization the Fletcher–Powell algorithm was used. The parameters are summarized in Table 3.14.2 as a function of the iteration steps. The correct values were achieved in 16 steps. ■

TABLE 3.14.2 Estimated parameters of the second-order generalized Wiener model during the iteration

Iter.	$\hat{c}_0(1+\hat{a}_1+\hat{a}_2)$ (0.4)	$\hat{b}_{10}$ (2)	$\hat{b}_{11}$ (1)	$\hat{b}_{20}$ (1)	$\hat{b}_{21}$ (0.5)	$\hat{a}_1$ (-1.5)	$\hat{a}_2$ (0.7)	σ_ε
0	1	1.5	1.5	1.5	0.6	-1.6	0.8	$2\cdot 10^4$
1	0.991	1.506	1.503	0.713	-0.008	-0.325	-0.534	$3\cdot 10^4$
2	0.997	1.502	1.501	1.242	0.399	-1.18	0.383	2965.1
3	0.998	1.502	1.5	1.29	0.448	-1.309	0.467	2690.3
4	0.988	1.523	1.501	1.007	0.466	-1.38	0.531	1769.5
5	0.964	1.561	1.51	0.993	0.53	-1.378	0.522	1740
6	0.067	3.715	1.57	1.049	0.569	-1.37	0.531	1036.3
7	-0.842	-1.79	0.274	0.825	0.474	-1.59	0.824	1548.6
8	0.702	7.76	6.665	1.267	0.826	-1.37	0.593	4241.5
9	0.155	1.359	1.97	0.998	0.303	-1.62	0.81	829.5
10	0.272	2.191	0.315	1.059	0.524	-1.459	0.648	170.3
11	0.398	1.842	0.612	1.02	0.47	-1.537	0.732	108.4
12	0.258	2.226	1.017	1	0.465	-1.508	0.702	20.5
13	0.382	2.03	0.958	0.999	0.499	-1.499	0.701	0.203
14	0.398	2.001	0.999	1	0.5	-1.5	0.699	$1\cdot 10^{-4}$
15	0.4	2	1	1	0.5	-1.5	0.7	$5\cdot 10^{-13}$
16	0.4	2	1	1	0.5	-1.5	0.7	$6\cdot 10^{-17}$

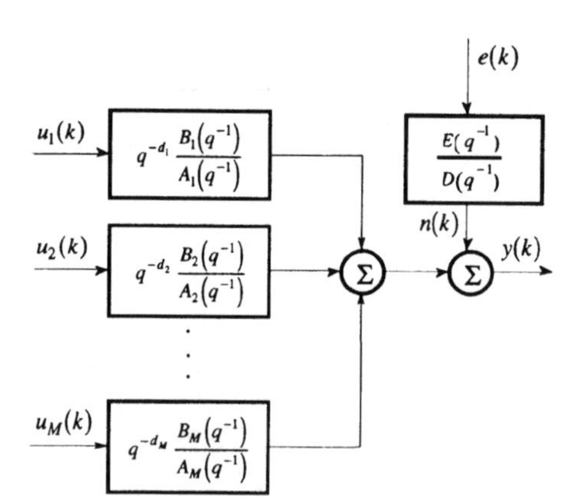

Fig. 3.14.2 Multi-input single-output linear dynamic model

2. Multi-input single-output equivalent nonlinear model

It is assumed that the model consists of parallel linear dynamic channels with nonlinear components of the input and possibly of the output signal as input signals. The model can be considered formally as a linear, multi-input single-output model (Figure 3.14.2).

Theoretically a reverse estimation of the channels can be formed according to the following algorithm.

Algorithm 3.14.1 The nonlinear model that is formally equivalent to a linear multi-input single-output model can be identified by a sequential iterative estimation of the channels as follows:

1. Assume an initial parameter vector for each subsystem except the ith $(0 \le i \le M)$.
2. Compute the output signals of all channels except i.
3. Compute the output signal of channel i as the difference of the measured output signal and the sum of the output signals of all submodels except the ith one.
4. Estimate the parameters of channel i from the measured input signal and from the computed output signal of this channel.
5. Compute the output signal of channel i.
6. Check whether the iteration has to be stopped. If not then first increase i by 1 and let $i ::= 0$ if $i > M$ and then go to Step 2. ■

Remark:
The same algorithm was used for two-input single-output Hammerstein model in Corlis and Luus (1969).

Example 3.14.3 *Identification of a quadratic, first-order simple Wiener model by the parametric Volterra model*

The parameters of the quadratic, first-order simple Wiener model were

$$v(t) + 10\dot{v}(t) = u(t), \qquad y(k) = 2 + v(k) + 0.5v^2(k)$$

TABLE 3.14.3 Estimated parameters of the parametric Volterra model during the iteration while identifying the simple Wiener model

Iter.	*Linear part*			*Channel with input $u^2(k)$*			*Channel with input $u(k)u(k-1)$*		
	$\hat{b}_{10}$	$\hat{a}_{11}$	σ_ε	$\hat{b}_{20}$	$\hat{a}_{21}$	σ_ε	$\hat{b}_{30}$	$\hat{a}_{31}$	σ_ε
1	0.3264	-0.8466	0.2056	0.01374	-0.9211	0.1942	0.01225	-0.9492	0.1880
2	0.2831	-0.8224	0.1799	0.02139	-0.9083	0.1659	0.02222	-0.9155	0.1594
3	0.2505	-0.8108	0.1639	0.02609	-0.8940	0.1490	0.02995	-0.8954	0.1445
4	0.2278	-0.8071	0.1540	0.02837	-0.8861	0.1405	0.03545	-0.8855	0.1381
5	0.2137	-0.8059	0.1498	0.02906	-0.8821	0.1368	0.03934	-0.8797	0.1351
6	0.2057	-0.8052	0.1479	0.02883	-0.8806	0.1352	0.04210	-0.8761	0.1349
7	0.2018	-0.8044	0.1470	0.02805	-0.8809	0.1343	0.04406	-0.8737	0.1347
8	0.2003	-0.8034	0.1464	0.02708	-0.8821	0.1337	0.04547	-0.8721	0.1345
9	0.2003	-0.8024	0.1461	0.02602	-0.8839	0.1332	0.04649	-0.8710	0.1344
10	0.2011	-0.8014	0.1459	0.02500	-0.8859	0.1327	0.04723	-0.8703	0.1343
11	0.2022	-0.8004	0.1458	0.02411	-0.8879	0.1324	0.04777	-0.8699	0.1342

The parametric Volterra model had the equation

$$y(k) = c_0 + \frac{b_{10}}{1 + a_{11}q^{-1}} u(k-1) + \frac{b_{20}}{1 + a_{21}q^{-1}} u^2(k-1)$$
$$+ \frac{b_{30}}{1 + a_{31}q^{-1}} u(k-1)u(k-2)$$

The test signal was a PRTS with maximum length 26, amplitude $\Delta U = \pm 2$ and mean value 1. The sampling time was $\Delta T = 2$ [s] and the minimum switching time of the PRTS was 5-times more. $N = 26 \cdot 5 = 130$ data pairs were used for the identification. Table 3.14.3 shows the parameters of the dynamic channels during the iterations. The input and output signal of the process and the computed output signals based on the estimated model are seen in Figure 3.14.3. The fit is quite good. ■

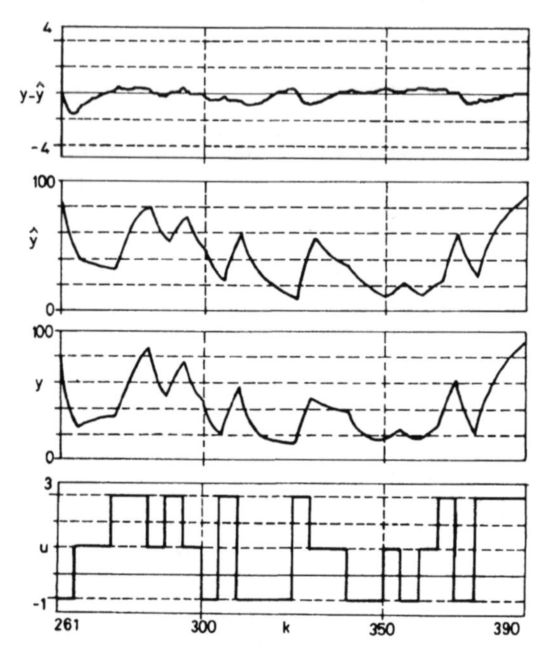

Fig. 3.14.3 Measured input (u) and output (y) signals of the simple Wiener model and the simulated output signal $(\hat{y})$ based on the estimated parametric Volterra model

3. Closed loop models

Two feedback schemes with single polynomial nonlinearity are drawn in Figure 3.14.4. A direct optimization without derivatives or a prediction error method can easily be derived. A correlation method was recommended by Billings and Fakhouri (1979d, 1979e). The method relies on the relation between the parts of the process and the linear and quadratic channels of the closed-loop system. Rosenthal and Beyer (1985) deals also with the identification of nonlinear closed systems.

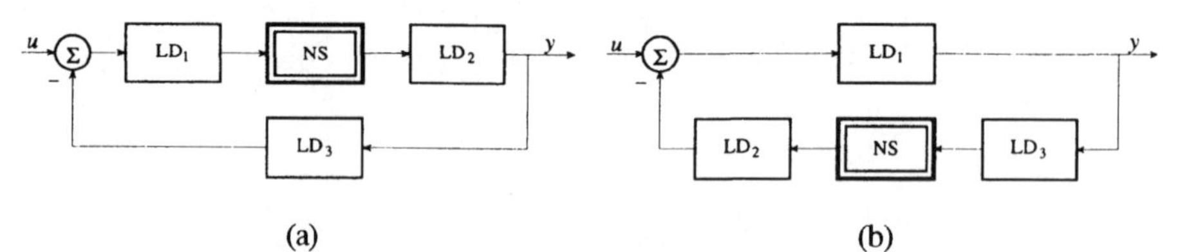

Fig 3.14.4 Closed loop models with single nonlinearity: (a) nonlinearity in the feedforward path; (b) nonlinearity in the feedback path

3.15 IDENTIFICATION OF LINEAR AND NONLINEAR MULTI-MODELS

A multi-model is a set of models that belongs to different values of a signal. This model belongs to the class of the gate models. A presentation of the gate models is in Section 1.7 and the identification method was given in Section 3.5.

Multi-models can be identified by a sequence of test signals exciting the domains of the process in turn or by a test signal that covers the whole input range. In the latter case the input and output signals are to be gated according to the different states of a signal or signals on which the parameters and/or structure of the multi-model depends.

From the regression point of view it is irrelevant whether the fitted model is linear or nonlinear. In most cases linear multi-models are used.

The identification of two processes with signal dependent parameters illustrates the method.

Example 3.15.1 *First-order quasi-linear system with output signal dependent static gain (Haber et al., 1986)*

The process is a noise-free first-order quasi-linear system with a signal dependent static gain

$$T\dot{y}(t) + y(t) = Ku(t)$$
$$T = T_0 = 10[\text{s}]$$
$$K(t) = K_0 + K_1 y(t) = 2 - 0.5y(t)$$

Figure 3.15.1 shows the dependence of the true continuous time parameters on the output signal. The input signal consists of four constant periods, each being 60 [s] long. The value of the input signal was

$$u(t) = \begin{cases} 1.0 & \text{if} \quad 0\,[\text{s}] \le t < 60\,[\text{s}] \\ 0.0 & \text{if} \quad 60\,[\text{s}] \le t < 120\,[\text{s}] \\ -0.5 & \text{if} \quad 120\,[\text{s}] \le t < 180\,[\text{s}] \\ 0.0 & \text{if} \quad 180\,[\text{s}] \le t < 240\,[\text{s}] \end{cases}$$

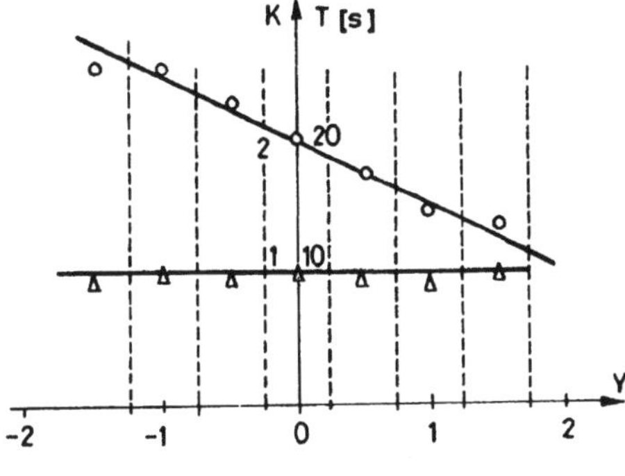

Fig. 3.15.1 True and estimated parameters of the first-order process with output signal dependent gain

The measured input and output signals and the change of the static gain during the experiment are plotted in Figure 3.15.2. The range of the output signal is about

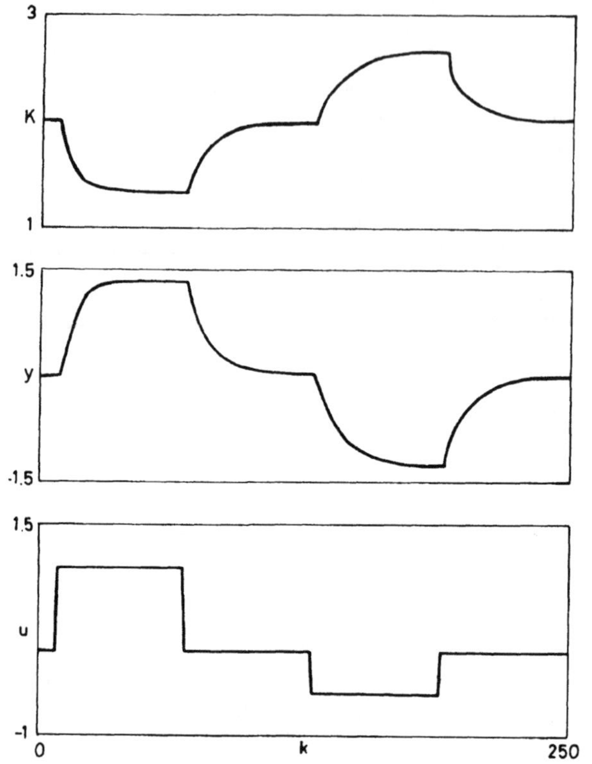

Fig. 3.15.2 Measured input (u), output (y) signals and the static gain (K)

TABLE 3.15.1 Estimated parameters of the domains at the identification of the system with output signal dependent gain

Domain	1	2	3	4	5	6	7
$Y_{min} \div Y_{max}$	-1.75÷-1.25	-1.25÷-0.75	-0.75÷-0.25	-0.25÷0.25	0.25÷0.75	0.75÷1.25	1.25÷1.75
$\frac{Y_{min}+Y_{max}}{2}$	-1.5	-1.0	-0.5	0.0	0.5	1.0	1.5
$\hat{a}_1$	-0.9492	-0.9524	-0.9519	-0.9518	-0.9506	-0.9499	-0.9508
$\sigma\{\hat{a}_1\}$	$6.363\ 10^{-3}$	$0.1705\ 10^{-3}$	$0.135\ 10^{-3}$	$0.334\ 10^{-3}$	$0.187\ 10^{-3}$	$0.255\ 10^{-3}$	$0.338\ 10^{-3}$
$\hat{b}_1$	0.1289	0.121	0.1091	0.09718	0.08488	0.07282	0.06587
$\sigma\{\hat{b}_1\}$	$121.1\ 10^{-3}$	$3.116\ 10^{-3}$	$1.815\ 10^{-3}$	$2.118\ 10^{-3}$	$2.049\ 10^{-3}$	$4.072\ 10^{-3}$	$6.481\ 10^{-3}$
σ_ε	$8.583\ 10^{-4}$	$5.496\ 10^{-4}$	$3.063\ 10^{-4}$	$3.661\ 10^{-4}$	$4.22\ 10^{-4}$	$7.449\ 10^{-4}$	$4.302\ 10^{-4}$
$\hat{K}$	2.538	2.544	2.271	2.015	1.717	1.455	1.338
K	2.75	2.5	2.25	2.0	1.75	1.5	1.25
$\hat{T}$	9.594	10.25	10.15	10.12	9.86	9.736	9.902
T	10.0	10.0	10.0	10.0	10.0	10.0	10.0

$-1.5 \le Y \le 1.5$. Therefore, the range (-1.75, 1.75) was divided into seven parts of equal widths ($\Delta Y = 0.5$). The domains are given in Table 3.15.1. The boundaries of the domains are also drawn in Figure 3.15.1. The sampling time of the identification was $\Delta T = 0.5$ [s]. Table 3.15.1 presents the results of the LS parameter estimation of first-order linear processes

$$\hat{y}(k) = \left[\hat{a}_1 \pm \sigma\{\hat{a}_1\}\right]\hat{y}(k-1) + \left[\hat{b}_1 \pm \sigma\{\hat{b}_1\}\right]u(k-1)$$

in each domain. The estimated static gains and time constants of each domain are drawn in the plot of the true parameters in Figure 3.15.1. As is seen, the coincidence is excellent. ■

Example 3.15.2 *First-order quasi-linear system with output signal-dependent time constant excited by small steps*

The process simulated is a noise-free first-order quasi-linear system with an output signal dependent time constant

$$T\dot{y}(t) + y(t) = Ku(t)$$
$$K(t) = K_0 = 2$$
$$T^{-1}(t) = \overline{T}_0^{-1} + \overline{T}_1^{-1}y(t) = 0.1 + 0.05y(t)\ \left[s^{-1}\right]$$

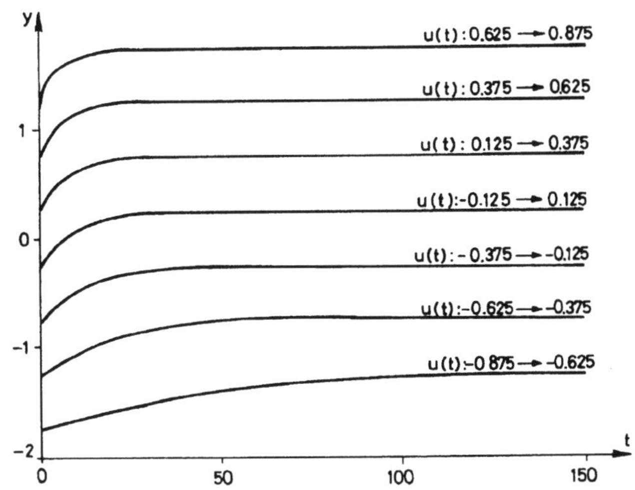

Fig. 3.15.3 Stepwise excitation of the first-order process with output signal dependent time constant

The process was excited by step signals with small amplitudes of $\Delta Y = 0.5$ in different working points between $y_{min} = -1.5$ and $y_{max} = 1.5$. The input signal consisted of stepwise changes with the following values: $-0.875 \rightarrow -0.625 \rightarrow\ -0.375 \rightarrow -0.125 \rightarrow$ $0.125 \rightarrow 0.375\ \rightarrow 0.625 \rightarrow 0.875$. Figure 3.15.3 shows the plot of the output signals. Each test took 150 [s]. The individual step responses were evaluated grapho-analytically as first-order systems. The static gain is the relation

$$K = \frac{y(\infty) - y(0)}{u(\infty) - u(0)}$$

and the time constant was read as the time belonging to $y_{63\%}$, the 63% value of the step response. The estimated parameters correspond to the middle value of the output signal, i.e., to the middle value of the boundaries of the domain. Table 3.15.2 summarizes the static gains and the time constants of each domain. The estimated values are drawn in

TABLE 3.15.2 Estimated static gains and time constants of the domains at the identification of the first-order system with output signal dependent time constant

No	$u_{min} \to u_{max}$	$y_{min} - y_{max}$	$\frac{y_{min}+y_{max}}{2}$	$\hat{K}$	K	$y_{63\%}$	$\hat{T}$	T	$\hat{a}_1$	$\hat{b}_1$
1	-0.875→-0.625	-1.75→-1.25	-1.5	2.0	2.0	-1.435	48.0	40	-0.9794	0.0412
2	-0.625→-0.375	-1.25→-0.75	-1.0	2.0	2.0	-0.935	22.5	20	-0.9565	0.0870
3	-0.375→-0.125	-0.75→-0.25	-0.5	2.0	2.0	-0.435	14.0	13.33	-0.9311	0.1378
4	-0.125→0.125	-0.25→0.25	0.0	2.0	2.0	0.065	10.5	10	-0.9092	0.1816
5	0.125→0.375	0.25→0.75	0.5	2.0	2.0	0.565	8.1	8	-0.8839	0.2322
6	0.375→0.625	0.75→1.25	1.0	2.0	2.0	1.065	6.6	6.67	-0.8594	0.2812
7	0.625→0.875	1.25→1.75	1.5	2.0	2.0	1.565	6.0	5.71	-0.8465	0.3070

the plot of the true parameters in Figure 3.15.4. The approximation of the true parameters is good. ■

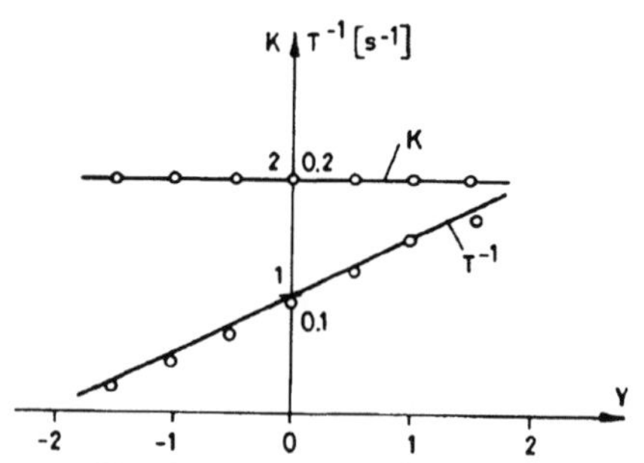

Fig. 3.15.4 True and grapho-analytically estimated parameters of the first-order process with output signal dependent time constant

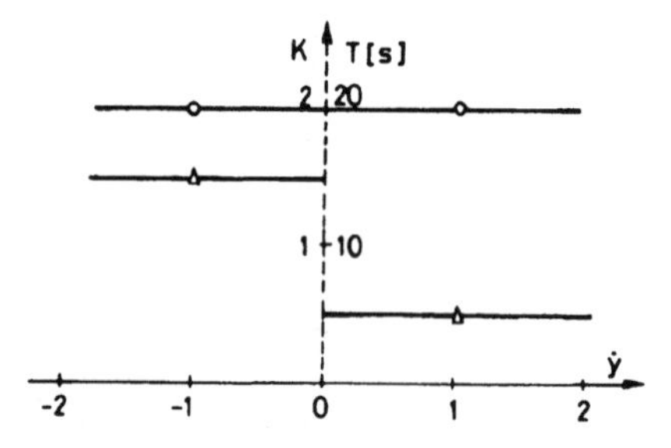

Fig. 3.15.5 True and estimated parameters of the first-order process with direction dependent time constant

Example 3.15.3 *First-order quasi-linear system with a direction dependent time constant (Haber et al., 1986)*

The process is a noise-free first-order quasi-linear system with a direction dependent time constant

$$T(t)\dot{y}(t) + y(t) = Ku(t)$$

$$K = K_0 = 2$$

$$T(t) = T_0 + T_1\mathrm{sgn}(\dot{y}(t)) = 10 - 5\mathrm{sgn}(\dot{y}(t)) \ [\mathrm{s}]$$

Figure 3.15.5 shows the dependence of the true continuous time parameters on the output signal. The input signal was the same as in Example 3.15.1. The measured input and output signals and the change of the time constant during the experiment are plotted in Figure 3.15.6. The input and output records are divided into two parts:

- part 1: where the output signal decreases;
- part 2: where the output signal increases.

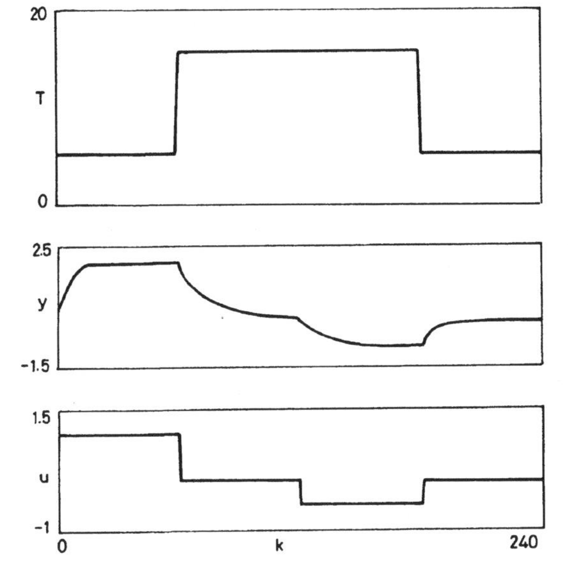

Fig. 3.15.6 Measured input (u) and output (y) signals and the time constant (T) of the first-order process with direction dependent time constant

Domain	1	2
$\mathrm{sgn}(\dot{y}(t))$	<0	>0
$\hat{a}_1$	-0.9355	-0.8187
$\sigma\{\hat{a}_1\}$	$0.137\ 10^{-4}$	$1.022\ 10^{-4}$
$\hat{b}_1$	0.129	0.3625
$\sigma\{\hat{b}_1\}$	$0.293\ 10^{-4}$	$1.915\ 10^{-4}$
σ_ε	$0.8235\ 10^{-4}$	$3.683\ 10^{-4}$
$\hat{K}$	2.0	2.0
K	2.0	2.0
$\hat{T}$	15.0	4.999
T	15.0	5.0

TABLE 3.15.3 Estimated static gains and time constants of the domains at the identification of the first-order system with direction dependent time constant

The domains are given in Table 3.15.3. The boundaries of the domains are also drawn in Figure 3.15.6. The sampling time of the identification was $\Delta T = 2$ [s]. Table 3.15.3 presents the results of the LS parameter estimation of first-order linear processes

$$\hat{y}(k) = \left[\hat{a}_1 \pm \sigma\{\hat{a}_1\}\right]\hat{y}(k-1) + \left[\hat{b}_1 \pm \sigma\{\hat{b}_1\}\right]u(k-1)$$

in each domain. The estimated static gains and time constants of both domains are drawn in the plot of the true parameters in Figure 3.15.5. As is seen, the coincidence is perfect. ■

Example 3.15.4 *First-order quasi-linear system with a direction dependent time constant excited by small steps*

The process is the same as in Example 3.15.3. Instead of performing a new test the exciting test given in Figure 3.15.6 can be evaluated grapho-analytically, since only the direction and not the size of the step is relevant.

- *part 1: where the output signal decreases:*
  These are the steps $u(t): 1 \to 0$ at $t = 60$ [s] and $u(t): 0 \to -0.5$ at $t = 120$ [s] The output signals decrease in both cases as the response of a first-order system with the parameters $\hat{K} = 2$ and $\hat{T} = 15$ [s].
- *part 2: where the output signal increases:*
  These are the steps $u(t): 0 \to 1$ at $t = 0$ [s] and $u(t): -0.5 \to 0$ at $t = 120$ [s]. The output signals increase in both cases as the response of a first-order system with the parameters $\hat{K} = 2$ and $\hat{T} = 5$ [s]. ∎

The output signal of the globally valid nonlinear model is

- either an output signal of one of the locally valid models (e.g., Pickhardt, 1997);
- or the weighted sum of the outputs of all locally valid submodels (e.g., Johansen and Foss, 1993,1995b; Nelles *et al.*, 1997).

Identification algorithms to find the local models automatically are reported in e.g., Nelles *et al.* (1997), Johansen and Foss (1995a), Pickhardt (1997), Hachino and Takata (1997).

Some examples for the identification of multi-models are given in the following papers:

- *distillation column* (Haber *et al.*, 1982; Haber and Zierfuss, 1991);
- *railway vehicle dynamics* (Broersen, 1973);
- *motor generator system* (Diekmann and Unbehauen, 1985);
- *air transportation system* (Diekmann and Unbehauen, 1985);
- *turbo-generator* (Jedner and Unbehauen, 1986);
- *fermentation process* (Johansen and Foss, 1995a);
- *exothermic chemical reaction in a continuous stirred tank reactor* (Johansen and Foss, 1993, 1995b);
- *pH-process* (Pottmann *et al.*, 1993);
- *transportation system for bulk goods* (Pickhardt, 1997);
- *heat exchanger* (Nelles *et al.*, 1997).

3.16 IDENTIFICATION OF QUASI-LINEAR MODELS WITH SIGNAL DEPENDENT PARAMETERS

A wide class of processes can be described by quasi-linear models with signal dependent parameters. This is based on the manifold of the imaginable signal dependencies. As control algorithms based on linear models can be easily applied to such models, makes the use of this model class very practical.

Two types of models can be used to describe processes with signal dependent parameters:

- models with analytical function of the signal dependence;
- models in tabular form of the signal dependence.

In this section models with analytical function of the signal dependence will be treated, models in tabular form of the signal dependence belong to the multi-models.

The identification methods applied to quasi-linear models with signal dependent parameters can be classified as follows:

1. two-step method: fitting a signal dependent model to a linear multi-model (to a set of piecewise linear models);
2. identification of continuous time signal dependent models in a form linear-in-parameters;
3. identification of continuous time signal dependent models in a form nonlinear-in-parameters
4. identification of discrete time signal dependent models in a form linear-in-parameters.

All methods can be performed from normal operating data. However, a linear multi-model, also called a set of piecewise linear models, can be more easily identified from small excitations in different working points. If step like excitations are applied the step responses can be evaluated grapho-analytically. Although a grapho-analytical method is less accurate than a statistical estimation, the step responses give an insight into the structure of the system.

3.16.1 Two-step method: fitting a signal dependent model to a linear multi-model

The procedure is very simple: a functional relationship has to be set up between the identified parameters of a set of multi-models (piecewise linear models). A graphical plot of the parameters against the signal the models depend on can help in choosing the correct function. There are some advantages of this method against the other methods:

- Parameter estimation methods applied to linear systems can be applied for determining the set of piecewise linear models;
- If applying special test signals grapho-analytical methods are sufficient;
- Instead of estimating parameters of a nonlinear dynamic model only the parameters of linear dynamic models and of nonlinear static models are sought;
- As a consequence of the above, fewer parameters have to be estimated than with the other methods.

Example 3.16.1 *First-order quasi-linear system with output signal dependent static gain identified from normal operating data (Continuation of Example 3.15.1)*

In Example 3.15.1 the linear multi-model of a noise-free first-order process with output signal dependent time constant

$$T\dot{y}(t) + y(t) = Ku(t)$$

$$T = T_0 = 10 \text{ [s]}$$

$$K(t) = K_0 + K_1 y(t) = 2 - 0.5y(t)$$

was identified. No special test signal was applied, the multi-model was determined in the range between the lowest and uppermost values of the input signal. The estimated parameters were given in Table 3.15.1. Both continuous and discrete time models can be set up. The estimated static gain $\hat{K}$ and time constant $\hat{T}$ were drawn in Figure 3.15.1. The following static functions were fitted to the estimated parameters:

$$\hat{b}_1(t) = (0.09711 \pm 0.00067) - (0.02212 \pm 0.00067)y(t)$$

$$\hat{a}_1 = -0.9509 \pm 0.00044$$

$$\hat{K}(t) = (1.983 \pm 0.034) - (0.4523 \pm 0.034)y(t)$$
$$\hat{T} = 9.945 \pm 0.09 \text{ [s]}$$

Figures 3.16.1 and 3.16.2 show the true and fitted parameter dependencies. The function of true value of the parameters of the difference equation was calculated as

$$a_1 = -\exp(-\Delta T/T) = -\exp(-0.5/10) = -0.9512$$
$$b_1 = [K_0 + K_1 Y](1 + a_1) = [2 - 0.5Y](1 - 0.9512) = 0.0976 - 0.0244Y$$

The coincidence between the true and the estimated parameters is good. ■

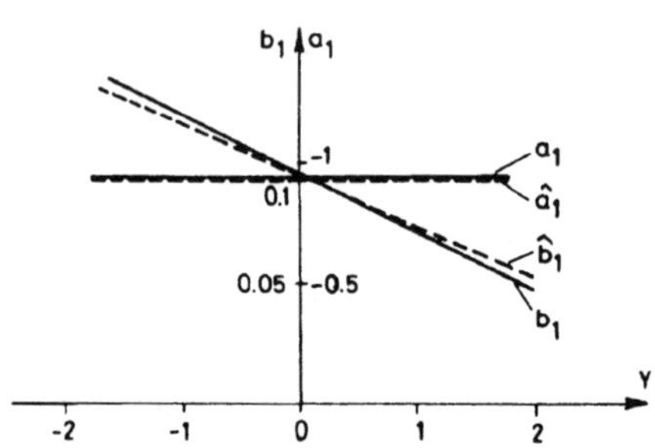

Fig. 3.16.1 True and estimated parameters of the difference equation of the first-order process with output signal dependent time constant identified from a set of small excitations

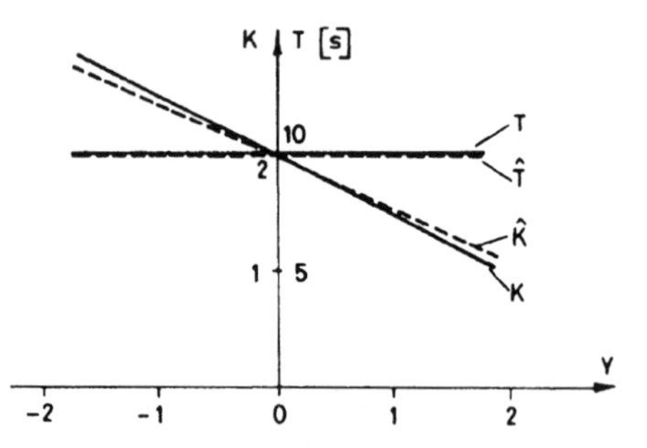

Fig. 3.16.2 True and estimated parameters of the differential equation of the first-order process with output signal dependent time constant identified from a set of small excitations

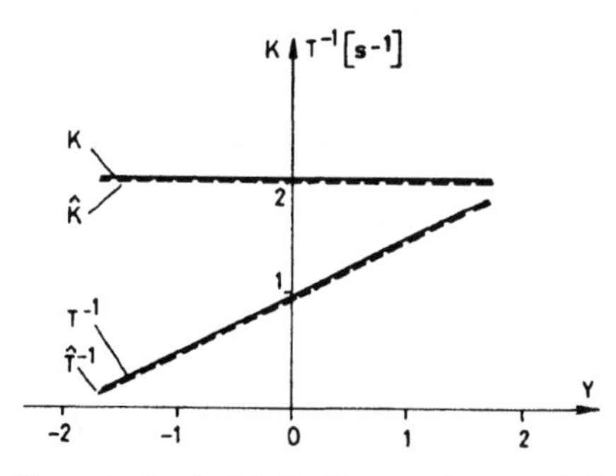

Fig. 3.16.3 True and estimated parameters of the differential equation of the first-order process with output signal dependent time constant

Example 3.16.2 *First-order quasi-linear system with output signal dependent time constant excited by small steps (Continuation of Example 3.15.2)*

In Example 3.15.2 a process with output signal dependent time constant was excited by step signals with small amplitudes in different working points. The plot of the measured signals was given in Figure 3.15.3. The individual step responses were evaluated grapho-analytically as first-order systems. Table 3.15.2 summarizes the static gains and the time constants of each domain. The estimated values were drawn in the plot of the true parameters in Figure 3.15.4. Based on this the following functions were fitted by the LS method

$$\hat{K} = 2.0$$

$$\hat{T}^{-1}(t) = (0.09623 \pm 0.00128) + (0.05026 \pm 0.00128) y(t) \left[\mathrm{s}^{-1}\right]$$

Figure 3.16.3 shows the plot of the true and the fitted values of the static gain and of the time constant. The coincidence is very good. ■

3.16.2 Identification of continuous time signal dependent models in a form linear-in-parameters

In Section 1.6 was shown that continuous time processes with signal dependent parameters can be modeled by difference equations linear-in-parameters if the Euler or the bilinear discretization transformation is used. Tables 1.6.5 to 1.6.10 summarize the difference equations of the first-order and second-order processes with a linear and inverse linear dependence of the static gain, time constant and damping factor (if exists). This method has:

1. *the advantage:*
   - the difference equation is linear in the parameters;
2. *the disadvantage:*
   - the number of parameters to be estimated is often more than the number of unknown parameters;
   - depending on the input signal and the signal dependence the model components may be strongly correlated.

Example 3.16.3 *First-order quasi-linear system with output signal dependent static gain identified from normal operating data (Continuation of Example 3.15.1)*

In Example 3.15.1 a noise-free first-order process with output signal dependent time constant was simulated with a test signal consisting of four steps with different magnitudes and directions. As the special form of the test signal will not be exploited, the measured data pairs are considered as normal operating data. Tables 1.6.5 and 1.6.8 give the difference equations of first-order system with linearly dependent static gain. The signal the parameters depend on now is the output signal. The input–output difference equation becomes:

- using Euler transformation:

$$y(k) = -a_1 y(k-1) + b_1 u(k-1) + c_1 y(k) u(k-1) \tag{3.16.1}$$

$$a_1 = \frac{\Delta T}{T_0} - 1 \qquad b_1 = K_0 \frac{\Delta T}{T_0} \qquad c_1 = K_1 \frac{\Delta T}{T_0}$$

- using bilinear transformation:

$$y(k) = -a_1 y(k-1) + b_1 \left[u(k) + u(k-1)\right] + c_1 y(k) \left[u(k) + u(k-1)\right] \tag{3.16.2}$$

$$a_1 = \frac{\Delta T - 2T_0}{\Delta T + 2T_0} \qquad b_1 = K_0 \frac{\Delta T}{\Delta T + 2T_0} \qquad c_1 = K_1 \frac{\Delta T}{\Delta T + 2T_0}$$

The parameters of the difference equations were estimated by the LS method.

TABLE 3.16.1 Estimated parameters of the first-order system with output signal dependent gain when fitting a model linear-in-parameters at the sampling time $\Delta T = 1\,[\mathrm{s}]$

Parameter (true value)		*Transformation*	
		Euler	*bilinear*
of the difference equation	a_1	-0.9043±0.00017	-0.9076±0.000975
	b_1	0.1911±0.000316	0.09318±0.000932
	c_1	-0.04753±0.000133	-0.02324±0.000385
of the differential equation	K_0 (2)	1.997	2.017
	K_1 (−0.5)	-0.497	-0.503
	T_0 (10)	10.499	10.323

Table 3.16.1 gives the true and the estimated parameters for the sampling time $\Delta T = 1$ [s]. The parameters of the continuous time process K_0, K_1 and T_0 can be calculated by the equations

- for the Euler transformation:

$$T_0 = \frac{\Delta T}{a_1 + 1} \qquad K_0 = b_1 \frac{T_0}{\Delta T} \qquad K_1 = c_1 \frac{T_0}{\Delta T}$$

- for the bilinear transformation:

$$T_0 = \frac{\Delta T}{2} \frac{1 - a_1}{1 + a_1} \qquad K_0 = b_1 \left[1 + 2\frac{T_0}{\Delta T}\right] \qquad K_1 = c_1 \left[1 + 2\frac{T_0}{\Delta T}\right]$$

TABLE 3.16.2 Estimated parameters of the first-order system with output signal dependent gain when fitting a model linear-in-parameters and using different sampling times

Transformation	*Parameter*	*Sampling time* [s]			
		0.5	1.0	2.0	4.0
Euler	K_0	1.996	1.997	1.997	1.994
	T_0	10.204	10.449	10.971	12.110
	T_1	-0.498	-0.497	-0.494	-0.490
bilinear	K_0	2.006	2.017	2.035	2.069
	T_0	10.123	10.323	10.765	11.723
	T_1	-0.501	-0.503	-0.506	-0.514

Figure 3.16.4 shows the time signals of the identification based on the bilinear approximation: measured input and output signals, the nonlinear component of the difference equation (3.16.2) $y(k)[u(k) + u(k+1)]$, the simulated output signals $\hat{y}_{\mathrm{L}}(k)$

and $\hat{y}_{NL}(k)$ based on the linear and nonlinear model, respectively. Table 3.16.2 summarizes the estimated parameters of the continuous time model using the sampling times $\Delta T = 0.5, 1, 2, 4$ [s]. The smaller the sampling time the better the fit to the true parameters.

Observe that the number of the parameters in the difference equation model linear-in-parameters is equal to the number of the parameters of the continuous time model. ■

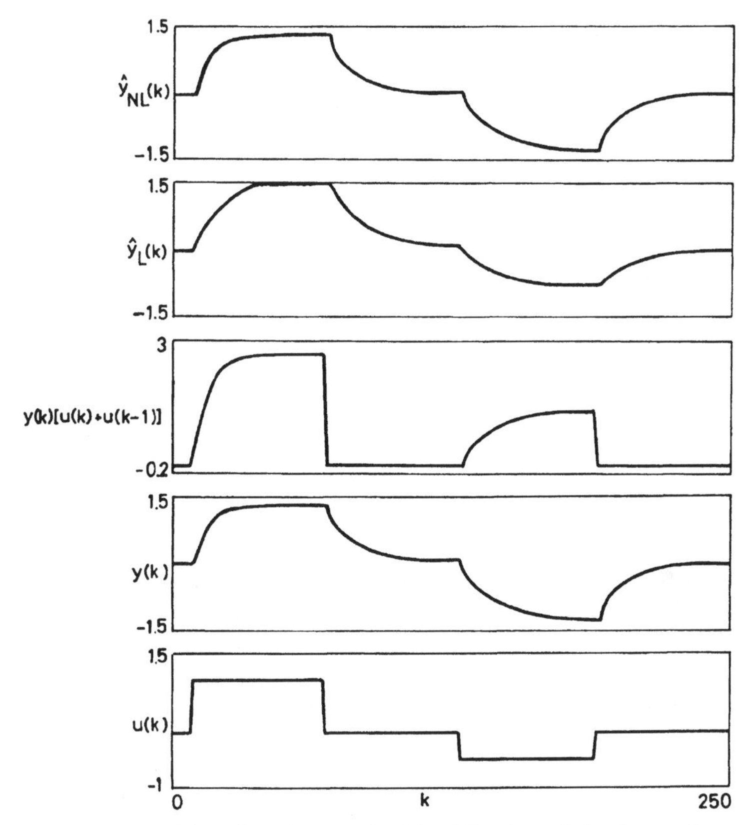

Fig. 3.16.4 Time functions at the identification of the first-order system with output signal dependent gain by a model linear-in-parameters

Example 3.16.4 *First-order quasi-linear system with output signal dependent time constant identified from normal operating data (Velenyak, 1981)*

The process is a noise-free first-order quasi-linear system with an output signal dependent time constant

$$T(t)\dot{y}(t) + y(t) = Ku(t)$$

$$K = K_0 = 2$$

$$T^{-1}(t) = \overline{T}_0^{-1} + \overline{T}_1^{-1}y(t) = 0.1 + 0.05y(t)\ \left[\mathrm{s}^{-1}\right]$$

Figure 3.16.5 shows the dependence of the true continuous time parameters on the output signal. The input signal consists of three constant periods, each being 60 [s] long. The value of the input signal was

$$u(t)=\begin{cases}1.0 & \text{if} \quad 0\,[\text{s}]\le t<60\,[\text{s}]\\ 0.0 & \text{if} \quad 60\,[\text{s}]\le t<120\,[\text{s}]\\ -0.5 & \text{if} \quad 120\,[\text{s}]\le t<180\,[\text{s}]\end{cases}$$

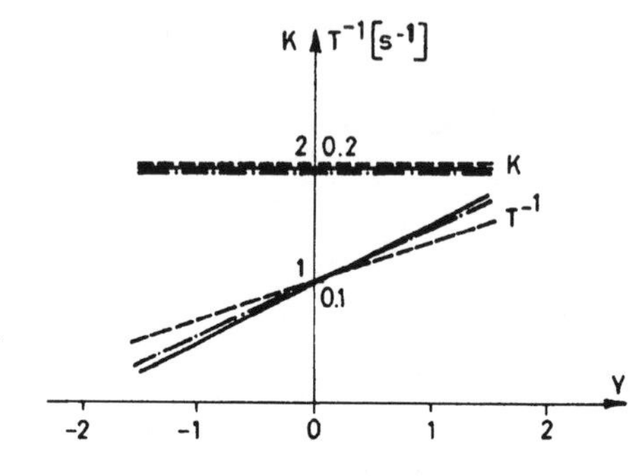

Fig. 3.16.5 True and estimated parameters of the differential equation of the first-order process with output signal dependent time constant: (true: —, Euler: -·-·-, bilinear: ---)

TABLE 3.16.3 Estimated parameters of the first-order system with output signal dependent time constant when fitting a model linear-in-parameters at the sampling time $\Delta T = 1\,[\text{s}]$

Parameter (true value)		*Transformation*	
		Euler	*bilinear*
of	a_1	-0.9050±0.000588	-0.8959±0.00510
the	b_1	0.1890±0.000902	0.1012±0.00390
difference	c_1	0.09119±0.000906	0.03256±0.00408
equation	d_1	-0.04565±0.000521	-0.01057±0.00233
of the	K_0 (2)	1.989	1.944
differential	$\bar{T}_0$ (10)	10.526	9.106
equation	$\bar{T}_1$ (20)	21.817, 21.906 } 21.862	28.303, 31.452 } 29.878

The measured input and output signals and the change of the time constant during the experiment are plotted in Figure 3.16.6. As the special form of the test signal will not be exploited, the measured data pairs are considered as normal operating data. Tables 1.6.5 and 1.6.8 give the difference equations of first-order system with an inverse linearly dependent time constant. The signal the parameters depend on is now the output signal $[x(t)=y(t)]$. The input–output difference equation becomes:

- using Euler transformation:

$$y(k)=-a_1y(k-1)+b_1u(k-1)+c_1x(k)u(k-1)+d_1x(k)y(k-1) \qquad (3.16.3)$$

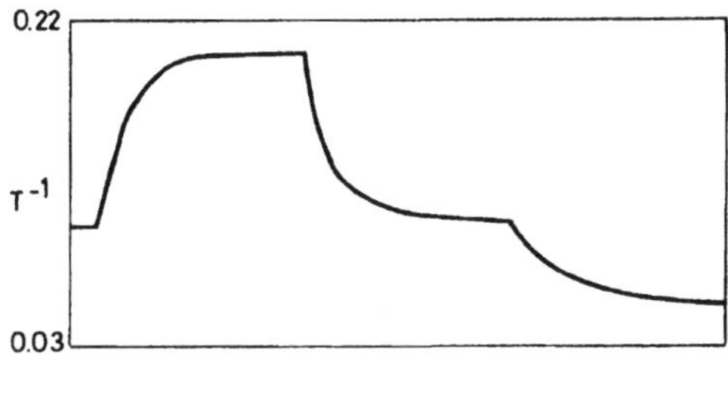

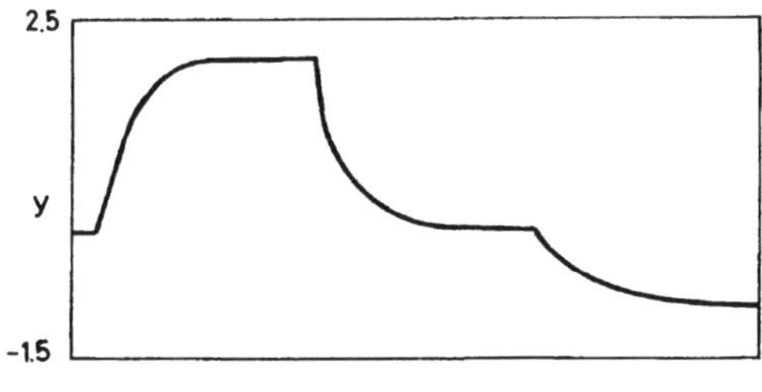

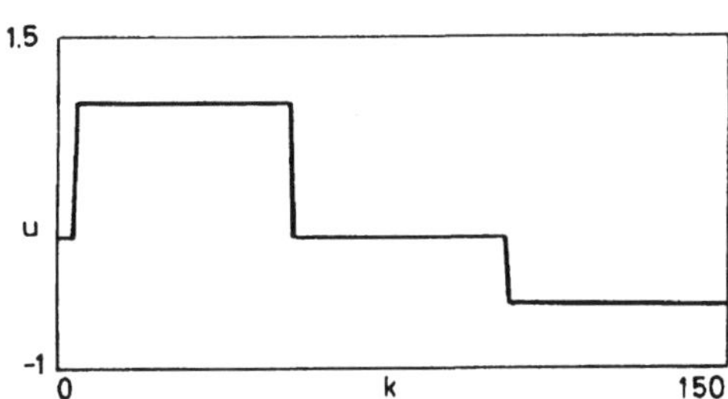

Fig. 3.16.6 Measured input (u), output (y) signals and the reciprocal time constant $\left(T^{-1}\right)$

$$a_1 = \frac{\Delta T}{\overline{T}_0} - 1 \quad b_1 = K_0 \frac{\Delta T}{\overline{T}_0} \quad c_1 = K_0 \frac{\Delta T}{\overline{T}_1} \quad d_1 = -\frac{\Delta T}{\overline{T}_1} \tag{3.16.4}$$

- using bilinear transformation:

$$\begin{aligned} y(k) = &-a_1 y(k-1) + b_1[u(k) - u(k-1)] + c_1 x(k)[u(k) - u(k-1)] \\ &+ d_1 x(k)[y(k) - y(k-1)] \end{aligned} \tag{3.16.5}$$

$$\begin{aligned} a_1 &= \frac{\Delta T - 2\overline{T}_0}{\Delta T + 2\overline{T}_0} \quad b_1 = K_0 \frac{\Delta T}{\Delta T + 2\overline{T}_0} \quad c_1 = K_1 \frac{\Delta T}{\Delta T + 2\overline{T}_0} \\ d_1 &= -\frac{\overline{T}_0}{\overline{T}_1} \frac{\Delta T}{\Delta T + 2\overline{T}_0} \end{aligned} \tag{3.16.6}$$

The parameters of the difference equations were estimated by the LS method. Table 3.16.3 gives the true and the estimated parameters for the sampling time $\Delta T = 1$ [s]. The parameters of the continuous time process K_0 and $\overline{T}_0, \overline{T}_1$ can be calculated by the equations

- for the Euler transformation:

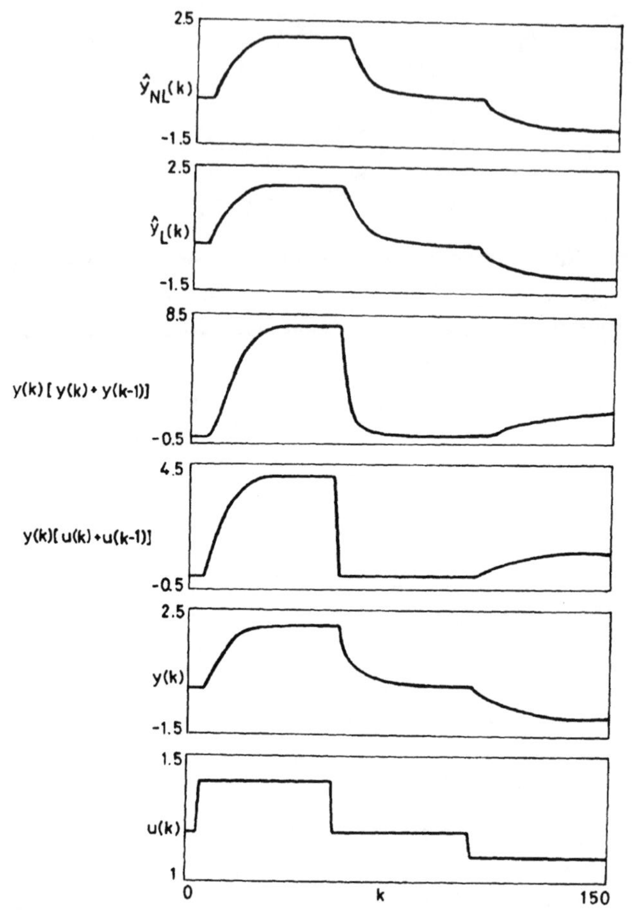

Fig. 3.16.7 Time functions at the identification of the first-order system with output signal dependent time constant by a model linear-in-parameters

$$\overline{T}_0 = \frac{\Delta T}{a_1 + 1} \quad K_0 = b_1 \frac{\overline{T}_0}{\Delta T} \quad \overline{T}_1 = \frac{K_0}{c_1 \Delta T} \quad \text{or} \quad \overline{T}_1 = -\frac{\Delta T}{d_1}$$

- for the bilinear transformation:

$$\overline{T}_0 = \frac{\Delta T}{2} \frac{1 - a_1}{1 + a_1} \qquad K_0 = b_1 \left(1 + 2 \frac{\overline{T}_0}{\Delta T} \right)$$

$$\overline{T}_1 = \frac{\overline{T}_0 K_0 \Delta T}{c_1 \left(\Delta T + 2\overline{T}_0 \right)} \quad \text{or} \quad \overline{T}_1 = -\frac{\overline{T}_0 \Delta T}{d_1 \left(\Delta T + 2\overline{T}_0 \right)}$$

Figure 3.16.7 shows the time signals of the identification based on the bilinear approximation: measured input and output signals, the nonlinear components of the difference equation (3.16.5) $y(k)[u(k)+u(k-1)]$ and $y(k)[y(k)+y(k-1)]$, the simulated output signals $\hat{y}_L(k)$ and $\hat{y}_{NL}(k)$ based on the linear and the nonlinear model, respectively. The time function of the two nonlinear components is very similar, this strong correlation has a negative effect on the parameter estimation.

TABLE 3.16.4 Estimated parameters of the first-order system with output signal dependent time constant when fitting a model linear-in-parameters and using different sampling times

Transformation	*Parameter*	*Sampling time* [s]			
		0.5	1.0	2.0	4.0
Euler	K_0	1.997	1.989	1.970	1.936
	$\overline{T}_0$	10.331	10.526	10.767	11.086
	$\overline{T}_1$	20.644	21.817	24.769	33.856
		20.619	21.906	25.284	36.101
bilinear	K_0	1.980	1.944		
	$\overline{T}_0$	9.710	9.106		
	$\overline{T}_1$	22.700	28.303		
		23.480	31.452		

Table 3.16.4 summarizes estimated parameters of the continuous time model for the sampling times $\Delta T = 0.5, 1, 2, 4$ [s]. Two values of $\overline{T}_1$ are given according to the two different calculation formulas. The bilinear discretization leads to better results and the smaller the sampling time the better the fit with the true parameters.

Observe that in this example the number of the parameters in the difference equation model linear-in-parameters is one more than the number of the parameters of the continuous time model. The equations are redundant; $\overline{T}_1$ can be calculated from two equations. With small sampling time and correct structure both defining equations result in the same $\overline{T}_1$.

Figure 3.16.5 shows the estimated parameter dependencies both for the Euler and the bilinear transformations for $\Delta T = 1$ [s]. ■

Example 3.16.5 *First-order quasi-linear system with input signal dependent time constant identified from normal operating data (Velenyak, 1981)*

The process is a noise-free first-order quasi-linear system with an input signal dependent time constant

$$T(t)\dot{y}(t) + y(t) = Ku(t)$$
$$K = K_0 = 2$$
$$T^{-1}(t) = \overline{T}_0^{-1} + \overline{T}_1^{-1}u(t) = 0.1 + 0.05u(t) \left[\mathrm{s}^{-1}\right]$$

(An output signal dependence of the parameters has more physical meaning than an input signal dependence. This case is treated here because some relation to Example 3.16.4 can be shown.) Figure 3.16.8 shows the dependence of the true continuous time parameters on the input signal. The input signal consists of four constant periods, each being 60 [s] long. The value of the input signal was

$$u(t) = \begin{cases} 1.0 & \text{if} \quad 0\ [\mathrm{s}] \le t < 60\ [\mathrm{s}] \\ 0.0 & \text{if} \quad 60\ [\mathrm{s}] \le t < 120\ [\mathrm{s}] \\ -0.5 & \text{if} \quad 120\ [\mathrm{s}] \le t < 180\ [\mathrm{s}] \\ 0.0 & \text{if} \quad 180\ [\mathrm{s}] \le t < 240\ [\mathrm{s}] \end{cases}$$

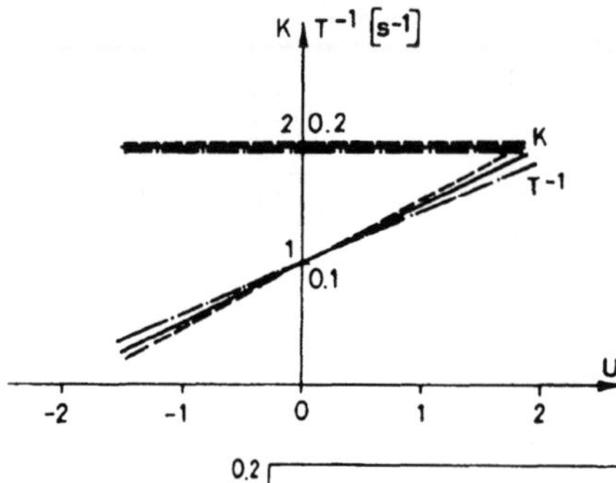

Fig. 3.16.8 True and estimated parameters of the differential equation of the first-order process with input signal dependent time constant (true: —, Euler: -·-·-, bilinear: ---)

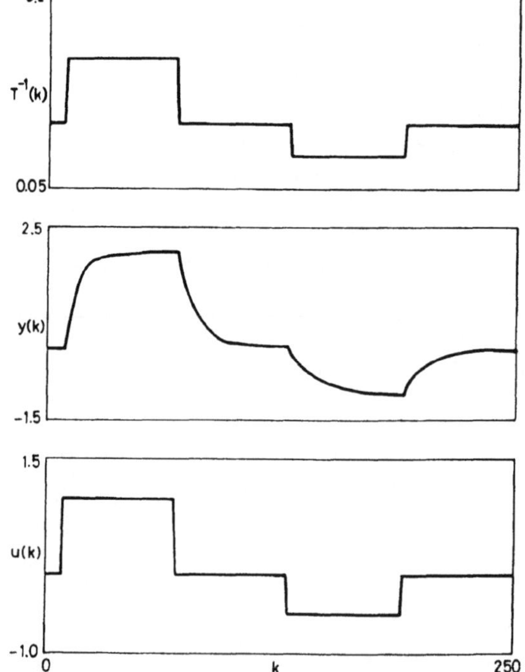

Fig. 3.16.9 Measured input (u), output (y) signals and the reciprocal time constant $\left(T^{-1}\right)$ of the first-order process with input signal dependent time constant

The measured input and output signals and the change of the time constant during the experiment are plotted in Figure 3.16.9. As the special form of the test signal will not be exploited, the measured data pairs are considered as normal operating data. The difference equations of a first-order system with an inverse linearly dependent time constant are given in (3.16.3) and (3.16.5). The signal the parameters depend on is now the input signal [$x(t) = u(t)$]

TABLE 3.16.5 Estimated parameters of the first-order system with output signal dependent time constant when fitting a model linear-in-parameters at the sampling time $\Delta T = 1\,[\mathrm{s}]$

Parameter		*Transformation*	
(true value)		*Euler*	*bilinear*
of	a_1	-0.9048±0.000170	-0.9085±0.000968
the	b_1	0.1897±0.000323	0.09292±0.000932
difference	c_1	0.08893±0.000609	0.04987±0.00190
equation	d_1	-0.04411±0.000343	-0.02559±0.00107
of the	K_0 (2)	1.993	2.031
differential	$\overline{T}_0$ (10)	10.504	10.429
equation	$\overline{T}_1$ (20)	22.407 22.671 } 22.539	19.432 18.645 } 19.039

The parameters of the difference equations were estimated by the LS method. Table 3.16.5 gives the true and the estimated parameters for the sampling time $\Delta T = 1$ [s]. The parameters of the continuous time process K_0 and $\overline{T}_0, \overline{T}_1$ can be calculated from equations (3.16.4) and (3.16.6).

Figure 3.16.10 shows the time signals of the identification based on the bilinear approximation: measured input and output signals, the nonlinear components of the difference equation (3.16.5) $u(k)[u(k)+u(k+1)]$ and $u(k)[y(k)+y(k+1)]$, the simulated output signals $\hat{y}_{\mathrm{L}}(k)$ and $\hat{y}_{\mathrm{NL}}(k)$ based on the linear and on the nonlinear model, respectively. Contrary to the case of Example 3.16.3, the time functions of the two nonlinear components differ from each other, which prophesies better results, as in Example 3.16.4.

Table 3.16.6 summarizes the estimated parameters of the continuous time model for the sampling times $\Delta T = 0.5, 1, 2, 4$ [s]. Two values of $\overline{T}_1$ are given according to the two different calculation formulas. The results are correct if the sampling time is small enough.

Similar to the case of the output signal dependence in Example 3.16.4, the difference equation is redundant in the original parameters.

Figure 3.16.8 shows the estimated parameter dependencies both for the Euler and the bilinear transformations for $\Delta T = 1$ [s]. ∎

Example 3.16.6 *First-order quasi-linear system with a direction dependent time constant (Continuation of Example 3.15.3)*

TABLE 3.16.6 Estimated parameters of the first-order system with input signal dependent time constant when fitting a model linear-in-parameters and using different sampling times

Transformation	*Parameter*	*Sampling time* [s]			
		0.5	1.0	2.0	4.0
Euler	K_0	10.225	10.504	11.038	12.162
	T_0	1.990	1.993	1.990	1.987
	$\bar{T}_1 = \bar{T}_1(c_1)$	21.186	22.407	25.318	31.421
	$\bar{T}_1 = \bar{T}_1(d_1)$	21.478	22.671	25.517	32.103
bilinear	K_0	2.013	2.031	2.067	2.136
	$\bar{T}_0$	10.188	10.429	10.912	11.803
	$\bar{T}_1 = \bar{T}_1(c_1)$	19.707	19.432	18.990	18.618
	$\bar{T}_1 = \bar{T}_1(d_1)$	19.351	18.645	17.472	15.968

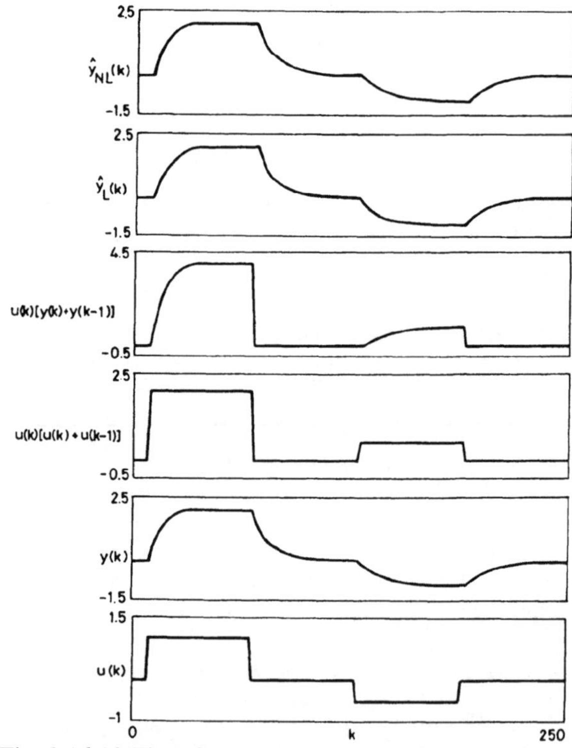

Fig. 3.16.10 Time functions at the identification of the first-order system with input signal dependent time constant by a model linear-in-parameters

The process is a noise-free first-order quasi-linear system with a direction dependent time constant

$$T(t)\dot{y}(t) + y(t) = Ku(t)$$
$$K = K_0 = 2$$
$$T(t) = T_0 + T_1 \text{sgn}(\dot{y}(t)) = 10 - 5\text{sgn}(\dot{y}(t)) \text{ [s]}$$

Figure 3.15.5 showed the dependence of the true continuous time parameters on the output signal. The input signal was the same as in Example 3.15.1. The measured input and output signals and the change of the time constant during the experiment were plotted in Figure 3.15.6. The nonlinear difference equation of the process can be set up based on Table 1.6.8. The signal the parameters depend on is $x(t) = \text{sgn}(\dot{y}(t))$. The input–output difference equation becomes:

- using Euler transformation:

$$y(k) = -a_1 y(k-1) + b_1 u(k-1) + c_1 x(k)[y(k) - y(k-1)]$$
$$a_1 = \frac{\Delta T}{T_0} - 1 \qquad b_1 = K_0 \frac{\Delta T}{T_0} \qquad c_1 = \frac{T_1}{T_0}$$

- using bilinear transformation:

$$y(k) = -a_1 y(k-1) + b_1 [u(k) - u(k-1)] + c_1 x(k)[y(k) - y(k-1)]$$
$$a_1 = \frac{\Delta T - 2T_0}{\Delta T + 2T_0} \qquad b_1 = K_0 \frac{\Delta T}{\Delta T + 2T_0} \qquad c_1 = -2\frac{T_1}{\Delta T + 2T_0}$$

TABLE 3.16.7 Estimated parameters of the first-order system with direction dependent time constant when fitting a model linear-in-parameters at the sampling time $\Delta T = 1$ [s]

Parameter (true value)		*Transformation*	
		Euler	*bilinear*
of the difference equation	a_1	-0.9048	-0.9092±0.00123
	b_1	0.1902	0.0913±0.00119
	c_1	-0.4755	0.474±0.0113
of the differential equation	K_0 (2)	1.999	1.918
	T_0 (10)	10.508	10.513
	T_1 (-5)	-4.997	-4.977

The parameters of the difference equations were estimated by the LS method. Table 3.16.7 gives the true and the estimated parameters for the sampling time $\Delta T = 1$ [s]. The parameters of the continuous time process K_0 and T_0, T_1 can be calculated by the equations

- for the Euler transformation:

$$T_0 = \frac{\Delta T}{a_1 + 1} \qquad K_0 = b_1 \frac{T_0}{\Delta T} \qquad T_1 = -T_0\, c_1$$

- for the bilinear transformation:

$$T_0 = \frac{\Delta T}{2}\frac{1-a_1}{1+a_1} \qquad K_0 = b_1\left[1 + 2\frac{T_0}{\Delta T}\right] \qquad T_1 = \frac{c_1}{2}\left[\Delta T + 2T_0\right]$$

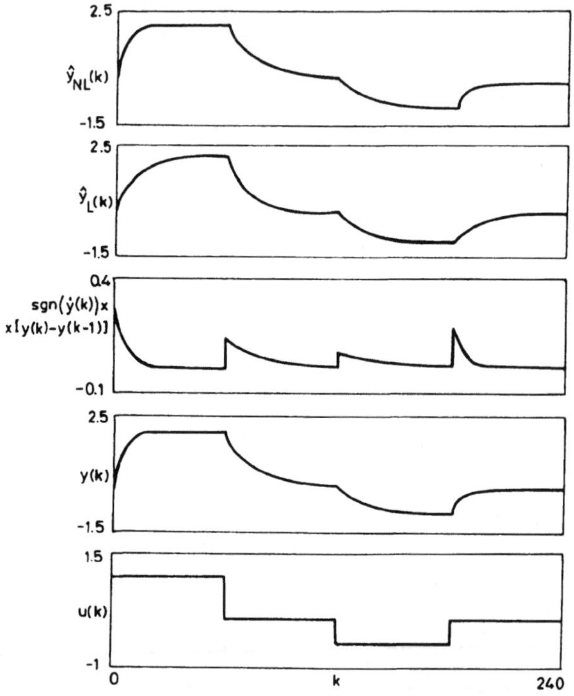

Fig. 3.16.11 Time functions at the identification of the first-order system with direction dependent time constant by a model linear-in-parameters

Figure 3.16.11 shows the time signals of the identification based on the bilinear approximation: measured input and output signals, the nonlinear component of the difference equation $\text{sgn}(\dot{y}(k))[y(k) - y(k-1)]$, the based on the linear and on the nonlinear model simulated output signals $\hat{y}_L(k)$ and $\hat{y}_{NL}(k)$, respectively. The coincidence both in the parameters and in the fit of the outputs is very good. ■

Remarks:

1. In some cases the number of the parameters in the difference equation model linear-in-parameters is more than the number of the parameters of the continuous time model. In other words, the equations are redundant.

2. In some cases the model components are too correlated, which fact has a negative influence on the accuracy of the parameter estimation.
3. The simulations presented in the examples illustrate Items 2 and 3.

3.16.3 Identification of continuous time signal dependent models in the form nonlinear-in-parameters

Redundancy can be overcome if only those parameters are estimated which describe the process. The equations are then usually nonlinear in the parameters. Several optimization procedures can be used to minimize the sum of residuals.

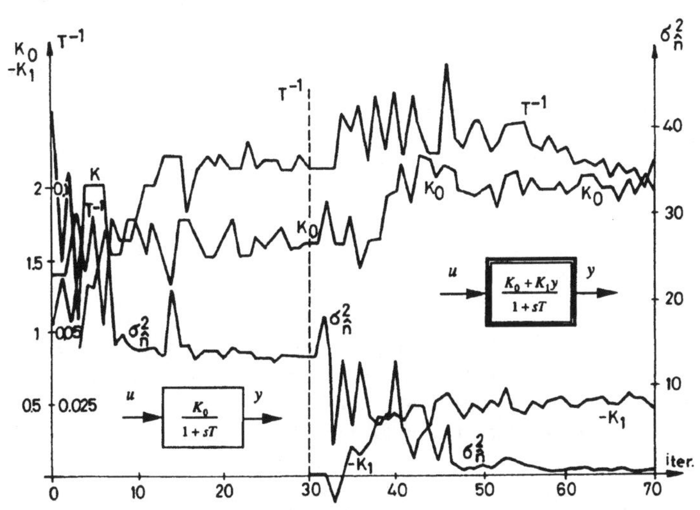

Fig. 3.16.12 Identification of the first-order system with output signal dependent gain by a model nonlinear-in-parameters

Example 3.16.7 *First-order quasi-linear system with output signal dependent static gain identified from normal operating data (Continuation of Example 3.15.1)*

The noise-free quasi-linear first-order process with output signal dependent time constant

$$T\dot{y}(t) + y(t) = K(t)u(t)$$

$$T = T_0 = 10 \text{ [s]}$$

$$K(t) = K_0 + K_1 y(t) = 2 - 0.5y(t)$$

was excited with a four-level test signal according to Figure 3.15.2. The parameters were estimated by the simplex method (Nelder and Mead, 1965). In the first 30 iteration steps only a linear model with the parameters K_0 and $T = T_0$ was estimated, then the nonlinear term K_1 was also estimated. Table 3.16.8 summarizes

- the initial values,
- the initial steps of the simplex, and
- the boundaries of the parameters.

Figure 3.16.12 shows the parameters and the normalized loss function $\left(\sigma_{\hat{n}}^2\right)$ defined as the mean square output error as a function of the iteration steps. The final parameters approximate well the true ones.

TABLE 3.16.8 Estimated parameters of the first-order system with output signal dependent gain when fitting a model nonlinear-in-parameters

Model	*Parameter*	*True value*	*Initial value*	*Initial step*	*Lower bound*	*Upper bound*	*Final estimate*
Linear	K_0	2	1	0.5	0.1	10.0	1.6
	K_1	-0.5					
	T^{-1}	0.1	0.07	0.02	0.02	1.0	0.106
Nonlinear	K_0	2	1.6	0.3	0.5	3.0	1.97
	K_1	-0.5	0.0	0.2	-2.0	2.0	-0.46
	T^{-1}	0.1	0.106	0.02	0.05	3.0	0.107

3.16.4 Identification of discrete time signal dependent models in the form linear-in-parameters

If the parameters of a quasi-linear differential equation depend on a signal, then the parameters of the difference equations depend on the same signal, as well. An assumed linear dependence of the coefficients of the difference equation makes the parameter estimation usually easier as the same assumption for the differential equation, because the discretization relationships are mostly nonlinear.

The quasi-linear process with the relative delay time d and of order n is given by

$$y(k) = c_0 + \frac{B\left(q^{-1}\right)}{A\left(q^{-1}\right)} u(k-d)$$

or

$$y(k) = c_0^* + \sum_{j=1}^{na} a_j y(k-j) + \sum_{i=0}^{nb} b_i u(k-d-i) \tag{3.16.7}$$

with

$$c_0^* = c_0 A(1)$$

A dependence, linear-in-parameters, of the parameters is

$$\theta_i(t) = \sum_{\ell=0}^{m_i} \theta_{i\ell} f_{i\ell}(x(t)) \tag{3.16.8}$$

where $x(t)$ is the signal the parameters depend on. (A functional dependence on more signals can also be formed but for simplicity not handled here.) The functions $f_{i\ell}(x(t))$ are assumed to be known and the coefficients $\theta_{i\ell}$ are searched.

Having no *a priori* knowledge, the first few terms of the Taylor series can be used in (3.16.8)

$$\theta_i(t)=\sum_{\ell=0}^{m_i^*}\theta_{i\ell}^* x^\ell(t) \tag{3.16.9}$$

Two identification methods can be used:

- two-step method: fitting the parameter dependence to a set of piecewise linear models having got from small excitations in different working points; or
- one-step method: identification of the globally valid model in one step.

1. Two-step method: evaluation of piecewise linear models

The coefficients of the quasi-linear model depend on the signal $x(t)$:

$$c_0^*(x(t))=\sum_{\ell=0}^{m_0} c_{0\ell}^* f_{0\ell}(x(t)) \tag{3.16.10}$$

$$a_j(x(t))=\sum_{\ell=0}^{m_j} a_{j\ell} f_{j\ell}(x(t)) \tag{3.16.11}$$

$$b_i(x(t))=\sum_{\ell=0}^{m_j} b_{i\ell} f_{i\ell}(x(t)) \tag{3.16.12}$$

The method uses the following procedure:

1. Divide the working domain of the process into several parts according to the values of the signal $x(t)$ the parameters depend on.
2. Identify in each working point a linear model for a small excitation.
3. Determine the parameters $c_{0\ell}^*, a_{j\ell}, b_{i\ell}$ of the equations (3.16.10) to (3.16.12) from the estimated parameters of the piecewise linear models.

Remarks:

1. The number of the parameters to be estimated simultaneously is few.
2. The structure of all linear models fitted has to be the same.

Example 3.16.8 *First-order quasi-linear system with output signal dependent time constant excited by small steps (Continuation of Example 3.15.2)*

In Example 3.15.2 a process with signal dependent time constant was excited by step signals with small amplitudes in different working points. The plot of the measured signals was given in Figure 3.15.3. The individual step responses were evaluated grapho-analytically as first-order systems. Table 3.15.2 summarizes the static gains and the time constants of each domain. The parameters of the pulse transfer functions for the sampling time $\Delta T=1$ [s] are calculated according to the formulas:

$$a_1=-\exp(-\Delta T/T)$$

$$b_1=K(1+a_1)$$

They are also given in Table 3.15.2. The following static functions were fitted by the LS method

$$\hat{a}_1=-(0.9094\pm 0.0013)+(0.04572\pm 0.0013)y(t)$$

$$\hat{b}_1 = (0.1811 \pm 0.0027) + (0.09144 \pm 0.0027)y(t)$$

The functions of the simulated parameters are

$$a_1 = -\exp\left\{-\Delta T\left[\overline{T}_0^{-1} + \overline{T}_1^{-1} y(t)\right]\right\} = -\exp\{-0.1 - 0.05y(t)\}$$
$$b_1 = K_0(1 + a_1) = 2\left[1 - \exp\{-0.1 - 0.05y(t)\}\right]$$

Figure 3.16.13 shows the plot of the true and the fitted values of the parameters a_1 and b_1. The coincidence is very good. ■

2. One-step method: identification of the globally valid model

Put (3.16.8) into (3.16.7)

$$y(k) = \sum_{\ell=0}^{m_0} c_{0\ell}^{*} f_{0\ell}(x(k)) + \sum_{j=1}^{na} \sum_{\ell=0}^{m_j} a_{j\ell} f_{j\ell}(x(k)) y(k-j)$$
$$+ \sum_{i=0}^{nb} \sum_{\ell=0}^{m_i} b_{i\ell} f_{i\ell}(x(k)) u(k-d-i) \qquad (3.16.13)$$

Equation (3.16.13) is linear in the parameters $c_{0\ell}^{*}, a_{j\ell}, b_{i\ell}$. Common parameter estimation algorithms can be used.

Remark:

The disadvantage of the method is that too many parameters have to be estimated.

Example 3.16.9 *First-order quasi-linear system with output signal dependent time constant identified from normal operating data (Continuation of Example 3.15.2)*

In Example 3.15.2 a noise-free first-order process with output signal dependent time constant was simulated with a test signal consisting of seven steps with different magnitudes. As the special form of the test signal will not be exploited, the measured data pairs are considered as normal operating data. From Example 3.16.8 we know the signal dependence of the parameters of the difference equation $a_1(x(t))$ and $b_1(x(t))$. Based on these the globally valid nonlinear difference equation of the process is

$$y(k) = -[a_{10} + a_{11}y(k)]y(k-1) + [b_{10} + b_{11}y(k)]u(k-1)$$
$$= -a_{10}y(k-1) - a_{11}y(k)y(k-1) + b_{10}u(k-1) + b_{11}y(k)u(k-1) \qquad (3.16.14)$$

Least squares parameter estimation of the whole input–output records resulted in the parameter estimates:

$$\hat{a}_{10} = -0.9067 \pm 0.00042 \quad ; \quad \hat{a}_{11} = 0.04628 \pm 0.00039$$
$$\hat{b}_{10} = 0.1862 \pm 0.00086 \quad ; \quad \hat{b}_{11} = 0.09271 \pm 0.00082 \qquad (3.16.15)$$

Equation (3.16.14) is not a recursive equation for $y(k)$ linear-in-parameters.

Substituting $y(k-1)$ instead of $y(k)$ into the parameter dependence leads to a recursive, difference equation linear-in-parameters

$$\begin{aligned} y(k) &= -\left[a_{10} + a_{11}y(k-1)\right]y(k-1) + \left[b_{10} + b_{11}y(k-1)\right]u(k-1) \\ &= -a_{10}y(k-1) - a_{11}y^2(k-1) + b_{10}u(k-1) + b_{11}y(k-1)u(k-1) \end{aligned} \quad (3.16.16)$$

An LS estimation leads to the parameters:

$$\begin{aligned} \hat{a}_{10} &= -0.9052 \pm 0.00043 \quad ; \quad \hat{a}_{11} = 0.0469 \pm 0.00040 \\ \hat{b}_{10} &= 0.1893 \pm 0.00087 \quad ; \quad \hat{b}_{11} = 0.09393 \pm 0.00084 \end{aligned} \quad (3.16.17)$$

The estimated parameters (3.16.15) and (3.16.17) are almost the same.

Figure 3.16.14 shows the plot of the true and the fitted values of a_1 and b_1 in (3.16.5) as a function of the output signal. The coincidence is very good. ■

Some examples for the identification of quasi-linear systems with signal dependent parameters are given in the following papers (among others):

- *electrically excited biological membrane* (Haber and Wernstedt, 1979)
- *concentration curve in furfural production* (Vuchkov *et al.*, 1985)
- *distillation column* (Haber *et al.*, 1982; Haber and Zierfuss, 1991).

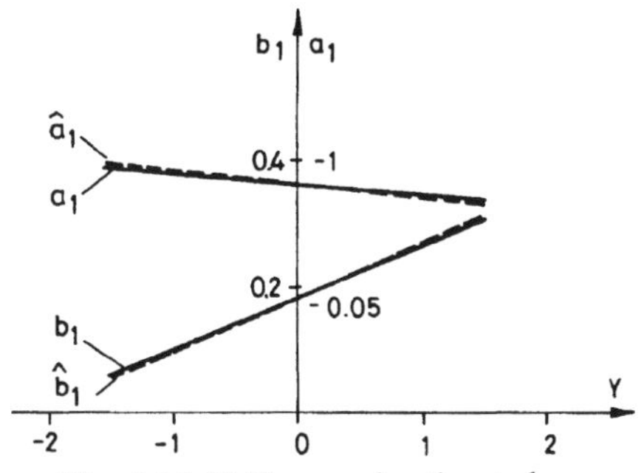

Fig. 3.16.13 True and estimated parameters of the difference equation of the first-order process with output signal dependent time constant estimated from a set of small excitations

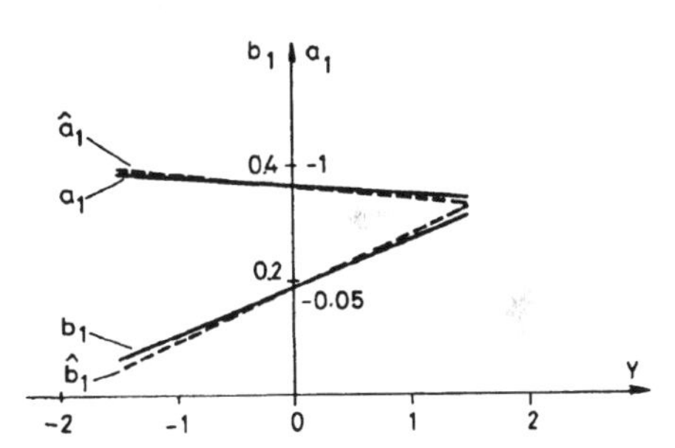

Fig. 3.16.14 True and estimated parameters of the difference equation of the first-order process with output signal dependent time constant estimated from normal operating record

3.17 IDENTIFICATION OF SPATIAL LINEAR AND NONLINEAR MODELS

With spatial models the model parameters depend not only on actual signal(s) but on the present and former input (and output) signals. Denote the signals the parameters depend on by $x_i(k)$. As mentioned in Sections 1.7 and 3.5 there are two types of these models:

- *nonparametric spatial models:* (Volterra or Zadeh models with gate functions)

$$x_1(k) = u(k), x_2(k) = u(k-1), \ldots, x_{m+1}(k) = u(k-m)$$

- *parametric spatial models:*

$$x_1(k) = u(k), x_2(k) = u(k-1), \ldots, x_{n+1}(k) = u(k-n)$$
$$x_{n+2}(k) = y(k-1), x_{n+3}(k) = y(k-2), \ldots, x_{2n+1}(k) = y(k-n)$$

Assume the signals are quantized in N_q levels. Then the number of domains are:

- *nonparametric spatial models:* N_q^{m+1}
- *parametric spatial models:* N_q^{2n+1}

Remarks:

1. The number of domains can be drastically reduced if there are only few levels of the input signal. This is the case, e.g., with PRBS and PRTS signals.
2. In the case of nonparametric models the size of the memory is about 10, which cause 100 domains even in the case of a PRBS excitation.
3. In the case of parametric models the size of the memory is usually not more than 3, but the output signal has to be quantized finely enough. Let, e.g., the input signal a PRBS, the order of the strictly causal process 2 and the number of levels of the output signals 5. Then the number of the domains is $2^2 \cdot 5^2 = 100$.
4. An other possibility to reduce the number of domains is to neglect some domains arbitrary (Pincock and Atherton, 1973).
5. Instead of taking the shifted input signals into account, the outputs of the Laguerre filters can also be considered (Pincock and Atherton, 1973).

   The identification procedure was described in Algorithm 3.5.1 in Section 3.5.
   Following authors presented the method in details:
   - *nonparametric model:*
     Roy and DeRusso (1962), Harris and Lapidus (1967a, 1967b), Miller and Roy (1964), Pincock and Atherton, (1973);
   - *parametric model:*
     Billings and Voon (1987).

Applications of the identification of spatial models with gate function method are reported in the following papers:

- *continuous stirred tank reactor with a single exothermic reaction* (Harris and Lapidus, 1967a, 1967b);
- *two phase induction motor* (Pincock and Atherton, 1973).

3.18 ESTIMATION OF THE STEADY STATE CHARACTERISTIC OF A PROCESS

Sometimes the task is to determine only the steady state characteristic of a process instead of estimating all parameters of the difference or differential equation. By means of it the optimal working point of a plant can be calculated. An example is the extremum control, where the minimum or maximum of the steady state characteristic is looked for. Such a task occurs, e.g., at the optimization of combustion conditions in furnaces, at the maximization of the capacity of grinding mills, etc.

3.18.1 Evaluation of steady state measurements

The steady state characteristic can be calculated from measurements taken in stationary states. The number of different working points has to be one more than the degree of the polynomial of the steady state characteristic. One has to wait until the transient responses to the excitations settle down and then the measurements in the steady state have to be averaged to filter the effect of the noises. The steady state characteristic can be determined

- by drawing the output values as a function of the input signals,
- by fitting a (usually a polynomial) functional relation between the input and output data pairs.

Example 3.18.1 *Stepwise excitation of a simple Hammerstein model*
The simple Hammerstein model consists of a static function

$$v(k) = 2 + u(k) + 0.5u^2(k) \tag{3.18.1}$$

followed by a linear term with

$$G(s) = \frac{1}{1+10s}$$

At the beginning the process was in the stationary state belonging to $U = 0$. Then the input signal was increased in the following steps $u(k): 0 \rightarrow 1 \rightarrow 2$. Before every new change of the input signal a stationary state was reached. The measured input and output signals are plotted in Figure 3.18.1. The steady state values of the input and the output signals in each working point are summarized in Table 3.18.1. The values belonging together give the characteristic in Figure 3.18.2 that coincides with Equation (3.18.1). A least squares estimation of the parameters of a quadratic function resulted in the same coefficients. ■

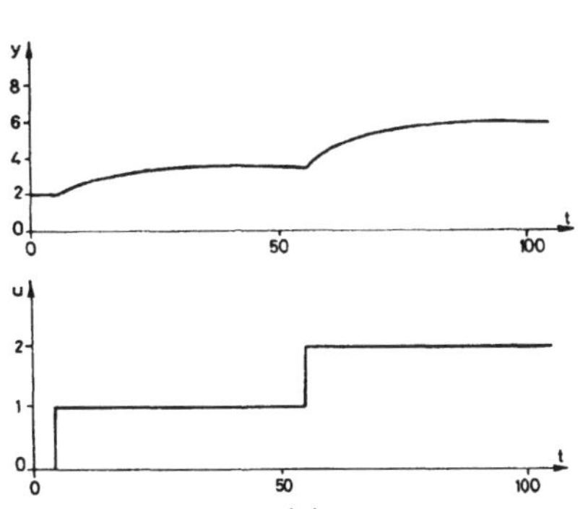

Fig. 3.18.1 Input (u) and output signals (y) at stepwise excitation of the simple Hammerstein model

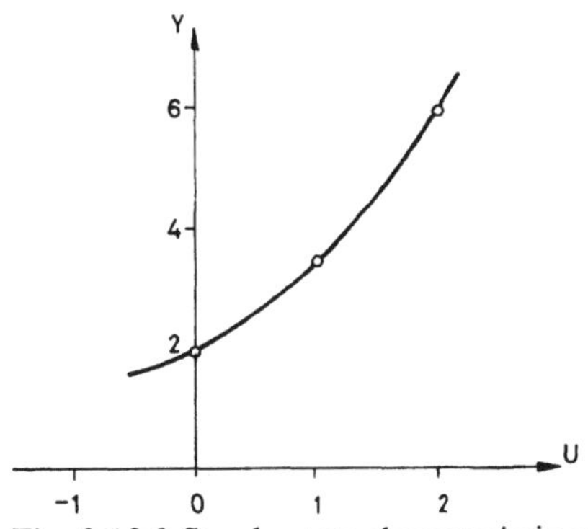

Fig. 3.18.2 Steady state characteristic of the simple Hammerstein model obtained from the stepwise excitation

Example 3.18.2 *Stepwise excitation of a simple Wiener model*
The simple Wiener model consists of a linear dynamic term with

$$G(s) = \frac{1}{1 + 10s}$$

followed by a static function

$$u(k) = 2 + v(k) + 0.5v^2(k) \tag{3.18.2}$$

At the beginning the process was in the stationary state belonging to $U = 0$. Then the input signal was increased in the following steps $u(k): 0 \to 1 \to 2$. Before every new change of the input signal a stationary state was reached. The measured input and output signals are plotted in Figure 3.18.3. The steady state values of the input and the output signals in each working point are summarized in Table 3.18.2. The values belonging together give the characteristic in Figure 3.18.4 that coincides with Equation (3.18.2) taking $v(\infty) = u(\infty)$ into account. A least squares estimation of the parameters of a quadratic function resulted in the same coefficients. ■

TABLE 3.18.1 Steady state values of the simple Hammerstein model obtained from the stepwise excitation

No	*U*	*Y*
1	0	2.0
2	1	3.5
3	2	6.0

TABLE 3.18.2 Steady state values of the simple Wiener model obtained from the stepwise excitation

No	*U*	*Y*
1	0	2.0
2	1	3.5
3	2	6.0

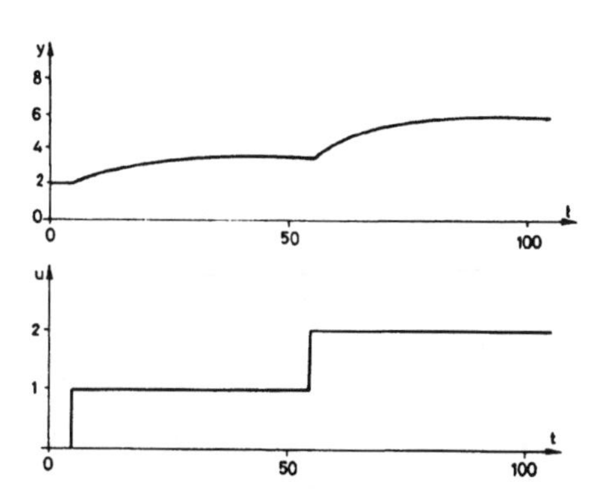

Fig. 3.18.3 Input (u) and output signals (y) at stepwise excitation of the simple Wiener model

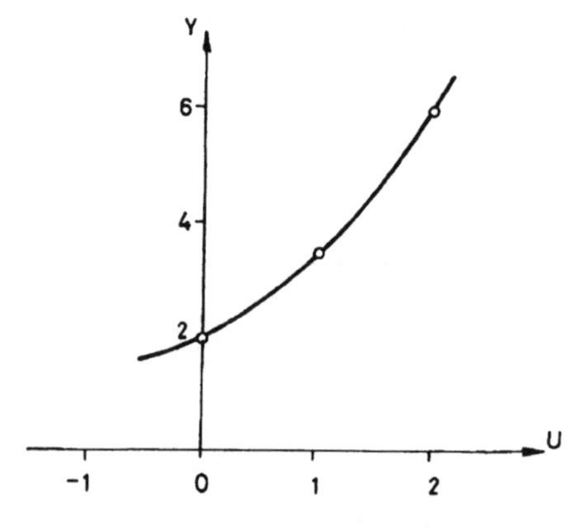

Fig. 3.18.4 Steady state characteristic of the simple Wiener model obtained from the stepwise excitation

3.18.2 Two-step method: fitting a model linear-in-parameters and the calculation the steady state relation

The method relies on the fact that the steady state equation can be obtained from a difference equation by replacing the delayed signals by their stationary values.

Algorithm 3.18.1 The two-step method of estimating the steady state equation of a process consists of the following steps:

1. Excite the process by a test signal whose
   - amplitude covers the range of the process, and
   - frequency range of the test signal covers the one of the process.
2. Estimate of the parameters of the discrete time dynamic model by regression.
3. Replace the delayed signals by their steady state values in the difference equation.
4. Replace $q = q^{-1} = 1$ in the difference equation.
5. Simplify the equation and express it to the steady state value of the output signal. ■

Remark:

If the model is a continuous time model, then Item 4 has to be changed to:

4. Replace $s = 0$ in the Laplace transformation of the differential equation.

Example 3.18.3 *Parametric correlation method applied to a simple Hammerstein model using a PRTS-excitation (Haber, 1979a, 1988)*

The simple Hammerstein model was simulated on the analog computer Dornier DO 80 at the Department of the Control and System Dynamics, University of Stuttgart. The parameters of the continuous time model were

$$\frac{B(s)}{A(s)} = \frac{1+2s}{(1+s)(1+3s)}, \qquad c_0 = 1.0, \quad c_1 = 2, \quad c_2 = -0.25$$

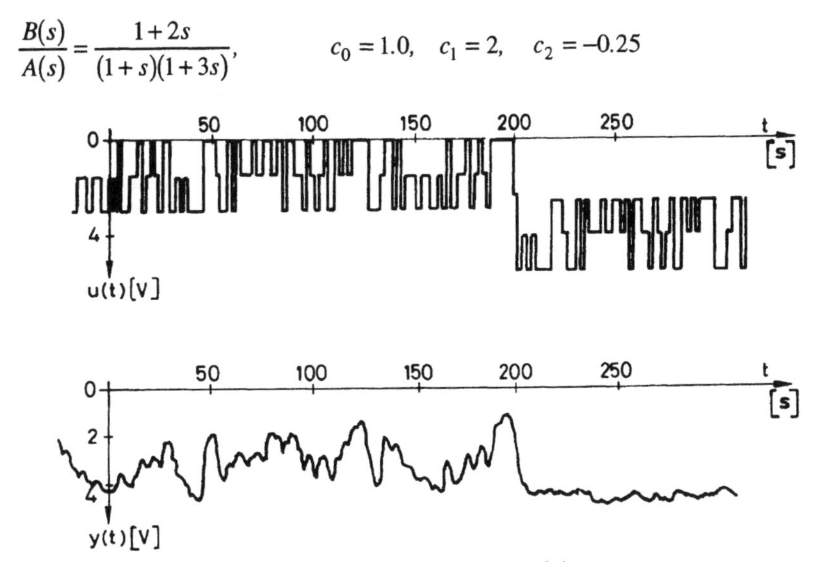

Fig. 3.18.5 Measured input (u) and output signals (y) at the identification of the simple Hammerstein model by the correlation method

The test signal generated by the process computer was a PRTS of maximal length and period of 80. The minimum and maximum values of the test signal were 0 and 3, respectively. Both the minimum switching time of the test signal and the sampling time were equal to 2 [s]. Ten percent Gaussian white noise was added to the output signal.

To compare the results of the identification with the true values of the simulated process, the equivalent pulse transfer function of the linear part of the process was calculated assuming a zero order holding device. The parameters are as follows

$$\frac{B(q^{-1})}{A(q^{-1})}=\frac{0.676q^{-1}-0.255q^{-2}}{1-0.649q^{-1}+0.069q^{-2}}$$

Without excitation the process output was $y_0=2$. The measured input and output signals are seen in Figure 3.18.5.

The process parameters were estimated by the two-step parametric correlation method: correlation analysis and least squares parameter estimation presented in Section 3.3. The multipliers were chosen as

$$x_1(k)=u(k)-\mathrm{E}\{u(k)\},\qquad x_2(k)=u^2(k)-\mathrm{E}\{u^2(k)\}$$

The correlation functions were evaluated in the shifting time domain $(-8\le\kappa\le 40)$ after having updated them during 1120 [s]. The cross-correlation functions $r_{x_1u}(\kappa)$ and $r_{x_1y}(\kappa)$ as well as $r_{x_2u}(\kappa)$ and $r_{x_2y}(\kappa)$ are drawn in Figure 3.18.6a and Figure 3.18.6b, respectively. The sum of the cross-correlation functions of the measured and the computed output signals based on the estimated model are compared in Figure 3.18.7.

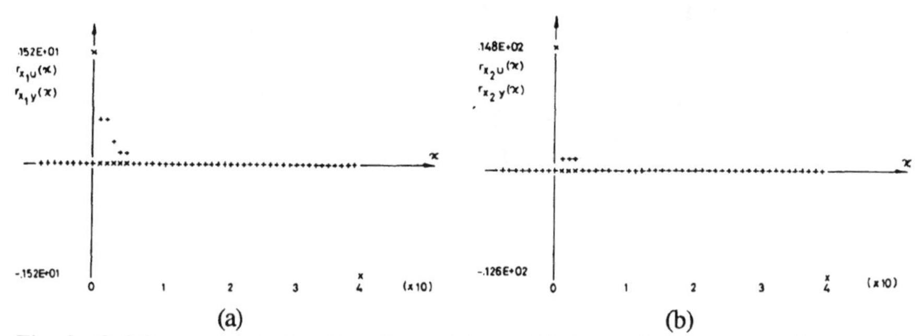

(a) (b)

Fig. 3.18.6 Cross-correlation functions of the multiplier and the measured input and output signals, respectively: (a) $r_{x_1u}(\kappa):*$, $r_{x_1y}(\kappa):+$; (b) $r_{x_2u}(\kappa):*$, $r_{x_2y}(\kappa):+$

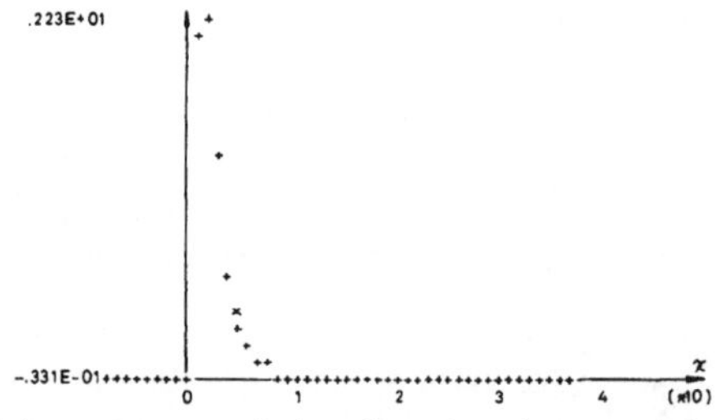

Fig. 3.18.7 Sum of the cross-correlation functions between the multipliers and the measured (*) and the computed (+) output signals based on the estimated model

The estimated process model without the constant term was

$$\hat{y}'(k) - 0.652\hat{y}'(k-1) + 0.063\hat{y}'(k-2)$$
$$= 0.5713u(k-1) + 0.2442u(k-2) - 0.07024u^2(k-1) - 0.03276u^2(k-2)$$

(y' means the output reduced by the constant term.) The estimated model approximates the simulated one well. The steady state equation without the constant term is

$$(1 - 0.652 + 0.063)\hat{Y}' = (0.5713 + 0.2442)U - (0.07024 + 0.03276)U$$
$$\hat{Y}' = 1.984U - 0.2506U^2 \qquad (3.18.3)$$

which almost completely coincides with the simulated values. In the knowledge of relationship (3.18.3) the input signal was set to the location of the extremum

$$U_{\text{extr}} = (1.984/2) \cdot 0.2506 = 3.9585$$

The value of the extremum is

$$Y_{\text{extr}} = y_0 + Y'_{\text{extr}} = 2.0 + 1.984 \cdot 3.9585 - 0.2506 \cdot \left(3.9585^2\right) = 5.936$$

■

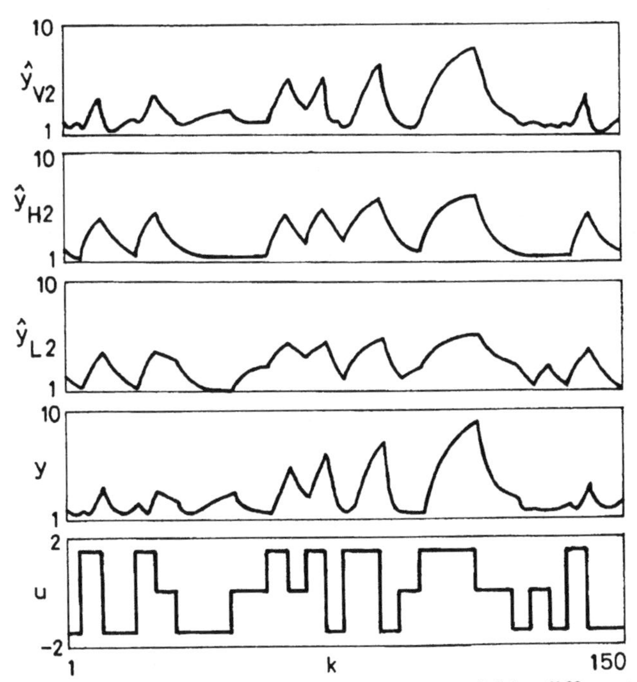

Fig. 3.18.8 Approximation of the simple Wiener model by different models linear-in-parameters (L2: linear, H2: Hammerstein, V2: parametric Volterra)

Example 3.18.4 *Least squares parameter estimation of a simple Wiener model using a PRTS excitation (Haber, 1985)*

The process was a simple Wiener model with the linear dynamic term

$$G(s) = \frac{2}{1+10s}$$

followed by the function (3.18.2). The input signal was a PRTS of maximal length, period 26, amplitude ±1.5 and mean value 0. The sampling time was $\Delta T = 2$ [s] and the minimum switching time of the PRTS was 5-times more. $N = 26 \cdot 5 = 150$ noise-free data pairs (Figure 3.18.8) were used for the parameter estimation. The following models were fitted:

- L1: first-order linear model;
- L2: second-order linear model;
- H1: quadratic, first-order generalized Hammerstein model;
- H2: quadratic, second-order generalized Hammerstein model;
- V2: quadratic, second-order parametric Volterra model.

TABLE 3.18.3 Estimated parameters at approximating the simple Wiener model by different models linear-in-parameters

Component	L1	L2	H1	H2	V2
$y(k-1)$	-0.8582 ±0.0232	-1.354 ±0.070	-0.850 ±0.021	-1.297 ±0.073	-1.460 ±0.026
$y(k-2)$		0.463 ±0.064		0.409 ±0.064	0.514 ±0.0223
1	0.4576 ±0.081	0.348 ±0.077	0.1455 ±0.091	0.1613 ±0.085	0.1717 ±0.029
$u(k-1)$	0.313 ±0.033	0.332 ±0.044	0.3247 ±0.03	0.3355 ±0.041	0.3401 ±0.014
$u(k-2)$		-0.1646 ±0.051		-0.1492 ±0.049	-0.2023 ±0.017
$u^2(k-1)$			0.2058 ±0.035	0.2132 ±0.052	0.06254 ±0.018
$u^2(k-2)$				-0.0943 ±0.0055	-0.29155 ±0.02
$u(k-1)u(k-2)$					0.3558 ±0.011
σ_ε	0.478	0.410	0.432	0.383	0.133
$Y = Y(U)$	3.096 $+2.118U$	3.183 $+1.532U$	0.968 $+2.161U$ $+1.3694U^2$	1.444 $+1.668U$ $+1.0646U^2$	2.1118 $+1.695U$ $+1.560U^2$
$\left.\frac{d\hat{Y}}{dU}\right\|_{U=0}$	2.118	1.532	2.161	1.668	1.695
$\frac{dY}{dU} = 0 \rightarrow \hat{U}_0$			-0.789	-0.783	-0.543

Table 3.18.3 shows the estimated parameters, standard deviations, the standard deviations of the residuals (σ_ε), the estimated steady state characteristic $Y(U)$, its

gradients at $U = 0$, the location and the values of the minima of the characteristics. The steady state characteristic of the simulated Wiener model is

$$Y = 2 + (2U) + 0.5(2U)^2 = 2 + 2U + 2U^2$$

which has a minimum in $U_{\text{extr}} = -1$ and the minimum value is $Y_{\text{extr}} = 1.5$. The gradient at $U = 0$ is 2.

The steady state characteristics were calculated from the estimated parameters of the dynamic model. In the case of the parametric Volterra model the equation of the model is

$$\hat{y}(k) = 1.460\hat{y}(k-1) - 0.541\hat{y}(k-2) + 0.1717 + 0.340u(k-1) - 0.2023u(k-2)$$
$$+0.06254u^2(k-1) - 0.29155u^2(k-2) + 0.3558u(k-1)u(k-2)$$

and the static relationship is

$$(1 - 1.46 + 0.541)Y = 0.1717 + (0.3401 - 0.2023)U + (0.06254 - 0.29155 + 0.3558)U^2$$

or

$$Y = 2.118 + 1.695U + 1.560U^2$$

The linear models approximate only the gradients of the steady state characteristic curve. Both the generalized Hammerstein and the parametric Volterra models approximate the convex feature of the static curve of the simple Wiener model. The parametric Volterra model gives the best fit (Figure 3.18.9). This can be seen also on the plot of the simulated output signals of the estimated models in Figure 3.18.8. ■

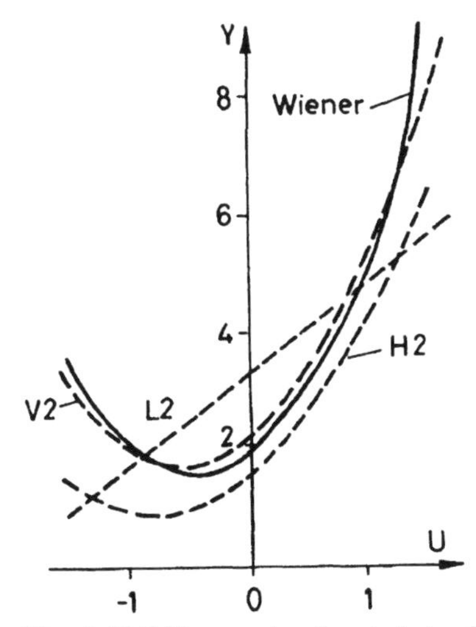

Fig. 3.18.9 True and estimated steady state characteristics of the simple Wiener model approximated by different models linear-in-parameters (L2: linear, H2: Hammerstein, V2: parametric Volterra)

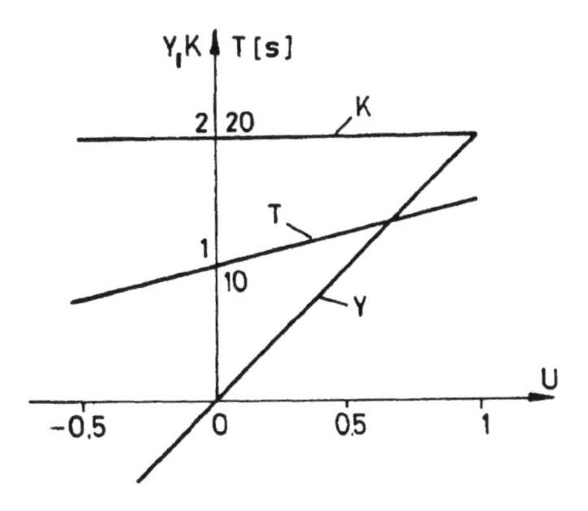

Fig. 3.18.10 True parameters of the first-order system with output signal dependent time constant

Example 3.18.5 *Two-step structure and parameter estimation of the steady state characteristic of a quasi-linear model with input signal dependent time constant (Haber et al., 1986)*

The process is a first-order quasi-linear system

$$T\dot{y}(t) + y(t) = Ku(t)$$

with constant gain and input signal dependent time constant

$$K(t) = K_0 = 2$$

$$T(t) = T_0 + T_1 u(t) = 10 + 5u(t) \text{ [s]}$$

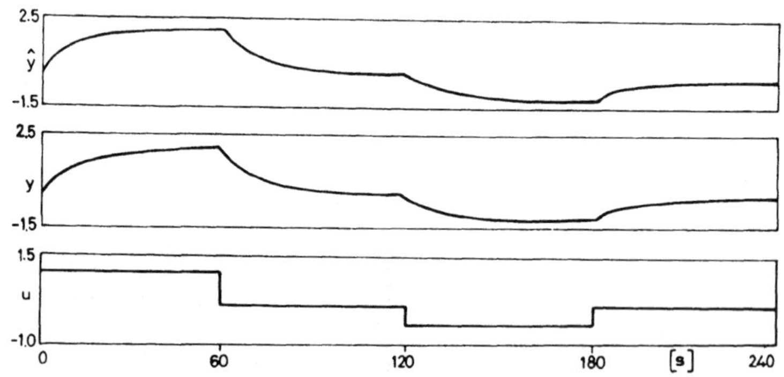

Fig. 3.18.11 Approximation of the first-order system with output signal dependent time constant by the best model linear-in-parameters

TABLE 3.18.4 Performance indices of different approximations linear-in-parameters of the first-order system with input signal dependent time constant

No	*Components*					*Leading term*			
	1	u	u^2	uy	y^2	$u(k)$:1, $u(k-1)$:0	σ_ε	$\sigma_{\hat{n}}$	$\sigma_{\hat{\theta}}$
1	0	1	0	0	0	1	0.0160	0.5	1
2	0	1	0	0	0	0	0.0419	0.5	1
3	**0**	**1**	**0**	**1**	**0**	**1**	**0.0142**	**1**	**1**
4	0	1	0	1	0	0	0.0417	0.5	0.5
5	0	1	1	1	1	1	0.0052	1	0.5
6	0	1	1	1	1	0	0.0401	0.5	0
7	1	1	1	1	1	1	0.0098	1	0
8	1	1	1	1	1	0	0.0400	0.5	0
9	1	1	0	1	0	1	0.1608	0.2	1
10	1	1	0	1	0	0	0.2140	0.2	0.5

The above relations are presented in Figure 3.18.10. Figure 3.18.11 shows the input signal and the measured output signal during the identification. The input signal consists of four constant periods, each being 60 [s] long. The values were

$$u(t)=\begin{cases}1.0 & \text{if} \quad 0\,[\text{s}]\le t< 60\,[\text{s}] \\ 0.0 & \text{if} \quad 60\,[\text{s}]\le t<120\,[\text{s}] \\ -0.5 & \text{if} \quad 120\,[\text{s}]\le t<180\,[\text{s}] \\ 0.0 & \text{if} \quad 180\,[\text{s}]\le t<240\,[\text{s}]\end{cases}$$

The whole record was sampled by 2 [s] and the LS method was applied to identify global valid linear-in-parameters models. Different components were considered as possible model components: $1, u(k), u^2(k), u(k)y(k), y^2(k)$. The results of the parameter estimation with ten different models are summarized in Table 3.18.4. Three features are given:

- σ_ε: standard deviation of the residuals;
- $\sigma_{\hat{n}}$: fitting of the noise-free model output to that measured;
- $\sigma_{\hat{\theta}}$: ratio of the estimated parameters to their standard deviations.

TABLE 3.18.5 True and estimated steady state values and static gains of the first-order system with input signal dependent time constant

U	Y	$\hat{Y}$	K	$\hat{K}$
-0.5	-1	-1.011	2	2.047
-0.25	-0.5	-0.501	2	1.028
0	0	0.0	2	1.985
0.25	0.5	0.492	2	1.949
0.5	1	0.975	2	1.914
0.75	1.5	1.448	2	1.880
1.0	2.0	1.915	2	1.847

The subjective measures were 1 if the estimation was perfect and 0 if it was bad. The equation of the best model is

$$\begin{aligned}\hat{y}(k) &= (0.838\pm 0.0034)\hat{y}(k-1)+(0.4075\pm 0.0156)u(k) \\ &-(0.08592\pm 0.161)u(k-1)-(0.05016\pm 0.0092)u(k)\hat{y}(k) \\ &+(0.04422\pm 0.0091)u(k-1)\hat{y}(k-1)\end{aligned}$$

The steady state equation is obtained by substituting $u(k)$ and $u(k-1)$ by U and $\hat{y}(k-1)$ by $\hat{Y}$:

$$\hat{Y}=1.985U-0.03664U\hat{Y} \tag{3.18.4}$$

$$\hat{K}=\frac{\partial \hat{Y}}{\partial U}=\frac{1.985}{(1+0.03664U)^2} \tag{3.18.5}$$

Since the test signal was in the range of $-0.5 \le u(t) \le 1$, the model is valid also only in this range. The second term in (3.18.4) and in the denominator of (3.18.5) can be neglected. The approximation of the linear steady state characteristic of the original model is very good as seen also from Table 3.18.5 and Figure 3.18.12. ■

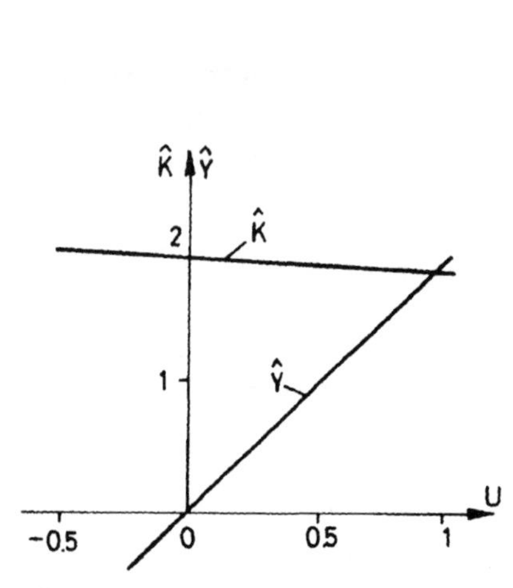

Fig. 3.18.12 Estimated parameters of the first-order system with output signal dependent time constant

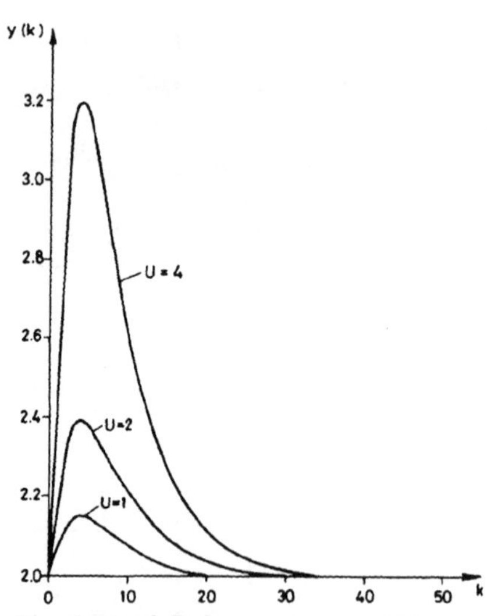

Fig. 3.18.13 Pulse responses of the simple Hammerstein model

3.18.3 Evaluation of finite-length excitations

If a process is excited by a test signal of finite length, the steady state characteristic can be computed by summing up the data vector over the whole time domain. To illustrate the method, a linear system will first be considered.

Assumption 3.18.1 The test signal has a finite length, i.e., $u(k) = 0$ if $k < 0$ and $k > N$. ■

Lemma 3.18.1

Excite the proportional linear system

$$y(k) = y_0 + \frac{B(q^{-1})}{A(q^{-1})} u(k-d) \tag{3.18.6}$$

by the test signal fulfilling Assumption 3.18.1. Define N that the output signal is also of finite length: $y(k) = 0$ if $k < 0$ and $k > N$.

The static gain is

$$K = \frac{\sum_{i=0}^{nb} b_i}{1 + \sum_{j=1}^{na} a_j} = \frac{\sum_{k=0}^{N} y'(k)}{\sum_{k=0}^{N} u(k)} \tag{3.18.7}$$

Proof. Denote the output signal reduced by the constant term y_0 by $y'(k)$.

$$y'(k) = y(k) - y_0 = y(k) - y(u(k))\big|_{u(k)=0}$$

The difference equation of the system is

$$\begin{aligned} &y'(k) + a_1 y'(k-1) + \ldots + a_{na} y'(k-na) \\ &= b_0 u(k-d) + \ldots + b_{nb} u(k-d-nb) \end{aligned} \tag{3.18.8}$$

Sum (3.18.8) from $k = 0$ to $k = N$

$$\begin{aligned} &\sum_{k=0}^{N} y'(k) + a_1 \sum_{k=0}^{N} y'(k-1) + \ldots + a_{na} \sum_{k=0}^{N} y'(k-na) \\ &= b_0 \sum_{k=0}^{N} u(k-d) + \ldots + b_{nb} \sum_{k=0}^{N} u(k-d-nb) \end{aligned} \tag{3.18.9}$$

Now

$$\sum_{k=0}^{N} y'(k) = \sum_{k=0}^{N} y'(k-1) = \ldots = \sum_{k=0}^{N} y'(k-na) \tag{3.18.10}$$

$$\sum_{k=0}^{N} u(k-d) = \ldots = \sum_{k=0}^{N} u(k-d-nb) \tag{3.18.11}$$

Substitute (3.18.10) and (3.18.11) into (3.18.9)

$$\left[\sum_{k=0}^{N} y'(k)\right]\left[1 + \sum_{j=1}^{na} a_j\right] = \left[\sum_{k=0}^{N} u(k)\right]\left[\sum_{i=0}^{nb} b_i\right] \tag{3.18.12}$$

which is equivalent to (3.18.7). ∎

Remarks:

1. The advantage of the method is that no parameter estimation is needed.
2. The method is not suitable for integrating or differentiating processes, at least not in the proposed form.

Corollary 3.18.1 Excite a linear process with a constant term $y_0 = y(u = 0)$ by a finite length excitation so, that the signal before and after the finite length excitation has a constant non-zero value $u(k) = u(\pm\infty)$, if $k < 0$ and $k > N$.

Choose N that the output signal is finite length within the interval $(0 \le k \le N)$, $y(k) = y(\pm\infty)$ if $k < 0$ and $k > N$. Define the deviations

$$u'(k) = u(k) - u(\pm\infty)$$
$$y'(k) = y(k) - y(\pm\infty)$$

Then the static gain is

$$K = \frac{\sum_{i=0}^{nb} b_i}{1 + \sum_{j=1}^{na} a_j} = \frac{\sum_{k=0}^{N} y'(k)}{\sum_{k=0}^{N} u'(k)} \tag{3.18.13}$$

or

$$K = \frac{\sum_{i=0}^{nb} b_i}{1 + \sum_{j=1}^{na} a_j} = \frac{y(\pm\infty) - y_0}{u(\pm\infty)} \tag{3.18.14}$$

Proof. (3.18.6) now becomes

$$y(k) = y_0 + \frac{B(q^{-1})}{A(q^{-1})}[u(\pm\infty) + u'(k-d)]$$

$$= y_0 + \frac{B(1)}{A(1)}u(\pm\infty) + \frac{B(q^{-1})}{A(q^{-1})}u'(k-d) \tag{3.18.15}$$

With

$$y(\pm\infty) = y_0 + \frac{B(1)}{A(1)}u(\pm\infty) \tag{3.18.16}$$

(3.18.15) becomes

$$y'(k) = \frac{B(q^{-1})}{A(q^{-1})}u'(k-d) \tag{3.18.17}$$

Applying Lemma 3.18.1 to (3.18.17) with (3.18.16) leads to (3.18.13) and (3.18.14). ■

Remarks:

1. The constant term can be measured by zero excitation $u(k) = 0$.
2. If the system has no constant term then (3.18.14) uses only the known values $u(\pm\infty)$ and $y(\pm\infty)$.

The steady state equation can be determined without parameter estimation for nonlinear process models linear-in-parameters, as well.

Theorem 3.18.1 Excite the non-integrating or differentiating nonlinear system

$$y'(k)+\sum_{j=1}^{na} a_j y'(k-j)=\sum_{i=1}^{M}\sum_{j=0}^{nb_i} b_{ij} f_i(k-d_i-j) \quad (3.18.18)$$

by a test signal fulfilling Assumption 3.18.1. Define N in such a way that also the reduced output signal is also of finite length: $y'(k)=0$, if $k<0$ and $k>N$.

The steady state equation is

$$Y'=\sum_{i=1}^{M} F_i \frac{\sum_{j=0}^{nb_i} b_{ij}}{1+\sum_{j=0}^{na} a_j} \quad (3.18.19)$$

where Y' and F_i are the stationary values

$$Y'=y(k)-y_0\big|_{k=\infty} \qquad F_i=f_i(k)\big|_{k=\infty}$$

The following relation exists between the coefficients of the steady state equation

$$\sum_{k=0}^{N} y'(k)=\sum_{i=1}^{M}\sum_{k=0}^{N} f_i(k)\frac{\sum_{j=0}^{nb_i} b_{ij}}{1+\sum_{j=0}^{na} a_j} \quad (3.18.20)$$

Proof. Sum up the output signal and all components from $k=0$ till $k=N$ in (3.18.18)

$$\left[\sum_{k=0}^{N} y'(k)\right]\left[1+\sum_{j=1}^{na} a_j\right]=\sum_{i=1}^{M}\left[\sum_{k=0}^{N} f_i(k)\right]\left[\sum_{j=0}^{nb_i} b_{ij}\right] \quad (3.18.21)$$

Equation (3.18.21) is equivalent to (3.18.20). ■

Algorithm 3.18.2 The steady state equation of the process (3.18.18) can be obtained as follows:

1. Excite the process by sufficient (not less than M) different finite length test signals $u_\ell(k)$. Denote the experiments by $\ell=1,\dots,M'$, $M'\geq M$. The number of experiments may be more than necessary.
2. Select N in such a way that the output signal and all components are zero for $k>N$ in each experiment.
3. Sum up the output signal and all components form $k=0$ till $k=N$ in each experiment, respectively.

4. Set up an equation system

$$\sum_{k=0}^{N} y'_\ell(k) = \sum_{i=1}^{M} \sum_{k=0}^{N} f_{\ell i}(k) \frac{\sum_{j=0}^{nb_i} b_{ij}}{1+\sum_{j=0}^{na} a_j} \qquad \ell = 1, \ldots, M' \tag{3.18.22}$$

where the subscript ℓ refers to the index of the test signal $u_\ell(k)$.

5. Solve the equation system with M' equations. If more tests than necessary were used, then a static regression by means of the least square method has to be applied.

■

Example 3.18.6 *Exciting the simple Hammerstein model by pulses*
The simple Hammerstein model consists of the nonlinear term (3.18.1) and of the aperiodic linear dynamic term

$$G(s) = \frac{1}{(1+5s)(1+10s)}$$

The sampling time was $\Delta T = 4$ [s]. The pulse transfer function obtained by SRE transformation is then

$$G(q^{-1}) = \frac{0.03286q^{-1} - 0.0269q^{-2}}{1 - 1.489q^{-1} + 0.5488q^{-2}}$$

The process was excited by pulses with length of one sampling time. The amplitudes were $U = 1$, 2 and 4, respectively. The pulse responses are seen in Figure 3.18.13.
The constant term $y_0 = 2$ was identified from a zero excitation.
The equation of the process between the input and the reduced output signal

$$y'(k) = y(k) - y_0$$

has the form

$$A(q^{-1})y'(k) = B_1(q^{-1})u(k-1) + B_2(q^{-1})u^2(k-1)$$

The steady state equation which is looked for is

$$Y' = K_1 U + K_2 U^2$$

The reduced output signal was summed from $k = 0$ to $k = 50$. The following static equations could be set up

$$1.498 = 1K_1 + 1K_2$$
$$2.995 = 2K_1 + 4K_2$$
$$11.989 = 4K_1 + 16K_2$$

The LS estimation leads to

$$K_1 = 0.9457 \pm 0.101554 \qquad K_2 = 0.5093 \pm 0.00431$$

The estimated values coincide with the static equation (3.18.1) of the process. ■

3.18.4 Correlation method

Different methods can be used:

- estimation of the difference equation of the process;
- correlation analysis with more multipliers.

1. Estimation of the difference equation of the process

Two step identification: correlation analysis and least squares parameter estimation can be applied to get the difference equation of the process. Then the steady state equation can be calculated by setting $q = q^{-1} = 1$.

2. Correlation analysis with more multipliers

Assumption 3.18.2 The following assumptions have to be met:

1. The process is given by its equation (3.18.18) linear-in-parameters .
2. There exists a multiplier to each model component that is uncorrelated to all other components (the number of components is M).
3. Both the auto-correlation and the existing cross-correlation function are of finite length in the shifting time domain $(\kappa_{min} \le \kappa \le \kappa_{max})$.
4. The cross-correlation function has already its steady state value at $\kappa = \kappa_{max}$. ■

Theorem 3.18.2 (Bamberger, 1978) Assume that Assumption 3.18.2 fulfills. The coefficients of the steady state Equation (3.18.18) can be obtained by

$$\frac{Y}{F_i} = \frac{\sum_{\kappa=\kappa_{min}}^{\kappa_{max}} r_{x_i y}(\kappa)}{\sum_{\kappa=\kappa_{min}}^{\kappa_{max}} r_{x_i f_i}(\kappa)} \tag{3.18.23}$$

If both correlation functions have a non zero value outside the shifting time domain $\kappa < \kappa_{min}$ and $\kappa > \kappa_{max}$ then besides (3.18.23) also

$$\frac{Y}{F_i} = \frac{r_{x_i y}(\kappa_{min})}{r_{x_i f_i}(\kappa_{min})} = \frac{r_{x_i y}(\kappa_{max})}{r_{x_i f_i}(\kappa_{max})} \tag{3.18.24}$$

is valid.

Proof. Let $x_i(k)$ the multiplier that is correlated only with the components $f_i(k-j)$. Correlate (3.18.18) with $x_i(k)$

$$r_{x_i y}(\kappa)+\sum_{j=1}^{na} a_j r_{x_i y}(\kappa - j)=\sum_{j=0}^{nb_i} b_{ij} r_{x_i f_i}(\kappa - j) \qquad (3.18.25)$$

(3.18.25) can be treated as a linear dynamic equation without a constant term, thus following Corollary 3.18.1 both (3.18.23) and (3.18.24) are valid. ∎

Example 3.18.7 *Correlation method applied to a simple Hammerstein model using a PRTS excitation (Continuation of Example 3.3.1)*

The simple Hammerstein model consists of a static function

$$v(k)=2+u(k)+0.5u^2(k)$$

and a linear dynamic term

$$\frac{B(q^{-1})}{A(q^{-1})}=\frac{0.065q^{-1}+0.043q^{-2}-0.008q^{-3}}{1-1.5q^{-1}+0.705q^{-2}-0.1q^{-3}}$$

The input–output difference equation of the output signal reduced by the constant term has the form of a generalized Hammerstein model that is a special case of (3.18.18) with $M=2$, $na=1$, $nb_1=nb_2=0$, $d_1=d_2=1$, $f_1(k)=u(k)$, $f_2(k)=u^2(k)$. The constant term was determined as the output signal at zero excitation

$$y=2.0$$

Two multipliers were used,

$$x_1(k)=u(k)-\mathrm{E}\{u(k)\}$$

$$x_2(k)=u^2(k)-\mathrm{E}\{u^2(k)\}$$

The plots of the correlation functions $r_{x_1 u}(\kappa)$, $r_{x_1 y}(\kappa)$ and $r_{x_2 u^2}(\kappa)$, $r_{x_2 y}(\kappa)$ were drawn in Figure 3.3.1. As the auto-correlation function of the PRTS with period 80 has two peaks, the boundaries of the shifting time domains were chosen as $\kappa_{min}=0$ and $\kappa_{max}=39$. The ratios given by (3.18.23) are

$$\frac{Y}{F_1}=\frac{\sum_{\kappa=0}^{39} r_{x_1 y}(\kappa)}{\sum_{\kappa=0}^{39} r_{x_1 u}(\kappa)}=\frac{0.1687}{0.1687}=1$$

$$\frac{Y}{F_2}=\frac{\sum_{\kappa=0}^{39} r_{x_2 y}(\kappa)}{\sum_{\kappa=0}^{39} r_{x_2 u^2}(\kappa)}=\frac{0.00703}{0.01406}=0.5$$

As both $r_{x_2 u^2}(\kappa)$ and $r_{x_2 y}(\kappa)$ have nonzero values at the boundaries, the ratio Y/F_2 can also be calculated as

$$\frac{Y}{F_2}=\frac{r_{x_2 y}(\kappa_{\min})}{r_{x_2 u^2}(\kappa_{\min})}=\frac{0.00017577}{0.00035156}=0.5$$

The steady state equation with the constant term is then

$$Y=2+U+0.5U^2$$

which is equal to what was simulated. ■

3.19 METHODS APPLIED TO SYSTEMS WITH NON-POLYNOMIAL NONLINEARITIES

The non-polynomial nonlinearities can be classified into two classes:

- single valued piecewise linear characteristics with breakpoints , and
- multivalued characteristics (hysteresis)

Common to both cases is that the nonlinear term can be described by a multi-model approach, i.e. by piecewise constant linear models. The parameters of the multi-model depend on the following signals:

- *static nonlinearity with single valued characteristics:* only on the input signal of the nonlinearity;
- *static nonlinearity and with double valued characteristics:* both on the input signal of the nonlinearity and on the sign of the derivation of the output signal

A common method for all types of nonlinearities is the direct search of the unknown parameters by derivative free methods. (See Section 3.4.) Following papers present examples for this method:

- *deadband plus saturation* (Jategaonkar, 1985);
- *hysteresis* (Androniku *et al.*, 1982; Jategaonkar, 1985).

There are basically two procedures in which form non-polynomial characteristics can be described and estimated:

- *approximation by a polynomial* (Goldberg and Durling, 1971);
- *piecewise linear description* (Butler and Bohn, 1966; Stone and Womack, 1970; Cantoni, 1971; Klein, 1973; Cooper and Falkner, 1975; Sehitoglu and Klein, 1975; Bai-Lan and Xiao-Lin, 1988; Maron, 1989).

The time domain parameter estimation algorithm can be of the following types:

- *optimization of all parameters* (e.g. Klein, 1973; Butler and Bohn, 1966);
- *successive optimization of parameter groups* (relaxation technique) (Chow and Chizeck, 1987; Vörös, 1994).

A typical dynamic nonlinear process is the simple Wiener–Hammerstein model (including the simple Wiener and the simple Hammerstein models). In addition to the direct optimization of a loss function there are three typical methods which can be used:

- *evaluation of repeated pulse tests* (Cooper and Falkner, 1975);
- *correlation method* (Billings and Fakhouri, 1978a, 1978b);
- *filtering of the input and output signals and graphical plotting* (Lammers *et al.*, 1979).

Both the correlation and the graphical plotting method works also for non-polynomial nonlinearities. The methods were presented in Section 3.11.

With a sinusoidal test signal the following method were applied:

- *describing function analysis* (Nordin and Bodin, 1995; Besancon–Voda and Drazdil, 1997)
- *linear and higher-order frequency response analysis* (Chen and Tomlinson, 1996)

Friction curves are often estimated with a constant velocity (Maron, 1989; Canudas de Wit and Lischinsky, 1997).

Some applications where non-polynomial or non-single valued nonlinearities were identified are listed below:

- *electromechanical positioning system* (Hoberock and Kohr, 1967) ;
- *visual system* (Troeslstra, 1969);
- *lateral plane of an aircraft* (Bell, 1974);
- *nonlinear actuator* (Moon, 1985);
- *positioning table* (Tomizuka *et al.*, 1985)
- *aircraft control surface actuator system* (Jategaonkar, 1985) ;
- *vibrating system* (Stanway and Mottershead, 1986);
- *spring-mass-damper mechanical system with friction and dead zone* (Isermann, 1988);
- *friction in a robot* (Dodds *et al.*, 1996);
- *hydraulic rotary servo system* (LeQuoc *et al.*, 1995).

3.20 REFERENCES

Alper, P. (1965). A consideration of the discrete Volterra series. *IEEE Trans. on Automatic Control, AC-10*, 3, pp. 322-327.

Androniku, A.M., G.A. Bekey and S.F. Masri (1982). Identification of nonlinear hysteretic systems using random search. *Prepr. 6th IFAC Symposium on Identification and System Parameter Estimation*, (Washington D.C.: USA), pp. 263–268.

Aracil, J. (1970). Measurements of Wiener kernels with binary random signals. *IEEE Trans. on Automatic Control*, AC–15, 1, pp. 123–125.

Åström, K.J. and T. Bohlin (1965). Numerical identification of linear dynamic systems from normal operating records. *IFAC Symposium*, (Teddington: UK).

Baeck, T. and H.P. Schwefel (1993). An overview of evolutionary algorithms for parameter optimisation. *Evolutionary Computation*, Vol. 1, pp. 1—23.

Baheti, R.S., R.R. Mohler and H.A. Spang III (1977). Second-order correlation method for bilinear system identification. pp. 1119–1125.

Bai-Lan, Li and T. Xiao-Lin (1988). On-line parameter identification for discontinuous nonlinear systems by pattern iteration recursive of partitioned data and expert system modification. *Prepr. 8th IFAC Symp.*

Identification and System Parameter Estimation, (Beijing: China), pp. 518–523.
Bányász, Cs., R. Haber and L. Keviczky (1973): Some estimation methods for nonlinear discrete time identification. *Prepr. 3rd IFAC Symposium on Identification and System Parameter Estimation*, (Hague: The Netherlands), pp. 793–802.
Bamberger, W. (1978). Methods for on-line optimization of the steady state behavior of nonlinear dynamic processes (in German). *PDV-Bericht, KfK-PDV* 159, pp. 173.
Bamberger, W. and R. Isermann (1978). Adaptive on-line steady state optimization of slow dynamic processes. *Automatica*, Vol. 14, pp. 223–230.
Bard, Y. and L. Lapidus (1970). Nonlinear system identification. *Ind. Eng. Chem. Fundam.*, Vol. 9, 4, pp. 628–633.
Bard, Y. (1974). *Nonlinear Parameter Estimation*. Academic Press. (New York: USA).
Barker, H.A., S.N. Obidegwu and T. Pradisthayon (1972). Performance of antisymmetric pseudorandom signals in the measurement of 2nd-order Volterra kernels by crosscorrelation. *Proc. IEE*, Vol. 119, 3, pp. 353–362.
Barker, H.A. and S.N. Obidegwu (1973a). Combined crosscorrelation method for the measurements of 2nd-order Volterra kernels. *Proc. IEE*, Vol. 120, 1, pp. 114–118.
Barker, H.A. and S.N. Obidegwu (1973b). Effects of nonlinearities on the measurement of weighting functions by crosscorrelation using pseudorandom signals. *Proc. IEE*, Vol. 120, 10, pp. 1293–1299.
Barker, H.A. and R.W. Davy (1978). Measurements of second-order Volterra kernels using pseudorandom ternary signals. *Int. Journal of Control*, Vol. 27, 2, pp. 277–291.
Baumgarten, S.L. and W.S. Rugh (1975). Complete identification of a class of nonlinear systems from steady state frequency response. *IEEE Trans. on Circuits and Systems, CAS-22*, pp. 753–759.
Beghelli, S. and R. Guidorzi (1976). Experimental results in the identification of a bilinear system: a test case. *Prepr. 4th IFAC Symp. Identification and System Parameter Estimation*, (Tbilisi: USSR), pp. 361–367.
Bell, D. (1974). Identification of nonlinear multi-variable systems subjected to random input signals. *Measurement and Control*, Vol. 7, 6, pp. 217–223.
Besancon–Voda, A. and P. Drazdil (1997). Estimation of a plant relay nonlinearity by nonlinear oscillations analysis, *Prepr. 4th European Control Conference*, (Brussels: Belgium), TU-A-I6, pp. 1–6.
Billings, S.A. and S.Y. Fakhouri (1977). Identification of nonlinear systems using the Wiener model. *Electronics Letters*, Vol. 13, 17, pp. 502–504.
Billings, S.A. and S.Y. Fakhouri (1978a). Identification of a class of nonlinear systems using correlation analysis. *Proc. IEE*, Vol. 125, Part D, 7, pp. 691–697.
Billings, S.A. and S.Y. Fakhouri (1978b). Correspondence to: identification of a class of nonlinear systems using correlation analysis. *Proc. IEE*, Vol. 125, Part D, 11, pp. 1307–1308.
Billings, S.A. and S.Y. Fakhouri (1979a). Identification of nonlinear S_m systems. *Int. Journal of Systems Science*, Vol. 10, 12, pp. 1401–1408.
Billings, S.A. and S.Y. Fakhouri (1979b). Identification of factorable Volterra systems. *Proc. IEE*, Part D, Vol. 126, 10, pp. 1018–1024.
Billings, S.A. and S.Y. Fakhouri (1979c). Nonlinear system identification using the Hammerstein model. *Int. Journal of Systems Science*, Vol. 10, 5, pp. 567–578.
Billings, S.A. and S.Y. Fakhouri (1979d). Identification of systems composed of linear dynamic and static nonlinear elements. *Prepr. 5th IFAC Symposium on Identification and Parameter Estimation*, (Darmstadt: FRG), pp. 493–500.
Billings, S.A. and S.Y. Fakhouri (1979e). Identification of nonlinear unity feedback systems. *Prepr. 17th IEEE Conf. on Decision and Control*, (San Diego: CA, USA), pp. 255–260.
Billings, S.A. and S.Y. Fakhouri (1980). Identification of nonlinear systems using correlation analysis and pseudorandom inputs. *Int. Journal of Systems Science*,, Vol. 11, 3, pp. 261–279.
Billings, S.A. (1980). Identification of nonlinear systems – A Survey. *Proc. IEE*, Vol. 127, Part D, 6, pp. 272–285.
Billings, S.A. and S.Y. Fakhouri (1982). Identification of systems containing linear dynamic and static nonlinear elements. *Automatica*, Vol. 18, 1, pp. 15–26.
Billings, S.A. and W.S.F. Voon (1984). Least squares parameter estimation algorithms for nonlinear systems. *Int. Journal of Systems Science*, Vol. 15, 6, pp. 601–615.
Billings, S.A. and M.B. Fadzil (1985). The practical identification of systems with nonlinearities. *Prepr. 7th IFAC/IFORS Symp. on Identification and System Parameter Estimation*, (York: UK), pp. 155–160.
Billings, S.A. and W.S.F. Voon (1987). Piecewise linear identification of nonlinear systems. *Int. Journal of Control*, Vol. 46, 1, pp. 215–235.
Billings, S.A., M.B. Fadzil, J.L. Sulley and P.M. Johnson (1988a). Identification of a nonlinear difference equation model of an industrial diesel generator. *Mechanical Systems and Signal Processing*, Vol. 2, 1, pp. 59–76.
Billings, S.A., M.J. Korenberg and S. Chen (1988b). Identification of nonlinear output–affine systems using an orthogonal least squares algorithm. *Int. Journal of Systems Science*, Vol. 19, 8, pp. 1559–1568.
Billings, S.A. and S. Chen (1989). Identification of nonlinear rational systems using a prediction error estimation algorithm. *Int. Journal of Systems Science*, Vol. 20, 3, pp. 467–494.
Billings, S.A., S. Chen and R.J. Backhouse (1989). The Identification of linear and nonlinear models of a turbocharged automotive diesel engine. *Mechanical Systems and Signal Processing*, Vol. 3, 2, pp. 123–142.

Broersen, P.M.T. (1973). Estimation of multivariable railway vehicle dynamics from normal operating records. *Prepr. 3rd IFAC Symp. on Identification and System Parameter Estimation*, (Hague: The Netherlands), pp. 425–433.

Butler, R.E. and E.V. Bohn (1966). An automatic identification technique for a class of nonlinear systems. *IEEE Trans. on Automatic Control*, AC–11, 2, pp. 292–296.

Cantoni, A. (1971). Optimal curve fitting with piecewise linear functions. *IEEE Trans. on Computers*, C-20, 1, pp. 59–67.

Canudas de Wit, C. and P. Lischinsky (1997). Adaptive friction compensation with partially known dynamic friction model, *Int. Journal of Adaptive Control and Signal Processing*, Vol. 11, pp. 65–80.

Carroll, C.W. (1961). The created response surface technique for optimizing nonlinear, restrained systems. *Operations Research*, Vol. 9, pp. 169–184.

Chang, F.H.I. and R. Luus (1971). A noniterative method for identification using Hammerstein model. *IEEE Trans. on Automatic Control*, AC-16, 5, pp. 464–468.

Chen, S. and S.A. Billings (1988). Prediction error estimation algorithm for nonlinear output–affine systems. *Int. Journal of Control*, Vol. 47, 1, pp. 309–332.

Chen, S., S.A. Billings and W. Luo (1989). Orthogonal least squares methods and their application to nonlinear system identification. *Int. Journal of Control*, Vol. 50, 5, pp. 1873–1896.

Chen, Q. and G.R. Tomlinson (1986). Parametric identification of systems with dry friction and nonlinear stiffness using a time series model, *Trans. ASME*, Vol. 118, pp. 252–263.

Chow, P-C. and H.J. Chizek (1987). Recursive identification of dynamic systems with varying memoryless nonlinearities having deadbands. *Proc. American Control Conference*, (Minneapolis: USA), pp. 1155–1157.

Cooper, B. and A.H. Falkner (1975). Identification of simple nonlinear systems. *Proc. IEEE*, Vol. 122, 7, pp. 753–755.

Corlis, R.G. and R. Luus (1969). Use of residuals in the identification and control of two-input, single-output systems. *I&EC Fundamentals*, Vol. 8, 5, pp. 246–253.

Diaz, H. and A.A. Desrochers (1988). Modeling of nonlinear discrete time systems from input–output data. *Automatica*, Vol. 24, 5, pp. 629–641.

Diekmann, K. and H. Unbehauen (1985). On-line parameter estimation in a class of nonlinear systems via modified least squares and instrumental variable algorithms. *Prepr. 7th IFAC Symp. on Identification and System Parameter Estimation*, (York: UK), pp. 149–153.

Diskin, M.H. and A. Boneh (1972). Determination of optimal kernels for second-order stationary surface run–off systems. *Water Resources Research*, Vol. 9, pp. 311–325.

Dodds, G., T. Ogasarawa, N. Glover and K. Kitagaki (1996). Telerobot control and real time simulation environment using parallel processing, *IEE Proc. Control Theory Applications*, Vol. 143, 6, pp. 543–550.

Draper, N.R. and H. Smith (1966). *Applied Regression Analysis*. John Wiley & Sons, Inc., (New York: USA).

Eykhoff, P. (1963). Some fundamental aspects of process parameter estimation. *IEEE Trans. on Automatic Control*, AC-8, pp. 347–357.

Eykhoff, P. (1974). *System Identification – Parameter and State Estimation*. John Wiley and Sons, (London: UK).

Fiacco, A.V. and G.P. McCormick (1964). The sequential unconstrained minimization technique for nonlinear programming: A primal-dual method. *Management Sci.*, Vol. 10, pp. 360–366.

deFigueiredo, R.J.P. (1984). A linear method for the design of optimal nonlinear filters for non-Gaussian processes. *Prepr. 9th IFAC World Congress*, (Budapest: Hungary), Vol. 14, pp. 33–35.

French, A.S. and E.G. Butz (1973). Measuring the Wiener kernels of a nonlinear system using the fast Fourier transform algorithm. *Int. Journal of Control*, Vol. 17, 3, pp. 529–539.

Gabr, M.M. and T.S. Rao (1984). On the identification of bilinear systems from operating records. *Int. Journal of Control*, Vol. 40, 1, pp. 121–128.

Gallman, P.G. (1976). A comparison of two Hammerstein model identification algorithms. *IEEE Trans. on Automatic Control*, AC-21, 1, pp. 124–126.

Gardiner, A.B. (1971). The identification testing of nonlinear systems, (manuscript).

Gardiner, A.B. (1973a). Frequency domain identification of nonlinear systems. *Prepr. 3rd IFAC Symp. on Identification and System Parameter Estimation*, (Hague: The Netherlands), pp. 831–834.

Gardiner, A.B. (1973b). Identification of processes containing single-valued nonlinearities. *Int. Journal of Control*, Vol. 18, 5, pp. 1029–1039.

Gautier, M., M. Monsion and J.P. Sagaspe (1976). Determination of a Volterra kernels representation for a nonlinear physiological system. *Prepr. 4th IFAC Symposium on Identification and Parameter Estimation*, (Tbilisi: USSR), pp. 468–474.

George, D.A. (1959). Continuous nonlinear systems. *Technical Report 355*, MIT Research Laboratory of Electronics, (Boston: MA, USA).

Gertner, A.G. and V.Ya. Zagurskii (1996). A behavioral model for nonlinear object. *Automatic Control and Computer Sciences*, Vol. 30, 3, pp. 10–19.

Goldberg, D.E. (1989). *Genetic algorithms in search, optimization and machine learning*, Addison–Wesley.

Goldberg, S. and A. Durling (1971). A computational algorithm for the identification of nonlinear systems. *Journal of the Franklin Institute*, Vol. 291, 6, pp. 427–447.

Gyftopoulos, E.P. and R.J. Hooper (1965). Signals for transfer function measurements in nonlinear systems. *Proc. Conf. Noise Analysis in Nuclear Systems*, (Gainsville: FL, USA), pp. 335–345.

Haber, R. and L. Keviczky (1974): The identification of the discrete time Hammerstein model. *Periodica*

Polytechnica – Electrical Engineering, Vol. 18, 1, pp. 71–84.
Haber, R. (1979a). Parametric identification of nonlinear dynamic systems based on correlation functions. *Prepr. 5th IFAC Symposium on Identification and Parameter Estimation,* (Darmstadt: FRG), pp. 515–522.
Haber, R. (1979b): An identification method for nonlinear dynamic models by means of process computers – modeling, theory, program package, testing (in German). *Report KfK-PDV-175,* Department of Control and Process Dynamics, Institute of Process and Steam Boilers Engineering. University of Stuttgart, (Stuttgart: FRG).
Haber, R. and J. Wernstedt (1979). New nonlinear dynamic models for the simulation of electrically excited biological membrane. *Int. Journal of Systems Science,* Vol. 5, 2, pp. 227–233.
Haber, R., I. Vajk and L. Keviczky (1982). Nonlinear system identification by 'linear' systems having signal dependent parameters. *Prepr. 6th IFAC Symposium on Identification and System Parameter Estimation,* (Washington, D.C.: USA), pp. 421–426.
Haber, R. (1985). Adaptive extremum control by the parametric Volterra model. *Proc. IFAC Conf. Digital Computer Applications to Process Control,* (Vienna: Austria), pp. 457–462.
Haber, R. and L. Keviczky (1985). Identification of 'linear' systems having signal dependent parameters. *Int. Journal of Systems Science,* Vol. 16, 7, pp. 869–884.
Haber, R., L. Keviczky and M. Hilger (1986). Process identification and control in the silicate industry. Part 4. Discrete time identification of dynamic processes: nonlinear systems (in Hungarian). *Scientific Publications of the Central Research and Design Institute for Silicate Industry,* No 85, (Budapest: Hungary).
Haber, R. (1988). Parametric identification of nonlinear dynamic systems based on nonlinear crosscorrelation functions. *Proc. IEE,* Vol. 135, Part D., 6, pp. 405–420.
Haber, R. and R. Zierfuss (1988). Identification of the simple Wiener model. *Report TUV-IMP-88/1,* Institute of Machine- and Processautomatisation, Technical University of Vienna, (Vienna: Austria).
Haber, R. and R. Zierfuss (1991). Identification of nonlinear models between the reflux flow and the top temperature of a destillation column. *Prepr. 9th IFAC/IFORS Symp. on Identification and System Parameter Estimation,* (Budapest: Hungary), pp. 486–491.
Hacjono, T. and H. Takata (1997). Identification of nonlinear systems by the automatic choosing function and the genetic algorithm, *12th IFAC Symp. on System Identification,* (Fukuoka: Japan), Vol. 1, pp. 69–74.
Harper, T.R. and W.J. Rugh (1976). Structural features of factorable Volterra systems. *IEEE Trans. on Automatic Control,* AC-21, 6, pp. 822–832.
Harris, G.H. and L. Lapidus (1967a). The identification of nonlinear reaction systems with two-level inputs. *A.I.Ch.E. Journal,* Vol. 13, 2, pp. 291–302.
Harris, G.H. and L. Lapidus (1967b). The identification of nonlinear systems. *Industrial and Engineering Chemistry,* Vol. 59, 6, 2, pp. 67–81.
Hatakeyama, T. (1997). Hydrological system model with Volterra series and M-P general inverse matrix, *12th IFAC Symp. on System Identification,* (Fukuoka: Japan), Vol. 3, pp. 1537–1541.
Hoberock, L.L. and R.H. Kohr (1967). An experimental determination of differential equations to describe simple nonlinear systems. *Trans. of the ASME, Journal of Basic Engineering,* Vol. 89, 2, pp. 393–398.
Hooper, R.J. and E.P. Gyftopoulos (1967). On the measurement of characteristic kernels of a class of nonlinear systems. *Proc. Symp. on Neutron Noise, Waves and Pulse Propagation,* (Oak Ridge: USA), pp. 343–356.
Hu, D.W. and Z.Z. Wang (1991). An identification method for the Wiener model of nonlinear systems, Proc. 30th IEEE Conf. on Decision and Control, (Brighton: UK), Vol. 1, pp. 783–787.
Hu, K.P. and Z.D. Yuan (1988). Identification of streamflow processes. *Prepr. 8th IFAC/IFORS Symp. on Identification and System Parameter Estimation,* (Beijing: China), pp. 1777–1781.
Hubbell, P.G. (1969). Identification of a class of nonlinear systems. *Prepr. 2nd Asilomar Conf. on Circuits and Systems,* (Pacific Grove: CA, USA), pp. 511–514.
Hung, G. and L. Stark (1977). The kernel identification method (1910–1970) – Review of theory, calculation, application an interpretation. *Mathematical Biosciences,* Vol. 37, pp. 135–190.
Hunter, I.W. and M.J. Korenberg (1986). The identification of nonlinear biological systems: Wiener and Hammerstein cascade models. *Biological Cybernetics,* Vol. 55, pp. 135–144.
Johansen, T.A. and B.A. Foss (1993). Constructing NARMAX models using ARMAX models, *Int. Journal of Control,* Vol. 58, pp. 1125–1153.
Johansen, T.A. and B.A. Foss (1995a). Semi-empirical modeling of nonlinear dynamic systems through identification of operating regimes and local models, *Modeling, Identification and Control,* Vol. 16, 4, 213–232.
Johansen, T.A. and B.A. Foss (1995b). Empirical modeling of a heat transfer process using local models and interpolation, *Proc. American Control Conference,* (Seattle: WA, USA), Vol. 5, pp. 3654–3658.
Isermann, R. (1988). *Identification of Dynamic Systems.* Springer Verlag, (Berlin: FRG).
Janiszowski, K. (1986). Determination of process models by means of linear transformation of the measured signals (in German). *Messen, Steuern und Regeln,* Vol. 29, pp. 29–33.
Jategaonkar, R.V. (1985). Parametric identification of discontinuous nonlinearities. *Prepr. 7th IFAC Symp. on Identification and System Parameter Estimation,* (York: UK), pp. 167–172.
Jedner, U. and H. Unbehauen (1986). Identification of a class of nonlinear systems by parameter estimation of a linear multi-model. *Prepr. IMACS/IFAC Symp. on Modeling and Simulation for Control of Lumped and Distributed Parameter Systems.* (Lille: France), pp. 287–290.

Jelonek, Z.J. and E. Economakos (1972). Complete identification of a nonlinear network with sinusoidal inputs (in Greek). *Technikon Hronikon*, pp. 21–27.
Kadri, F.L. (1971). Nonlinear plant identification by crosscorrelation. *Prepr. IFAC Symp. Digital Simulation of Continuous Processes*, (Györ: Hungary), pp. K1/1–9.
Kadri F.L. (1972). A method for determining the crosscorrelation functions for a class of nonlinear systems. *Int. Journal of Control*, Vol. 15, 4, pp. 779–783.
Kadri F.L. and J.D. Lamb (1973). An improved performance criterion for pseudorandom sequences in the measurements of 2nd order Volterra kernels by crosscorrelation. *Prepr. 3rd IFAC Symp. on Identification and System Parameter Estimation*, (Hague: The Netherlands), pp. 843–846.
Keviczky, L. (1976). Nonlinear dynamic identification of a cement mill to be optimized. *Prepr. 4th IFAC Symp. on Identification and System Parameter Estimation*, (Tbilisi: USSR), pp. 388–396.
Klein, R.E. (1973). Identification of multivalued parameters in a class of noisy dynamic processes. *Prepr. 3rd IFAC Symp. on Identification and System Parameter Estimation*, (Hague: The Netherlands), pp. 851–856.
Klippel, W. (1996). Nonlinear system identification for horn loudspeakers, *Journal of Audio Engineering Society*, Vol. 44, pp. 811–820.
Korenberg, M.J. (1973). A new statistical method of nonlinear system identification. *Prepr. 16th Midwest Symp. on Circuit Theory*, XVIII., pp. 1.1–1.10.
Korenberg, M.J. (1985). Orthogonal identification of nonlinear difference equation models. *Prepr. Midwest Symp. on Circuit Theory*, pp. 90–94.
Korenberg, M.J. and I.W. Hunter (1986). The identification of nonlinear biological systems: LNL cascade models. *Biological Cybernetics*, Vol. 55, pp. 123–134.
Korenberg, M., S.A. Billings, Y.P. Liu and P.J. McIlroy (1988a). Orthogonal parameter estimation algorithm for nonlinear stochastic systems. *Int. Journal of Control*, Vol. 48, 1, pp. 193–210.
Korenberg, M.J., A.S. French and S.K.L. Voo (1988b). White noise analysis of nonlinear behavior in an insect sensory neuron: Kernel and cascade approaches. *Biol. Cybern.* Vol. 58, pp. 1110–1120.
Korenberg, M.J. (1989). A robust orthogonal algorithm for system identification and time series analysis. *Biological Cybernetics*, Vol. 60, pp. 267–276.
Korenberg, M.J. and I.W. Hunter (1990). The identification of nonlinear biological systems: Wiener kernel approach. *Annals of Biomedical Engineering*, Vol. 18, pp. 629–654.
Kortmann, M.J. and H. Unbehauen (1986). Application of a recursive prediction error method to the identification of nonlinear systems using the Wiener model. *Prepr. IMACS Symp. on Modeling and Simulation for Control of Lumped and Distributed Parameter Systems*, (Villeneuve d'Asq: France), pp. 281–285.
Kortmann, M.J. and H. Unbehauen (1987). A new algorithm for automatic selection of optimal model structure in the identification of nonlinear systems (in German). *Automatisierungstechnik*, Vol. 35, 12, pp. 491–498.
Kortmann, M.J. (1988). The identification of nonlinear single- and multi-variable systems by means of nonlinear models (in German). *VDI-Fortschrittberichte, Series 8: Measurement and Control*, No 177, VDI Verlag, (Düsseldorf: FRG).
Kortmann, M.J., K. Janiszowski and H. Unbehauen (1988). Application and comparison of different identification schemes under industrial conditions. *Int. Journal of Control*, Vol. 48, 6, pp. 2275–2296.
Kortmann, M.J. and H. Unbehauen (1988). Two algorithms for model structure Determination of Nonlinear Dynamic Systems with Applications to Industrial processes, *Prepr. 8th IFAC/IFORS Symp. on Identification and System Parameter Estimation*, (Beijing: China), pp. 939–946.
Koukoulas, P. and Kalouptsidis (1995). Nonlinear system identification using Gaussian inputs. *IEEE Trans. on Signal Processing*, Vol. 43, 8, pp. 1831–1841.
Krempl, R. (1973a). Application of three-level pseudo-random signals for parameter estimation of nonlinear systems. *Prepr. 3rd IFAC/IFORS Symp. on Identification and System Parameter Estimation*, (Hague: The Netherlands), pp. 835–838.
Krempl, R. (1973b). Application of three-level pseudo-random signals for identification of nonlinear control systems. *Dissertation* (in German), Faculty of Mechanical Engineering, Ruhr-University, (Bochum: FRG).
Kurth, J. (1996). Reduced Volterra series – a new model approach for identification of nonlinear systems (in German), *Automatisierungstechnik*, Vol. 44, 6, pp. 265–273.
Kurth, J. and H. Rake (1994). Identification of nonlinear systems with reduced Volterra series, *Postprint IFAC Symp. System Identification*, (Copenhagen: Denmark), Vol. 1, pp. 143–150.
Lammers, H.C., H.B. Verbruggen and E. de Boer (1979). An identification method for a combined Wiener–Hammerstein filter describing the encoding part of the cochlear system. *Prepr. 5th IFAC Symposium on Identification and System Parameter Estimation*, (Darmstadt: FRG), pp. 485–491.
Lee, Y.W. and M. Schetzen (1965). Measurements of the Wiener kernels of a nonlinear system by crosscorrelation. *Int. Journal of Control*, Vol. 2, pp. 237–254.
Leontaritis, I.J. and S.A. Billings (1985). Input–output parametric models for nonlinear systems. Part II: Stochastic nonlinear systems. *Int. Journal of Systems Science*, Vol. 41, 2, pp. 329–341.
Leontaritis, I.J. and S.A. Billings (1988). Prediction error estimation for nonlinear stochastic systems. *Int. Journal of Systems Science*, Vol. 19, 4, pp. 519–536.
LeQuoc, S., Y.F. Xiong and R.M.H. Cheng (1995). Design of nonlinear controller for hydraulic rotary servo system, *Proc. 3rd European Control Conference*, (Rome: Italy), pp. 336–341.
Li, Y. and S. Chen (1988). Identification of model for a cement rotary kiln. *Prepr. 8th IFAC/IFORS Symp.*

on Identification and System Parameter Estimation, (Beijing: China), pp. 711–716.
Liu, Y.P., M.J. Korenberg, S.A. Billings and M.B. Fadzil (1987). The nonlinear identification of a heat exchanger. *Proc. 24th Conf. on Decision and Control*, (Los Angeles: CA, USA), pp. 1883–1888.
Marmarelis, P.Z. and K.-I. Naka (1972). White noise analysis of a neuron chain: an application of the Wiener theory. *Science*, Vol. 175, 3, pp. 1276–1279.
Marmarelis, P.Z. and K.-I. Naka (1973). Nonlinear analysis and synthesis of receptive-field responses in the catfish retina. *Journal of Neurophysics*, Vol. 36, pp. 605–648.
Marmarelis, V.Z. (1978). Random versus pseudorandom test signals in nonlinear system identification. *Proc. IEE*, Vol. 125, 5, pp. 425–428.
Marmarelis, V.Z. (1979). Error analysis and optimal estimation procedures in identification of nonlinear Volterra systems. *Automatica*, Vol. 15, pp. 161–174.
Maron, J.C. (1989). Identification and adaptive control of mechanical systems with friction, *Proc. IFAC Symp. on Adaptive Systems in Control and Signal Processing*, (Glasgow: UK), pp. 325–330.
Marquardt, D.W. (1963). An algorithm for least squares estimation of nonlinear parameters. *SIAM Journal*, Vol. 11, pp. 431–441.
Mathews, V.J. (1995). Orthogonalization of correlated Gaussian signals for Volterra system identification, *IEEE Signal Processing Letters*, Vol. 2, 10, pp. 188–190.
Miller, R. W. and R. Roy (1964). Nonlinear process identification using decision theory. *IEEE Trans. on Automatic Control*, AC-9, 10, pp. 538–540.
Moon, S.F. (1985). Parameter identification of processes with nonlinear actuators. *Proc. American Control Conference*, (Boston: USA), pp. 1704–1706.
Moore J.V. (1982). Global convergence of output error recursions in colored noise, *IEEE Trans. on Automatic Control*, Vol. 27, 6, pp. 1189–1199.
Mosca, E. (1970). System identification by reproducing kernel Hilbert space methods. *Prepr. 2nd IFAC Symp. on Identification and System Parameter Estimation*, (Prague: Czechoslovakia), pp. 1.1/1–8.
Mosca, E. (1971). A deterministic approach to a class of nonparametric system identification problems. *IEEE Trans. on Information Theory*, IT-17, 6, pp. 686–695.
Mosca, E. (1972). Determination of Volterra kernels form input–output data. *Int. Journal of Systems Science*, Vol. 3, 4, pp. 357–374.
Narendra, K.S. and P.G. Gallman (1966). An iterative method for the identification of nonlinear systems using a Hammerstein model. *IEEE Trans. on Automatic Control*, AC-11, 3, pp. 546–550.
Nelder, J.A. and R. Mead (1965). A simplex method for function minimization. *Computer Journal*, Vol. 7, pp. 308–313.
Nelles, O., O. Hecker and R. Isermann (1997). Automatic model selection in local linear model trees (LOLIMOT) for nonlinear system identification of a transport delay process, *12th IFAC Symp. on System Identification*, (Fukuoka: Japan),Vol. 2, pp. 727–732.
Nordin, M., and P. Bodin (1995). A backlash gap estimation method, Proc. 3rd European Control Conference, (Rome: Italy), pp. 3486–3491.
Pajunen, G.A. (1985a). Identification and adaptive control of Wiener type nonlinear processes. *Prepr. 7th IFAC Conf. on Digital Computer Applications to Process Control*, (Vienna: Austria), pp. 559–606.
Pajunen, G.A. (1985b). Recursive identification of Wiener type nonlinear systems. *Prepr. American Control Conference*, (Boston: USA), pp. 1365–1370.
Pickhardt, R. (1997). Adaptive control using a multi-model approach applied to a transportation system for bulk goods (in German), *Automatisierungstechnik*, Vol. 45, 3, pp. 113–120.
Pincock D.G. and D.P. Atherton (1973) The identification of a two phase induction machine with SCR speed control. *Prepr. 3rd IFAC Symp. on Identification and System Parameter Estimation*, (Hague: The Netherlands), pp. 505–512.
Pottmann, M., H. Unbehauen and D.E. Seborg (1993). Application of a general multi-model approach for identification of highly nonlinear processes – a case study, *Int. Journal of Control*, Vol. 57, 1, pp. 97–120.
Ream, N. (1970). Nonlinear identification using inverse-repeat m sequences. *Proc. IEE*, Vol. 117, 1, pp. 213–218.
Root, W.L. (1971). On the structure of a class of system identification problems. *Automatica*, Vol. 7, pp. 219–231.
Rosen, J.B. (1960). The gradient projection method for nonlinear programming: I. Linear constraints. *SIAM Journal*, Vol. 8, pp. 181–217.
Rosen, J.B. (1961). The gradient projection method for nonlinear programming: II. Nonlinear constraints. *SIAM Journal*, Vol. 9, pp. 514–532.
Rosenthal, B. and J. Beyer (1985). Investigations for describing nonlinear dynamic systems by means of simple model structures (in German). *30. International Scientific Kolloquium*, (Ilmenau: GDR), pp. 275–278.
Roy, R.J. and P.M. DeRusso (1962). A digital orthogonal model for nonlinear processes with two level inputs. *IRA Trans. on Automatic Control*, pp. 93–101.
Roy, R.J. and J. Sherman (1967a). System identification and pattern recognition. *Prepr. 1st IFAC/IFORS Symp. on Identification and System Parameter Estimation*, (Prague: Czechoslovakia), pp. 1.6/1–9.
Roy, R.J. and J. Sherman (1967b). A learning technique for Volterra series representation. *IEEE Trans. on Automatic Control*, AC-12, 6, pp. 761–766.
Rugh, W.J. (1981). *Nonlinear System Theory*. The Volterra/Wiener approach. The John Hopkins University Press, (Baltimore: USA).

Sandberg, A. and L. Stark (1968). Wiener G-function analysis as an approach to nonlinear characteristics of human pupil light reflex. *Brain Res.*, Vol. 11, pp. 194–211.
Schetzen, M. (1965a). Measurements of the kernels of nonlinear systems of finite order. *Int. Journal of Control,* Vol. 1, 3, pp. 251–263.
Schetzen, M. (1965b). Determination of optimum nonlinear systems for generalized criteria based on the use of gate functions. *IEEE Trans. on Information Theory*, IT-11, pp. 117–125.
Schetzen, M. (1974). A theory of nonlinear system identification. *Int. Journal of Control*, Vol. 20, 4, pp. 577–592.
Schetzen, M. (1980). *The Volterra and Wiener Theories of Nonlinear Systems*. John Wiley & Sons, (New York: USA).
Schweizer, J. (1987). A contribution to the identification of nonlinear controlled systems by means of correlation analysis. *Dissertation*, Faculty of Mechanical Engineering, University of Kaiserslautern, (Kaiserslautern: FRG).
Sehitoglu, H. and R.E. Klein (1975). Identification of multivalued and memory nonlinearities in dynamic processes. *Simulation*, Vol. 25, 3, pp. 86–92.
Stanway, R. and J.E. Mottershead (1986). Identification of combined viscous and Coulomb friction – a numerical comparison of least squares algorithms. *Trans. Inst. M.C.*, Vol. 8, 1, pp. 9–16.
Stark, L.W. (1969). The pupillary control system: its nonlinear adaptive and stochastic engineering design characteristics. *Automatica*, 5, pp. 655–676.
Stoica, P. (1981). On the convergence of an iterative algorithm used for Hammerstein systems identification. *IEEE Trans. on Automatic Control*, AC-26, 4, pp. 967–969.
Stoica, P. and T. Söderström (1982). Instrumental-variable methods for identification of Hammerstein systems. *Int. Journal of Control*, Vol. 35, 3, pp. 459–476.
Stone, J.J. and B.F. Womack (1970). Identification of a class of nonlinear systems by use of piecewise continuous expansions. *SWIEEECO Record*, pp. 198–200.
Sutter, E. (1987). A practical nonstochastic approach to nonlinear time domain analysis. In: Marmaleris V.Z. (editor). Advance Methods of Physiological System Modeling. *USC Biomedical Simulations Resource*, Vol. 1, (Los Angeles: USA), pp. 303–315.
Talmon, J.L. (1971). Approximated Gauss–Markov estimators and related schemes. *Report 71-E-17*, Eindhoven University of Technology, Department of Electrical Engineering, (Eindhoven: The Netherlands).
Taylor, L.W. and A.V. Balakrishnan (1967). Identification of human response models in manual control systems. *Prepr. 1st IFAC Symp. on Identification and Parameters Estimation*, (Prague: Czechoslovakia), pp. 1.7/1–8.
Tomizuka, M., A. Jabbari, R . Horowitz, D.M. Auslander and M. Denome (1985). Modeling and identification of mechanical systems with nonlinearities. *Prepr. 7th IFAC Symp. on Identification and Parameters Estimation*, (York: UK), pp. 845–850.
Troelstra, A. (1969). System identification with threshold measurements. *IEEE Trans. on Systems Science and Cybernetics*, SSC-5, 4, pp. 313–321.
Tuis, L. (1975). Application of multilevel pseudorandom signals to identification of nonlinear control systems. *Dissertation* (in German), Faculty of Mechanical Engineering, Ruhr–University, (Bochum: FRG).
Tuis, L. (1976). Identification of nonlinear systems by means of multilevel pseudo-random signals applied to a water-turbine unit. *Prepr. 4th IFAC Symp. on Identification and Parameters Estimation*, (Tbilisi: USSR), pp. 569–579.
Velenyak, J. (1981). Identification of systems with signal dependent parameters. *Diploma Thesis* (in Hungarian), Department of Automation, Technical University of Budapest, (Budapest: Hungary).
Velev, K.D. and I.N. Vuchkov (1986). Identification and control of parametrically dependent stochastic processes. *Prepr. 2nd IFAC Symp. on Stochastic Control*, (Vilnius: USSR), pp. 126–131.
Vörös, J. (1994). Identification of discontinuous block oriented nonlinear dynamic systems, *Prep. 1st IFAC Workshop on New Trends in Design of Control Systems*, (Smolenice:Slovak Republic), pp. 189–194.
Vörös, J. (1985). Nonlinear system identification with internal variable estimation. *Prepr. 7th IFAC Symp. Identification and System Parameter Estimation*, (York: UK), pp. 439–443.
Vörös, J. (1995). Identification of nonlinear dynamic systems using extended Hammerstein and Wiener models, *Control Theory and Advanced Technology*, Vol. 10, 4, Part 2, 103–12!.
Vuchkov, I.N., K.D. Velev and V.K. Tsochev (1985). Identification of parametrically dependent processes with applications to chemical technology. *Prepr. 7th IFAC/IFORS Symp. on Identification and System Parameter Estimation*, (York: UK), pp. 1089–1093.
Zheng, Q. and E. Zafiriou (1995). Nonlinear system identification for control using Volterra–Laguerre expansion, *Proc. American Control Conference*, (Seattle: WA, USA), pp. 2195–2199.
Zhu, Q.M. and S.A. Billings (1994). Identification of polynomial and rational NARMAX models. *11th IFAC Symp. System Identification*, (Copenhagen: Denmark), Vol.1, pp. 295—300.
Westenberg, J.Z. (1969). Some identification schemes for nonlinear noisy processes. *TH Report* 69–2–09, Eindhoven University of Technology, (Eindhoven: The Netherlands).
Wiener, N. (1958). *Nonlinear Problems in Random Theory*. MIT Press, (Cambridge: MA, USA).
Wysocki, E.M and W.J. Rugh (1976). Further results on the identification for the class of nonlinear systems. *IEEE Trans. on Circuits Systems*, CAS-23, pp. 664–670.

MATHEMATICAL MODELLING:
Theory and Applications

1. M. Křížek and P. Neittaanmäki: *Mathematical and Numerical Modelling in Electrical Engineering.* Theory and Applications. 1996
ISBN 0-7923-4249-6
2. M.A. van Wyk and W.-H. Steeb: *Chaos in Electronics.* 1997
ISBN 0-7923-4576-2
3. A. Halanay and J. Samuel: *Differential Equations, Discrete Systems and Control.* Economic Models. 1997 ISBN 0-7923-4675-0
4. N. Meskens and M. Roubens (eds.): *Advances in Decision Analysis.* 1999
ISBN 0-7923-5563-6
5. R.J.M.M. Does, K.C.B. Roes and A. Trip: *Statistical Process Control in Industry.* Implementation and Assurance of SPC. 1999
ISBN 0-7923-5570-9
6. J. Caldwell and Y.M. Ram: *Mathematical Modelling.* Concepts and Case Studies. 1999 ISBN 0-7923-5820-1
7. 1. R. Haber and L. Keviczky: *Nonlinear System Identification - Input-Output Modeling Approach.* Volume 1: Nonlinear System Parameter Identification. 1999 ISBN 0-7923-5856-2; ISBN 0-7923-5858-9 Set

   2. R. Haber and L. Keviczky: *Nonlinear System Identification - Input-Output Modeling Approach.* Volume 2: Nonlinear System Structure Identification. 1999 ISBN 0-7923-5857-0; ISBN 0-7923-5858-9 Set

KLUWER ACADEMIC PUBLISHERS – DORDRECHT / BOSTON / LONDON